AF526221

Nanoscale Materials and Modeling—Relations Among Processing, Microstructure and Mechanical Properties

MATERIALS RESEARCH SOCIETY
SYMPOSIUM PROCEEDINGS VOLUME 821

Nanoscale Materials and Modeling—Relations Among Processing, Microstructure and Mechanical Properties

Symposium held April 13–16, 2004, San Francisco, California, U.S.A.

EDITORS:

Peter M. Anderson
Ohio State University
Columbus, Ohio, U.S.A.

Timothy Foecke
National Institute of Standards and Technology
Gaithersburg, Maryland, U.S.A.

Amit Misra
Los Alamos National Laboratory
Los Alamos, New Mexico, U.S.A.

Robert E. Rudd
Lawrence Livermore National Laboratory
Livermore, California, U.S.A.

Materials Research Society
Warrendale, Pennsylvania

This material is based upon work supported by the National Science Foundation under Grant Number CMS-0414166. Any opinions, findings, and conclusions or recommendations expressed in this material are those of the author(s) and do not necessarily reflect the views of the National Science Foundation.

Single article reprints from this publication are available through University Microfilms Inc., 300 North Zeeb Road, Ann Arbor, Michigan 48106

CODEN: MRSPDH

Published by:

Materials Research Society
506 Keystone Drive
Warrendale, PA 15086
Telephone (724) 779-3003
Fax (724) 779-8313
Web site: http://www.mrs.org/

Manufactured in the United States of America

CONTENTS

PART I: PROCESSING—VAPOR AND ELECTRO-DEPOSITION

A. EXPERIMENTS

B. MODELING

PART II: PROCESSING—MECHANICAL DEFORMATION

A. EXPERIMENTS

B. MODELING

PART III: PROCESSING—SINTERING, SOLIDIFICATION, AND GEL TECHNIQUES

PART IV: INTERNAL STRESS AND STRENGTHENING MECHANISMS

A. EXPERIMENTS

*Invited Paper

B. MODELING

*Invited Paper

PART V: DEFORMATION, TEXTURE DEVELOPMENT, AND DUCTILITY

A. EXPERIMENTS

B. MODELING

*Invited Paper

PART VI: FRACTURE, FATIGUE, AND DEGRADATION

A. EXPERIMENTS

PREFACE

This volume contains a portion of the papers presented in Symposium P, "Nanoscale Materials and Modeling—Relations Among Processing, Microstructure and Mechanical Properties," held April 13–16 at the 2004 MRS Spring Meeting in San Francisco, California. As indicated by the title, the symposium highlighted structure-processing-mechanical property relations in nanoscale materials. In all, nine sessions were organized primarily around specific systems such as single and multilayer thin films, nano-dots, nano-porous and nanocomposite materials, and nanostructured metals and alloys. A focus session on theory, modeling, and simulation of mechanical properties at the nanoscale and another on grain boundaries and interfaces in nanoscale materials were held, the latter as a joint session with Symposium N. Among the several student participants were MRS Graduate Student Award winners M. Buehler, A. Dalis and S. Welz (gold) and B. McCaughey (silver).

The contributions to the proceedings reflect many of the themes of the Symposium. However, the presentation sequence has been updated from the original symposium, in order to emphasize the processing-structure-mechanical properties theme. The first three parts are devoted to processing methods for nanostructured materials, including vapor and electro-deposition techniques, mechanical deformation such as ball milling and ECAP, and elevated temperature techniques such as sintering and solidification. The fourth part is devoted to internal stress and the physics of strengthening mechanisms at the nanoscale. The remaining two parts highlight mechanical properties such as stress-strain and nanoindentation response, texture development, ductility, fracture, and fatigue. Typically, each part is ordered with experimental contributions appearing first, followed by papers describing related modeling and simulation. The first three parts tend to emphasize processing-structure and processing-mechanical properties relations, while the latter three parts tend to emphasize structure-mechanical properties relations. It is expected that this organization will foster the theme of the symposium.

A successful symposium and proceedings requires several aspects of support. We are grateful to the Mechanics and Materials Program at the National Science Foundation and also Hysitron, Inc. for financial support that enabled several participants to attend. Our appreciation is extended to T.J. Balk, R.C. Cammarata, M. Dao, H. Gao, W.W. Gerberich, R.G. Hoagland, E. Ma, A.K. Mukherjee, T.G. Nieh, W.D. Nix, F. Spaepen, Z. Suo, H. Van Swygenhoven, J.R. Weertman, J.M.K. Wiezorek, and Y.T. Zhu for their role as invited speakers. We also thank session chairs C. Barbour, V. Bulatov, W. Gerberich, R. Hoagland, N. Moody, A. Mukherjee, V. Randle, H. Van Swygenhoven, C. Volkert, and Y. Zhu for their management of time and technical discussion points, and also MRS Meeting Chairs I. Baumvol, J. De Yoreo, S. Mantl, and T. Neenan for their support and adoption of our Symposium into the Meeting. Special appreciation is given to the MRS staff for their technical assistance.

Peter M. Anderson
Timothy Foecke
Amit Misra
Robert E. Rudd

August 2004

MATERIALS RESEARCH SOCIETY SYMPOSIUM PROCEEDINGS

Volume 782— Micro- and Nanosystems, D. LaVan, M. McNie, A. Ayon, M. Madou, S. Prasad, 2004, ISBN: 1-55899-720-2
Volume 783— Materials, Integration and Packaging Issues for High-Frequency Devices, P. Muralt, Y.S. Cho, J-P. Maria, M. Klee, C. Hoffmann, C.A. Randall, 2004, ISBN: 1-55899-721-0
Volume 784— Ferroelectric Thin Films XII, S. Hoffmann-Eifert, H. Funakubo, A.I. Kingon, I.P. Koutsaroff, V. Joshi, 2004, ISBN: 1-55899-722-9
Volume 785— Materials and Devices for Smart Systems, Y. Furuya, E. Quandt, Q. Zhang, K. Inoue, M. Shahinpoor, 2004, ISBN: 1-55899-723-7
Volume 786— Fundamentals of Novel Oxide/Semiconductor Interfaces, C.R. Abernathy, E. Gusev, D.G. Schlom, S. Stemmer, 2004, ISBN: 1-55899-724-5
Volume 787— Molecularly Imprinted Materials—2003, P. Kofinas, M.J. Roberts, B. Sellergren, 2004, ISBN: 1-55899-725-3
Volume 788— Continuous Nanophase and Nanostructured Materials, S. Komarneni, J.C. Parker, J. Watkins, 2004, ISBN: 1-55899-726-1
Volume 789— Quantum Dots, Nanoparticles and Nanowires, P. Guyot-Sionnest, N.J. Halas, H. Mattoussi, Z.L. Wang, U. Woggon, 2004, ISBN: 1-55899-727-X
Volume 790— Dynamics in Small Confining Systems—2003, J.T. Fourkas, P. Levitz, M. Urbakh, K.J. Wahl, 2004, ISBN: 1-55899-728-8
Volume 791— Mechanical Properties of Nanostructured Materials and Nanocomposites, R. Krishnamoorti, E. Lavernia, I. Ovid'ko, C.S. Pande, G. Skandan, 2004, ISBN: 1-55899-729-6
Volume 792— Radiation Effects and Ion-Beam Processing of Materials, L. Wang, R. Fromknecht, L.L. Snead, D.F. Downey, H. Takahashi, 2004, ISBN: 1-55899-730-X
Volume 793— Thermoelectric Materials 2003—Research and Applications, G.S. Nolas, J. Yang, T.P. Hogan, D.C. Johnson, 2004, ISBN: 1-55899-731-8
Volume 794— Self-Organized Processes in Semiconductor Heteroepitaxy, R.S. Goldman, R. Noetzel, A.G. Norman, G.B. Stringfellow, 2004, ISBN: 1-55899-732-6
Volume 795— Thin Films—Stresses and Mechanical Properties X, S.G. Corcoran, Y-C. Joo, N.R. Moody, Z. Suo, 2004, ISBN: 1-55899-733-4
Volume 796— Critical Interfacial Issues in Thin-Film Optoelectronic and Energy Conversion Devices, D.S. Ginley, S.A. Carter, M. Grätzel, R.W. Birkmire, 2004, ISBN: 1-55899-734-2
Volume 797— Engineered Porosity for Microphotonics and Plasmonics, R. Wehrspohn, F. Garcial-Vidal, M. Notomi, A. Scherer, 2004, ISBN: 1-55899-735-0
Volume 798— GaN and Related Alloys—2003, H.M. Ng, M. Wraback, K. Hiramatsu, N. Grandjean, 2004, ISBN: 1-55899-736-9
Volume 799— Progress in Compound Semiconductor Materials III—Electronic and Optoelectronic Applications, D. Friedman, M.O. Manasreh, I. Buyanova, F.D. Auret, A. Munkholm, 2004, ISBN: 1-55899-737-7
Volume 800— Synthesis, Characterization and Properties of Energetic/Reactive Nanomaterials, R.W. Armstrong, N.N. Thadhani, W.H. Wilson, J.J. Gilman, Z. Munir, R.L. Simpson, 2004, ISBN: 1-55899-738-5
Volume 801— Hydrogen Storage Materials, M. Nazri, G-A. Nazri, R.C. Young, C. Ping, 2004, ISBN: 1-55899-739-3
Volume 802— Actinides—Basic Science, Applications and Technology, L. Soderholm, J. Joyce, M.F. Nicol, D. Shuh, J.G. Tobin, 2004, ISBN: 1-55899-740-7
Volume 803— Advanced Data Storage Materials and Characterization Techniques, J. Ahner, L. Hesselink, J. Levy, 2004, ISBN: 1-55899-741-5
Volume 804— Combinatorial and Artificial Intelligence Methods in Materials Science II, R.A. Potyrailo, A. Karim, Q. Wang, T. Chikyow, 2004, ISBN: 1-55899-742-3
Volume 805— Quasicrystals 2003—Preparation, Properties and Applications, E. Belin-Ferré, M. Feuerbacher, Y. Ishii, D. Sordelet, 2004, ISBN: 1-55899-743-1
Volume 806— Amorphous and Nanocrystalline Metals, R. Busch, T. Hufnagel, J. Eckert, A. Inoue, W. Johnson, A.R. Yavari, 2004, ISBN: 1-55899-744-X

MATERIALS RESEARCH SOCIETY SYMPOSIUM PROCEEDINGS

Volume 807— Scientific Basis for Nuclear Waste Management XXVII, V.M. Oversby, L.O. Werme, 2004, ISBN: 1-55899-752-0

Volume 808— Amorphous and Nanocrystalline Silicon Science and Technology—2004, R. Biswas, G. Ganguly, E. Schiff, R. Carius, M. Kondo, 2004, ISBN: 1-55899-758-X

Volume 809— High-Mobility Group-IV Materials and Devices, M. Caymax, E. Kasper, S. Zaima, K. Rim, P.F.P. Fichtner, 2004, ISBN: 1-55899-759-8

Volume 810— Silicon Front-End Junction Formation—Physics and Technology, P. Pichler, A. Claverie, R. Lindsay, M. Orlowski, W. Windl, 2004, ISBN: 1-55899-760-1

Volume 811— Integration of Advanced Micro- and Nanoelectronic Devices—Critical Issues and Solutions, J. Morais, D. Kumar, M. Houssa, R.K. Singh, D. Landheer, R. Ramesh, R. Wallace, S. Guha, H. Koinuma, 2004, ISBN: 1-55899-761-X

Volume 812— Materials, Technology and Reliability for Advanced Interconnects and Low-k Dielectrics—2004, R. Carter, C. Hau-Riege, G. Kloster, T-M. Lu, S. Schulz, 2004, ISBN: 1-55899-762-8

Volume 813— Hydrogen in Semiconductors, N.H. Nickel, M.D. McCluskey, S. Zhang, 2004, ISBN: 1-55899-763-6

Volume 814— Flexible Electronics 2004—Materials and Device Technology, B.R. Chalamala, B.E. Gnade, N. Fruehauf, J. Jang, 2004, ISBN: 1-55899-764-4

Volume 815— Silicon Carbide 2004—Materials, Processing and Devices, M. Dudley, P. Gouma, P.G. Neudeck, T. Kimoto, S.E. Saddow, 2004, ISBN: 1-55899-765-2

Volume 816— Advances in Chemical-Mechanical Polishing, D. Boning, J.W. Bartha, G. Shinn, I. Vos, A. Philipossian, 2004, ISBN: 1-55899-766-0

Volume 817— New Materials for Microphotonics, J.H. Shin, M. Brongersma, F. Priolo, C. Buchal, 2004, ISBN: 1-55899-767-9

Volume 818— Nanoparticles and Nanowire Building Blocks—Synthesis, Processing, Characterization and Theory, O. Glembocki, C. Hunt, C. Murray, G. Galli, 2004, ISBN: 1-55899-768-7

Volume 819— Interfacial Engineering for Optimized Properties III, C.A. Schuh, M. Kumar, V. Randle, C.B. Carter, 2004, ISBN: 1-55899-769-5

Volume 820— Nanoengineered Assemblies and Advanced Micro/Nanosystems, J.T. Borenstein, P. Grodzinski, L.P. Lee, J. Liu, Z. Wang, D. McIlroy, L. Merhari, J.B. Pendry, D.P. Taylor, 2004, ISBN: 1-55899-770-9

Volume 821— Nanoscale Materials and Modeling—Relations Among Processing, Microstructure and Mechanical Properties, P.M. Anderson, T. Foecke, A. Misra, R.E. Rudd, 2004, ISBN: 1-55899-771-7

Volume 822— Nanostructured Materials in Alternative Energy Devices, E.R. Leite, J-M. Tarascon, Y-M. Chiang, E.M. Kelder, 2004, ISBN: 1-55899-772-5

Volume 823— Biological and Bioinspired Materials and Devices, J. Aizenberg, C. Orme, W.J. Landis, R. Wang, 2004, ISBN: 1-55899-773-3

Volume 824— Scientific Basis for Nuclear Waste Management XXVIII, J.M. Hanchar, S. Stroes-Gascoyne, L. Browning, 2004, ISBN: 1-55899-774-1

Volume 825E—Semiconductor Spintronics, B. Beschoten, S. Datta, J. Kikkawa, J. Nitta, T. Schäpers, 2004, ISBN: 1-55899-753-9

Volume 826E—Proteins as Materials, V.P. Conticello, A. Chilkoti, E. Atkins, D.G. Lynn, 2004, ISBN: 1-55899-754-7

Volume 827E—Educating Tomorrow's Materials Scientists and Engineers, K.C. Chen, M.L. Falk, T.R. Finlayson, W.E. Jones Jr., L.J. Martinez-Miranda, 2004, ISBN: 1-55899-755-5

Prior Materials Research Society Symposium Proceedings available by contacting Materials Research Society

Part I: Processing—Vapor and Electro-Deposition

A. Experiments

Mat. Res. Soc. Symp. Proc. Vol. 821 © 2004 Materials Research Society P2.8

Structure-property relationships of Au films electrodeposited on Ni

C. San Marchi, N.R. Moody, M.J. Cordill**, G. Lucadamo, J.J. Kelly, T. Headley*, N. Yang
Sandia National Laboratories, Livermore, CA 94550
*Sandia National Laboratories, Albuquerque, NM 87185
**University of Minnesota, Minneapolis MN 55455

Thin gold films and coatings on metal have long constituted an important technology for the microelectronics industry and will continue to be important for microdevices such as contact springs. The properties of these materials may be highly processing dependent, particularly when the gold is deposited by electrochemical means. In this study, we characterize gold electrodeposited on Ni substrates from two bath chemistries: hard Au sulfite with proprietary hardening additive and soft Au cyanide. TEM and SEM show that the bath chemistry alters the microstructure and the resulting surface of the electrodeposits. Nanoindentation techniques were used to determine the elastic and plastic properties of the Au electrodeposits as a function of the specifics of processing. Soft Au electrodeposits have a grain size of on the order of 300 nm and a hardness of about 1 GPa. Hard Au electrodeposits produced from the sulfite bath feature grain sizes as small as 30 nm, some twinning, and fine porosity uniformly distributed both within the grains and at grain boundaries. The hardness is about 2 GPa, approaching the hardest values reported for sputtered gold films. The effect of the hardening agent on the microstructure of electrodeposits from the Au sulfite bath was also investigated and found to significantly refine the grain size at concentrations of at least 4 mL/L, although little additional refinement was found at higher concentrations.

INTRODUCTION

Au electrodeposits are ubiquitous in the microelectronics industry serving numerous technical needs [1-2]. Pure or soft Au deposits are used for their high electrical conductivity, for corrosion resistance and for joining. Soft gold easily deforms, as the name implies, and while useful for bonding and joining, for contact applications hard gold is preferred for improved wear resistance. Hard gold is typically achieved with the addition of transition metal complexes and result in very fine-grained microstructures. Paradoxically, the fine grain size results in a fast diffusional conduit for transporting the transition metals to the free surface where they oxidize and increase electrical contact resistance. As localized heating is common in electrical contacts, especially at high current densities, thermal stability is an important consideration for transition metal-hardened Au electrodeposits and for thin Au electrodeposits on transition metals [3]. In this study, the microstructure and basic mechanical properties (modulus and hardness) of Au electrodeposits from Au cyanide and Au sulfite bath chemistries are compared. In addition, the effect of a proprietary hardening additive on the character of the Au sulfite electrodeposits is investigated.

EXPERIMENTAL DETAILS

Two basic chemistries were used for electrodeposition of gold and are referred to as hard Au sulfite and Au cyanide throughout this text. The Au sulfite is a commercial electrolyte of sodium auro sulfite called Technigold 25E (Technic, Rhode Island) and used at pH 6.5 [4]. This solution has a proprietary additive (12 mL/L) used to refine the grain size of the deposit and to produce high hardness. The Au cyanide electrolyte is a phosphoric acid chemistry discussed previously in the literature [5]. All deposits were made at 50°C and 2 mA/cm^2.

For mechanical analysis, a thick Ni substrate was prepared by electroplating nickel from a nickel sulfamate bath onto wrought Ni discs 2 cm in diameter. The Ni electrodeposit was on the order of 100 μm thick and subsequently polished to a mirror finish. The nickel was cleaned with a commercial cleaning solution and activated in HCl. The final processing steps were varied. In one case, gold was electrodeposited directly onto the cleaned nickel surface from the hard Au sulfite electrolyte (S1). In the other cases, a Ni Wood's strike (a thin layer of nickel electrodeposited from an all chloride bath) was deposited prior to electrodeposition of the gold from the hard Au sulfite bath (S2 and S3), or the Au cyanide bath (C1). In the case of S2, there was a delay before the plating current was applied which reduces the adhesion of the Au electrodeposit. The process steps used to produce the Au electrodeposits are summarized in Table 1.

For microstructural analysis, gold was electrodeposited to thicknesses on the order of 10 μm from the Au baths on free standing Ni electrodeposits about 100 μm in thickness. In addition to varying the final processing steps, the effect of the hardening additive on the microstructure was investigated. Although the vendor specifies 12 mL/L of additive, electrodeposits were also produced from Au sulfite baths with 0, 2, 4, and 6 mL/L of the hardening additive, Table 1.

Nanoindentation was performed on a Nano Indenter XP™ with a Berkovich diamond indenter. An excitation frequency of 45 Hz was used with a displacement of 2 nm for continuous stiffness measurement. Transmission electron microscopy (TEM) was used to analyze the fine microstructure of the Au electrodeposits. Three millimeter disks were punched from the Au coated thin Ni substrates. This usually caused partial debonding of the gold from the nickel such that the gold could be peeled from the Ni substrate resulting in free-standing Au foils. These foils were then ion milled from both sides to achieve perforation near the middle of the foil thickness. An Au grid was similarly milled to confirm that observed features were not, in fact, milling artifacts. Standard θ–2θ X-ray diffraction (XRD) was used to determine the texture of the gold and the Au (111) reflection was used to estimate the crystallite size. Pole figures for the Au (111), (200), (220) and (311) directions were also collected. In a few cases, the Ni substrate could be observed; weak Ni (111) peak appeared in the Au (200) at an angle of 54°which is consistent with a Ni (200) grains exhibiting strong out-of-plane texture. Given the (200) out-of-

Table 1. Processes used to produce Au electrodeposits.

Sample	Process
For mechanical analysis (1-2μm thick gold)	
C1	Clean, HCl, Au cyanide
S1	Clean, HCl, hard Au sulfite (12mL/L of hardening additive)
S2	Clean, HCl, Ni Wood's strike, hard Au sulfite COLD* (12mL/L additive)
S3	Clean, HCl, Ni Wood's strike, hard Au sulfite (12mL/L additive)
For microstructural analysis (10 μm thick gold)	
S3-0	as S3 except no additive
S3-2	as S3 except 2mL/L of hardening additive
S3-4	as S3 except 4mL/L of hardening additive
S3-6	as S3 except 6mL/L of hardening additive
S3-12	Same as S3 except deposited thickness is greater

* COLD indicates a delay before the current was applied to the Au bath

plane texture of the Ni substrate, the Ni substrate does not interfere with the features observed in the other pole figures for Au.

RESULTS AND DISCUSSION

Effect of electrodeposition process and chemistry

The electrodeposits from the Au cyanide bath (C1) have a matte appearance and coppery color, while the electrodeposits from the hard Au sulfite chemistry (S3) have a mirror-like, polished appearance and a true Au color. Comparison of these surfaces at high magnification, Figure 1, shows the presence of surface features on the same scale as the grain size and that the grain size is an order of magnitude greater in C1 than S3, ~300 nm and ~30 nm respectively. The crystallite size could not be quantitatively determined for C1 from the XRD analysis, although the data indicate sizes greater than 100 nm. For sample S3, a crystallite size of 32 nm was determined from Au (111) peak broadening. A majority of the grains feature twins when deposited from Au cyanide (C1), while significantly fewer twins are present in material from the hard Au sulfite chemistry (*e.g.*, S3), Figure 2.

Substantial porosity is also evident in these electrodeposits as shown in Figure 2. The pores are nearly spherical and believed to be the consequence of hydrogen evolution during deposition [6]. In C1, the pores vary in size from a few nanometers to as large as 20 nm in diameter. These pores exist predominantly at grain boundaries and twin boundaries, especially the largest pores. Intragranular pores also exist, however these tend to have a smaller size distribution. The porosity in S3 is similar in nature, but with significantly higher density. In this instance, the pores are more uniformly distributed, both at boundaries and within the grains. The pores in S3 are also smaller, typically around 5 nm in diameter, and more uniform in size than in C1.

An out-of-plane fiber texture was observed for both C1 and S3. Pole figures show that C1 has a bimodal (111)/(200) texture oriented out-of-plane. The central intensity of the (111) pole is moderate and greater than the (200) pole intensity. In comparison, S3 exhibits weak (111) out-of-plane texture.

The thin Au electrodeposits on thick Ni substrates appear quite robust except for S2, where gold spalls off in localized regions of a few millimeters in size. The majority of S2, nevertheless, appears well adhered. There is no evidence of spalling or failure at the interface between the gold and Ni substrate as a consequence of indentation testing in any of the thin (1 μm) electrodeposits (C1, S1, S2 and S3).

Figure 1. SEM images of nanoindentation marks in (a) C1 and (b) S3. The relatively coarse grain size in sample C1 is apparent from the relatively rough as-deposited surface compared to sample S3.

The elastic modulus and hardness data are shown in Table 2. The quoted values are average values at depths between 100 and 200 nm. This range avoids surface effects and influence of the substrate; the thickness of these Au electrodeposits is 1-2 μm. Hardness data are shown in Figure 3 as a function of displacement into the surface; these data represent the average of at least 14 indents. The measured modulus is near the bulk modulus of gold (80 GPa) despite the elastic anisotropy of gold [7] and the texture of these electrodeposits. The comparatively rough surface of C1, Figure 1, may account for the relatively large uncertainty in those measurements. The lower hardness of C1 is due to its coarser grain size compared to the other electrodeposits. The hard Au electrodeposits display a hardness of about 2 GPa, which is in the range reported for sputter deposited gold films [8,9]. S1, S2 and S3 have similar properties indicating that the Ni Woods strike does not have a significant effect on the inherent mechanical properties of the Au electrodeposit. It remains to be seen whether the adhesion of the gold to the nickel is improved by the Ni strike, but it is clear the application of the current adversely affects the adhesion.

The effect of the nickel substrate is observed at relatively low displacements into the surface and dominates at depths over 1000 nm. For all materials the modulus converges to about 150 GPa for displacements greater than 1500 nm and the hardness converges to values near 3 GPa. These values approach the values obtained for the Ni substrate in S2 in a region where the Au had spalled off: 180 GPa and 3.8 GPa for the modulus and hardness respectively.

Effect of hardening additive

The thickness of the Au electrodeposits from baths with varying amounts of hardening agent is significantly greater than the samples used for nanoindentation (on the order of 10 μm compared 1 μm) resulting in residual stresses that are greater in these electrodeposits. This may account for the spontaneous spallation observed for samples S3-6 and S3-12. Indeed, the spallation and fracture of the thick (10 μm) Au electrodeposit was so severe that TEM foils could not be prepared for thick versions of S3 (*i.e.*, S3-12).

The samples with no additive (S3-0) and the lowest concentration (S3-2) have similar microstructures: for both electrodeposits the grain size is on the order of 1 μm and significant twinning is apparent, Figure 4. The texture, however, is somewhat different. S3-0 has a strong (111) out-of-plane texture with a secondary preferred (200) orientation that is also rather strong. The out-of-plane texture in S3-2 is moderate in comparison, but still (111), while the secondary preferred orientation is (311). Fine porosity on the order of a few nanometers in diameter exists in these two samples and increases with additive concentration.

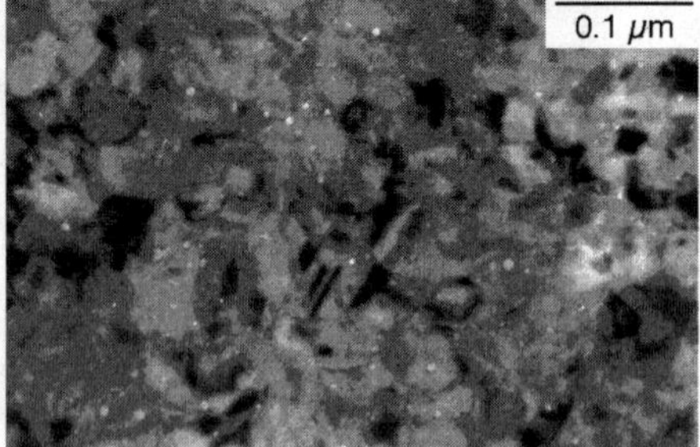

Figure 2. TEM micrographs of (a) C1 and (b) S3. The white circular features are pores.

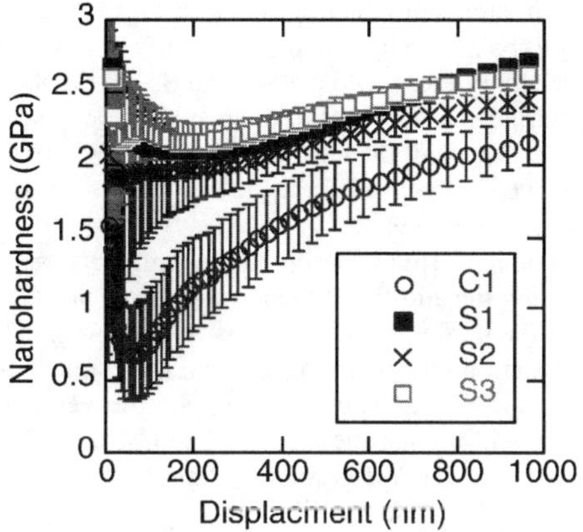

Figure 3. Nanohardness as a function of displacement into the surface.

Table 2. Modulus and hardness data from nanoindentation testing, average values at depths between 100 to 200 nm for at least 14 indents. The standard deviation in these measurements is quoted as the uncertainty. (S1, S2 and S3 were deposited from electrolyte containing 12 mL/L of hardening additive).

	Modulus (GPa)	Hardness (GPa)
C1	73 ±16	1.0 ±0.3
S1	79 ± 3	2.1 ±0.1
S2	74 ± 7	1.9 ±0.3
S3	80 ± 4	2.2 ±0.2
Ni substrate	181 ±11	3.8 ±0.5

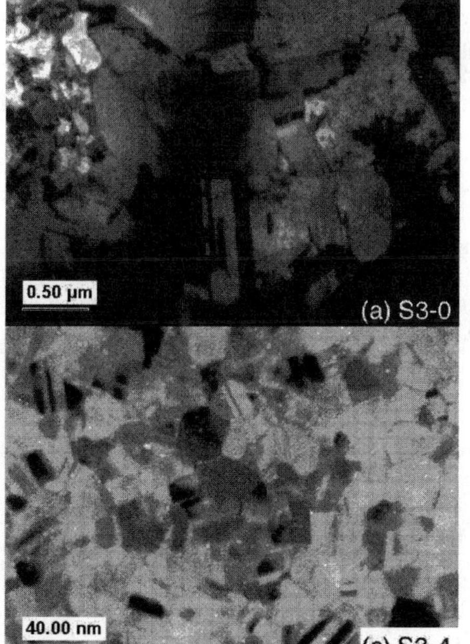

Figure 4. TEM images showing microstructure of samples as a function of concentration of hardening additive; (a) S3-0, Au sulfite with no hardening additive; (b) S3-2, with 2 mL/L of hardening additive; (c) S3-4, with 4 mL/L of additive; and (d) S3-6, with 6 mL/L of additive.

Materials S3-4, S3-6 and S3-12 have approximately equivalent grain size, Figure 2 and 4; the crystallite size was determined from XRD data to be 27, 25 and 32 nm respectively. S3-4 did not display any texture, the grain structure is essentially random, while S3-6 and S3-12 display weak (111) out-of-plane texture. The porosity also continues to increase in concentration and size as the additive concentration increases. As might be expected, the current efficiency is observed to decrease as the concentration of hardening additive is increased since the low efficiency is associated with the evolution of hydrogen on the Au surface.

Nanoindentation of these samples has not been performed. Based on the similarity in crystallite size for S3-4, S3-6 and S3-12, increasing the additive concentration to greater than 4 mL/L is not expected to have a large effect on the hardness. A high density of nano-sized pores can provide effective hardening by pinning dislocation motion in metals, and may contribute to some difference in the hardness of S3-4 compared to S3-12. The present data, however, indicate that hardness near 2 GPa can be anticipated for S3-4. The hardness of S3-0 and S3-2 will certainly be much lower because of their comparatively coarse grain size, although, the stiffness of these samples may be greater due to their strong (111) texture [7].

CONCLUSIONS

Hard Au electrodeposits from Au cyanide chemistry have hardness in excess of 1 GPa and grain size of about 300 nm. Electrodeposits from Au sulfite with 12 mL/L of proprietary hardening additive have hardness of 2 GPa and a grain size near 30 nm. The hardening additive has little effect on grain size at a concentration of 2 mL/L, but at a concentration of 4 mL/L the additive eliminates texture formation and refines the grain size to about 30 nm. Higher concentrations of additive cause significant increases in porosity, but have little effect on the grain size and presumably hardness.

ACKNOWLEDGEMENTS

The authors wish to thank Mark Rodriguez of Sandia National Laboratories, Albuquerque, NM for performing the diffraction measurements. Sandia is a multiprogram laboratory operated by Sandia Corporation, a Lockheed Martin Company, for the United States Department of Energy under contract DE-AC04-94AL85000.

REFERENCES

1. I.R. Christie and B.P. Cameron, *Gold Bulletin* **27**, 12 (1994).
2. Y. Okinaka and M. Hoshino, *Gold Bulletin* **31**, 3 (1998).
3. M. Antler, *Plat. Surf. Finish.* **85** (no. 12), 85 (1998).
4. R.J. Morrissey, *Plat. Surf. Finish.* **80** (no. 4), 75 (1993).
5. M. Bick, J.A. Lochet, US patent no. 3,878,068 (1975).
6. J.J. Kelly, N. Yang, T. Headley and J. Hachman, *J. Electrochem. Soc.* **150**, C445 (2003).
7. J.D. Kiely and J.E. Houston, *Phys. Rev. B* **57**, 12588 (1998).
8. N.R. Moody, D.P. Adams, D. Medlin, T. Headley, N. Yang, *Int. J. Fracture* **119/120**, 407 (2003).
9. N.R. Moody, D.P. Adams, D. Medlin, T. Headley, N. Yang, *Mater. Sci. Forum* **426-432**, 3403 (2003).

Mat. Res. Soc. Symp. Proc. Vol. 821 © 2004 Materials Research Society P3.22

Effect of Substrate Bias on Structure and Properties of W incorporated Diamond-like Carbon Films

Ai-Ying Wang and Kwang-Ryeol Lee
Future Technology Research Division, Korea Institute of Science and Technology,
P.O.Box 131, Cheongryang, Seoul, 130-650, South Korea

ABSTRACT

W incorporated diamond-like carbon (W-DLC) films were deposited on silicon (100) wafers by a hybrid deposition method combining ion beam deposition of carbon with DC magnetron sputtering of tungsten. During the films deposition, a wide range of negative bias voltage from 0 to −600 V was applied. W concentration in the film could be controlled by varying the Ar/C_6H_6 ratio in the supplying gas. In the present experimental condition, WC_{1-x} nano-sized particles were not observed in the amorphous carbon matrix. Regardless of the W concentration in the film, it was found that the G-peak position of the Raman spectra had a lowest value at a bias voltage of −200 V, which represents the highest sp^3 bond fraction in the film. The highest residual stress, hardness and Young's modulus were also observed when the bias voltage was −200 V. This result shows that the mechanical properties of W-DLC films were mainly dependent on the atomic bond structure of carbon. On the other hand, the electrical resistivity significantly decreased by the W incorporation.

INTRODUCTION

After Aisenberg *et al.* firstly reported the synthesis of diamond-like (DLC) carbon films using a low energy carbon ion beam [1], significant progress in the understanding the growth process of DLC films has been achieved in the last three decades. One noteworthy feature is that the incorporation of metallic components (Ti, W, Ag, Cu, Si, Au etc.) or other third elements (N, Ar, B) is used to overcome the shortcomings of DLC films such as high residual compressive stress, poor thermal stability and ductility [2-12]. Earlier studies on the third element addition showed a considerable improvement of the physical properties of the metal containing DLC films (Me-DLC). Recently, nanocomposite films composed of nano-sized particulates dispersed in the matrix phase drew much attention due to its potential of unique properties. Nevertheless, the size, volume fraction and distribution of the nanometer metallic particulates formed in the films must be optimized carefully [13].

In the present work, we report the deposition behavior of W incorporated DLC (W-DLC) films. It is widely accepted that the Me-DLC films consist of the embedded nanometer metallic particulates (if metal formed carbides) and amorphous carbon or hydrocarbon matrix. However, we focus on the deposition behavior of the films at low W concentration where WC_{1-x} particulate phase does not form. The dependence of the structure and properties of the deposited film on the negative bias voltage are studied by a comprehensive set of experiments. We confirm that the W incorporation does not change the deposition mechanism of the DLC films. This result is useful for understanding the structure and the properties of WC-C nanocomposite films of high W concentrations [14].

EXPERIMENTAL

W-DLC films were prepared by a novel hybrid deposition system consisting of a DC magnetron sputter gun of W target (high purity 99.95 %) and an end-Hall type ion gun. C_6H_6 was supplied to the ion gun for the carbon film deposition. W concentration in the film was varied by changing the Ar fraction of the reaction gas in the range of 65~85%. The base pressure was kept less than 2×10^{-4} Pa, but the deposition pressure ranged from 0.08 to 0.133 Pa depending on the Ar fraction. A p-type Si (100) wafer of thickness 500±10 μm was used as the substrate. A thin Si wafer of thickness 200±10 μm was also used to measure the residual compressive stress of the film. The deposition rate of the film decreased with increasing Ar fraction. The deposition time was adjusted to obtain films of thickness 350±60 nm. A substrate bias voltage ranging from 0 to −600 V in DC was applied by a medium frequency (250 KHz) power supply. For ease of comparison, a series of pure DLC films were also deposited by the ion beam method without operating the sputter gun.

The W concentration in the film was evaluated by RBS using a collimated $^4He^{2+}$ ion beam of 2 MeV. The film thickness was measured by an Alphastep-200 profilometer. The microstructures of the films were characterized by micro-Raman spectroscope using a 514.5 nm Ar^+ laser. The mechanical properties were measured by the nano-indentation technique in a continuous stiffness measurement (CSM) mode with a maximum indentation depth of 350 nm. The characteristic hardness of the film was chosen in the depth where the measured value is not affected by the soft Si substrate. The residual stress of the film was calculated from the curvature of the film/substrate composite using the Stoney equation. The electrical resistivity of the film was measured by a four-point probe system (1.0 mm equal probe spacing).

RESULTS AND DICCUSSION

Typical RBS spectra of the W-DLC films deposited at the substrate bias voltage of –200 V are shown in figure 1. The spectra show that the W concentration increased with increase of Ar fraction. The W concentration in the film increased from 2.9 to 8.5 at.% as the Ar fraction in the supplying gas increased from 65% to 85% (figure 2). In the present concentration range, the WC_{1-x} crystalline phase was not observed in the film as confirmed by the grazing incidence x-ray diffraction (GIXRD) method. It must be noted that the W concentration was mainly dependent on the Ar fraction. Even if the applied bias voltage slightly varied the W concentration, the variation was limited to ±0.5 at.% in the range of the bias voltage from 0 to –600 V.

Figure 3 shows the residual compressive stress of the film as a function of the bias voltage. W incorporation into DLC films considerably reduced the residual compressive stress, especially

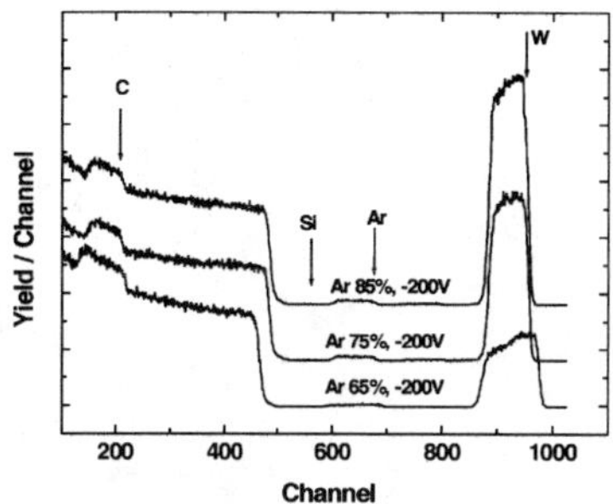

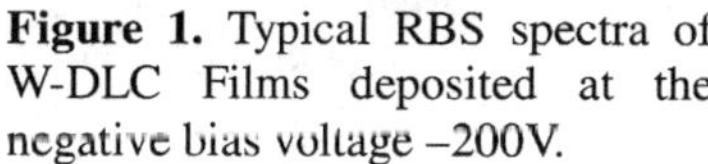

Figure 1. Typical RBS spectra of W-DLC Films deposited at the negative bias voltage –200V.

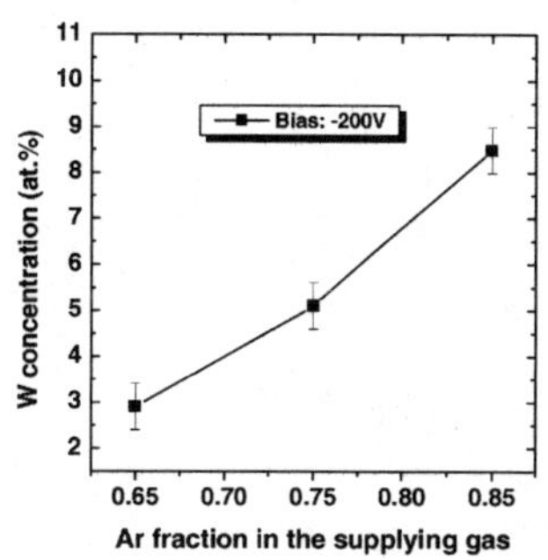

Figure 2. Dependence of the W concentration on the Ar fraction in the reaction gas.

when the negative bias voltage was small. The residual compressive stress decreased from 2.27 to 0.58 GPa when the negative bias voltage was 0 V. As the negative bias voltage increased, the reduction in the residual stress became less significant. When the negative bias voltage was –600V, the reduction in the residual stress caused by the W incorporation was not observed. As in the pure DLC film, the residual stress of the W-DLC has a maximum at –200 V for all W concentrations. This behavior can be explained by a simple model by Davis for the ion bombardment phenomena during thin film deposition [15]. This model suggests that the compressive stress is mainly balanced by the knock-on processes causing the stress formation and the thermal spike effect causing the stress relaxation in the film. A maximum residual stress, as observed in the present work, was successfully predicted by the kinetic model.

Mechanical properties of the film exhibited similar behavior. The dependence of the film hardness on the bias voltage is summarized in figure 4. The maximum hardness was observed at a bias voltage of –200 V regardless of the W concentration. The plane strain modulus of the film exhibited the same dependence on the negative bias voltage. These results show that the mechanical properties and the residual compressive stress of the W-DLC film are determined by the same structural factors. It must be noted that when W is incorporated in the film without forming WC_{1-x} particulates, the structure-property relationship was essentially the same as that in pure DLC film.

Figure 5 shows the representative Raman spectra of the deposited WC-C films as a function of the bias voltage when the W concentration was 2.9 at.%. In the range of 1100~1700 cm^{-1}, there is a broad asymmetric Raman scattering band, representing the typical characteristic of DLC film. We found that the intensity of the Raman spectra decreased with increasing bias voltage. However, the reason for this phenomenon is yet to be resolved. The Raman spectrum analysis of amorphous carbon includes deconvolution of the spectrum with two Gaussian peaks: the G and D peaks. The atomic bond structure of the film is characterized by the intensity ratio,

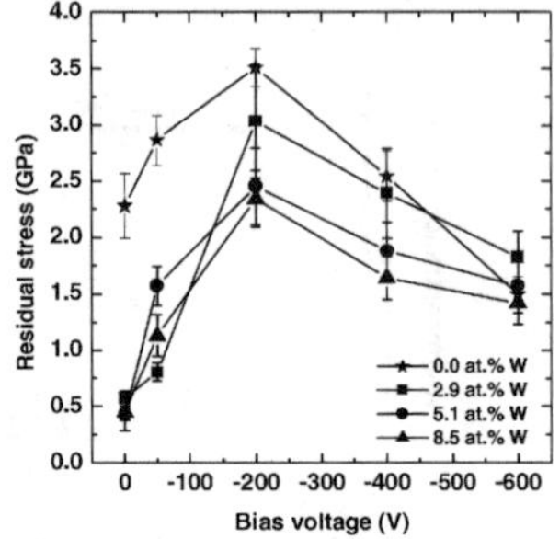

Figure 3. The residual compressive stress of the films as a function of the bias voltages.

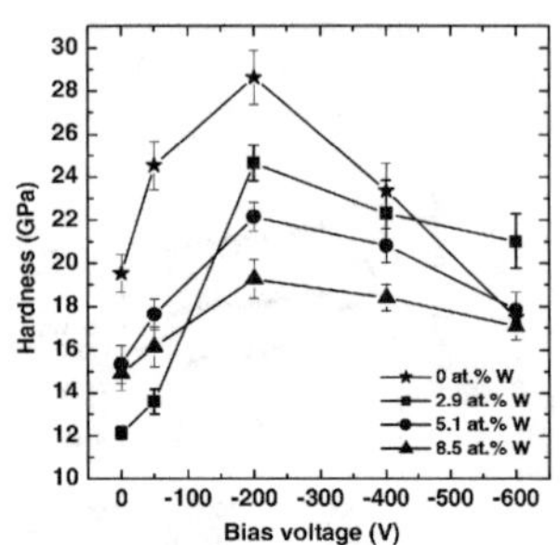

Figure 4. The hardness of the films as a function of the bias voltages.

the full width at half maximum (FWHM) values, or the position of each peak. In the present work, G-peak position was used to qualitatively compare the atomic bond structure, since the error in G-peak position was much smaller than other characteristics. The G-peak positions as a function of bias voltage are shown in figure 6. It was reported that the G-peak position shifted upward due to the residual compressive stress by 4.1 cm^{-1}/GPa [17]. G-peak position in figure 6 was after the correction of the residual stress effect. It is obvious that the G-peak position runs through a minimum at –200 V and then increases with further increment of the bias voltage. In addition, increasing W incorporation also resulted in an increase of the G-peak position. Since the lower G-peak position represents a higher fraction of sp^3 carbon bonding [17, 18], it can be said that the film deposited at the bias voltage of –200 V has the highest fraction of sp^3 carbon

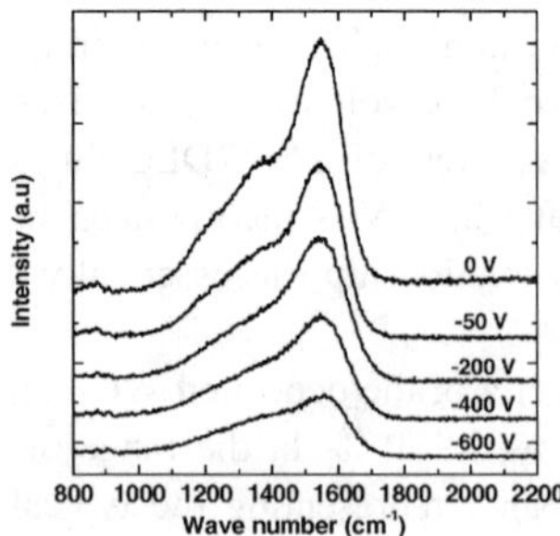

Figure 5. Representative Raman spectra of the deposited films as a function of bias voltage.

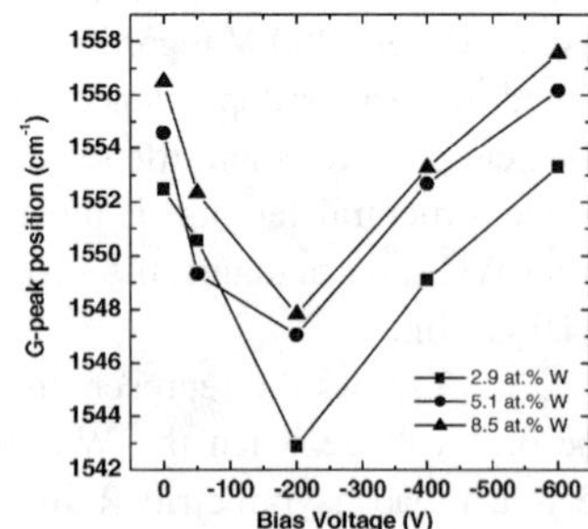

Figure 6. Dependence of G-peak position on the negative bias voltage for various W concentrations.

bond. This result shows that the residual compressive stress and the mechanical properties are closely related to the sp^3 hybridized bond. The mechanical properties of carbon materials are dependent on the degree of 3-dimensional interlinks of the atomic bond network. For example, atomic bonds in diamond are fully interlinked in 3-dimensions while for graphite 2-dimensional graphitic planes are bonded by a weak Van der Waals force. High content of sp^3 hybridized bond would increase the 3-dimensional interlinks, resulting in the high hardness and elastic modulus. However, increasing 3-dimensional interlinks in amorphous carbon also increases the distortion of the carbon bond, which causes the high residual stress because the covalent bond is highly directional. It is worth noting that the carbon bond network mainly determined the mechanical properties of W-DLC film even if a significant amount of W was incorporated.

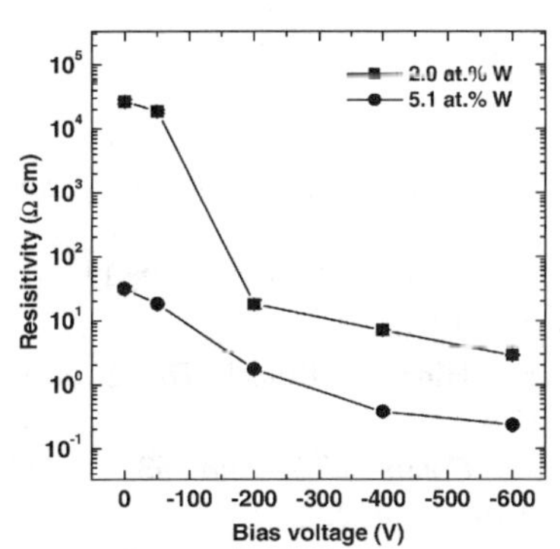

Figure 7. Electrical resistivity of the deposited WC-C films as a function of the bias voltage.

However, W incorporation significantly decreased the electrical resistivity as shown in figure 7. The resistivity of the film decreased by a factor of 3~5 due to W incorporation. (The resistivity of pure DLC films is in the order of 10^7 Ω cm.) This result shows that the W incorporation changed the electronic band structure. It is also noted that the G-peak position shown in figure 6 shifted upward with increasing W concentration. Hence, the W incorporation also increased the sp^2 hybridized bond that can increase unpaired π electrons. At a fixed W concentration, increasing the bias voltage further decreased the resistivity. This decrease would be due to the change in the atomic bond structure from the diamond-like phase to the graphitic one. This result shows that the W incorporation can control the electrical properties without causing considerable changes in the carbon bond structure or the mechanical properties.

CONCLUSIONS

W-DLC films were deposited using a novel hybrid deposition system composed of DC magnetron sputtering and an end-Hall ion gun source. The dependence of the atomic bond structure and the mechanical properties on the negative bias voltage and W concentration were comprehensively studied. For all negative bias voltages, increasing W concentration decreased the content of sp^3 hybridized bonding. However, the Raman analysis proposed that the highest content of sp^3 hybridized bond was obtained at a negative bias voltage of –200 V, where the maximum in the residual compressive stress, the hardness and the elastic modulus of the film were also observed. Despite a significant addition of W, the mechanical properties of the deposited W-DLC films are mainly dependent on the changes in the atomic bond structure of carbon. The mechanical properties of WC-C nanocomposite film are discussed in terms of the present experimental results [14]. W incorporation made it possible to control the electrical resistivity without a considerable change in the mechanical properties of the W-DLC films.

ACKNOWLEDGEMENTS

This research was performed with the financial support of the Center for Nanostructured Materials Technology under the 21st Century Frontier R&D Program of the Ministry of Science and Technology, Korea. The authors would like to thank Dr. Jun-Hee Han for the mechanical properties measurements and Miss Min-Kyung Choi for Raman spectrum analysis.

REFERENCES

1. S. Aisenberg, R. Chamot, *J. Appl. Phys.* **42**, 2953 (1971).
2. C.-S. Lee, K.-R. Lee, K. Y. Eun, K.-H. Yoon, J.-H. Han, *Diamond Relat. Mater.* **11**, 198 (2002).
3. S. Zhang, Y. Q. Fu, H. J. Du, X. T. Zeng, Y. C. Liu, *Surf. Coat. Technol.* **162**, 42 (2002).
4. W. J. Meng, B. A. Gillispie, *J. Appl. Phys.* **84**, 4314, (1998).
5. K. I. Schiffmann, M. Fryda, G. Goerigk, R. Lauer, P. Hinze, A. Bulack, *Thin Solid Films*, **347**, 60 (1999).
6. V. I. Ivanov-Omskii, A. B. Lodygin, S. G. Yastrebov, *Comp. Sci. Technol.* **63**, 1193 (2003).
7. S. H. Yang, H. Kong, K. R. Lee, S. J. Park, D. E. Kim, *Wear*, **252**, 70 (2002).
8. W. Zhang, A. Tanaka, K. Wazumi, Y. Koga, *Thin Solid Films,* **416**, 145 (2002).
9. K. Oguri, T. Arai, *Surf. Coat. Technol.* **47**, 710(1991).
10. S.-J. Park, K.-R. Lee, D. H. Ko, K. Y. Eun, *Diamond Relat. Mater.* **11**, 1747 (2002).
11. X. Nie, J. C. Jiang, L. D. Tung, L. Spinu, E. I. Meletis, *Thin Solid Films,* **415**, 211 (2002).
12. Y. Y. Chang, D. Y. Wang, W. T. Wu, *Thin Solid Films*, **420-421**, 241 (2002).
13. S. Zhang, D. Sun, Y. Q. Fu, H. J. Du, *Surf. Coat. Technol.* **167**, 113 (2003).
14. A. Y. Wang, T. Y. Kim, J.-H. Han, K.-R. Lee (to be published).
15. C. A. Davis, *Thin Solid Films*, **226**, 30 (1993)
16. J. Robertson, *Mater. Sci. Eng. R*, **37**, 129(2002)
17. J.-K. Shin, C.-S. Lee, K.-R. Lee, K. Y. Eun, *Appl. Phys. Lett.* **78**, 631 (2001)
18. A. C. Ferrari, J. Robertson, *Phy. Rev. B*, **61**, 14095 (2000)

Mat. Res. Soc. Symp. Proc. Vol. 821 © 2004 Materials Research Society P3.1

Effect of Mold Dimensions on Temperature Distribution of Sintered Body during Plasma Activated Sintering

Gil-Geun Lee
Div. of Materials Science and Engineering, Pukyong National University,
San 100, Yongdang-Dong, Nam-Gu, Busan 608-739, Korea

ABSTRACT

In the present study, the focus is on the analysis of the effect of the mold dimensions on the temperature distribution of a sintered body during plasma activated sintering. The temperature distribution of a cylindrical mold with various dimensions was measured using K-type thermocouples. The temperature homogeneity of the mold was studied based on the direction and dimensions of the mold. A temperature gradient existed in the radial direction of the mold during the plasma activated sintering. Also, the magnitude of the temperature gradient was increased with increasing sintering temperature. In the longitudinal direction, however, there was no temperature gradient. The temperature gradient of the mold in the radial direction decreased with increasing inner diameter of the mold and its length.

INTRODUCTION

The plasma activated sintering process has been used for preparing various functional materials, including nanostructured materials, composite materials, magnetic materials, and others [1]. The thermal energy for plasma activated sintering is obtained from an external electric current. In powder sintering, the use of an electric current is known to enhance the sintering kinetics, and therefore, high densities are achieved at low temperature or in shorter times compared to conventional sintering methods. In general, the external electric current is applied to the punch and passes through the powder and die. The passage of the electric current provides resistive heating of the powders and die by the Joule effect; that is, the electric power is used as the heating source for consolidating the compact. The powder characteristics; i.e., conductivity, size distribution and morphology, affect the Joule effect. Also, the degree of the Joule effect depends not only on the powder characteristics but also the punch and die; i.e., material, size and morphology, and other factors [2].

A homogeneous temperature distribution in the compact during plasma activated sintering is very important to obtain high performance functional materials with large dimensions. The temperature distribution of the compact when applying the external electric current depends on the Joule effect of the powder and die. During plasma activated sintering, however, it is difficult to maintain a homogeneous temperature distribution in the large compact due to short sintering time [3]. Until now, the determination of mold dimensions for plasma activated sintering has been carried out in light of experience. For practical plasma activated sintering applications, the relationship between the temperature distribution and mold dimensions should be verified.

In the present study, to obtain a basic guiding principle for the design of the mold for the plasma activated sintering, the homogeneity of the temperature distribution in the mold during plasma activated sintering is studied using a cylindrical mold with various dimensions.

EXPERIMENTAL DETAILS

Table 1 shows the dimensions of the die and punch of the cylindrical molds for the plasma activated sintering. Graphite has high thermal and electric conductivities, therefore, this material was used for the mold material. Figure 1 shows a schematic diagram of the mold for measuring the temperature distribution of the mold. To measure the inside temperature distribution of the mold, temperature measuring holes were made that had a diameter of 2 mm as shown in figure 1. The holes were made at intervals of 3 mm from the inner surface of the die in the radial direction. In the longitudinal direction of the die, the holes were made at intervals of 10 mm from the center of the die. The holes in the longitudinal direction were made at a distance of 3 mm from the inner surface of the die. The same amount of carbon black powder having an average particle size of 0.5 μm was used as the compact material to minimize the compact material effect or powder effect. The temperature of the mold compacted with carbon black powder increased from room temperature to 1000℃ in a 1×10^{-3} Torr atmosphere using the plasma activated sintering apparatus. The temperature control of the mold during heating was carried out using a pyrometer by measuring the temperature on the outside surface of the die. The temperature distribution of the die was measured using K-type thermocouples inserted into the temperature measuring holes. The measured temperature distributions of the dies were studied as functions of the direction and dimensions of the mold.

RESULTS AND DISCUSSION

Figure 2 shows the changes in temperature of the die with heating time for the various distances from the inner surface of the die in the case of mold I. The surface temperature is the controlled and measured temperature on the outside of the surface of the die by the pyrometer. The die temperature increased with decreasing distance from the inner surface of the die. Also, the temperature difference between the surface temperature and measured temperature at a

Table 1. Dimensions of the cylindrical molds.

	Die dimensions (mm)			Punch dimensions (mm)	
	Outer Diameter (D_o)	Inner Diameter (D_i)	Length (L)	Diameter (D)	Length (l)
I	50	12.5	50	12.5	30
II	50	22.5	50	22.5	30
III	50	30.5	50	30.5	30
IV	50	22.5	40	22.5	25
V	50	22.5	30	22.5	20

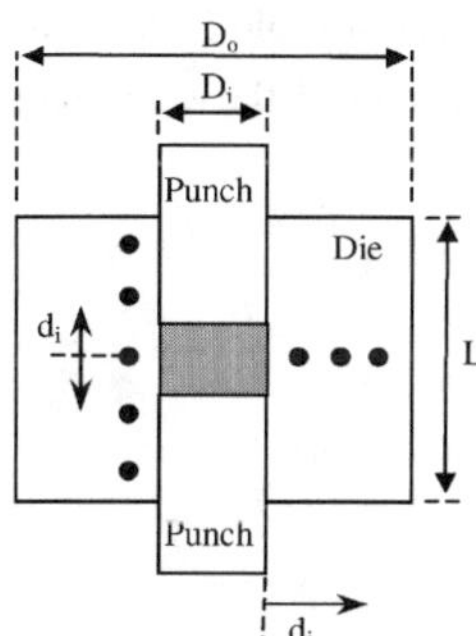

● : Temperature measuring hole
d_i : Temperature measuring distance

Figure 1. Schematic diagram of the mold for measuring temperature distribution in the mold.

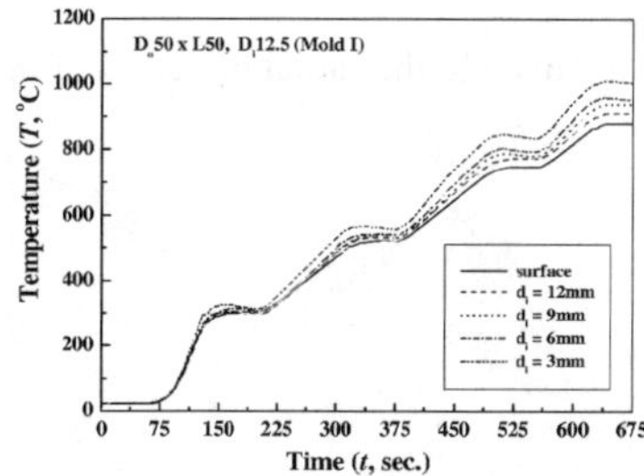

Figure 2. Changes in temperature of the die with heating time for various distances from the inner surface of the die for mold I.

specific distance increased with increasing distance between the measuring points of these temperatures. This means that the center of the die had a higher amount of heat generation by the Joule effect from the external electric current than that generated at the periphery of the die.

Figure 3 shows the changes in the temperature of the die with distance in the radial direction of the die for various surface temperatures in the case of mold I. The temperature of the die simultaneously decreased with the increasing distance in the radial direction of the die. This means that a temperature gradient existed in the radial direction of the die during plasma activated sintering. Also, the degree of the temperature gradient increased with the increasing controlled temperature or surface temperature as shown in figure 3. There is a good possibility that the powder compact has a temperature gradient in the radial direction during the plasma activated sintering, and the degree of temperature nonuniformity was significant with increasing sintering temperature and the dimensions.

Figure 4 shows the changes in the temperature of the die with distance in the longitudinal direction of the die for various surface temperatures in the case of mold I. In the horizontal axis of figure 4, the zero, positive and negative values mean the center of the die height, the upper

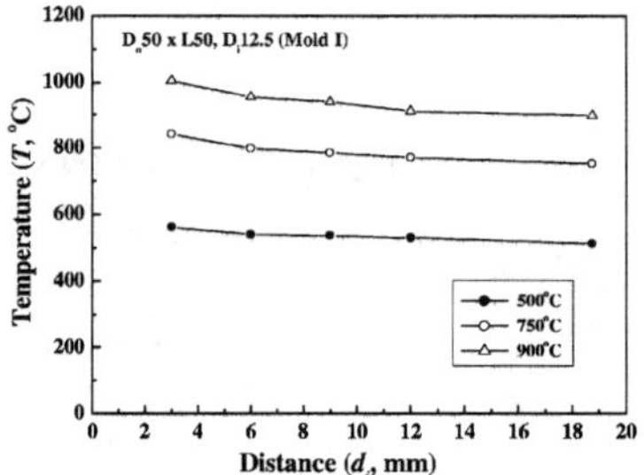

Figure 3. Changes in the temperature of the die with distance in the radial direction of the die for various surface temperatures for mold I.

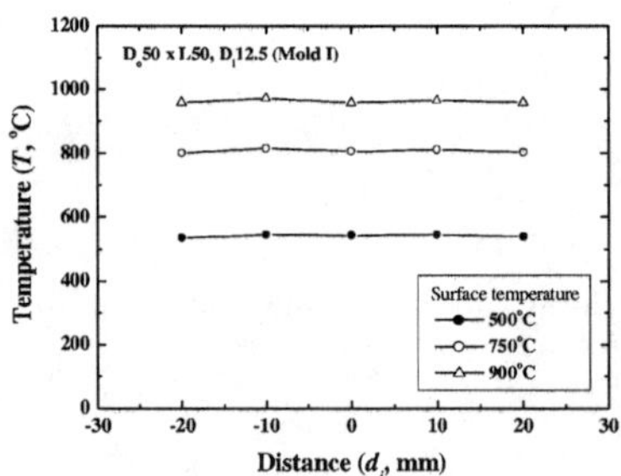

Figure 4. Changes in the temperature of the die with distance in the longitudinal direction of the die for various surface temperatures for mold I

part of the die and the lower part of the die, respectively. There is no difference in the temperature with distance, and the temperature distribution does not change with surface temperature or control temperature. This means that there is no temperature gradient in the longitudinal direction during the plasma activated sintering. Also, the powder compact has a homogeneous temperature distribution in the thickness direction. It is known that the temperature homogeneity of the powder compact or die is mainly dependent on the radial directional temperature distribution.

Figure 5 shows the changes in the temperature of the die with specific distance in the radial direction for the various inner diameters of the die (D_i) at a surface temperature of 900℃. The specific distance means the relative distance of the temperature measuring distance (d_i) to the die thickness (d_o). It is possible to make an exception of die thickness effect on the temperature distribution in the radial direction for specific distances. The temperature gradient decreased with the increasing inner diameter of die or diameter of the punch as shown in figure 5. Figure 6 shows the changes in the temperature of the die with specific distance for the various die lengths (L) in the radial direction. The temperature gradient of the die in the radial direction decreased with the increasing die length. It is known that the temperature distribution in the die changed with changing dimensions of the die: inner diameter and length. This means that the energy conversion behavior of the external electric energy by the Joule effect during the plasma activated sintering depends on the mold dimensions. There is a good possibility that the temperature uniformity of the powder compact has a very strong dependence on the die and punch dimensions.

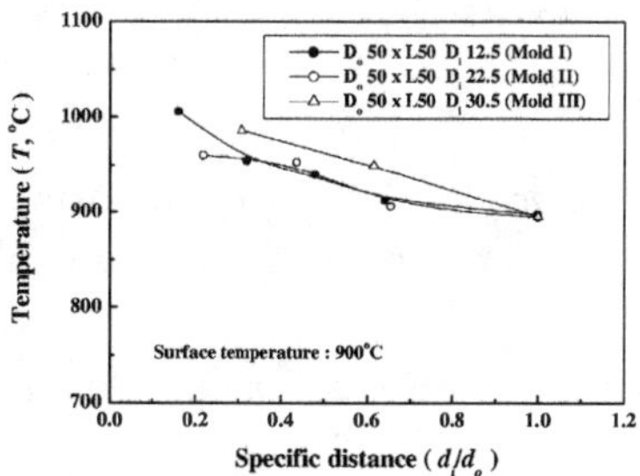

Figure 5. Changes in the temperature of the die with specific distance in the radial direction for various inner diameters.

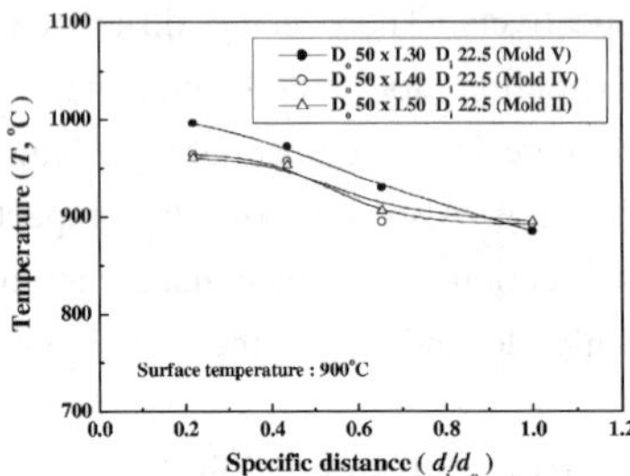

Figure 6. Changes in the temperature of the die with specific distance in the radial direction for various die lengths.

CONCLUSIONS

A temperature gradient existed in the radial direction of the mold during plasma activated sintering. The degree of the temperature gradient increased with increasing sintering temperature. There is a good possibility that the powder compact has a temperature gradient in the radial direction during plasma activated sintering, and the degree of temperature nonuniformity is significant with increasing sintering temperature and dimensions. In the longitudinal direction of the mold, however, there was no temperature gradient, and the temperature distribution did not change with the sintering temperature. The temperature gradient of the mold in the radial direction decreased with the increasing inner diameter of the mold and length. It was shown that the temperature uniformity of the powder compact strongly depends on the mold dimensions.

REFERENCES

1. J. R. Groza, "Field-Activated Sintering," *ASM Handbook Vol. 7 Powder Metal Technologies and Applications*, ed. ASM International Handbook Committee (ASM International, New York, 1998) pp. 583-589.
2. K. Matsugi, G. Sasaki and O. Yanagisawa, *J. Japan Society of Powder and Powder Metallurgy* **50** (3), 238-247 (2003).
3. S. Kihara, H. Ishida and S. Morioka, *J. Japan Society of Powder and Powder Metallurgy* **49** (6), 503-509 (2002).

Mat. Res. Soc. Symp. Proc. Vol. 821 © 2004 Materials Research Society P3.19

Synthesis and Study of Nanostructures Via Microwave Heating

Oxana V. Kharissova [1], Eder Zavala [1], Ubaldo Ortíz [2],
Jorge L. Hernández-Piñero [3], Stanislav Soloviev [4]

[1] Facultad de Ciencias Físico Matemáticas, Universidad Autónoma de Nuevo León, México
okhariss@ccr.dsi.uanl.mx , ederz@hotmail.com
[2] Facultad de Ingeniería Mecánica y Eléctrica, Universidad Autónoma de Nuevo León, México
uortiz@ccr.dsi.uanl.mx
[3] Facultad de Ciencias Biológicas, Universidad Autónoma de Nuevo León, México
[4] University of South Caroline, USA
soloviev@engr.sc.edu

ABSTRACT

This work is devoted to microwave heating of graphite, sucrose, calcined sucrose, and a mixture of graphite with sucrose to produce carbon nanotubes (CNT's). The samples were submitted to microwave radiation (power 800W, frequency 2.45 GHz) in air and high vacuum (10^{-5} Torr) for 30 – 60 min. The oven temperature was approximately 1200°C. After vaporization the condensed material was collected on various fused silica targets (different morphologies were used). The samples were found to contain a significant proportion of nanotubes, nanoparticles and fibers (1-2.8 micrometers), which appeared to be highly graphitized and helical structured. After deposition, the morphology of carbon nanotubes was studied with SEM, TEM and AFM techniques. It was observed that multi-walled nanotubes (MWNT's) were produced by this method. The morphology of fused silicon based substrates (SiO_2, SiC) was studied as an important factor for the growth of carbon nanotubes. Many aspects as the size and shape of the obtained nanotubes on different substrates (porous and non-porous fused silicon substrates) were achieved, as well as the concentration of them across the substrate and other properties.

INTRODUCTION

The science of carbon nanotubes has received a big enhancement in last years, especially since the development of standardized methods for its production in laboratories. The most recent efforts have been focused mainly in the reaching of a way for producing carbon nanotubes at a bigger scale, i.e., quantities bigger enough for being able to experiment on them easily and consequently getting a better quality of results on researches about properties and technologic applications. Numerous potential applications, such as flat panel displays [1], chemical sensors [2], hydrogen storage [3], etc., had been proposed. A number of methods such as arc discharge [4], laser vaporization [5], gas-phase pyrolysis [6,7], plasma-enhanced [8,9] or thermal chemical vapor deposition (CVD) [10,11], had been developed for the production of CNT's. The synthesis of CNT's is often accompanied by the formation of other forms of carbon, such as fullerenes, polyhedral particles and amorphous forms of carbon. In many instances, purification of CNT's is necessary [12].

The microwave (MW) irradiation technique is widely applied in some areas of chemistry [13] and technology to produce or degrade various materials and chemical compounds [14], as well as in the study of chemical processes. A synthesis of fullerenes method using microwave-induced naphthalene-nitrogen plasma at atmospheric pressure was reported in 1995 [15]. Some positive results were reported in 1999 by MW-heating of chloroform in presence of argon [16].

The objective of this work is to obtain nanotubes using MW radiation by heating graphite, sucrose and perform their characterization by Scanning Electron Microscopy (SEM), Transmission Electron Microscopy (TEM) and Atomic Force Microscopy (AFM). It is also offered the alternative of using silicon carbide substrates that have received a special treatment for modifying its morphology and by consequence increasing the concentration of nanotubes on its surface. More ahead we will see that the use of these new substrates not only has modified the concentration of nanotubes on the samples [17] but also many other interesting characteristics that will be detailed next.

EXPERIMENTAL

Preparation of nanotubes was carried out in a domestic MW-oven (power 800 W and frequency 2.45 GHz) [18,19]. The MW-action allows the graphite to be heated without direct contact with the energy source; the process control is achieved varying the power and heating time from 60 min. The samples were prepared from powdered graphite (99 %) with sucrose; and calcined sucrose. The mixtures of each one composition were contained under vacuum atmosphere (10^{-5} Torr) into a quartz capsule. The obtained samples were characterized by SEM, TEM and AFM. The AFM images were recorded in the contact mode in air at room temperature.

Inside each quartz bubble they were introduced several small slides of silicon carbide (SiC) with different morphologies; they were used non-porous (smooth) quartz slides (NPQS) as well as porous SiC slides (PSCS). The porosity of the SiC slides was achieved after submitting them to a chemical treatment [20] that caused the creation of tiny cavities as the ones that can be seen in Fig.1 taken with AFM and with SEM. After putting away the quartz bubbles of the microwave exposure the SiC slides were cautiously extracted and next they were analyzed with the SEM and TEM techniques.

RESULTS

It is well-known that materials absorb microwaves in different manners. In particular, graphite does not need a preliminary heating to absorb microwaves, so it can be directly MW-heated [18]. As it is shown in Table I, the condensation was carried out through carbon vaporization and the process was completed in 60 min.

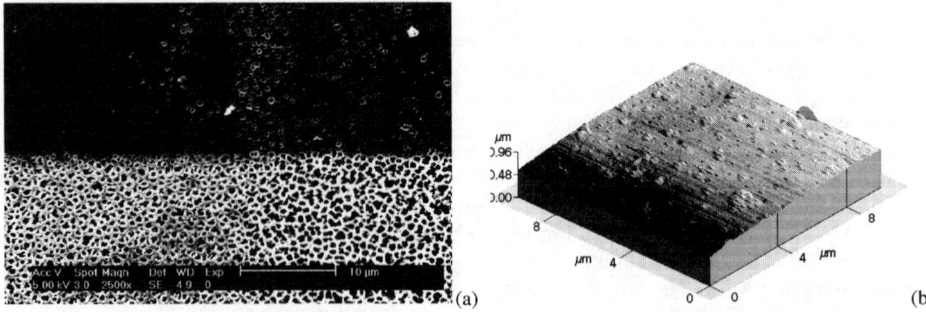

Figure 1. SEM and AFM images of the porous SiC slides Avg. pore diameter: 1µm (SEM Mag. 2.5Kx)

Table I. Microwave heating time of the samples at 800w, 2.45GHz.

Sample number	Time (min)	Target	Observation
1	60	Graphite	Nanotubes
2	60	Sucrose	No heating
3	60	Calcined Sucrose	No heating
4	60	Sucrose and graphite	Nanotubes, nanofibers

The SEM observation of *graphite samples*, allowed to conclude that the size of the nanotubes was increased with heating time. The diameter of formed CNT's is about 100 nm; others investigators have reported that when using a coating with Co/SiO_2, the diameter may reach up to 0.5 µm [18]. In our experiments, a catalyst is not employed and the diameters of nanotubes vary from 100 nm to 200 nm (Fig.2-a). The diameter of nanotubes slightly increases with time in the range 30-60 min. Most of the CNT's are surrounded by carbonaceous material and smaller particles. The TEM images are coincident with the images of SEM. TEM images of the samples were used to characterize the structure of the CNT's, which exhibit a multi-walled bamboo-like structure (Fig.2-b). Some of the tips are open, most are closed, or end on massive carbonaceous particles.

To investigate the internal structure of nanotubes obtained by microwave heating, the samples were also analyzed by AFM. According to AFM studies (Fig.3), carbon vaporization via microwave treatment produces multi-layer nanotubes (thickness 117 Å). The tube contains approximately 20 wall layers. This result agrees with these of Zhang et.al.[21].

An easy way for improving the purity of nanotubes was reported using a domestic multimode microwave oven [22]. In the present research, we noted that the CNT's appeared to be highly graphitized. The nanotubes obtained by this technique do not require further purification.

The SEM observation of the samples with *calcined sucrose and graphite mixture* allowed to conclude that the amount of CNT's on them (and its size), are smaller than the ones obtained in the only-graphite sample, but nanofibers are much more abundant (Fig.4-a). The TEM and SEM analysis show the presence of typical nanofibers (Fig. 4-b), enlarged CNT's (Fig.5-a) and some growing CNT's just after the MW-heating started (Fig. 5-b).

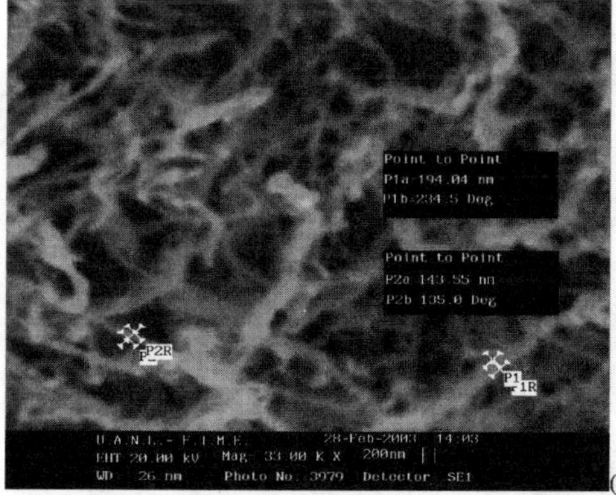

(a)

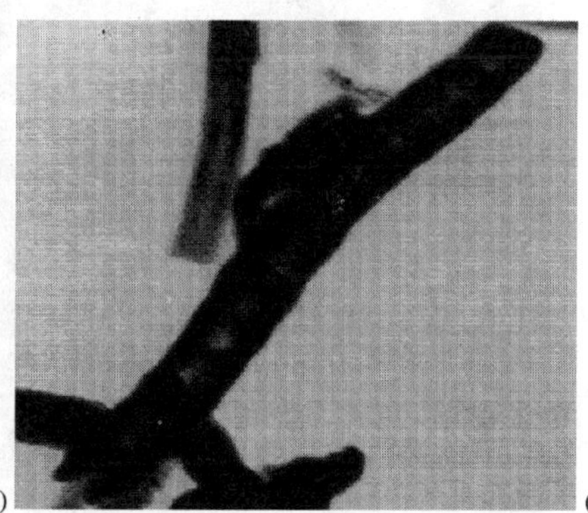
(b)

Figure 2. (a) SEM image of MWCNT's on the graphite target heated for 60 min without catalyst; (b) TEM image of bamboo-like MWCNT's on the graphite target heated for 60 min without catalyst (20-50nm between layers).

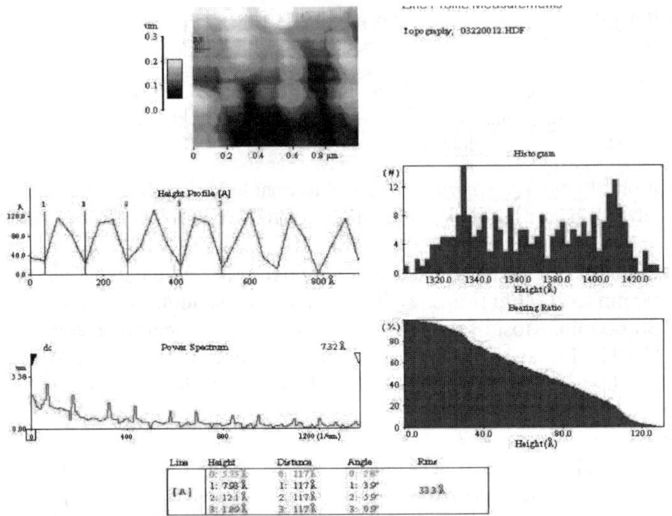

Figure 3. AFM transversal image of a multi-walled CNT from a graphite target.

The presence of sucrose, generated filled nanotubes, and nanofibers. The size of nanofibers is smaller than the nanotubes on the graphite sample.

The SEM examination showed without doubt a much bigger concentration of carbon nanotubes bonded over the porous substrate, till the degree of forming fibrous agglomerations across all the surface (Fig.4-a), which are much more abundant than those obtained of the samples were NPQS were used.

On the other hand, pictures obtained with TEM show additional characteristics of the

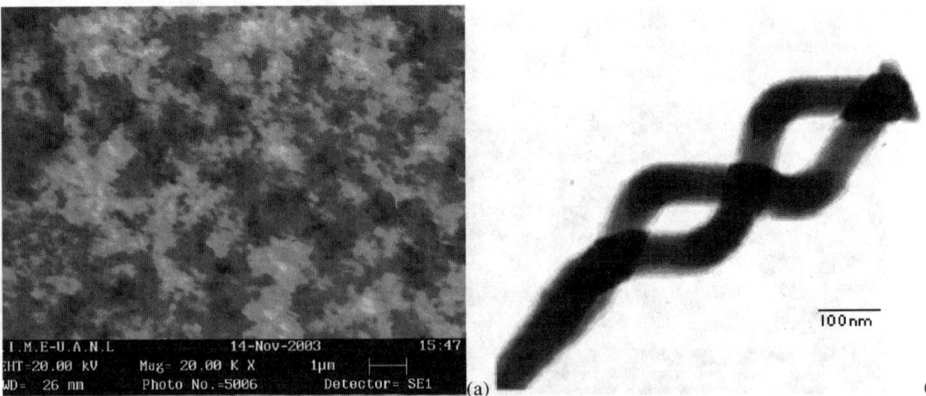

Figure 4. (a) SEM image of nanofibers from a calcined sucrose and graphite mixture without catalyst, after heating 60 min. Mag. 20Kx (b) TEM image of a nanofiber from calcined sucrose and graphite sample heated 60 min without catalyst.

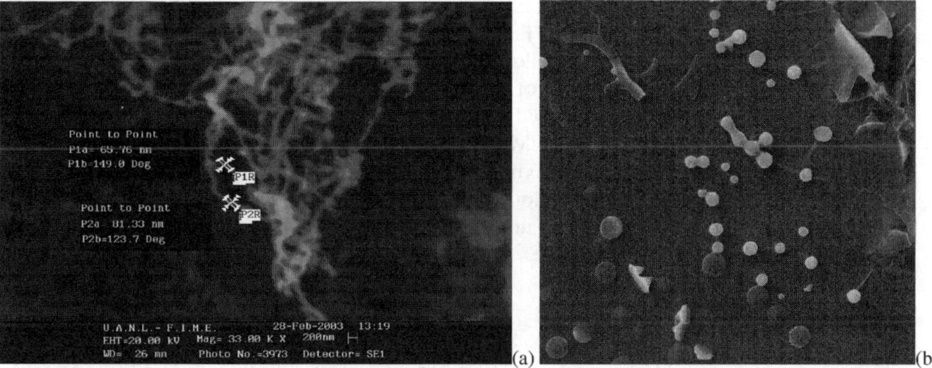

Figure 5. (a) SEM image of CNT's on graphite target (b) CNT's growth just after MW-heating started (sucrose-graphite mixture). Mag.1.6Kx. Diameter: 2-10μm approximately.

nanotubes obtained with the porous substrate, specifically the carbon nanotubes show themselves as considerable larger than in non-porous, and they also show a better bonding to the porous SiC surface, as well as suspects of a certain uniformity or growth pattern of nanotubes on regions were its growing reached only its 50% approximately. It were also registered the different growing stages of nanotubes on the sample's substrate. It was observed a great amount of proto – nanotubes which remained forming clusters or that also corresponded to a part of another well defined and well formed nanotubes. The images showed both SWNT's as well as MWNT's.

CONCLUSIONS

We can produce nanotubes by vaporization using microwave heating. The optimal time was 60 min. The presence of calcined sucrose diminished the quantity and size of nanotubes but increased that of nanofibers and filled nanotubes. The use of graphite-only targets produced a great amount of multi-walled bamboo-like nanostructures. As a contribution of this method, aligned multi-layer carbon nanotubes with diameters from 50 to 400 nm and lengths from 1 μm to 10 μm were obtained, their layer size is about 117 Å.

Small CNT's have bamboo-like structure with empty closed compartments inside. When calcined sucrose was added, CNT's contained other nanotubes or nanofibers inside, and both samples contain fibers with diameters from of 5 to 12 nm.

The reason why a bigger concentration of nanotubes fixed over porous SiC than over smooth quartz was observed is simple. By the fact of being a bigger superficial area of substrate, it is offered to nanotubes a bigger region for its bonding and subsequent growing. Nanotubes doesn't show a fixation criteria, i.e. a bonding preference over hills or valleys on the porous surface but instead the simple existence of a bigger area allows a bigger concentration of them all across the substrate. The enormous agglomeration of nanotubes over the PSCS shows with clearness the increment of its concentration judged against those from NPQS, particularly because the smooth quartz and porous SiC pieces that were analyzed with microscopes hailed (each couple) from the same bubble where graphite was heated.

The morphology of the porous-kind substrate offers the advantage over other substrates of allowing the atomic layers of graphite to get a quicker fixation over the irregular substrate surface. On that trail is that the nanotube quickly erect a solid base over which it can start its growing. The reason why we found large nanotubes that grew over PSCS is due to the quick raise of the nanotube's base which allows that the remaining microwave energy conferred to the sample is driven to the sole purpose of completing the nanotube development instead of keep on building a strong enough base for holding itself, which is the case of those that grew on NPQS.

It can also be observed from SEM images that the partial development of nanotubes show a certain degree of uniformity, this maybe due to the fact that nanotubes haven't grew up enough to the point of making a chaotic agglomeration or cluster that put out of sight its bonding with the porous substrate. The suggested uniformity seen from the images is directly related to the pore distribution across the substrate.

ACKNOWLEDGMENTS

The authors are very grateful to the CONACyT-Mexico (grant No. I39207U), to the Universidad Autónoma de Nuevo León (Monterrey, Mexico, grant No.CA804-02) and to the Facultad de Ciencias Físico Matemáticas for financial support, as well as M.S. Claudia López González (FIME, UANL) for technical assistance.

REFERENCES

1. W.B.Choi, Y.W.Jin, H.Y.Kim, S.J.Lee, M.J.Yun, J.H.Kang, Y.S.Choi, N.S.Park, N.S.Lee, J.M. Kim, *Appl. Phys. Lett.* 78 (2001)1547
2. C.Liu, Y.Y.Fan, M.Liu, H.T.Cong, H.M.Cheng, M.S.Dresselhaus, *Science* 286 (1999) 1127.
3. J.Kong, N.R.Franklin, C.W.Zhou, M.G.Chapline, S.Peng, K.J.Cho, H.J.Dai, *Science* 287(2000) 622.
4. D.S.Bethune, C.H.Kiang, M.S.de Vries, G.Gorman, R.Savoy, J.Vázquez, R.Beyers, *Nature* 363 (1993) 605.
5. A.Thess, R.Lee, P.Nicolaev, H.Dai, P.Petit, J.Robert, C.Xu, Y.H.Lee, S.G.Kim, A.G.Rinzler, D.T.Colbert, G.E.Scuseria, D.Tomanek, J.E.Fisher, R.E.Smalley, *Science* 273 (1996) 483.
6. M.Terrones, N.Grobert, J.P.Zhang, H.Terrones, J.Olivares, W.K.Hsu, J.P.Hare, A.K.Cheetham, H.W.Kroto, D.R.M.Walton, *Chem. Phys. Lett.* 285 (1998) 299.
7. M.Nath, B.C.Satishkumar, A.Govindaraj, C.P.Vinod, C.N.R.Rao, *Chem. Phys. Lett.* 322 (2000) 33.
8. Jin Lee, Cheol and Park, Jeunghee, School of Electrical Engineering, *Kunsan National University, Kunsan,* (2000) 573-701.
9. Byszewsri, P. and Klusek, Z., *Optoelectronics Review*, 9 (2001) 203-210.
10. X.Ma, E.G.Wang, *Appl. Phys. Lett.* 78 (2001) 978
11. S.L.Sung, S.H.Tsai, C.H.Tseng, F.K.Chaiiang, X.W.Liu, H.C.Shih, *Appl. Phys. Lett.* 78 (1999) 197
12. E.G.Rakov, *Russ. Chem. Rev.* 70(10) (2001) 827-863
13. Kharissova, O.V., Ortiz, U., Hinojosa, M., *Mat. Res. Soc.*, 654 (2001) AA.3.14.1-6.
14. Teooreanu I.F., *Revue Roum. Chim.*, 40(10) (1995) 965.
15. Tesuya Ikeda, Toshiihiro Kamo, Minoru Danno, *Appl. Phys. Lett*, 67(7) (1995) 900.
16. Su-yuan Xie, Rong-bin Huang, La-jia Yu, Jie Ding, and Lan-sun Zheng, *Applied Physics Letters*, 75(18) (1999) 2764-2766.
17. Kawasaki, S.; Komiyama, S.; Ohmori, S.; Yao, A.; Okino, F.; Touhara, H. Preparation of Carbon Nanotubes by Using Mesoporous Silica. Mat. Res. Soc. Symp. Proc. 2001, *675*.
18. Kharissova O.V., Nieto, I., Ortiz U., Aguilar J.A.,Hinojosa M. *Mat.Res.Soc.,* 740**,** (2003) I7.25.1-6
19. Kharissova, O.V.; Nieto, I.; Ortíz, U.; Aguilar, J.A.; Hinojosa, M. Condensation of Carbon Vapor in the Microwave Oven. MRS Proceedings 2002, *740. Materials Research Society*, 2002 Fall Meeting, Boston, MA.
20. Soloviev, S.; Das, T.; Sudarshan, T.S. Structural and Electrical Characterization of Porous Silicon Carbide Formed in n-6H-SiC Substrates. Electrochemical and Solid-State Letters 2003, *6* (2), 1-3
21. W.D.Zhang, J.T.L.Thong, W.C.Tjiu, L.M.Gan, *Diamond and Related Materials*, **11**(2002) 1638-1642.
22. R.Avetik, Pradhan Bhabendra K, Chang Jiping, Chen Gugang, Eklunnd Peter C, *J. Phys. Chem. B*, 106(34) I(2002) 8671-8675

B. Modeling

Mat. Res. Soc. Symp. Proc. Vol. 821 © 2004 Materials Research Society P8.16

Modeling the Stress Evolution of Ion Beam Synthesized Nanocrystals

D. O. Yi[1,3], I. D. Sharp[2,3], Q. Xu[2,3], C. Y. Liao[2,3], J. W. Ager III[3], J. W. Beeman[3], Z. Liliental-Weber[3], K. M. Yu[3], D. Zakharov[3], E. E. Haller[2,3], and D. C. Chrzan[2,3]
[1]Applied Science and Technology Group, Univ. of Calif., Berkeley, CA 94720, U.S.A.
[2]Dept. of Materials Science and Engineering, Univ. of Calif., Berkeley, CA 94720, U.S.A.
[3]Materials Sciences Division, Lawrence Berkeley Nat. Lab, Berkeley, CA 94720, U.S.A.

ABSTRACT

Under certain conditions, nucleation and growth can lead to substantial stresses in nanocrystals embedded in a host matrix. These stresses may be relaxed through subsequent annealing treatments. A model is presented for the relaxation of these stresses via diffusive processes within the matrix. The model reflects the effects of surface tension, potential phase transformations at or near the processing temperature, and differential thermal expansion. It is demonstrated that the model describes well the stress relaxation of ion beam synthesized Ge nanocrystals embedded in a silica matrix.

INTRODUCTION

In the past decade, a number of techniques have been developed to create nanometer-sized crystalline semiconductors embedded in a host matrix. These nanocrystals are potential building blocks for photonic and opto-electronic applications and their properties must be well understood. One interesting property that is not yet understood quantitatively is the stress state of ion beam synthesized (IBS) nanocrystals. These nanocrystals are typically found to be under compressive stresses of a few GPa[1–3]. After thermal annealing, however, these stresses may decrease. Several reports have described qualitatively the existence of the compressive stresses and the subsequent relaxation of stress after thermal processing[1–3]. Here, a quantitative model of the relaxation process is presented.

BACKGROUND

The experiments under consideration[4] involve the implantation of multi-energy Ge ions into a thin film (500 nm) of silicon dioxide (figure 1). The implanted samples are annealed at 900°C for one hour to produce Ge nanocrystals. The resulting 'as-grown' nanocrystals are, on average, 5 nm in diameter.

The stress state of the nanocrystals is characterized using Raman spectroscopy. The effects of phonon confinement produce a red shift of the peak frequency from the bulk value of the characteristic wavenumber, as measured by Raman spectroscopy at room temperature. However, Raman measurements performed on the embedded, as-grown, Ge nanocrystals show a blue shift from the bulk value. This blue shift is attributed to the presence of compressive stress on the nanocrystal[5]. When additional Raman

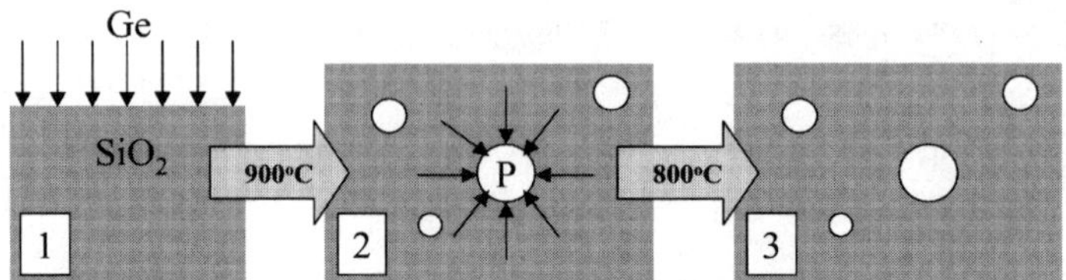

Figure 1: (1)Ge ions are implanted into silica. (2) Nucleation and growth at 900°C results in nanocrystals under pressure. (3)Annealing at 800°C relaxes as-grown pressure.

measurements are performed on nanocrystals that have been liberated from their host matrix[6], the expected red shift due to phonon confinement is observed.

After the initial Raman measurements, the as-grown samples are annealed further at a temperature of 800°C for time intervals from 30 minutes to 48 hours, and quenched to room temperature for measurement. For longer anneals, the Raman signals show increasing red shifts as compared to the as-grown value. The resulting decrease in the as-grown pressure is 1.6 GPa.

Several effects contributing to the stress state of the nanocrystal are considered: the surface tension between the matrix and nanocrystal, the differential thermal expansion of Ge and silica, and the equilibrium size of the matrix cavity in which the nanocrystal resides. For nanometer-sized particles, the pressure due to the interface energy is inversely proportional to the nanocrystal radius. In order to account for the temperature changes during processing and characterization, one must consider the thermal expansion coefficients of silica and Ge, the latter being an order of magnitude higher than the former. The differential expansion contributes to a tensile stress that reduces slightly the compressive stress state of the nanocrystal when quenched to room temperature. However, considering all these sources of stress, one may conclude that the observed pressure is mainly attributable to a volumetric difference: the nanocrystal resides in a matrix cavity that is too small. This volumetric difference may arise from the matrix atoms not being able to move rapidly enough to accomodate the nanocrystal as it is growing, or from a liquid to solid phase transition (liquid Ge has higher density than solid Ge). In subsequent annealing steps, the volume of the silica cavity increases, thereby decreasing the compressive stress on the nanocrystal.

Modeling the mechanisms involved in establishing the stress during the initial growth stage is difficult, given the dynamics of the matrix atoms and nucleating atoms. However, modeling the relaxation of stress during subsequent thermal treatment is straightforward.

THEORY

The model of the nanocrystal-matrix system is based upon the following assumptions. A spherical, isotropic, linear-elastic nanocrystal is considered to be embedded in an infinite, isotropic, linear-elastic matrix. The nanocrystal surface is welded to the matrix. Assuming that the nanocrystal radius is larger than the matrix cavity radius, one can obtain the radial stress fields and strained radii for the composite system. Imposing stress continuity at the boundary and equal strained matrix cavity and nanocrystal radii, one obtains the pressure, $P(t)$, on the nanocrystal as a function of nanocrystal and matrix cavity radii:

$$P(t) = \frac{12\, B_{nc}\, \mu_{matrix}\, (R^0_{nc} - R_{matrix}(t)) + 3\, B_{nc}\, R_{matrix}(t)\, \frac{2\gamma}{R^0_{nc}}}{4\, \mu_{matrix} R^0_{nc} + 3\, B_{nc}\, R_{matrix}(t)}. \tag{1}$$

The initial matrix cavity radius R^0_{matrix} becomes a function of initial nanocrystal pressure P^0_{nc} and initial nanocrystal radius R^0_{nc}:

$$R^0_{matrix} = \frac{1 - \frac{P^0_{nc}}{3\, B_{nc}}}{1 + \frac{1}{4\, \mu_{matrix}} \left(P^0_{nc} - \frac{2\gamma}{R^0_{nc}} \right)} R^0_{nc}. \tag{2}$$

R^0_{nc} and R^0_{matrix} are taken to be the radii of a sphere and spherical cavity, respectively, under no stresses, and are termed 'equilibrium' radii in this paper. In equations (1) and (2), B_{nc} and μ_{matrix} are the bulk modulus of the nanocrystal and shear modulus of the matrix, respectively, and γ is the surface tension of the nanocrystal/matrix interface.

The pressure in the nanocrystal acts as a driving force for an interstitial diffusive process that takes matrix atoms away from the nanocrystal interface and allows the matrix cavity to expand. Imposing a condition of local thermodynamic equilibrium at the nanocrystal/matrix interface, the pressure is related to the concentration of matrix interstitials[7]. In amorphous silica, tetragonal silica molecules are randomly distributed, allowing interstitial diffusion through open spaces in the matrix. As the nanocrystal volume expands, there is a decrease in free energy, $\mathrm{d}F_{nc}$ in the nanocrystal:

$$\mathrm{d}F_{nc} = -P\,\Omega + \frac{2\gamma}{R_{nc}}\,\Omega, \tag{3}$$

where R_{nc} is the equilibrium radius of the nanocrystal at temperature T, P is the pressure in the nanocrystal, and Ω is the volume of a matrix molecule. At thermodynamic equilibrium, the decrease in free energy of the nanocrystal $\mathrm{d}F_{nc}$, is balanced by a corresponding increase in free energy in the matrix $\mathrm{d}F_{matrix}$:

$$\begin{aligned} \mathrm{d}F_{matrix} &= -\mu_{SiO_2} + 2\,\mu_O \quad +\mu_{Si} \\ &= \underbrace{-\mu_{SiO_2} + 2\,\mu_O + E_{Si}} \quad +k_B\, T\, \ln(c_{Si}) \\ &\equiv \qquad \bar{E} \qquad +k_B\, T \ln(c_{Si}). \end{aligned} \tag{4}$$

The free energy increase is written in terms of μ_{SiO_2}, μ_{Si} and μ_O, the chemical potentials of SiO_2, Si and O, respectively. Rewriting μ_{Si} in terms of E_{Si}, the energy needed to create a Si interstitial and $k_B\, T\, \ln(c_{Si})$, the configurational entropy of Si interstitials, $\bar{E}$ is defined as the effective temperature-dependent energy barrier to create a Si matrix interstitial. Then, imposing local thermodynamic equilibrium at the boundary and solving for the atomic concentration of Si interstitials, c_{Si}:

$$c_{Si}\left(R_{matrix}\right) = \exp\left(-\frac{\bar{E}}{k_B T}\right) \exp\left(\left[P - \frac{2\gamma}{R_{nc}}\right] \frac{\Omega}{k_B T}\right). \tag{5}$$

As the pressure in the nanocrystal drives the creation of matrix interstitials at the interface, a higher concentration of Si interstitials accumulates at the matrix-nanocrystal boundary. Because oxygen is a fast diffuser, only Si accumulates at the boundary. Assuming the growth of the matrix cavity is slow compared to the rate of change in the effective concentration of interstitials, one can use steady-state diffusion to solve for the spatial distribution of interstitials, with the concentration at the nanocrystal-matrix interface given by (5):

$$c_{Si}(r) = \exp\left(-\frac{\bar{E}}{k_B T}\right)\left(1 + \frac{R_{matrix}}{r}\left(\exp\left(\left[P - \frac{2\gamma}{R_{nc}}\right] \frac{\Omega}{k_B T}\right) - 1\right)\right). \tag{6}$$

The gradient of the concentration at the interface is related to the flux of Si interstitials, which is, in turn, related to the rate of change in the matrix cavity radius. Given these relationships, one arrives at an expression for the evolution of the matrix cavity radius:

$$\frac{dR_{matrix}}{dt} = \frac{D_0\, e^{-\frac{E}{k_B T}}}{R_{matrix}(t)}\left(\exp\left(\left[P\left(t, R_{matrix}\right) - \frac{2\gamma}{R_{nc}}\right] \frac{\Omega}{k_B T}\right) - 1\right), \tag{7}$$

where D_0 is the prefactor of the effective interstitial diffusivity. E is the effective energy barrier which includes $\bar{E}$ and the migration energy. Note that $\bar{E} = \bar{E}(T)$. In equation (7), the matrix cavity increases when the nanocrystal pressure exceeds the pressure due to interface tension. The pressure is given by equation (1) and decreases when the matrix cavity grows. Ultimately, the pressure will reach a final value equal to the pressure due to surface tension alone.

The time-dependent nanocrystal pressure is determined by equations (1), (2) and (7). To take into account nanocrystal growth at 900°C, rate of change in matrix cavity radius during the 800°C annealing step, and Raman measurements taken at room temperature, the expression for the equilibrium radii are modified by:

$$R(t, \Delta T) = R(t)\,\left[1 + \alpha\,\Delta T\right], \tag{8}$$

where α is the thermal expansion coefficient of the matrix or nanocrystal and ΔT is the temperature change. For a decrease in temperature, the nanocrystal contracts causing a tensile contribution to the stress on the nanocrystal. This effect offsets the as-grown pressure by approximately 400 MPa, not enough to counteract the compressive stress imposed by the matrix.

DISCUSSION

The time evolution of the nanocrystal stress state can be computed numerically with equations (1), (2) and (7). There are several unknown parameters in these equations: P^0_{nc}, γ and $D_0 \exp^{-E/k_B T}$, that can be calculated given the experiments described above. These equations are fitted to the experimental data and the comparison between the fitted curve and the data are shown in figure 2.

Experimentally, the difference between Raman measurements of Ge nanocrystals embedded in silica and of Ge nanocrystals liberated from the matrix results in the relative pressure as shown in figure 2. The stress state of embedded nanocrystals, will, after a long-time anneal at 800°C, reduce to a stress state due to surface tension alone. The surface tension γ is taken to be the same value for the nanocrystal/matrix and the nanocrystal/air interfaces, because, experimentally, there are no appreciable differences in the Raman measurements between a long-time annealed embedded nanocrystal and a liberated nanocrystal. Correspondingly, the theoretical model reflects the relative pressure, $P_{relative}$, as the difference between the nanocrystal pressure given by equation (1) and the pressure due to surface tension, $2\gamma/R_{nc}$.

The corresponding fitting paramaters are given by:

$$\begin{aligned} P_{as-grown} &= 1.2\ GPa \\ D_0 \exp^{-E/k_B T} &= 1.3 \times 10^{-20}\ cm^2/s, \end{aligned} \tag{9}$$

where $P_{as-grown}$ is the relative initial nanocrystal pressure, given by the difference between the initial nanocrystal pressure P^0_{nc} and the pressure due to the surface tension γ.

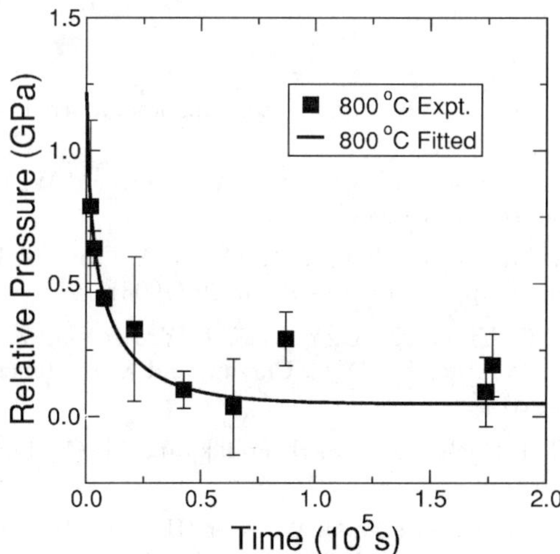

Figure 2: Relative nanocrystal pressure as a function of annealing time, fit to experimental data[4]. Error bars are the standard deviation from multiple Raman measurements on each sample.

The experimentally measured as-grown pressure is 1.6 ± 0.2 GPa , higher than the fitting parameter. The first experimental point is taken after half an hour of annealing. During this time interval, the silica may relax at a faster rate, allowing a greater decrease in the stress state of the nanocrystal. For the temperature considered, the effective interstitial diffusivity is of the same order of magnitude as the diffusivity of Si in silica[8, 9]. Overall, the experimental data is described well by the theory.

CONCLUSIONS

A quantitative model for the relaxation of pressure in a nanocrystal embedded in a matrix was developed. A parametric fit to experimental data for Ge in silica yields very good agreement between experiment and theory.

ACKNOWLEDGEMENTS

The authors acknowledge insightful discussions with W. G. Wolfer. D.C.C. gratefully acknowledges the support of the Miller Institute for Basic Research. This work is supported in part by the Director, Office of Science, Office of Basic Energy Sciences, of the U.S. Department of Energy under contract No. DE-AC03-76SF00098, and by U.S. NSF Grants No. DMR-0109844 and EEC-0085569.

REFERENCES

[1] M. Fujii, S. Hayashi, and K. Yamamoto, Japanese Journal of Applied Physics **30**, 687 (1991).

[2] M. Dubiel, H. Hofmeister, E. Schurig, E. Wendler, and W. Wesch, Nuclear Instruments Methods B **166-167**, 871 (2000).

[3] A. Wellner, V. Paillard, C. Bonafos, H. Coffin, A. Claverie, B. Schmidt, and K. H. Heinig, Journal of Applied Physics **94**, 5639 (2003).

[4] I. D. Sharp, D. O. Yi, Q. Xu, C. Y. Liao, J. W. Beeman, Z. Liliental-Weber, K. M. Yu, D. Zhakarov, J. W. Ager III, D. C. Chrzan, and E. E. Haller, submitted to Phys. Rev. Let. (unpublished).

[5] F. Cerdeira, C. J. Buchenauer, F. H. Pollak, and M. Cardona, Physical Review B **5**, 580 (1972).

[6] I. D. Sharp, Q. Xu, C. Y. Liao, J. W. Ager III, J. W. Beeman, K. M. Yu, D. Zhakarov, Z. Liliental-Weber, and E. E. Haller, Materials Research Society Proceedings **777**, T7.6.1 (2003).

[7] G. W. Greenwood, A. J. E. Foreman, and D. E. Rimmer, Journal of Nuclear Materials **4**, 305 (1959).

[8] T. Takahashi, S. Fukatsu, K. M. Itoh, M. Uematsu, A. Fujiwara, H. Kageshima, Y. Takahashi, and K. Shiraishi, Journal of Applied Physics **93**, 3674 (2003).

[9] D. Tsoukalas, C. Tsamis, and P. Normand, Journal of Applied Physics **89**, 7809 (2001).

Mat. Res. Soc. Symp. Proc. Vol. 821 © 2004 Materials Research Society P2.10

Modeling the Elastic Fields in Epitaxially Grown Multilayers

Ganesh Vanamu[1], Tariq A. Khraishi[2], Abhaya K. Datye[1] and Saleem H. Zaidi[3]
[1]Department of Chemical and Nuclear Engineering, University of New Mexico, Albuquerque, NM-87131, U.S.A.
[2]Mechanical Engineering Department, University of New Mexico, Albuquerque, NM-87131, U.S.A.
[3]Gratings, Inc., 2700 B Broadbent Pkwy NE, Albuquerque, NM 87107.

ABSTRACT

A model was developed to calculate the elastic fields, including strain energy density, in multilayers grown epitaxially on a planar substrate. This model works well for compliant and non-compliant substrates. In particular we illustrate the model for four layer heterostructure and apply it for graded Ge (Si_xGe_{1-x}) grown on a planar silicon substrate. Using the equations for static equilibrium and Hooke's law for isotropic materials under a plane stress condition, the elastic fields associated with each layer were calculated. The strain partitioning in this model reduces to the limiting case of a two-layer structure available in the literature. As it turns out here, strain partitioning is a function of the bulk unstrained lattice parameters, elastic constants and thicknesses of the layers. The model was qualitatively verified by comparing the strain energy density with the dislocation density away from a relatively thick substrate. This model helps shed some light on the factors important in achieving defect free multilayers for optoelectronic devices.

INTRODUCTION

In recent years there has been great interest in the study of multilayer epitaxial structures (such as Si-Ge alloys) with varying composition (compositionally graded buffer layers). Graded buffer-layer composition has offered significant improvement in the epilayer quality [1] and hence was used for many advanced electronic and optoelectronic applications [2-5]. These different applications have different defect density tolerances. Although many applications can be explored with moderate defect density material, the ultimate implementation of these materials into products requires the threading dislocation densities to be lower than some critical amount in order to integrate III/V devices on silicon. In the case of field effect transistors (FETs) on silicon utilizing SiGe alloy buffers, the defect density tolerance is on the order of 5×10^5-10^7 cm^{-2}. In the case of light-emitting diodes (LEDs) and photovoltaics on silicon, the defect densities need to be on the order of 10^5- 10^6 cm^{-2}, whereas for monolithic lasers, photodetectors and solar cells on silicon, the densities should not exceed 10^4 cm^{-2}. In principle the dislocations would be formed when the strain energy density exceeds a certain critical amount equal to the energy of formation of a single dislocation [6]. This last reference modeled a two-layer only heterostructure in which the authors calculated the film critical thickness based on strain energy considerations. The ultimate problem is thus to calculate the strain energy per unit volume in each layer.

A model of strain partitioning among the layers can be obtained using simple elasticity theory. Strain partitioning for a two-layer heterostructure (composed of a substrate and one growth epilayer) has been done using both isotropic [6-10] and anisotropic [11] elasticity models. We have extended such models to four- and five-layers. The aim of this work is to find the variation of strain energy density with the thickness/number of top epilayers from a planar substrate. Here, the strain partitioning is a function of the bulk lattice constant, elastic constants and thickness of all layers.

We also verified the developed model experimentally. We have applied the model to the specific case of compositionally graded germanium (Si_xGe_{1-x}) on a silicon planar substrate. The growth of the multilayers of graded germanium were done by Metal Organic Chemical Vapor Deposition (MOCVD) on a planar silicon substrate.

THEORETICAL MODEL

Consider a planar heterostructure made up of four layers with cubic unit cells as shown in figure 1. The figure on the left describes the unstrained bulk layers before the growth, and the one on the right describes the structure after the growth. The grown layers strain one another reaching some equilibrium configuration. The strained layers are assumed to only have normal in-plane stresses (perpendicular to the growth direction). The two orthogonal normal stresses, σ_{xx} and σ_{yy}, are equal in magnitude. There is no stress in the out-of-plane direction (growth direction) assuming plane-stress conditions for the relatively thin structure. As a consequence, the strains in the two in-plane directions are also equal in magnitude but the out-of-plane strain is different due to Poisson's effect. For a planar interface, it is assumed that the lattice constant strains uniformly everywhere in the sample including across the interface as shown in the figure. In this heterostructure, the normal in-plane strain in each layer (substrate is treated as layer number one and the first graded layer as number two and so on) is defined by,

$$\varepsilon^1 = \frac{a^p - a^1}{a^1}, \; \varepsilon^2 = \frac{a^p - a^2}{a^2}, \; \varepsilon^3 = \frac{a^p - a^3}{a^3}, \; \varepsilon^4 = \frac{a^p - a^4}{a^4} \tag{1}$$

where a^p is the strained lattice parameter of the layers of the heterostructure as shown in figure1. a^1 is the unstrained lattice parameter of layer number one and so forth. For the case of isotropic materials under plane stress, the stress-strain equations in each layer (including substrate) are given according to Hooke's law as [10],

$$\sigma^1 = B^1\,\varepsilon^1\,,\; \sigma^2 = B^2\,\varepsilon^2,\; \sigma^3 = B^3\,\varepsilon^3,\; \sigma^4 = B^4\,\varepsilon^4 \tag{2}$$

where $B^1 = Y^1/(1-\nu^1)$, Y^1 is the Young's modulus, ν^1 is Poisson's ratio and σ^1 is the bi-axial stress (stress along the x and y- directions) in the first layer and so forth.

Using the static equilibrium criterion, strain partitioning can be determined in each layer as follows. The condition of static equilibrium requires that the forces in all the layers balance each other. The equilibrium equation can be written as follows [10],

$$\sigma^1 h^1 + \sigma^2 h^2 + \sigma^3 h^3 + \sigma^4 h^4 = 0, \tag{3}$$

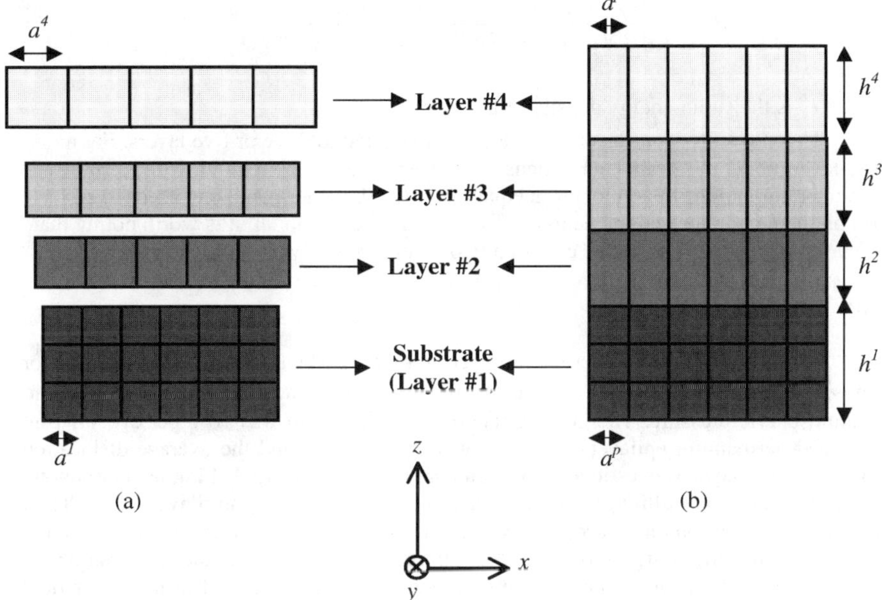

Figure 1. (a) A schematic cross-sectional illustration of unstrained layers, (b) strained layers.

where h^1 is the thickness of first layer, this is here the substrate, h^2 is the thickness of the second layer or the first epilayer and so forth.

Equations 1-3 can be solved to find an expression for the strained lattice parameter of the layers. The expression is as follows,

$$a^p = \frac{h^1B^1 + h^2B^2 + h^3B^3 + h^4B^4}{h^1B^1/a^1 + h^2B^2/a^2 + h^3B^3/a^3 + h^4B^4/a^4} \qquad (4)$$

Combining equations 1 and 4 again yields the strain in each layer,

$$\varepsilon^i = \frac{\sum_{j=1}^{4} h^j B^j}{a^i \sum_{k=1}^{4}(h^k B^k / a^k)} - 1 \qquad (i = 1, 2, 3 \text{ or } 4) \qquad (5)$$

The elastic strain energy density (energy per unit volume) in each layer can be expressed as follows,

Strain energy density $= U_o^i = \dfrac{(\sigma^i)^2}{B^i} = B^i(\varepsilon^i)^2 \ (i = 1, 2, 3 \text{ or } 4)$

Therefore $$U_o^i = B^i\left(\frac{h^1B^1 + h^2B^2 + h^3B^3 + h^4B^4}{a^i(h^1B^1/a^1 + h^2B^2/a^2 + h^3B^3/a^3 + h^4B^4/a^4)} - 1\right)^2 \quad (6)$$

where U_o^i is the strain energy density of layer i.

This model for four layers can be easily extended to three or five layers. For a three-layer model the strain energy density in layer i is given by removing one term from the numerator and the denominator in equation 6. On the other hand, one needs to add a term on the numerator and the denominator for a five-layer model. It is worth noting that the current model reduces to the case of a two-layer model given in the literature [6].

EXPERIMENT

Four layers of graded germanium were grown by Metal Organic Chemical Vapor Deposition (MOCVD) on a 4"-diameter planar silicon substrate at 20A°/s at 900°C and 0.2 atmospheric pressure. The grading rate of the germanium was 10% per every 1 μm. So graded germanium epilayers over silicon were analyzed and the average dislocation density in each layer was estimated. We prepared the samples (polishing using diamond lapping film and ion milling) to analyze the cross-sections of the multilayers using TEM (transmission electron microscopy). We performed the TEM characterization of the dislocations in different regions of the samples to obtain qualitative variation of dislocation density in the layers. The dislocations were characterized using bright field and dark field TEM. We qualitatively observed the variation of dislocation density as a function of the thickness of the graded growth film. The experimental results were compared with the developed model and good qualitative agreement was found (see **RESULTS** below).

RESULTS

The strain energy density in each layer was calculated using equation 6 with the help of linear interpolation of the material constants (such as the lattice parameters and elastic constants, see table I). The variation of the strain energy density across the heterostructure with the thickness/number of top epilayers was plotted in figure 2 (with the strain energy density in J/m^3 plotted along the *y*-axis and the layer number plotted along the *x*-axis). It is clearly seen that the strain energy density increases as the distance from the substrate increases. It is known from thin film/dislocation theory that the strain energy density and the dislocation density are correlated with one another [6]. Figure 3 shows that the substrate is essentially free of dislocations and that the dislocation density increases with the film thickness away from the substrate. Therefore, the current model qualitatively agrees with the experimental results.

CONCLUSIONS

We have calculated the elastic fields (stresses, strains and strain energy densities) in multilayers grown epitaxially on a planar substrate using an isotropic elasticity model.

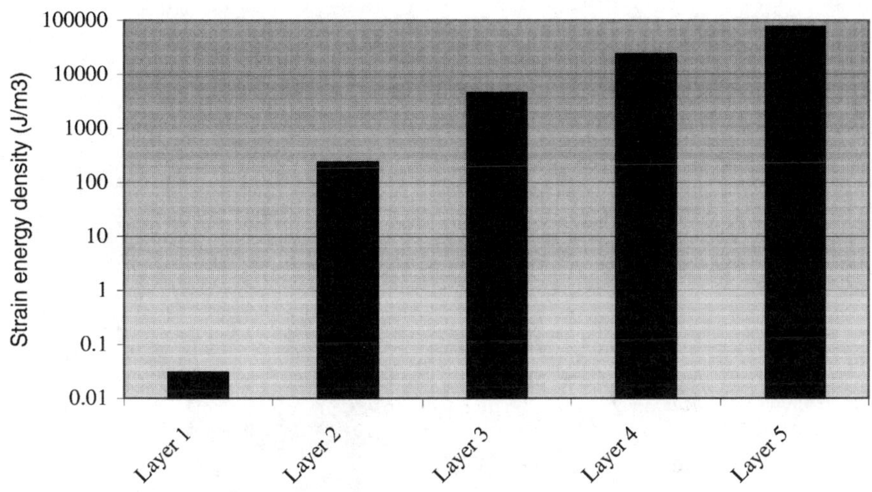

Figure 2. Variation of strain energy density in the heterostructure layers. Strain energy density in a layer directly correlates with its dislocation content.

An increase in the strain energy density with distance from the substrate was found which is due to the introduction of additional strain by growing a new layer over the previous layer. The strain energy density of the substrate in the particular Si_xGe_{1-x}/Si heterostructure considered here is negligible compared to that of the epilayers due to the fact that the substrate is non-compliant (i.e. much thicker than the epilayers). The model was compared with experiment for graded SiGe grown over a planar silicon substrate. The results revealed qualitative agreement between the calculated strain energy density and the experimentally-observed dislocation density in the layers.

Table I. The linear interpolation of the properties of the layered materials

Layer #	Thickness of layer (μm)	Fraction Ge (%)	Fraction Si (%)	Young's modulus (Pa)	Poisson's ratio	Unstrained lattice constant (Å)
Layer 1	2000	0	100	9.8×10^{10}	0.28	5.431
Layer 2	1	10	90	9.57×10^{10}	0.278	5.455
Layer 3	1	20	80	9.34×10^{10}	0.276	5.478
Layer 4	1	30	70	9.11×10^{10}	0.274	5.502
Layer 5	1	40	60	8.88×10^{10}	0.272	5.525

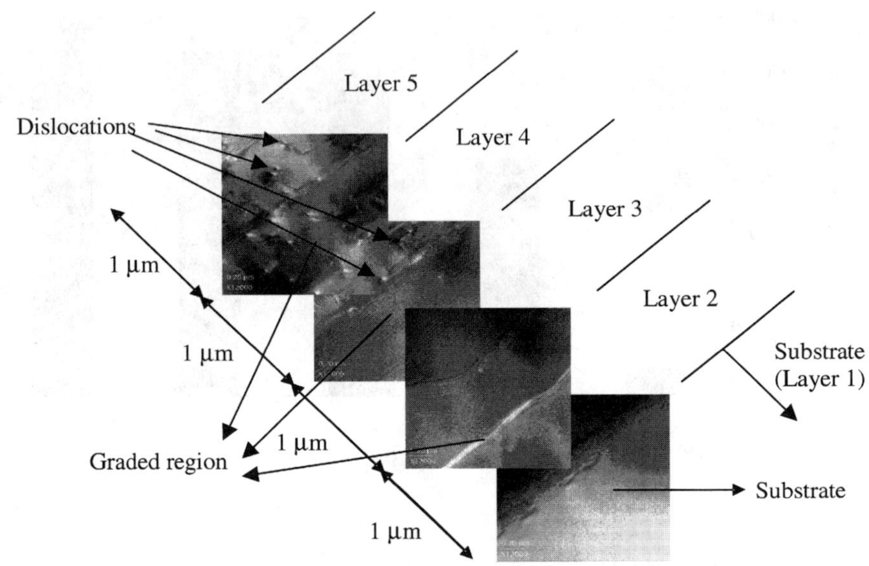

Figure 3. Cross-sectional TEM image of a planar structure of Si_xGe_{1-x} grown using MOCVD showing an increase in the dislocation density with distance from the silicon substrate. The dislocations are the lines and dots in the TEM image.

ACKNOWLEDGMENTS

This work is supported by the Ceramic and Composite Materials Center, through NSF grant EEC 99-08205 and the members of the industrial advisory board is gratefully acknowledged.

REFERENCES

1. S. B. Samavedam, E. A. Fitzgerald, Journal of Applied Physics **81**, 3108 (1997).
2. E. A. Fitzgerald, Y. H. Xie, D. Monroe, P. J. Silverman, J. M. Kuo, A. R. Kortan, F. A. Thiel and B. E. Weir, J. Vac. Sci. Technol. B **10**, 1807 (1992).
3. F. Schaffler, D. Tobben, H. R. Herzog, G. Abstreiter, and B. Hollander, Semicond. Sci. Technol. **7**, 260 (1992).
4. S. S. Iyer, G. L. Patton, J. M. C. Stork, B. S. Meyerson, and D. L. Harame, IEEE Trans. Electron Devices **36**, 2043 (1989).
5. R. People, IEEE J. Quantum Electron **QE-22**, 1696 (1986).
6. D. Zubia, S. D. Hersee, T. A. Khraishi, Appl. Phys. Lett. **80**, 740 (2002).
7. Y. H. Lo, Appl. Phys. Lett **59**, 2311 (1991).
8. D. Teng and Y. H. Lo, Appl. Phys. Lett. **62**, 43 (1993).
9. L. B. Freund and W. D. Nix, Appl. Phys. Lett. **69**, 173 (1996).
10. J. P. Hirth and A.G. Evans, J. Appl. Phys. **60**, 2372 (1986).
11. P. Wakeland, T. A. Khraishi, D. Zubia, Journal of Electronic Materials **32,** 836 (2003).

Mat. Res. Soc. Symp. Proc. Vol. 824 © 2004 Materials Research Society P3.39

Finite Element Analysis of Elastic and Plastic Deformation during Growth of $Si_{1-x}Ge_x$/Si Heterostructures

F. Sahtout Karoui, A. Karoui, G. Rozgonyi

Materials Science and Engineering Department,
North Carolina State UniversityRaleigh, NC 27695-7916, USA.

ABSTRACT

The elastic and plastic stress distribution in strained $Si_{1-x}Ge_x$ heterostructure (x=0.2 and x=0.5), was investigated during the growth process using a nonlinear transient finite element modeling. The material plastic behavior is described by the von Mises yield criteria coupled with isotropic work hardening conditions. The calculated stress distribution during growth of the SiGe cap layer shows that the von Mises stress fluctuates strongly within the layers and at the interfaces. The surface of constant composition $Si_{1-x}Ge_x$ layer is found under compressive stress in both cases. Within that layer the normal stress in the growth direction remains compressive for x=0.2, while it changes from compressive to tensile for x=0.5. In the graded layer the stress goes from tensile to compressive for x=0.5 and in the opposite way for x=0.2. High plastic deformation is observed in the layers, with the maximum von Mises plastic stress being higher for x=0.5 and localized in the SiGe graded region. The plastic strain vanishes monotonically up to 8 μm deep into the Si bulk substrate, in agreement with TEM images that revealed dislocation loops penetrating into the substrate. The time dependent analysis shows that elastic and plastic deformation appear almost instantaneously in the sublayers, while in the Si substrate it is delayed up to 300 s.

INTRODUCTION

Heteroepitaxial strained $Si_{1-x}Ge_x$ layers on Si substrates have been used to increase charge carrier mobility[1] in bipolar transistors and to adjust the bandgap of modern optoelectronic devices. Their electronic properties depend on germanium content and stress. However, improperly designed film-induced local stresses may degrade the structural integrity and electronic properties of the strained Si film. The evolution of surface morphology during strained layer heteroepitaxy is generally known to serve as one potential strain relaxation mechanism in the epitaxial film. Relaxation by misfit dislocation introduction is a competing relaxation mechanism at epitaxial layer thickness greater than the critical layer thickness, see Fig. 1 and 2. Thin films are typically deposited under severe kinetic constraints, resulting in highly non-equilibrium microstructures for a wide range of thickness-layer composition as shown in Fig. 2. These films often exhibit stress levels far in excess of the bulk yield strength, which degrade the structural integrity and the electronic properties of the strained silicon film. The evolution of the film stress during deposition is poorly controlled and its impact, in most cases, is severe. A method to quantify and understand the interplay of stresses and defect formation during $Si_{1-x}Ge_x$ layers growth are essential to lessen subsequent dislocation thread propagation into the device zone.

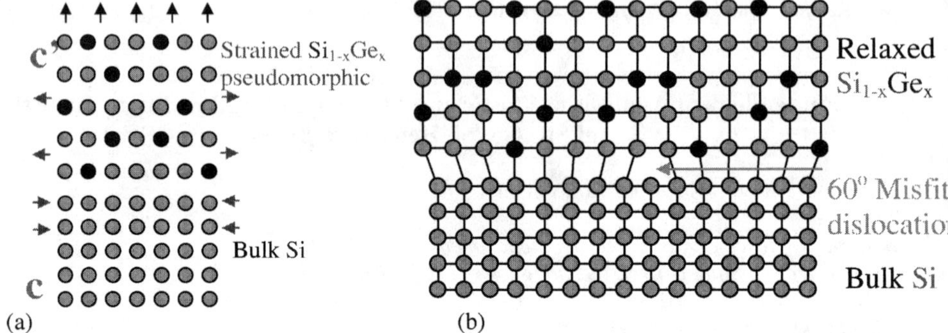

Figure 1. Model of stress relaxation in SiGe heterostructures via misfit dislocations: (a) Strained pseudomorphic $Si_{1-x}Ge_x$ on silicon, (b) Relaxed $Si_{1-x}Ge_x$ layer on silicon.

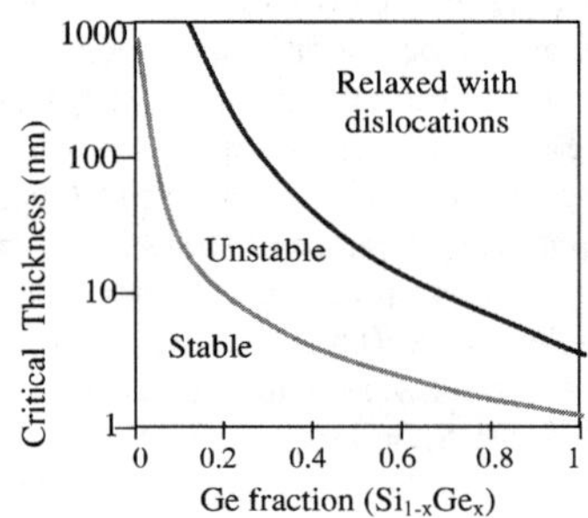

Figure 2. SiGe layer behavior as a function of layer thickness and Ge content.

The aim of this work is to investigate the elastic and plastic strain/stress distribution during the growth process, in strained $Si_{1-x}Ge_x$ epilayers on thick silicon 001 substrates. For this purpose, we employed a nonlinear transient finite element (FE) modeling implemented within ANSYS package. We present in this paper the calculated stress distribution during the growth of the $Si_{1-x}Ge_x$ constant layer (1050°C for 500 s), for both cases: x=0.2 and x=0.5.

SYSTEM MODEL

The studied sample shown in Fig. 3, has a SiGe relaxed graded buffer, considered here as 'engineered substrate' for the dislocation free top strained Si layer which will be used for device fabrication. The relaxed graded layer, which has Ge-rich top surface, introduces 4% lattice mismatch between Si and Ge. The fact that this layer is gradual, results in a disperse three-dimensional misfit dislocation network. At high temperature, high level of stress is generated in the sublayers due to thermal expansion mismatch, which can exceed the yield stress, leading to the onset of plastic deformation. Indeed, between growth temperature, 1050°C (our case) and room temperature, the thermal expansion coefficient of Si, α_{Si}, varies from 4.9×10^{-6} to 2.57×10^{-6} K^{-1}, and α_{Ge} varies from 8.55×10^{-6} to 5.9×10^{-6} K^{-1}. [2]

FINITE ELEMENT MODEL

We have considered a 2D plane strain analysis, see Fig. 3 (b), assuming no displacement in the Y direction because much larger than the wafer thickness. Reflecting boundary conditions were applied on each side (left and right) of the sample to insure continuity of the sample along x-axis. Since these surfaces are not free surfaces, they are not exposed to external stresses and the solution shall repeat itself to infinity. A nonlinear transient analysis has been carried out on the structure (without the strained Si layer) subjected to a thermal load of 1050°C for 500s.

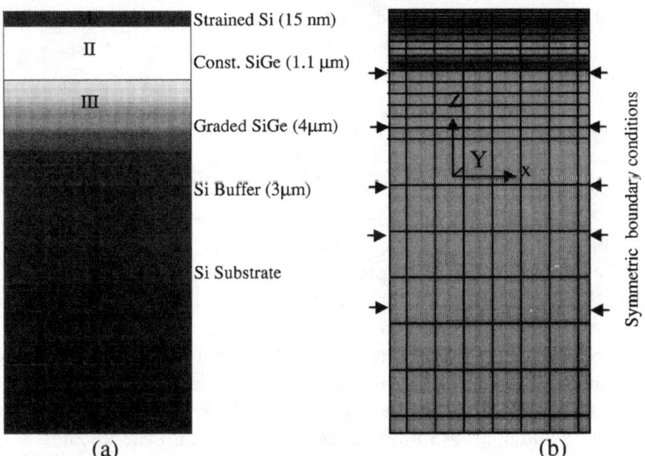

Figure 3. (a) Strained $Si/Si_{1-x}Ge_x$ layer on a Si(001) substrate, (b) Finite element model.

The elastic behavior follows Hooke's law while the material plastic behavior is described by the von Mises yield criteria coupled with isotropic work hardening conditions.[3] Under thermal load the residual strain is expressed as a function of the thermal expansion coefficient and is given by:

$$\varepsilon^{tot} = \varepsilon_x^{el} + \alpha \Delta T$$

Where ε^{tot} is the total strain, ε_x^{el} is the elastic strain, α the coefficient of thermal expansion and ΔT the temperature change. The stress tensor, assuming a plane strain isotropic condition, is:

$$\begin{bmatrix} \sigma_{xx} \\ \sigma_{zz} \\ \tau_{xz} \end{bmatrix} = \frac{E}{(1+\nu)(1-2\nu)} \begin{bmatrix} 1-\nu & \nu & 0 \\ \nu & 1-\nu & 0 \\ 0 & 0 & \frac{1}{2}-\nu \end{bmatrix} \cdot \begin{bmatrix} \varepsilon_{xx}^{el} \\ \varepsilon_{zz}^{el} \\ \gamma_{xz}^{el} \end{bmatrix}$$

Where E is the Young modulus, ν the poisson ratio, σ_{xx}, σ_{zz} the normal stresses in x and z directions, respectively. The material plastic behavior is described by a bilinear stress-strain curve. A yield stress (σ_Y) of 210 MPa has been chosen for Si and $Si_{1-x}Ge_x$ in accordance to published experimental data from Ref. 4 and 5. An important factor governing the plastic behavior of a material is the phenomenon of strain-hardening, also referred to as work-hardening. The simplest example of hardening is linear strain-hardening, which is actually a reasonable approximation used in our calculation. In this case, we have associated the gradient in the elastic region with the standard Young modulus E and the gradient in the plastic region with the elastic/plastic modulus. From these moduli and the initial yield stress, a linear relationship between the yield stress σ_Y and the plastic strain is derived. Note that temperature is an important factor when describing strain hardening. The relaxation phenomenon occurs by plastic deformation at nanometer scale. As more dislocations are generated, their interaction probability increases. Basically, the larger the number of produced dislocations, the more they interlock themselves and hence the larger the stresses required for the material to yield.

RESULTS AND DISCUSSION

The data presented here are those obtained at the end of the growth process before cooling. These include, the time and depth dependent elastic and plastic strain/stress distributions within each layer which causes slip and dislocations. Plot in Fig. 4 shows the normal stress (σ_{zz}) in the growth direction. As we can see, the constant composition $Si_{1-x}Ge_x$ layer (II) in both samples is essentially under compressive stress. Within that layer the stress remains compressive for x=0.2 while it changes from compressive to tensile at the SiGe/SiGe-graded interface for x=0.5. Within the graded layer the stress goes from tensile to compressive for x=0.5 and in the opposite way for x=0.2. An inversion of the stress sign occurs about the center of that layer, resulting in opposite stresses at its interfaces. This phenomenon is also observed in the Si buffer layer. In both samples, the stress along the z-axis in the cap layer (II) is compressive. The 50% Ge sample produced five times higher stress in that layer than the 20% Ge sample. The large difference in the residual stress in that layer will result in different states of the strain in the device zone right above the cap.

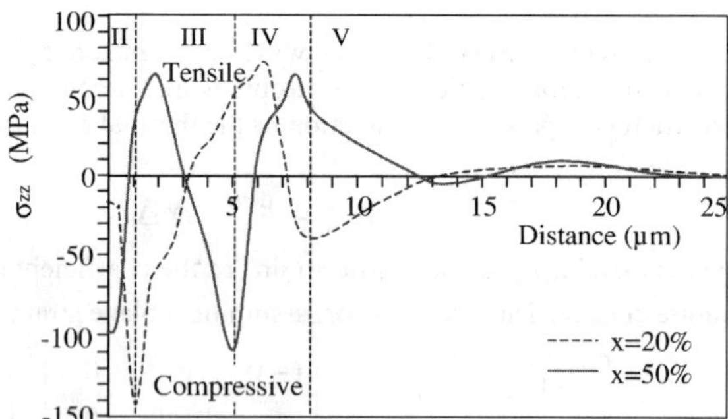

Figure 4. Calculated stress in the [001] growth direction.

As shown in Fig. 5, high plastic deformation is observed in all layers underneath the strained silicon top layer. The maximum plastic stress is localized in the SiGe-graded region and depends on the Ge content. In that region, high dislocation density is expected as observed by TEM imaging.[6] The strain profile in the graded layer appeared in good agreement with the anisotropic etching rate profile obtained for that structure in our laboratory.[7] The plastic strain vanishes monotonically up to 8 µm into the Si bulk, in agreement with TEM images that revealed dislocation loops penetrating into the substrate.[8]

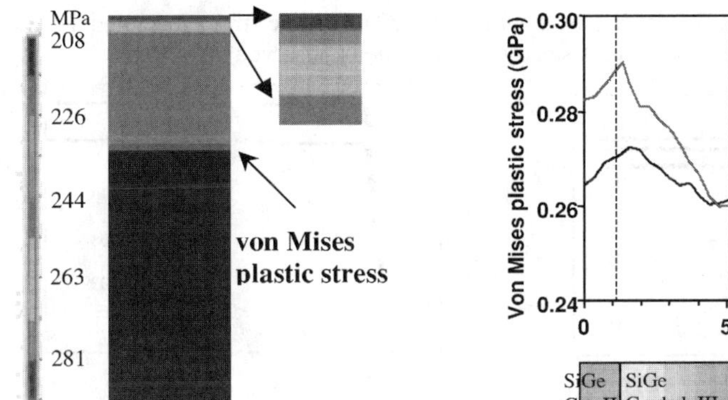

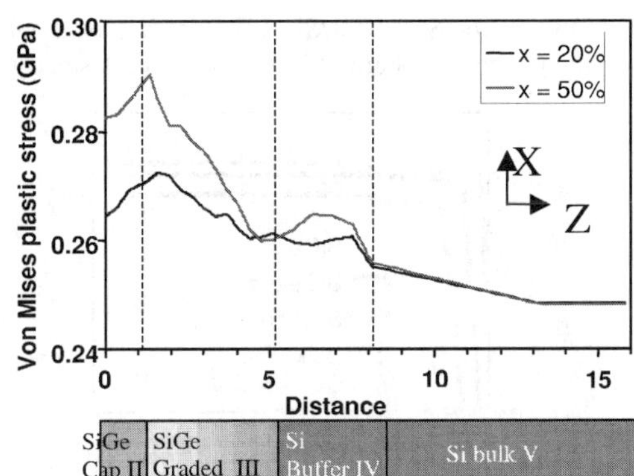

Figure 5. Von Mises plastic stress generated through the structure after 1050°C for 500s.

Figure 6 shows the evolution of elastic strain and layers displacements during the growth process of the SiGe cap. While the elastic strain appears almost instantaneously in the thin layers and fluctuates all along the process time, it is delayed up to 250-300 s for the Si substrate, then increasing abruptly reaching a maximum value within 100 s. The displacements along the growth direction are slightly delayed for the sublayers under the cap and increase linearly with time, while for the Si bulk the displacement occurs after 300 s.

Figure 7 illustrates the variations of the plastic strain as a function of time for nodes chosen on the layer centers. The plastic flow occurs first in the SiGe cap followed by the SiGe graded and Si buffer. The plastic deformation in the Si bulk occurs also after 300 s. The

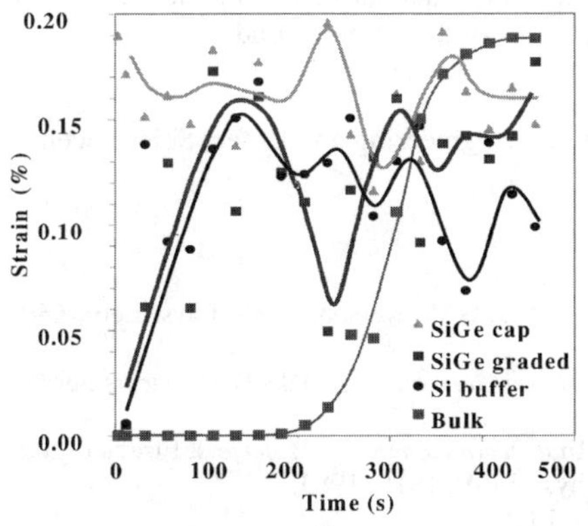

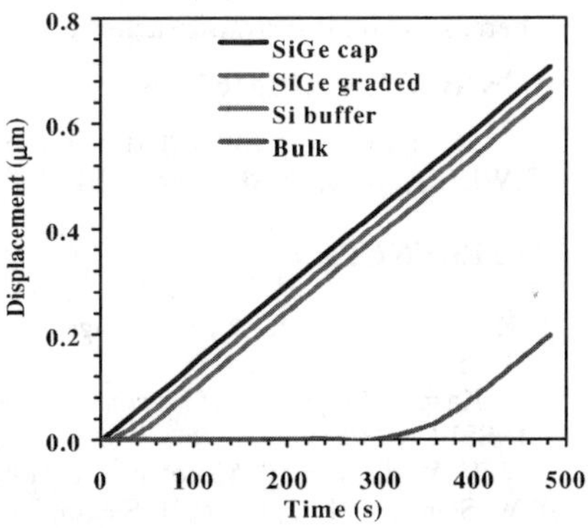

Figure 6. Dynamics of the elastic strain and displacements during SiGe cap growth, x = 0.2.

calculated plastic strain due to thermal mismatch saturates after 10, 30, 50, and 450 s for the sublayers and substrate, respectively. It varies from 3.4×10^{-3} to 4.1×10^{-3} in the Si-buffer, SiGe graded layer and SiGe cap, which might result in high density of dislocations. The plastic strain close to the Si-buffer/Si-bulk interface is about 2.5×10^{-3}. The plastic strain generated depends on the Ge content being higher for 50%. A steady state regime is observed after 160 s.

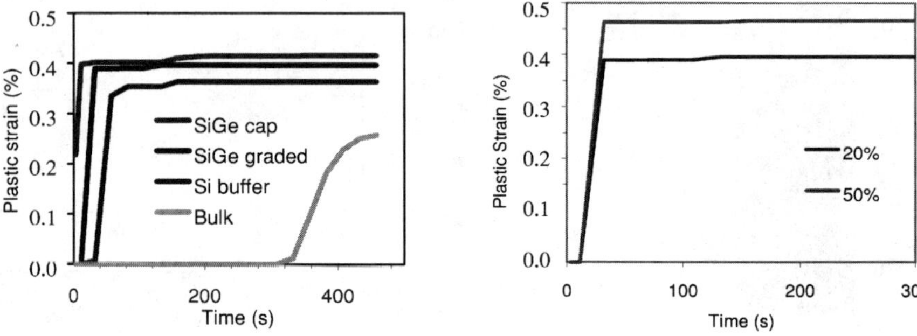

Figure 7. (a) Time dependent plastic strain within each layer for the 20% Ge sample; (b) Plastic strain in the SiGe constant layer for the two samples.

CONCLUSION

The residual elastic and plastic strain/stress generated in the structure during the Si_xGe_{1-x} constant layer growth have been investigated using finite element analysis. Beside the elastic regime, plastic deformation has been identified. We found that the interface SiGe(cte)/SiGe(graded) endures maximum plasticity resulting in a high density of dislocations in the graded layer. The process of relaxation largely extends into the Si bulk. Our simulation showed that the residual stress increases with the Ge content. The non-linear transient analysis showed that strain appears almost instantaneously in the thin layer, while in the substrate it is delayed up to 300 s. The elastic component of the strain fluctuates all along the process time, whereas the plastic strain reaches a steady state behavior after few seconds.

ACKNOWLEDGEMENTS

This work is supported by the Silicon Wafer Engineering and Defect Science center (SiWEDS), under NSF Grant # EEC-9726176.

REFERENCES

1. K. Ismail, M.Arafa, K. L. Saenger, J. O. Chu, and B. S. Meyerson, Appl. Phys. Lett., **66(9)**, 1077 (1995).
2. E. Kasper, *Properties of Strained and Relaxed Silicon Germanium*, EMIS Datareviews Series No. 12, INSPEC (1995)
3. V. V. Vasiliev, E. V. Morozov, *Mechanics and Analysis of Composite Materials*, Elsevier (2001).
4. W. Schroter, H. G. Brian, H. Siethoff, J. Appl. Phys., **54(4)**, 1816 (1983).
5. D. Lowney, et al, Semicond. Sci. Technol. **17**, 1081 (2002).
6. M. T. Currie, S. B. Samavedam, T. A. Langdo, C. W. Leitz, and E. A. Fitzgerald, Appl. Phys. Lett., **72 (14)**, 1718 (1998).
7. K. Bray et al., Interfaces in Electronic Materials Symposium, ECS meeting, October 2003, Orlando,FL.

Mat. Res. Soc. Symp. Proc. Vol. 821 © 2004 Materials Research Society P3.31

Instability Induced by Near-substrate Electric Field

David Salac and Wei Lu
Department of Mechanical Engineering, University of Michigan, Ann Arbor, MI 48109, USA

ABSTRACT

This paper reports experimental work using an electric field close to a substrate to induce morphological patterns in a thin polymer film. The film was spin-coated onto a glass wafer, and subsequently heated to above its glass transition temperature to allow viscous flow. An electric field was applied by two parallel electrodes spacing 10 mm apart. The initially flat polymer/air interface lost stability and formed islands. The self-assembled islands exhibited a narrow size distribution and demonstrated spatial ordering. We attribute the pattern formation to the minimization of combined interface energy and electrostatic energy.

INTRODUCTION

Polymeric thin films enable manufacture of a wide variety of novel, nanostructured devices and templates from which other nanofeatured materials can be created. Examples include block co-polymer self-assembly [1], nanoimprint lithography [2], and soft lithography [2].

One promising fabrication technique discovered recently is lithographically induced self-assembly (LISA) [3-6]. LISA is a process that utilizes the concept of electrohydrodynamic (EHD) instabilities of polymers when exposed to an electric field perpendicular to the surface. In instances of LISA, a flat, thin polymer layer was first applied to a substrate using a technique such as spin coating. An upper mask is then set at a fixed distance (a few hundred nanometers) above the film, forming a small air gap. Between the substrate and the mask an electric field perpendicular to the film surface is developed from either image charges [3,4] or external power supply [5,6]. This causes pressure buildup at the polymer/air interface, and subsequent destabilization of the thin film. As a result, columns are formed towards the upper mask. It is also possible to direct this assembly using a topographically structured top mask [6], thus reproducing the mask pattern. Various orderings of columns have been constructed between the substrates and the masks, for example, use of polymer/polymer/air trilayers have yielded structures comprised of cylinders having polymer cores surrounded by another polymer [6]. Removal of the core via selective etching results in an ordered structure comprised of hollow cylinders.

Thus, in the LISA process, polymer structures are determined by the topographical patterns on the mask, making the technique well-suited to pattern replication. However, the requirement of close placement of upper mask relative to substrate complicates production of large or curved surfaces. Specifically, the mask surface must be treated to prevent polymer adherence, so that the fine structures induced are not damaged during mask removal.

Here, we describe our recent experiment, in which we have used an electric field applied in the plane of the air/polymer interface, instead of perpendicular to the interface. This setup greatly simplifies production, since no mask is required to produce fine features; instead, the morphological features self-assemble.

EXPERIMENTAL DETAILS

A schematic of the experimental setup is show in Fig. 1. It consists of a pedestal, a pair of electrodes, a Pyrex riser, a substrate and the polymer film. The pedestal was constructed from virgin electrical grade polytetrafluoroethylene (PTFE). Besides providing support for the substrate and electrodes, it insulates the platform from other components. PTFE has high electrical resistivity, high dielectric strength and can withstand high temperature. On the pedestal surface, two grooves were machined 10mm apart to hold the electrodes tightly. The electrodes also had two grooves on the inner surface to secure the Pyrex riser, which raised the sample above the pedestal to allow rapid cooling. The electrodes were connected to a Glassman high-voltage power supply, capable of up to 125 kV. Thus the average electric field strength could reach 1.25×10^7 V/m.

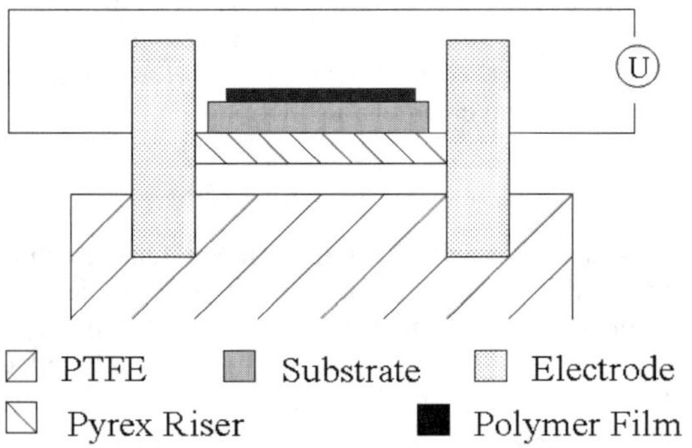

Figure 1 A schematic of the experimental setup.

Each sample consisted of a substrate and a spin-coated thin film. The substrates were made of Pyrex glass wafer, of dimensions of 10mm by 30mm. They were thoroughly cleaned with Toluene to remove dust and any organic compounds before spin coating. The setup in Fig. 1 was enclosed in an aluminum oven equipped with a front observation window. The outside of the aluminum enclosure was covered with a layer of thermal insulating material. A temperature controller and a thermocouple controlled the temperature inside. The operational temperature of the oven ranges between 300 K (room temperature) and 600 K. The aluminum enclosure protected the environment from the high voltage inside.

The polymer used in our experiment was Polystyrene (PS). The properties of PS are well known, and those relevant the experiment include: molecular weight: 108 kg/mol; glass transition temperature: 100 °C; surface tension: 30 mN/m; and dielectric constant: 2.5. PS can be easily spin-coated onto a substrate. Each prepared sample was heated to 160°C, above the glass transition temperature of PS. An electrical voltage of 9.5kV was applied, resulting in an electric field with an average strength of 10^6 V/m. The sample was left in these conditions for times ranging from 6 to 96 hours. After annealing, the sample was cooled to below the glass transition temperature and then the electric field was removed. During each experiment, a control sample was also utilized. A control sample experienced the same temperature and time condition, but was not exposed to any electric field. The samples were observed under both an optical microscope and an atomic force microscope (AFM) to investigate the overall patterns as well as the detailed features.

RESULTS AND DISCUSSION

Two typical experimental results are shown in Fig. 2 and Fig. 3. In the first experiment, the substrate was spin-coated with a 10% PS solution. The second experiment used a 5% PS solution. They were both annealed for 96 hours. Figure 2 is an optical micrograph showing overall morphology. It was observed that patterns consisting of islands formed, in both cases. No structure formations were observed in any control sample. Thus, we concluded that the electric field is the driving force behind this formation.

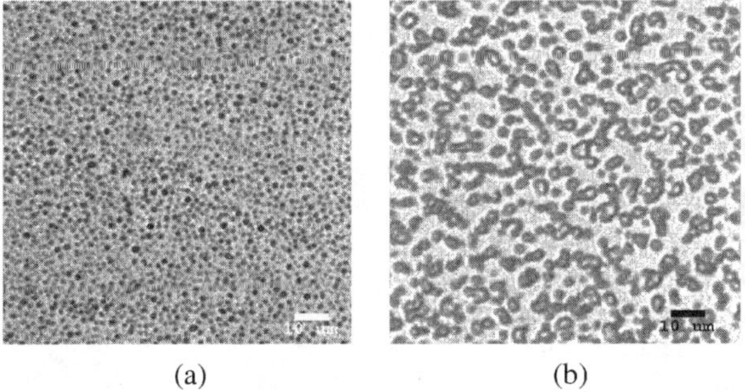

(a) (b)

Figure 2 Two optical images showing the overall morphology. (a) the polymer thin film is prepared from a 10% solution (b) the polymer thin film is prepared from a 5% solution.

A difference in morphology can be observed in the two images in Fig.2. Figure 2a (10% solution) shows a close-packed hexagonal pattern, with a narrow size distribution of islands. Figure 2b (5% solution) has a larger characteristic feature size. Also, some islands seem to have agglomerated into strips. Figure 3 (10% solution) shows a physical characterization of the surface structures, obtained using atomic force microscopy (AFM) in tapping mode. The diameters of most islands ranged from 1.2μm to 1.6μm. Some smaller islands have diameters of about 300nm. The height of the islands averaged about 300nm. The 5% solution shows an average diameter of features of approximately 2.5 μm and an average height of roughly 400 nm.

The cause of the observed differences in structure density, diameters, heights and distributions is not yet known, though we postulate that the difference in feature size is due to the film thickness and non-linearity in the late evolution process. A 5% solution produces a thinner polymer film, and thus less material per unit area, than a film produced from a 10% solution. Imagine a small perturbation for two film thicknesses. The initial stage of evolution is quite similar in terms of linear stability analysis. However, the formation of islands in a thinner film depletes neighboring materials and starts nonlinear evolution earlier in the process. In the final stage, morphology significantly changes the local electric field distribution. The islands in a thinner film do not have a chance to refine as much as they do in a thicker film. We plan to conduct experiments and nonlinear simulations as part of future research to better understand this phenomenon.

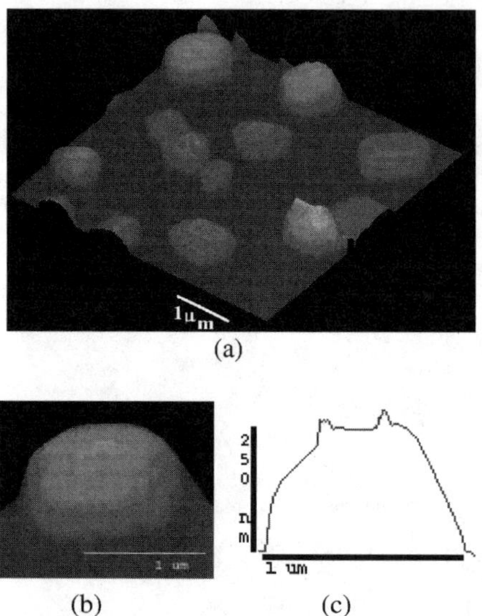

Figure 3 The AFM images showing the morphology (a) 5μm by 5μm area. (b) a closer image of an island (c) height profile. 10% solution.

Theoretically, the origin of the film instability can be understood by considering the competition of forces acting at the polymer/air interface [7-9]. The surface tension tends to minimize the polymer/air interface and stabilize a flat film. By contrast, the electric field induces an electrostatic pressure at the interface that destabilizes the film. This balance of forces can be written in terms of pressures, as

$$p = p_0 - \gamma \frac{\partial^2 h}{dx^2} + p_{el}, \tag{1}$$

where p_0 is the atmospheric pressure, $h(x)$ the film thickness, γ the surface tension, and p_{el} the electrostatic pressure. The pressure induces a Poiseuille flow [10]

$$j = \frac{h^3}{3\mu}\left(-\frac{\partial p}{\partial x}\right), \tag{2}$$

where μ is the temperature-dependent viscosity of the polymer melt. Mass conservation for the incompressible polymer melt requires

$$\frac{\partial j}{\partial x} + \frac{\partial h}{\partial t} = 0 \tag{3}$$

The pattern formation process at the polymer/air interface is described by Eqs. (1-3). A linear perturbation analysis gives the fastest growing wavelength by $\lambda_m = 2\pi\sqrt{2\gamma \Big/ \left(-\frac{\partial p_{el}}{\partial h}\right)}$. Thus an increase of electric field will reduce the feature size. With an in-plane electric field, the morphological change at one point will influence the entire domain. This global influence is desired as it allows structures at long distance sense each other, which may lead to patterns with better uniformity.

CONCLUSIONS

We applied an electric field close to a substrate to induce morphological patterns in a thin polymer film. The initially flat polymer/air interface lost stability and formed islands. The self-assembled islands exhibited a narrow size distribution and demonstrated spatial ordering. We expect that a larger electric field may reduce the island size into sub-100nm scale.

ACKNOWLEGEMENTS

The authors would like to acknowledge the support by the University of Michigan Rackham Research Grant and U.S. National Science Foundation Career Award No. DMI-0348375.

REFERENCES

1 T. B. Liu, C. Burger, and B. Chu, *Prog Polym Sci.* **28**, 5 (2003)
2 Y. Xia, J. A. Rogers, K. E. Paul and G. M. Whitesides, *Chem. Rev.* **99**, 1823 (1999)
3 S.Y. Chou and L. Zhuan, *J. Vac. Sci Technol.* **17**, 3197 (1999).
4 S. Y. Chou, L. Zhuang, and L. J. Guo, *Appl. Phys. Lett.* **75**, 1004 (1999).
5 E. Schäffer, T. Thurn-Albrecht, T. P. Russell, and U. Steiner, *Nature* **403**, 874 (2000)
6 M. D. Morariu, N. E. Voicu, E. Schaffer, Z.Q. Lin, T.P. Russell, and U. Steiner, *Nature Materials* **2**, 48 (2003).
7 E. Schäffer, T. Thurn-Albrecht, T. P. Russell, and U. Steiner, *Europhys. Lett.* **53**, 518 (2001).
8 Z. Suo and J. Liang, *Appl. Phys. Lett.* **78**, 3971 (2001)
9 L. Wu and S. Y. Chou, *Appl. Phys. Lett.* **82**, 3200 (2003).
10 A. Vrij, *Faraday Soc* **42**, 23 (1966).

Part II: Processing—
Mechanical Deformation

A. Experiments

Mat. Res. Soc. Symp. Proc. Vol. 821 © 2004 Materials Research Society

Structures and Mechanical Properties of ECAP Processed 7075 Al Alloy upon Natural Aging and T651 Treatment

Yonghao Zhao[1], Xiaozhou Liao[1], Ruslan Z. Valiev[2] and Yuntian T. Zhu[1]
[1]Materials Science and Technology Division, Los Alamos National Laboratory, Los Alamos, NM-87545, U.S.A.
[2]Institute of Physics of Advanced Materials, Ufa State Aviation Technical University, 12 K. Marx Street, 450000 Ufa, Russian Federation

ABSTRACT

Equal-channel angular pressing (ECAP) processed ultrafine grained (UFG) and coarse grained (CG) 7075 Al alloys were treated by natural aging and T651 temper (annealed at 120 °C for 48 h in Ar atmosphere), respectively. Mechanical tests showed that for the UFG sample, the natural aging resulted in the highest strength (the ultimate tensile strength is 720 MPa). In contrast, for the CG sample, the T651 treatment resulted in a higher strength (the ultimate strength is 590 MPa) than the natural aging (530 MPa). Microstructural analyses indicated that the enhanced strength of the T651 treated CG sample was mainly caused by high densities of G-P zones and metastable η' precipitates. The enhanced strength of the naturally aged UFG sample was mainly caused by the high densities of G-P zones and dislocations. Upon T651 treatment, the dislocation density of the UFG sample deceased significantly, overcompensating the precipitation strengthening.

INTRODUCTION

In the last decade, equal channel angular pressing (ECAP) technique has been widely used to produce bulk ultrafine grained (UFG) metal and alloys, such as Al and its alloys, Cu, Ni, Ti and its alloys, and steels. During an ECAP process, the sample is deformed by being pressed through a die containing two channels, equal in cross section and intersecting at a certain angle. The bulk UFG materials are thought to have considerable potential for industrial applications due to their high strength with good ductility, superplasticity at moderate temperatures and high strain rates, etc. [1].

For Al alloys, most ECAP effort has focused on work strengthening Al-Mg alloys [2-5]. Much less attention has been paid to the precipitate hardening Al-Zn-Mg 7000 series alloys [6,7], which show the highest strength of all commercial Al alloys and are widely used for structural applications in military and civil aircraft as well as sporting goods. Conventional cold working, which usually improves the strength of metals and alloys, has been found to be ineffective in improving the strength of 7000 series Al alloys [8]. It is of great interest to investigate if the strengthening effect observed in ECAP-processed UFG materials can be added to the precipitation hardening effect in the 7000 series Al alloys. If these two strengthening effects can be made additive, it will be possible to significantly improve the strength of 7000 Al alloys, making them much more attractive in high strength structural applications.

In the literature, the highest strength of the 7000 Al alloys was obtained by a T651 temper (annealing at 120 °C for 24 h after solution treatment). T651 treatment produced high densities of coherent spherical-shaped G-P zones and semicoherent intermediate metastable η' phase, which increase the resistance to dislocation movement arising from the strong atomic bonds in the zones [8]. The objective of this work was to investigate the microstructures and mechanical properties of naturally aged (NA) and T651 treated UFG 7075 Al alloy produced by ECAP process. For comparison, coarse-grained (CG) 7075 samples, subjected to the same natural aging and T651 temper, were also characterized.

EXPERIMENTAL DETAILS

Commercial 7075 Al alloy was homogenized by solution treatment for 5 h at 480 °C and quenched to room temperature. The initial grain size is approximately 40 μm. An ECAP die with an intersecting channel angle of 90° and an outerarc angle of 45° was used. The sample was processed for 2 passes by route B_c in which the work piece was rotated 90° along its longitudinal axis between adjacent passes. The ECAP processed UFG and initial solution-treated CG 7075 Al alloys were naturally aged at room temperature for one month and aged at 120 °C for 24 h (T651 temper), then were cut into small pieces in an orientation perpendicular to the pressing direction. The surfaces of these thin pieces were polished for subsequent characterization.

Tensile tests were carried out using a Shimadzu Universal Tester. The samples were cut and polished into a 2.0 mm×1.4 mm cross section and a gauge length of 18.0 mm for tensile tests (in an orientation parallel to the pressing direction) at a displacement rate of 1×10^{-2} mm/s. Microhardness measurements were carried out on Buehler Micromet® Hardness Tester with a load of 500 g and loading time of 15 s. The indenter is the Vickers diamond pyramid. For each sample, at least 20 points were measured to obtain an average value with a typical uncertainty of ±2 %.

Quantitative x-ray diffraction (XRD) measurements were performed on a Scintag x-ray diffractometer, which was equipped with a Cu target operating at 1.8 kW and a secondary monochromator to select the Cu K_α radiation. θ-2θ scans with a step size of $2\theta = 0.02°$ and a counting time of 10 s were performed at room temperature. Pure Al powder (99.999%) was annealed at 200 °C in Ar and used as a XRD reference. Transmission electron microscopy (TEM) and high-resolution transmission electron microscopy (HREM) were performed using a Phillips CM30 microscope operating at 300 kV and a JEOL 3000F microscope operating at 300 kV, respectively. The TEM and HREM samples were prepared by mechanical grinding of the 7075 Al alloy disks to a thickness of about 10 μm. Further thinning to a thickness of electron transparency was carried out using a Gatan Dual Ion Milling System with an Ar+ accelerating voltage of 4 kV and liquid nitrogen for cooling the specimens.

EXPERIMENTAL RESULTS

The engineering stress-strain curves of the UFG and CG samples are shown in figure 1. Upon natural aging, the tensile yield strength, σ_{ys}, and ultimate strength, σ_{us}, of the UFG sample are 650 MPa and 720 MPa, respectively, which are about 103% and 35% higher, respectively, than those of the CG sample (320 MPa and 530 MPa). The elongation to failure of the UFG sample (8.4%) is smaller than that of the CG sample (20.5%). Upon T651 temper, the σ_{ys} and σ_{us} of the UFG sample decreased to 550 MPa and 600 MPa, respectively, while the σ_{ys} and σ_{us} of the CG sample increased to490 MPa and 590 MPa, respectively. The present mechanical properties of the CG sample agree well with the literature [9]. The Vickers microhardness of the natural aged UFG sample (201 H_V) is about 48% larger than that of CG sample (136 H_V). Upon T651 treatment, the hardness of the UFG sample decreased to 178 H_V, while that of the CG sample increased to 174 H_V.

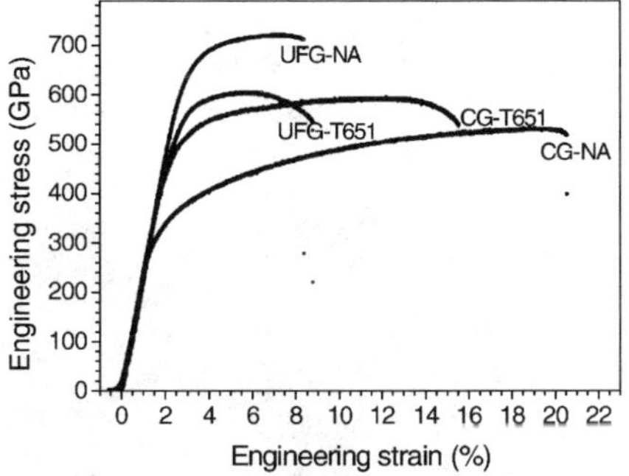

Figure 1. The engineering stress-strain curves of the UFG and CG samples treated by natural aging (NA) and T651 temper.

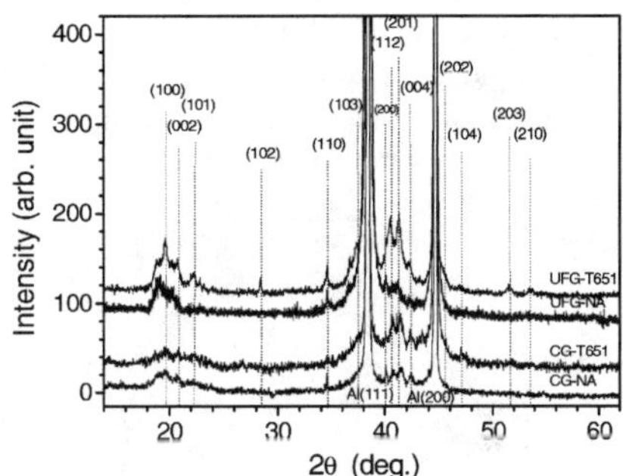

Figure 2. XRD patterns of the UFG and CG samples treated by natural aging (NA) and T651 temper.

Table I. The tensile yield stresses (σ_{ys}), ultimate stresses (σ_{us}), ultimate elongation (ε_{ue} and Vickers microhardnesses (DPH, H_V) of the naturally aged and T651 treated UFG and CG 7075 Al alloys.

	UFG-NA	UFG-T651	CG-NA	CG-T651
σ_{ys} (MPa)	650	550	320	490
σ_{us} (MPa)	720	600	530	590
ε_{ue} (%)	8.4	8.8	20.5	15.5
H_V/0.5, DPH	201	178	136	174

The XRD patterns of the UFG and CG samples treated by natural aging and T651 temper are shown in figure 2. The indexes of diffraction planes of the metastable hexagonal η' were also indicated in the figure. The UFG and CG samples have similar variations in the x-ray patterns upon natural aging and T651 temper. For the naturally aged UFG and CG samples, besides Al reflections, there appeared a broad peak at about $2\theta = 20°$ and some other weak peaks. The broad peak at about 20° corresponds to the G-P zones [10], and the other weak peaks are from the metastable η' phase. Upon T651 temper, there precipitated large amount of η' phase coexisted with G-P zones. Because the XRD sample has the same area involved in the reflection, the intensity of the XRD patterns can be compared. From figure 2, the intensity of the G-P zone broad peak of the naturally aged UFG sample is larger than that of the naturally aged CG sample, indicating that the volume fraction of the G-P zones in the naturally aged UFG sample is larger than that in the naturally aged CG sample. Similarly, the volume fractions of the G-P zones and metastable η' phase in the T651 treated UFG sample are larger than those in the T651 treated CG sample.

The above-mentioned phase precipitations in the UFG and CG samples can also be observed by TEM and HREM, as shown in figures 3 (a-d). The HREM pictures were observed from a

[110] direction of the Al matrix. For the naturally aged UFG and CG samples, both spherical G-P zones (as indicated by arrows in the inset in figure 3a) and plate-shaped η′ (see the inset in figure 3c) are observed. The volume fraction of the G-P zones in the naturally aged UFG sample is larger than that in the CG sample by numerous observations in different areas, which agrees with the XRD results. After T651 temper (figures 3b and 3d), a large amount of plate-shaped η′ phase precipitated out and mixed with the G-P zones. The volume fractions of the G-P zones and metastable hexagonal η′ phase in the UFG sample are larger than those in the CG sample, agreeing with XRD results.

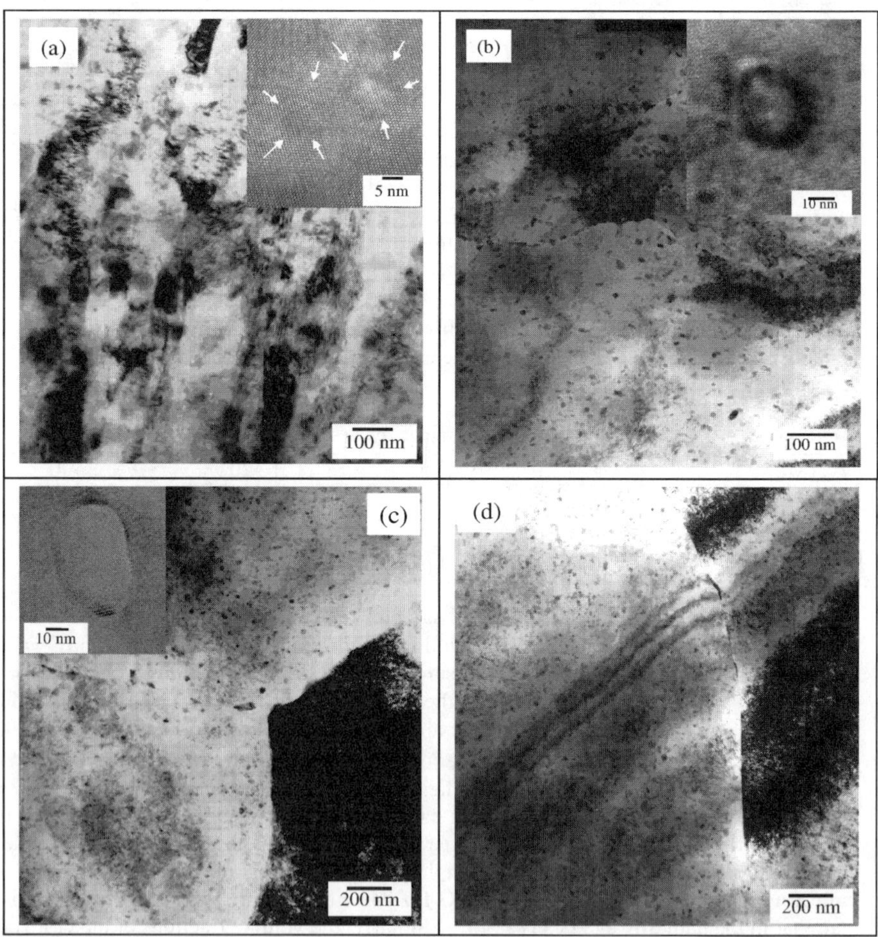

Figure 3. TEM and HREM pictures of the UFG (a,b) and CG (c,d) samples treated by natural aging (a,c) and T651 temper (b,d).

From TEM pictures, one can obtain the average grain size (see Table II). TEM showed that the naturally aged UFG sample was composed of lamella grains with a length of about 430 nm and a width of about 150 nm, which was produced by shear deformation during the ECAP process (figure 3a). Upon T651 temper, the lamella-grains grew into equiaxed grains with a size of about 2 μm. The grain sizes of the naturally aged and T651 treated CG samples were about 40 and 50 μm, respectively. The grain size and microstrain can also be calculated from XRD broadening peaks by the Scherrer and Wilson method [11], as listed in Table II. The XRD calculated grain size is smaller than that from TEM, because XRD gives the scattering domain size or sub-grain size. From the XRD results, the grain size of the naturally aged UFG sample is about 70 nm. The microstrain of the naturally aged UFG sample was 0.54 %, and it decreased significantly to 0.25 % upon T651 temper. The microstrains of the naturally aged and T651 treated CG samples were about 0.05 %.

Table II. The average grain sizes calculated from TEM pictures (D), microstrains ($<\varepsilon^2>^{1/2}$) and dislocation densities (ρ) of the naturally aged and T651 treated UFG 7075 Al alloy and the CG 7075 Al counterpart.

Samples	UFG-NA	UFG-T651	CG-NA	CG-T651
D (nm)	290	1,900	42,000	49,000
$<\varepsilon^2>^{1/2}$ (%)	0.54	0.25	0.05	0.08
ρ (10^{15} m^{-2})	0.94	0.02	0.0	0.0

The dislocation densities in the UFG and CG samples can be calculated from the grain size and microstrain [12,13], as listed in Table II. The dislocation density of the naturally aged UFG sample was about 0.94×10^{15} m^{-2}, and it decreased significantly to about 0.02×10^{15} m^{-2} upon T651 temper. For the naturally aged and T651 treated CG samples, the dislocation densities remained at about zero.

DISCUSSION

The strength of the naturally aged UFG 7075 alloy is significantly higher than its naturally-aged CG counterparts, indicating that it is possible to add strengthening from ECAP processing to that from precipitation hardening.

The significantly high strength of the UFG 7075 Al alloy may be attributed to (i) solid solution strengthening, (ii) grain refinement strengthening, (iii) dislocation strengthening and (iv) precipitation strengthening. The first two effects appear to have less contribution compared to the last two effects, which were verified elsewhere [14]. The XRD analysis revealed high dislocation density in the naturally aged UFG sample, suggesting that dislocation strengthening has a significant contribution to the strength enhancement. The dislocation networks and tangles within grains and near grain or sub-grain boundaries make dislocation glide more difficult. The precipitation strengthening results from the precipitate's ability to impede dislocation motion by forcing dislocations to either cut through or circumvent the fine precipitates. In either case, higher density of precipitates leads to higher strength. For naturally aged 7075 Al alloy, the dislocation cutting of the G-P zones and η′ fine plates is the main strengthening mechanism, because the strong atomic bonds in the zones can effectively hinder dislocation motion [8]. Both XRD and HREM revealed that the volume fraction of the G-P zones in the UFG sample is larger than that in the CG sample. Therefore, the cooperative interaction of the high densities of the G-P zones and dislocations has resulted in the observed significantly higher strength of the naturally aged UFG 7075 Al alloy. The enhanced strength of the T651 treated CG sample was mainly caused by high densities of G-P zones and metastable η' precipitates. Upon T651 treatment, the dislocation density of the UFG sample deceased significantly, overcompensating the precipitation strengthening.

CONCLUSION

A combination of ECAP processing and precipitation hardening has the potential to render Al 7000 series alloys significantly stronger than those processed by either technique alone. The high densities of G-P zones and dislocations are primarily responsible for the high strength of naturally-aged UFG 7075 alloy. A T651 temper resulted in the high strength of the CG sample, caused by high densities of G-P zones and metastable η' precipitates. However, a T651 temper decreased the strength of the UFG sample, because the T651 treatment deceased the dislocation density significantly, overcompensating the precipitation strengthening.

REFERENCES

1. R.Z. Valiev, R.K. Islamgaliev and I.V. Alexandrov, Prog. Mater. Sci. **45**, 103 (2000).
2. J. Wang, Y. Iwahashi, Z. Horita, M. Furukawa, M. Nemoto, R.Z. Valiev and T.G. Langdon. Acta mater. **44**, 2973 (1996).
3. M. Furukawa, A. Utsunomiya, K. Matsubara, Z. Horita and T.G. Langdon, Acta mater. **49**, 3829 (2001).
4. S. Lee, A. Utsunomiya, H. Akamatsu, K. Neishi, M. Furukawa, Z. Horita and T.G. Langdon, Acta mater. **50**, 553 (2002).
5. D.G. Morris and M.A. Munoz-Morris, Acta mater. **50**, 4047 (2002).
6. Z. Horita, T. Fujinami, M. Nemoto and T.G. Langdon, Metall. Trans. **31A**, 691 (2000).
7. L.J. Zheng, C.Q. Chen, T.T. Zhou, P.Y. Liu and M.G. Zeng, Mater. Character. **49**, 455 (2003).
8. W.F. Smith, *Structure and Properties of Engineering Alloys*, (McGraw-Hill: New York, 1993) Chap. 5-9, p. 214.
9. Y. Murakami, Aluminum-Based Alloys, in *Processing Metals and Alloys*, edited by R.W. Cahn (VCH: Weinheim, New York, c1991), p. 253.
10. Y.H. Zhao, X.Z. Liao, Z. Jin, R.Z. Valiev and Y.T. Zhu, in *Ultrafine-Grained Materials III*, edited by Y.T. Zhu, T.G. Langdon, R.Z. Valiev, S.L. Semiatin, D.H. Shin and T.C. Lowe (The Minerals, Metals & Materials Society, 2004), p. 511.
11. Y.H. Zhao, K. Zhang and K. Lu, Phys. Rev. B, **56**, 14322 (1997).
12. Y.H. Zhao, H.W. Sheng and K. Lu, Acta mater. **49**, 365 (2001).
13. Y.H. Zhao, K. Zhang and K. Lu, Phys. Rev. B, **66**, 085404-01 (2002).
14. Y.H. Zhao, X.Z. Liao, Z. Jin, R.Z. Valiev and Y.T. Zhu, to be published in Acta mater. (2004).

Mat. Res. Soc. Symp. Proc. Vol. 821 © 2004 Materials Research Society P9.5

Microstructural Refinement in Steels by Machining

S. Swaminathan, C. Swanson, T. L. Brown, R. F. Kezar, S. Chandrasekar, W. D. Compton and K. P. Trumble
Center for Materials Processing and Tribology,
Schools of Engineering,
315 N. Grant Street, Purdue University,
West Lafayette, IN, 47907-2023

ABSTRACT

A study has been made of microstructure and hardness of machining chips created from commercially pure iron and carbon steels. Large shear strains imposed during chip formation in machining are found to produce significant microstructure refinement in the chips, resulting in higher hardness compared to the bulk. Transmission electron and scanning electron microscopy have shown the chips to consist entirely of ultra-fine grain structures with ferrite grain sizes in the range of 100-800 nm. With high carbon steels, the microstructure of the bulk material prior to machining is also seen to have a significant influence on the characteristics of the chip.

INTRODUCTION

It has been known for some time now that significant microstructure refinement can be realized in metals and alloys by the application of large strain deformation [1,2]. Examples of this approach can be found in the use of high-energy ball milling to create nano-particles [3]; and the use of severe plastic deformation (SPD) processes such as equal channel angular extrusion (ECAE) [4] and high pressure torsion (HPT) [5] to prepare bulk and thin samples, respectively, with nanocrystalline microstructure. It has been shown, recently, that chip formation by machining provides a simple configuration in which large shear strains of known magnitude can be imposed in metals and alloys in a single stage of deformation [6,7]. Furthermore, the method is equally applicable to materials of both low and high initial strength. The creation of nanocrystalline microstructures in a variety of non-ferrous materials and commercially pure iron was demonstrated [6,8].

The present study was undertaken with the objective of studying the effects of large strain deformation, imposed in a single stage of deformation, on the microstructure of steels of varying carbon content. Preliminary experiments have also been carried out to explore the effect of initial microstructure of steel on microstructural changes arising from the large deformation.

BACKGROUND

Machining is a process in which a hard, wedge-shaped tool removes material (chip) from the surface of a solid (bulk) by very large strain deformation (Fig. 1a)). Chip formation occurs by concentrated shear along a narrow deformation zone called the shear plane (Figs. 1a) and 1b)). The effective plastic strain imposed at the shear plane can be varied by changing the tool rake angle and the friction between tool and chip. Shear strains in the range of 2-10 and strain rates of up to 1×10^6 s^{-1} are common features of machining [9]. The temperature in the deformation zone may be systematically varied by changing the cutting velocity.

EXPERIMENTAL PROCEDURE

Chips from commercially pure iron (99.8%), plain carbon steels, 52100 steel and stainless steel (316L) were produced by lathe machining. The ferrite grain size in the bulk material was greater than 10 μm in all the cases. The cutting conditions were selected so as to ensure that the temperature rise due to shear deformation at the shear plane (Fig. 1a)) was minimal. The shear strain imposed in the chip was approximately 4 in all the cases. Typical dimensions of chip samples were 200-3000 μm in width, 100-1000 μm in thickness and a few millimeters in length. Hardness and microstructure of the bulk and chip samples were characterized. Hardness measurements were made using Vickers indentation taking care to ensure that the size of an indentation (>25 μm) was kept approximately the same in the bulk and the chip samples. At least 20 indentations were made in each sample to estimate the mean hardness and associated standard deviation. Microstructure was characterized by a combination of optical microscopy, scanning electron microscopy (SEM) and transmission electron microscopy (TEM). Samples were metallographically prepared and etched with 2% Nital for optical microscope and SEM observations. For the TEM observations, disk specimens were thinned from ~100 μm thick chip samples by electrolytic twin-jet polishing on a Struers Tenupol-2. The electron transparent samples were observed in a JEOL-2000-FX-TEM operating at 200 kV. Bright-field and dark-field images were obtained, together with selected area electron diffraction (SAD) patterns from the imaged regions.

The effect of initial microstructure on microstructure refinement arising from large strain deformation, was examined in a set of machining experiments carried out with 1080 steel samples. These samples were heat-treated to have different initial (bulk) microstructures. For this set of experiments, three 1080 steel rods were initially austenitized at 900°C for two hours in a furnace under a controlled argon atmosphere (95%Ar and 5%H_2 by volume). One rod was annealed (furnace cooled) to produce a coarse pearlite; the second rod was normalized in air to create fine pearlite; and the third rod was quenched in cold water (25°C) to form a fully martensitic structure. Chips were produced from each of these rods by machining; the hardness and microstructure of the chips were characterized as before. Additionally, chip samples created from the annealed and normalized 1080 steels were subjected to heat treatment at different temperatures in the range of 300°C to 600°C for 30 minutes in a controlled argon atmosphere. Hardness and microstructure of the heat treated samples were analyzed.

RESULTS

Characteristics of steel chips

Figure 1b) shows an optical micrograph of a partially formed iron chip, as yet not separated from the bulk, produced in a specially devised "quick stop" experiment. Also, shown superimposed on the microstructure are average Vickers hardness values of the bulk and the chip. The hardness values show a steep increase over a narrow deformation zone (shear plane) separating the chip from the bulk. The presence of flow lines and lack of visible grain structure in the chip suggest the grain size to be in the sub-micrometer range.

Figure 2 summarizes the Vickers hardness values measured on the chips and the bulk of different steels. The chips in each case are seen to be significantly harder than the corresponding bulk samples. The increase in hardness is about 200% in the case of the iron chips, 100% for the 1018 steel chips, 50% for the 1080 chips and 150% for the stainless steel chips. Furthermore, the measurements showed no evidence of hardness anisotropy in the chips.

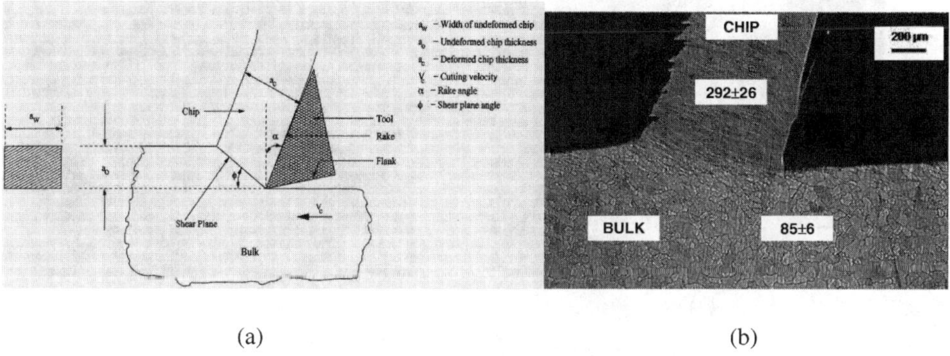

(a) (b)

Figure 1. (a) Schematic of 2-D plane strain machining. The shear plane angle (ϕ) together with the rake angle (α) determine the average shear strain in the chip (b) microstructure of partially formed iron chip as yet not separated from the bulk that was produced in a specially devised "quick stop" experiment. The superimposed Vickers hardness values show a significant increase in hardness across a narrow deformation zone (shear plane) in going from the bulk into the chip.

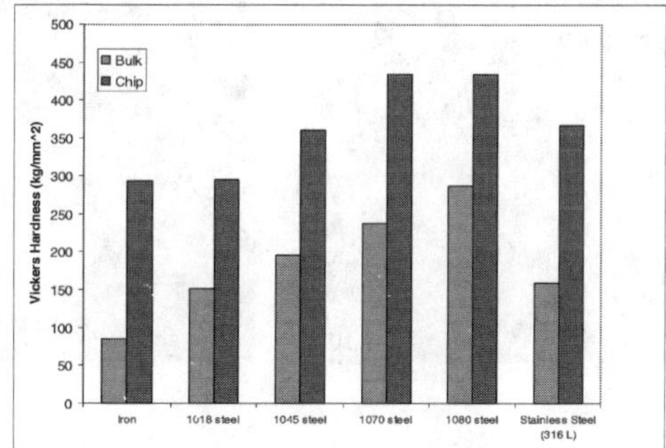

Figure 2. Hardness of the bulk and chip samples. The chips have significantly higher hardness compared to the bulk. The plain carbon steels were normalized prior to machining.

Figures 3a) and 3b) show TEM images characteristic of the pure iron and 52100 chip steel samples, respectively. Also, shown in the inset of the figures are SAD patterns with spotted rings characteristic of fine-grained, polycrystals. The iron chip (Fig. 3a)) shows numerous grains with elongated morphologies, with an average grain size of 175 ± 100 nm in the width and 685 ± 190 nm in the length direction. Furthermore, the grains appear to be severely deformed with a high density of dislocations. The microstructure of the 52100 steel chip (Fig. 3b)) is seen to consist of equiaxed ferrite grains and carbides, with grain sizes in the range of 100 to 500 nm. Figures 4a) and 4b) show SEM images of lightly etched 1045 steel bulk and chip samples, respectively. The elongation of the cementite lamellae in the chip suggests directionality of flow. Nanostructured, equiaxed ferrite grains may be seen interspersed between the lamellae.

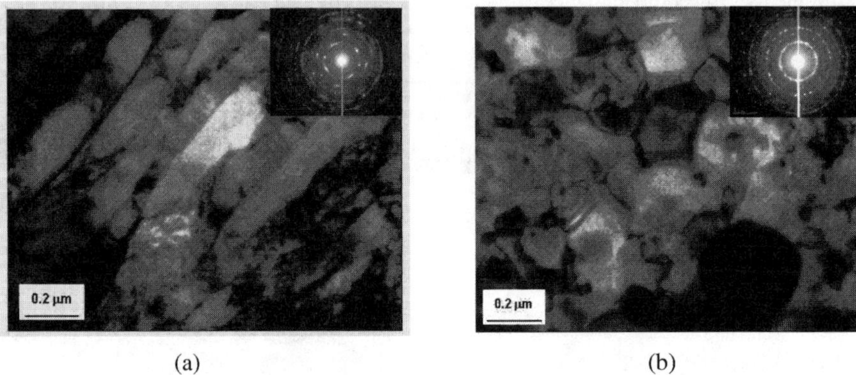

(a) (b)

Figure 3. Bright-field TEM images of chip samples, along with SAD patterns, showing nanocrystalline microstructure (a) iron chip showing elongated grains (b) 52100 steel chip showing equiaxed ferrite grains and carbide particles.

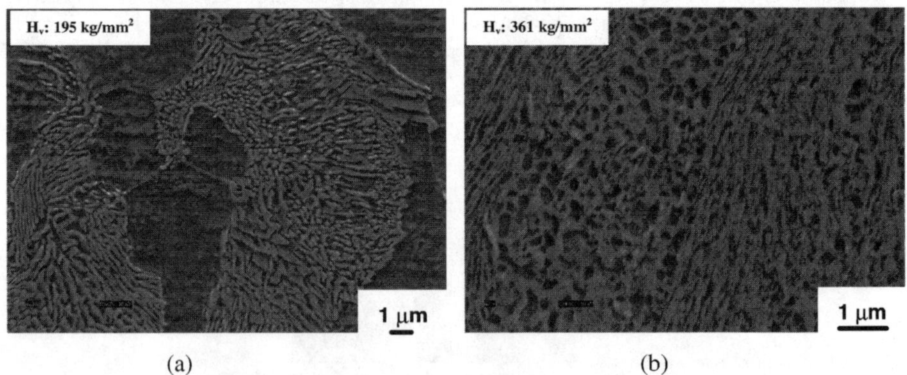

(a) (b)

Figure 4. SEM images of 1045 steel prepared metallographically and etched lightly with 2% Nital (a) normalized bulk showing large ferrite grains and pearlite colonies (b) chip showing equiaxed ferrite grains between cementite lamellae. The lamellae appear to show flow in a particular direction. Vickers hardness (H_v) values for the samples are given in the figure.

Effect of initial microstructure in 1080 steel

Figures 5a) and 5b) show SEM images of 1080 bulk steel (annealed) and chip samples, respectively. Distinct colonies of cementite lamellae, characteristic of coarse pearlite, can be seen in the bulk (Fig. 5a)). The cementite lamellae appear to be broken up in the chip (Fig. 5b)), but their distribution is suggestive of directional flow. The hardness of the chip is seen to be twice that of the bulk. The chips machined from normalized and quenched bulk 1080 samples were also observed to undergo significant microstructure refinement, as only flow lines were visible in the etched chip microstructure. However, the hardness increase in going from the bulk to the chip was 46% in case of the normalized steel and only 7% in case of the quenched steel. These observations indicate that the initial microstructure of the bulk steel sample significantly influences the properties of resulting chips. TEM studies of the 1080 steel chips are underway to elucidate the details of the microstructure.

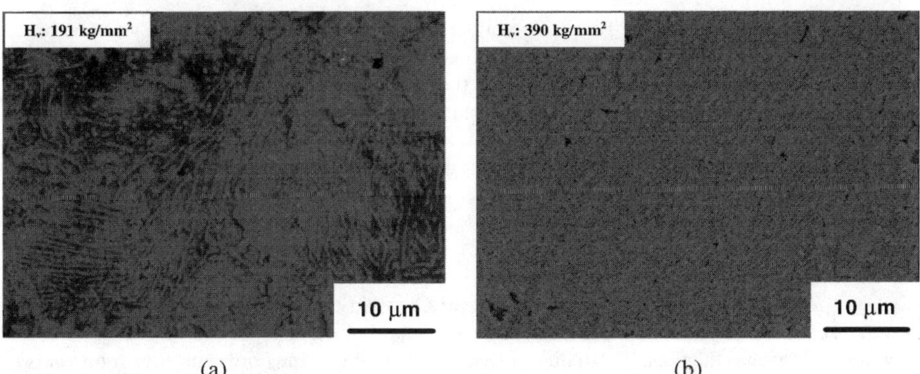

(a) (b)

Figure 5. SEM images of 1080 steel, which was furnace cooled (annealed) after austenitization to produce coarse pearlite (a) bulk sample showing randomly oriented pearlite colonies and (b) chip sample showing mostly broken cementite. There is distinct flow pattern associated with the broken cementite lamellae in the chip. The hardness is seen to increase by 100% from bulk to the chip.

Heat treatment of 1080 steel chips

Table 1. Vickers hardness as a function of annealing temperature for chips produced from the annealed and normalized 1080 steel bulk samples.

Temperature (°C)	Hardness (annealed bulk) (kg/mm^2)	Hardness (normalized bulk) (kg/mm^2)
Chip	390±25	422±22
300	434±24	467±32
400	440±20	466±22
500	364±10	422±15
600	244±8	353±8

Table 1 summarizes the hardness values of steel chips after heat treatment at different temperatures. The hardness of chips machined from both annealed and normalized bulk samples is seen to increase after heat treatment in the range of 300°C and 400°C. However, the hardness decreases when heat treatment temperatures are greater than 400°C. It is possible that the increase in hardness noted between 300°C and 400°C is due to precipitation hardening. Dissolution of cementite lamellae in the ferrite matrix could have occurred due to the large strain imposed in the chip during machining. Subsequent heat treatment at higher temperatures will cause re-precipitation of cementite, leading to an overall increase in the hardness of the chips. This is similar to what has been observed in annealing of heavily drawn pearlitic steel wires [10]. At heat treatment temperatures greater than 400°C, the growth of ferrite grains will lead to a decrease in the hardness of chips.

CONCLUSIONS

The creation of ultra-fine grained microstructure in pure iron and steels by the process of chip formation in machining has been demonstrated. The ultra-fine grained chips have been found to have significantly higher hardness than the bulk material in most of the steels. The results indicate that the process of chip formation can be used to study microstructure refinement and related phenomena arising from large strain deformation in steels of both low and high initial strength. Lastly, the machining approach suggests a low-cost process for making nanostructured steels in large volume.

ACKNOWLEDGEMENTS

We would like to thank the Ford Motor Company, State of Indiana 21st Century Research and Technology Fund and DOE/Oak Ridge National Laboratory (ORNL) for support of this work. Additional thanks are also due to Drs. Andrew Sherman (Ford) and Ray Johnson (ORNL) for their encouragement of this work.

REFERENCES

1. G. Langford and M. Cohen, *Trans. ASME* **62**, 623 (1969).
2. J. D. Embury and R. M. Fisher, *Acta. Metall.* **14**, 147 (1966).
3. H. J. Fecht, E. Hellstern, Z. Fu and W. L. Johnson, *Metall. Trans.* **A21(9)**, 2333 (1990).
4. V. M. Segal, V. I. Reznikov, A. E. Drobyshevskiy and V. Kopylov, *Russ. Metall.* **1**, 99 (1981).
5. R. Z. Valiev, R. K. Islamgaliev, and I. V. Alexandrov, *Prog. Mater. Sci.* **45**, 103 (2000).
6. T. L. Brown, S. Swaminathan, S. Chandrasekar, W. D. Compton, A. H. King and K. P. Trumble, *J. Mater. Res.* **17(10)**, 2484 (2002).
7. T. L. Brown, S. Swaminathan, B. C. Rao, R. F. Kezar, S. Chandrasekar, W. D. Compton, K. P. Trumble and A. H. King in *Ultrafine Grained Materials III* edited by Y. T. Zhu, T. G. Langdon, R. Z. Valiev, S. L. Semiatin, D. H. Shin and T. C. Lowe, (TMS (The Minerals, Metals and Materials Society, 2004)) pp 167-172.
8. S. Swaminathan, T. L. Brown, M. Ravi Shankar, B. C. Rao, K. P. Trumble, W. D. Compton and S. Chandrasekar in *Ultrafine Grained Materials III* edited by Y. T. Zhu, T. G. Langdon, R. Z. Valiev, S. L. Semiatin, D. H. Shin and T. C. Lowe, (TMS (The Minerals, Metals and Materials Society, 2004)) pp 161-166.
9. M. C. Shaw, *Metal Cutting Principles*, (Oxford University Press, 1984).
10. J. Languillaume, G. Kapelski and B. Baudelet, *Mater. Lett.* **33**, 241 (1997).

Mat. Res. Soc. Symp. Proc. Vol. 821 © 2004 Materials Research Society P8.10

Unconfined Twist: a Simple Method to Prepare Ultrafine Grained Metallic Materials

Yonghao Zhao, Xiaozhou Liao and Yuntian T. Zhu*
Materials Science and Technology Division, Los Alamos National Laboratory,
Los Alamos, NM-87545, U.S.A.

ABSTRACT

A new simple method – unconfined twist was employed to prepare ultrafine grained (UFG) Fe wire. A coarse grained (CG) Fe wire with a diameter of 0.85 mm was fixed at one end, and twisted at the other end. After maximum twist before fracture, in the cross-sectional plane, concentrically deformed layers with a width of several micrometers formed surrounding the center axis of the wire. The near-surface deformed layers consist of lamella grains with a width in submicrometer range. In the longitudinal plane, deformed bands (with a width of several micrometers) formed uniformly, which were composed of lamella crystallites (with a width in submicrometer range). The tensile yield strength and ultimate strength of the twisted Fe wire are increased by about 150 % and 100 % compared with the values of its CG counterpart.

INTRODUCTION

Ultra-fine grained (UFG) metals and alloys have been the subject of considerable research in recent years. Such interest has been spurred by the recognition that these materials possess some appealing mechanical properties, such as high yield and fracture strengths, superior wear resistance, increased resistance to tribological and environmentally-resisted damage, increased strength and/or ductility with increasing strain rate and potential for enhanced superplastic deformation at lower temperature and higher strain rates. Nevertheless, the industrial application of the UFG materials has not been realized yet because of the difficulty to fabricate them into dense and bulk dimensions. Among the numerous UFG material synthesis methods, such as mechanical alloying and compaction [1,2], inert-gas condensation and consolidation [3], electrodeposition [4], crystallization of amorphous solids [5,6] and severe plastic deformation (SPD) [7], the SPD method, such as equal channel angular pressing (ECAP) [8,9] is especially attractive because it can economically produce relatively large bulk dense UFG samples without changing their cross-sectional dimensions. Another SPD method, high-pressure torsion (HPT) [7], can effectively refine coarse grains into several tens nanometers, but it can only process very thin disk (about 1 mm thick). Moreover, both HPT and ECAP methods need very high pressure (up to several GPa). Such high pressure brought difficulties for the industrial applications of these SPD preparations. Therefore, simpler SPD method would be attractive.

In the present work, we developed a new simple SPD method – unconfined twist to process UFG metallic materials. In this method, an initial rod-shaped sample was fixed at one end, and was twisted at the other end. During the twist, severe plastic deformation occurred within the sample and formed the UFG structures. The twisted UFG sample is found to have greatly enhanced mechanical strength compared with its coarse grained (CG) counterpart.

*corresponding author: yzhu@lanl.gov

EXPERIMENTAL DETAILS

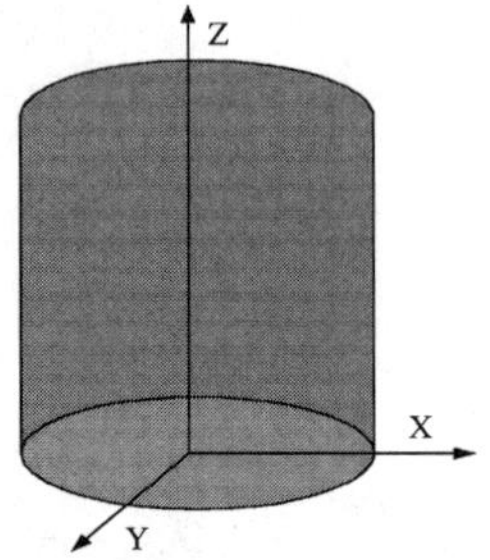

Figure 1. Schematical representation of the Fe wire.

Commercial Fe wire (with a diameter of 0.85 mm and a length of 12 mm) was fixed on a twisting machine and twisted for 180 rotations clockwise at a speed of 30 rpm. Tensile tests of the twisted and untwisted samples were carried out using a Shimadzu Universal Tester at a displacement rate of 1×10^{-2} mm/s. The morphology of the fracture surface was observed by JEOL Scanning Microscope 6300 FXV operating at an accelerate voltage of 30 kV.

The twisted UFG and initial CG Fe wires were then cut and polished cross-sectionally (X-Y plane in figure 1) and longitudinally (Y-Z plane). The polished specimens were etched using a nital etchant (mixture of 1-ml ethanol and 10-ml nitric acid) for scanning electron microscopy (SEM) charaterizations.

Transmission electron microscopy (TEM) was performed using a Phillips CM30 microscope operating at 300 kV. The TEM samples were prepared by mechanical grinding of the twisted and untwisted Fe wires cross-sectionally and longitudinally to a thickness of about 10 μm. Further thinning to a thickness of electron transparency was carried out Gatan Precision Ion Milling System with an Ar+ accelerating voltage of 4 kV.

EXPERIMENTAL RESULTS

The engineering stress-strain curves of the UFG and CG samples are shown in figure 2. The UFG sample has a much higher strength than its CG counterpart, as listed in Table I. The tensile yield strength (σ_{ys}) and the ultimate strength (σ_{us}) of the UFG sample are 690 MPa and 750 MPa, respectively, which are about 130 % and 103 % higher than those of the CG sample (300 MPa and 370 MPa). The elongation to failure of the UFG sample (2.5 %) is much smaller than that of the CG sample (15.8 %).

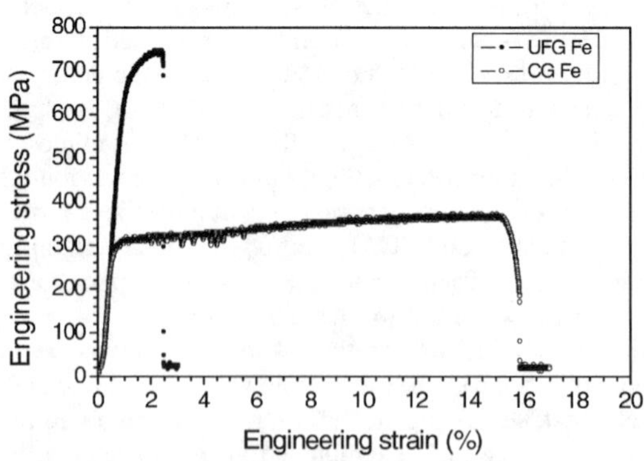

Figure 2. The engineering stress-strain curves of the twisted UFG and initial CG Fe samples.

Table I. The tensile yield strength (σ_{ys}), ultimate strength (σ_{us}) and ultimate elongation (ε_{ue}) of the twisted UFG and initial CG Fe wires.

Samples	UFG	CG
σ_{ys} (MPa)	690	300
σ_{us} (MPa)	750	370
ε_{ue} (%)	2.5	15.8

Figures 3(a-d) show the SEM pictures of the UFG and CG Fe wires. Figures 3(a) and 3(c) were taken from the cross-sectional plane (X-Y), and figures 3(b) and 3(d) were taken from the longitudinal plane (Y-Z). The CG Fe sample was composed of equi-axed grains (with an average size of about 15 μm) in X-Y plane and lamella grains (with an average size of about 25 μm) in Y-Z plane, as shown in figures 3(a) and 3(b). The micro-holes are carbide. After twist, in the cross-sectional X-Y plane, concentrically deformed layers with a thickness of about several micrometers formed surrounding the center axis of the wire, as shown in figure 3(c). The crystallites at the outer part were elongated (see positions XY-A and XY-B in figure 3c), while the grains at the center of the Fe wire remains unchanged (see position XY-C in figure 3c). In the longitudinal Y-Z plane, deformed bands with a width of about several micrometers formed, as shown in figure 3(d).

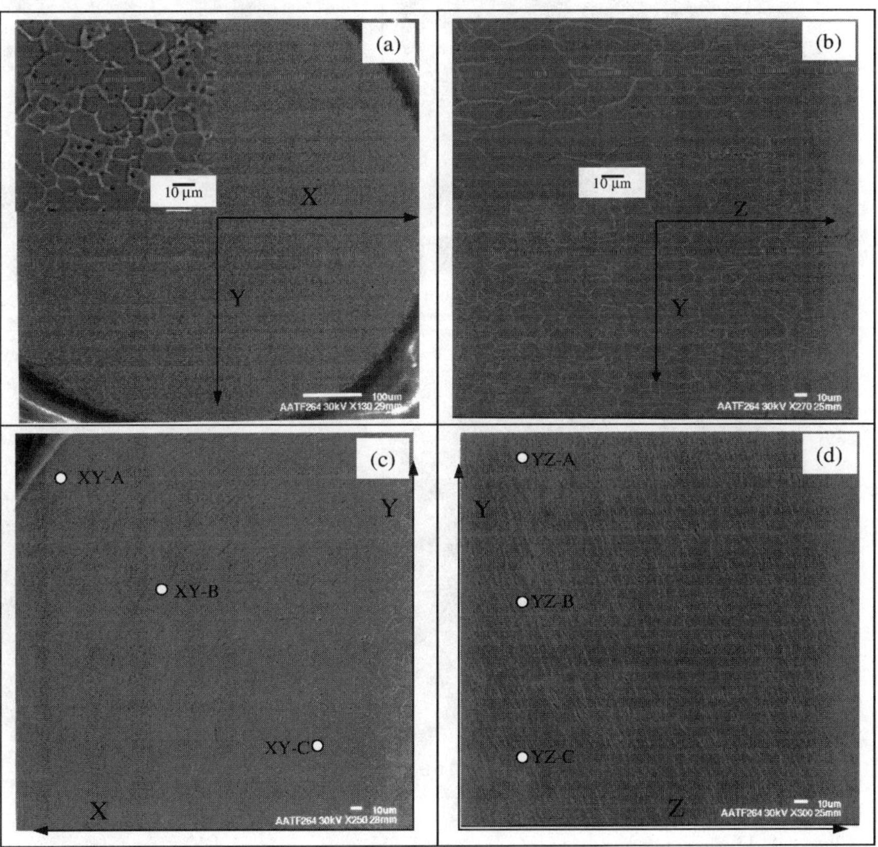

Figure 3. SEM pictures of the CG (a,b) and UFG (c,d) samples. Figures 3(a) and 3(c) were taken from the cross-sectional X-Y plane, and figures 3(b) and 3(d) were taken from the longitudinal Y-Z plane.

To study the detailed structures within the deformed layers and bands of the twisted UFG Fe wire, TEM was used to observe different positions (such as XY-A, XY-B, YZ-A, YZ-B and YZ-C in figures 3c and 3d) in the UFG sample. Figures 4(a-d) show the TEM pictures taken from positions XY-A, XY-B, YZ-A and YZ-C, respectively. One can see that near the surface position XY-A in X-Y plane, the deformed layers were composed of lamella grains with a width of about several hundred nanometers. The longitudinal direction of the lamella grains is parallel to that of

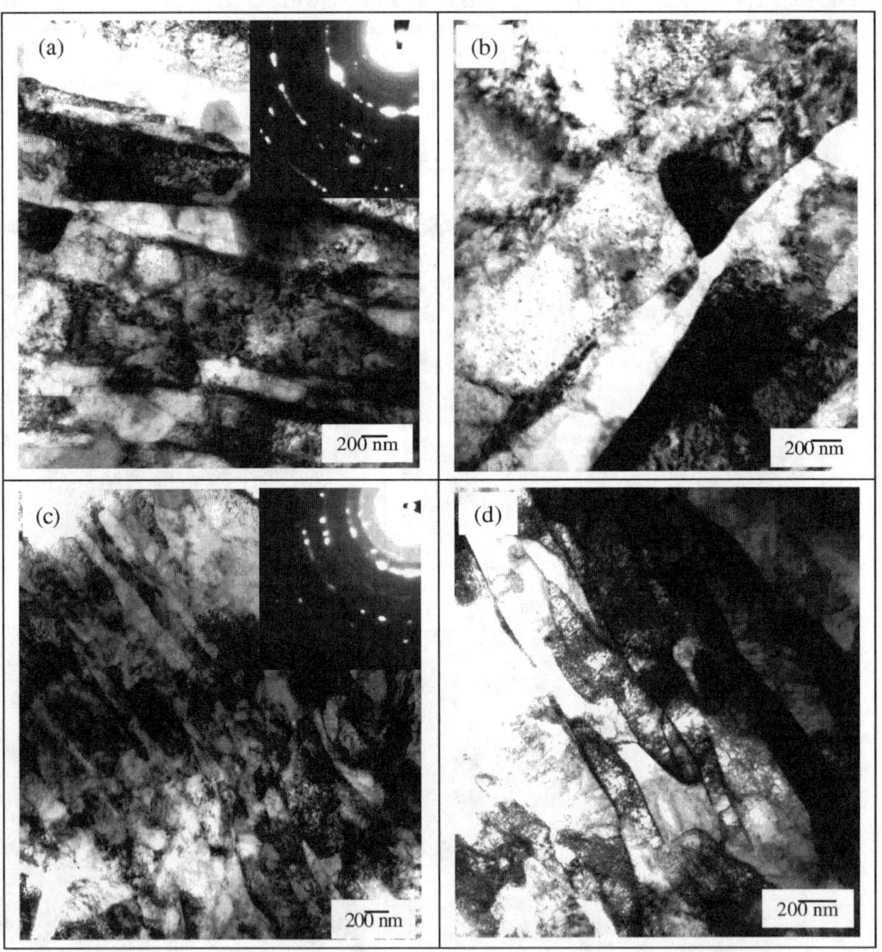

Figure 4. TEM pictures of the UFG Fe sample. Figures 4(a) and 4(b) were taken from positions XY-A and XY-B, respectively, and figures 4(c) and 4(d) from positions YZ-A and YZ-C, respectively.

the deformed layers. The deformed layers in position XY-B contain the evidently larger lamella grains than those in XY-A position, as shown in figure 4(b). The selected area diffraction (SAD) indicates the crystallographic orientations of the crystallites within the deformed layers are mainly small-angle grain boundaries, as shown the inset in figure 4(a).

From figures 4(c) and 4(d), the deformed bands in positions YZ-A and YZ-C contain lamella grains with a width of about 100-200 nm. The longitudinal direction of the lamella grains is parallel to that of the deformed bands, which has a certain angle with Z-axis. The crystallographic orientations of the crystallites within the deformed bands are mainly small-angle grain boundaries (see the inset in figure 4c).

From the TEM pictures, there are many networks within the crystallites, which are corresponding to the high density of dislocations. Moreover, the grain boundaries of the twisted sample are straight and sharp, different from the high-energy non-equilibrium boundary configurations observed in the ECAP processed UFG materials [10, 11]. This difference was caused by the different deformation mechanisms of the different SPD methods.

Figures 5(a,b) show the morphologies of the surface fracture for the tensile UFG and CG Fe samples. The fracture surface of the CG sample contains the spherical dimples, of which the size ranges from several micrometers to about 10 micrometers. For the UFG sample (see figure 5b), the fracture surface contains the concentrical-circle layers (with a thickness of several micrometers) caused by the deformed layers. Within the concentrical layers (see the inset in figure 5b), there observed the elongated dimples with a width size of about 1 μm, which were caused by the submicrometer grains within the deformed layers.

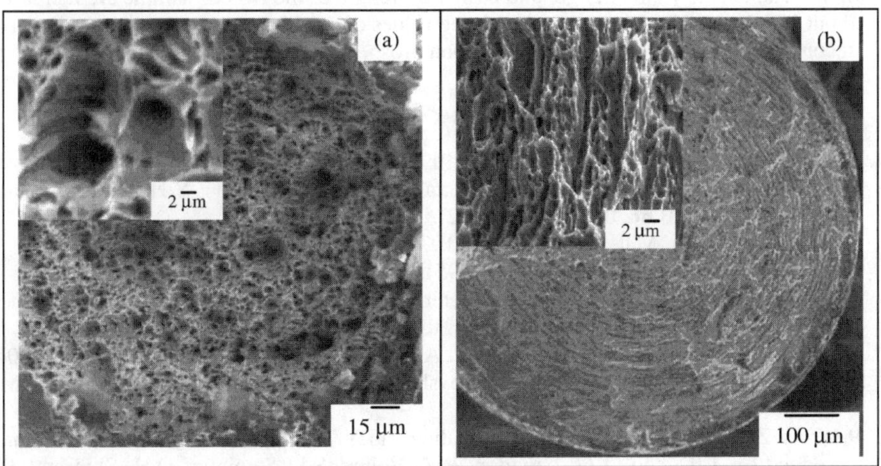

Figure 5. Morphologies of the fracture surface for the CG (a) and UFG (b) samples.

DISCUSSION

Based on above results, the twist process introduced unique UFG structures into the Fe wire. Upon twist, the deformed layers and bands with sizes of several micrometers were formed in the X-Y and Y-Z planes, respectively. Both the deformed layers and bands contain lamella grains with a width in submicrometer range. Near the surface of the Fe wire, the coarse crystallites were refined into plate shaped grains with a thickness of about 100-200 nm along Z-axis and a length of several hundred nanometers along X(Y)-axis. At the center of the Fe wire, the crystallites are coarse grained in X-Y plane, but are about 100-200 nm thick along Z-axis.

The mechanical strength of the twisted sample was enhanced by about 100 % compared with the untwisted sample. The strength enhancement was mainly caused by grain refinement strengthening and dislocation strengthening. The grain refinement strengthening is generally described by a Hall-Petch equation [12,13]. The dislocation networks and tangles within grains (as shown in figures 4a-d) will improve the resistance to dislocation movements when the moving dislocations are forced to cut through these regions of high dislocation density.

Compared to the other SPD methods, such as ECAP and HPT, the unconfined twist is simple and easy to operate. However, the processed sample must have enough plasticity to be deformed into UFG structures before fracture. For the metals and alloys lacking of enough plasticity, the twist process can be performed at an elevated temperature.

CONCLUSION

A new simple method – unconfined twist was found to effectively introduce ultrafine grained structures into the CG Fe wire. Upon twist, the deformed layers and bands were formed in the cross-sectional plane and the longitudinal plane, respectively. Both the near-surface deformed layers and the deformed bands are composed of lamella grains with a width in submicrometer range. The tensile yield strength and ultimate strength of the twisted sample are increased by about 150 % and 100 % compared with the values of its CG unprocessed counterpart. The unconfined twist SPD method shows a good prospect for industrial application.

REFERENCES

1. C.C. Koch, Nanostruct. Mater. **9**, 13 (1997).
2. Y.H. Zhao, H.W. Sheng and K. Lu, Acta Mater. **49**, 365 (2001); Y.H. Zhao, K. Zhang and K. Lu, Phys. Rev. B **66**, 085404-01 (2002).
3. H. Gleiter, Prog. Mater. Sci. **33**, 223 (1989).
4. U. Erb, Nanostruct. Mater. **6**, 533 (1995).
5. A. Inoue, T. Zhang, T. Masumoto, Mater. Trans. Jpn **31**, 177 (1990).
6. Y.H. Zhao, K. Zhang and K. Lu, Phys. Rev. B **56**, 14322 (1997).
7. R.Z. Valiev, R.K. Islamgaliev and I.V. Alexandrov, Prog. Mater. Sci. **45**, 103 (2000).
8. V.M. Segal, V.I. Reznikov, A.E. Drobyshevskiy and V.I. Kopylov, Russ. Metall. (Metally) **1**, 99 (1981).
9. Y.H. Zhao, X.Z. Liao, Z. Jin, R.Z. Valiev and Y.T. Zhu, in *Ultrafine-Grained Materials III*, edited by Y.T. Zhu, T.G. Langdon, R.Z. Valiev, S.L. Semiatin, D.H. Shin and T.C. Lowe (The Minerals, Metals & Materials Society, 2004), p. 511.
10. R.Z.Valiev, Nanostruct. Mater. **6**, 73 (1995); R.Z. Valiev, Mater. Sci. Eng. **A234**, 236 (1997).
11. J.Y. Huang, Y.T. Zhu, H. Jiang, T.C. Lowe, Acta Mater. **49**, 1497 (2001); J.Y. Huang, X.Z. Liao, Y.T. Zhu, F. Zhou, E.J. Lavernia, Phil. Mag. **83**, 1407 (2003).
12. E.O. Hall, Proc. Phys. Soc. B **64**, 747 (1951).
13. N.J. Petch, J. Iron Steal Inst. **174**, 25 (1953).

Mat. Res. Soc. Symp. Proc. Vol. 821 © 2004 Materials Research Society P8.27

Microstructural Refinement in Single-Phase Copper Solid Solutions by Machining

S. Swaminathan, S. Chandrasekar, W. D. Compton, K. P. Trumble and A. H. King
Schools of Engineering,
Purdue University,
315 N. Grant Street,
West Lafayette, IN, 47907-2023

ABSTRACT

A study has been made of the effect of solute (Mn, Al, Ni) additions on microstructure refinement due to large strain deformation in single phase, copper solid solutions. The solutes were specifically selected for their influence on stacking fault energy (SFE) of copper, and the large strain deformation was imposed by chip formation in machining. The microstructure of Cu-0.7at%Ni chip consists of elongated, sub-micrometer sized grains while Cu-7at%Al chip is made up of long, thin microbands with twins. The microstructure of the chip changes as the SFE of the material varies. With all of the solid solutions studied, the hardness of the chips is found to be significantly greater than that of the bulk material. Recrystallization temperature of solid solution chips is found to be higher than those of pure copper chips.

INTRODUCTION

Recent interest in ultra-fine grained (UFG) materials with enhanced mechanical properties has accentuated interest in severe plastic deformation (SPD) processes by which the grain size of metals and alloys can be made sub-microcrystalline. The SPD processes that have attracted most interest in this regard are: equal channel angular extrusion (ECAE) [1] and high pressure torsion (HPT) [2], to produce bulk and thin samples, respectively. Recently, it has been demonstrated that large plastic strains imposed in a chip, in a single stage of deformation during machining, results in significant microstructure refinement [3-5]. Machining is an attractive method for studying microstructural changes and related phenomena arising from large deformation as the conditions can be readily altered, *e.g.*, to introduce a wide range of strain at different temperatures and strain rates. Additionally, machining is applicable to a variety of materials including high strength steels and alloys.

The objective of this study was to understand the effects of solute additions to copper on the properties of resulting chips. Machining has been successfully used to create nanocrystalline copper chips with enhanced hardness [3]. The presence of a solute in copper markedly affects the stacking fault energy (SFE) of the resulting copper solid solution [6-8]. Increasing solute content in an alloy increases the dislocation density [8] and hence, the stored energy corresponding to a given state of strain. The addition of solutes is also likely to have a major influence on time-temperature behavior of an alloy in any post-deformation heat treatment. Solutes retard recovery, recrystallization and grain growth [9]. This should facilitate retention of the ultra-fine grained microstructure, and associated enhancements in mechanical properties, to much higher temperatures. Microstructure refinement and associated mechanical properties of chips were explored in single-phase copper alloys that incorporated three different solute additions, namely, manganese (SFE negligibly influenced) [6], nickel (SFE increased) [7] and aluminum (SFE decreased) [8].

BACKGROUND

Machining is a process in which a hard, sharp wedge-shaped tool moves relative to a workpiece (bulk) such that a layer of metal, the chip, is removed by large strain deformation. Figure 1 shows a schematic of the machining process and the associated geometric parameters. The tool surface along which the chip flows is known as the rake face and that surface ground back to clear the new or machined workpiece surface is known as the flank. The angle between the rake face of the tool and the normal to the work surface is called the rake angle (α). The width of the undeformed chip into the plane is a_w. The amount of interference between the tool and the workpiece is the undeformed chip thickness or depth of cut (a_o), the deformed chip thickness is a_c and the relative velocity between the tool and the workpiece is the cutting velocity (V_c). When the tool cutting edge is perpendicular to the cutting velocity and a_o is small compared to the cutting edge length and a_w, a state of 2-D plane strain deformation prevails during chip formation. This is the preferred configuration for investigations of machining.
Chip formation in machining occurs by concentrated shear along a plane called the shear plane (Fig. 1) with the uniform shear strain (γ) imposed at the shear plane during chip formation given by

$$\gamma = \frac{\cos(\alpha)}{\sin(\phi)\cos(\phi - \alpha)} \qquad (1)$$

The shear plane angle (ϕ) can be obtained from a measurement of a_o and a_c (Fig. 1) as

$$\tan(\phi) = \frac{\frac{a_o}{a_c}\cos(\alpha)}{1 - \frac{a_o}{a_c}\sin(\alpha)} \qquad (2)$$

The geometry of the deformation zone is completely determined by the shear plane angle. An approximate estimate of the shear strain can also be made from a measurement of inclination of flow lines (Fig. 2) in the chip with respect to the shear plane [10].

Equation 1 suggests that the strain (γ) can be varied over a range by varying the rake angle between positive and negative values. Additionally, the friction at the tool-chip interface also influences γ *via* its effect on the shear plane angle (ϕ). The cutting speed affects the temperature at the shear plane. Effective plastic strains in the range of 1-15, strain rates of up to 1×10^6 per second and shear plane temperatures of up to 0.7 T_m can be realized under appropriate machining conditions [11]. It should be noted that these ranges are greater than can be realized uniformly in the usual SPD processes.

EXPERIMENTAL DETAILS

Copper solid solutions were prepared by induction melting of 99.99% pure copper in a graphite crucible with appropriate amount of solute followed by casting into 25 mm diameter rods. The following compositions by atomic percent were prepared: Cu-0.7at%Al, Cu-0.7at%Ni, Cu-0.7at%Mn, Cu-7at%Al, Cu-7at%Ni and Cu-7at%Mn. For comparison, a pure copper rod was prepared by melting and casting. The cast rods were homogenized at 950°C for 36 hours in a controlled argon atmosphere (95% Ar and 5% H_2 by volume). All rods, except the Cu-Ni alloys, were repeatedly cold forged and annealed. The annealing conditions were 650°C for Cu, Cu-0.7at%Al and Cu-0.7at%Mn, and 750°C for Cu-7at%Al and Cu-7at%Mn, for two hours in a controlled argon atmosphere. To remove porosity arising from casting, Cu-Ni alloys were repeatedly hot forged at 1000°C. All the materials were then machined into rods of 15.9 mm diameter. The initial grain size in all the rods was greater than 50 μm. Chips were machined from these rods using a high speed steel tool with rake angle of +20°. The cutting speed was varied between 0.7 m/min and 49.8 m/min to introduce a range of temperatures at the shear plane during chip formation.

The hardness and microstructure of the chip and host material prior to machining (bulk) were characterized. The samples were metallographically polished and at least twenty hardness measurements were made using Vickers indentation taking care to ensure that the size of an indentation was kept about the same in the bulk and the chip. Microstructure was characterized by a combination of optical microscopy and transmission electron microscopy (TEM). The metallographically prepared samples were electro-etched using 77% phosphoric acid solution in water for optical microscopy observations. For the TEM observations, disk specimens were thinned from ~100 μm thick chip samples by electrolytic twin-jet polishing using a solution of 77% phosphoric acid and 23% water on a Struers Tenupol-5. The electron transparent samples were observed in a JEOL-2000-FX TEM operating at 200 kV. Bright-field images were obtained, together with selected area electron diffraction (SAD) patterns from the imaged regions. To study recovery and recrystallization, chip samples were annealed at different temperatures for one hour in a controlled argon atmosphere followed by quenching in distilled water; the hardness and microstructure of the resulting chips were characterized.

RESULTS

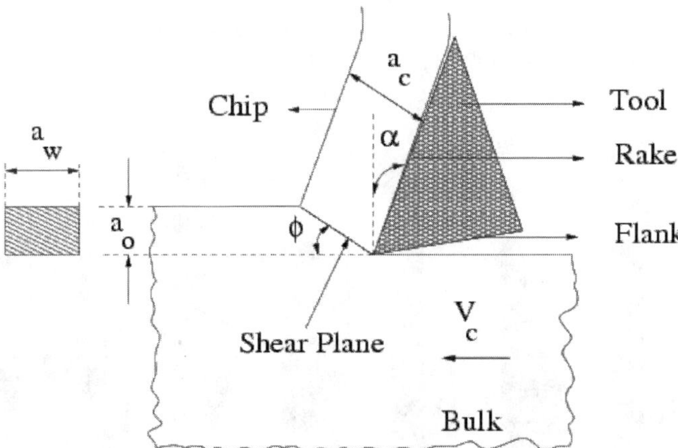

Figure 1. Schematic of 2-D plane strain machining. a_w is the width of the undeformed chip, a_o the undeformed chip thickness, a_c the deformed chip thickness and V_c the cutting velocity. The shear plane angle (ϕ) together with the rake angle (α) determine the average shear strain (γ) in the chip.

Figure 2 shows an optical micrograph of Cu-0.7at%Ni alloy produced in a specially devised "quick stop" experiment. Using, the values of a_c and a_o, along with the rake angle (α), measured from the figure gives a shear plane angle (ϕ) of 10.2° (eqn. 2). The resulting shear strain (γ) is 5.3 as determined using eqn. 1. Also, shown superimposed in the Fig. 2 are average Vickers hardness values for the bulk and the chip. The hardness is seen to increase by 100% in going from the bulk to the chip over a narrow deformation zone (shear plane). The presence of only flow lines in the chip with no visible grain structures suggests that the chip is composed of sub-micrometer sized grains.

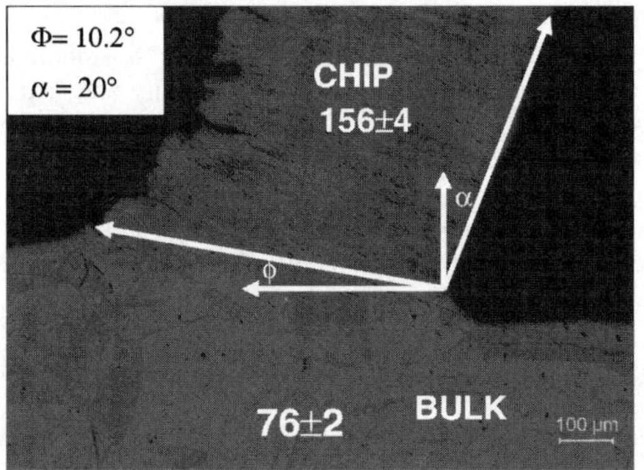

Figure 2. Optical micrograph of partially formed Cu-0.7at%Ni chip as yet not separated from the bulk that was produced in a specially devised "quick stop" experiment.

Figure 3 shows a comparison of hardness between the bulk and the chip samples of copper and its solid solutions. In all cases the chip is seen to be significantly harder than the bulk. The measurements did not show any hardness anisotropy in the chips.

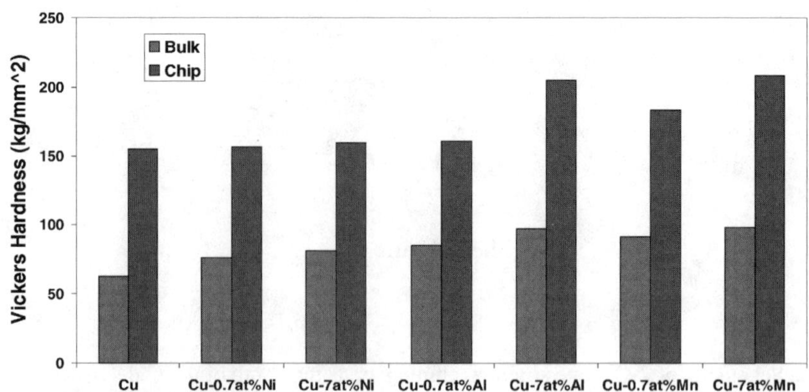

Figure 3. Hardness of the bulk and chip samples.

Figures 4a) and b) show TEM images of Cu-7at%Al and Cu-0.7at%Ni chips, respectively. Low stacking-fault Cu-7at%Al shows long thin microbands (Fig. 4a)) with an average band size (width) of 70 nm. Preliminary analysis using high resolution TEM indicated the presence of twins with an average width of 5 nm within these bands. Furthermore, SAD patterns revealed the chip to be highly textured. In the case of Cu-0.7at%Ni, where the SFE is slightly higher than that of pure copper, elongated grains with sizes under a micrometer can be seen (Fig. 4b)). The grains appear to be severely deformed with a high density of dislocations.

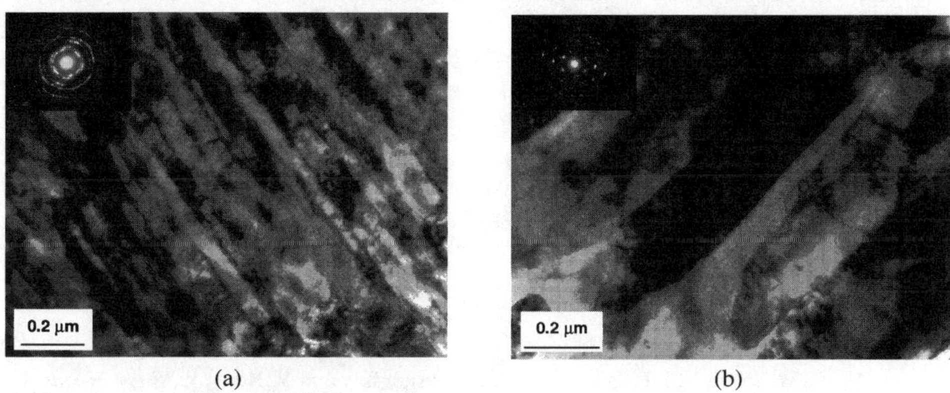

Figure 4. Bright-field TEM images of chip samples, along with SAD patterns (a) Cu-7at%Al showing ultra-fine microbands and (b) Cu-0.7at%Ni showing elongated nanocrystals.

Figure 5 shows the recovery and recrystallization behavior of chips in the form of hardness versus temperature curves. The addition of solutes is seen to delay the recrystallization of Cu chips to higher temperatures with manganese retarding recrystallization to temperatures as high as 300°C. It is interesting to note that in Cu-Mn alloys, a ten fold increase in solute concentration increased the recrystallization temperature by only 50°C.

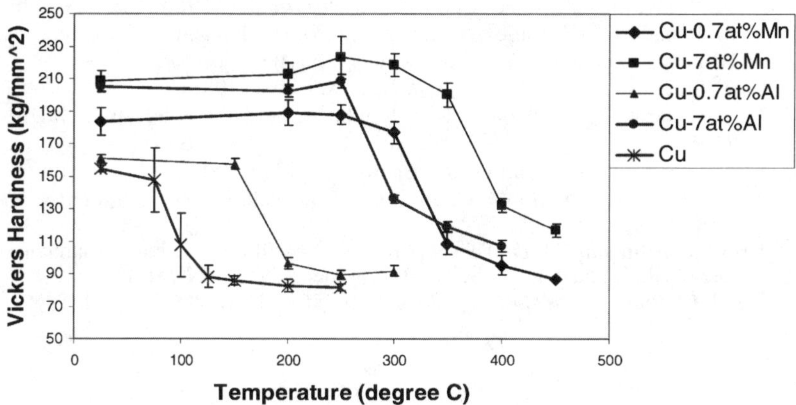

Figure 5. Variation of Vickers hardness with temperature for Cu and Cu alloy chips. As the solute concentration increases, the recrystallization is seen to occur at higher temperatures.

CONCLUSION

The microstructure of machining chips produced by large strain deformation in copper solid solutions was found to be influenced by the SFE of the material. The hardness of the chips was greater than that of the bulk in all of the materials studied. Furthermore, the recrystallization temperature of the chips was found to increase with increasing solute content. Ongoing work is directed towards obtaining a better quantitative understanding of the role of solutes in influencing microstructure and mechanical property changes arising from large strain deformation.

ACKNOWLEDGEMENTS

We would like to thank the Ford Motor Company, State of Indiana 21st Century Research and Technology Fund and DOE/Oak Ridge National Laboratory (ORNL) for support of this work. Additional thanks are also due Drs. Andrew Sherman (Ford) and Ray Johnson (ORNL) for their encouragement of this work.

REFERENCES

1. V. M. Segal, V. I. Reznikov, A. E. Drobyshevskiy and V. Kopylov, *Russ. Metall.* **1**, 99 (1981).
2. R. Z. Valiev, R. K. Islamgaliev, and I. V. Alexandrov, *Prog. Mater. Sci.* **45**, 103 (2000).
3. T. L. Brown, S. Swaminathan, S. Chandrasekar, W. D. Compton, A. H. King and K. P. Trumble, *J. Mater. Res.* **17**, 2484 (2002).
4. S. Swaminathan, T. L. Brown, M. Ravi Shankar, B. C. Rao, K. P. Trumble, W. D. Compton and S. Chandrasekar in *Ultrafine Grained Materials III* edited by Y. T. Zhu, T. G. Langdon, R. Z. Valiev, S. L. Semiatin, D. H. Shin and T. C. Lowe, (TMS (The Minerals, Metals and Materials Society, 2004)) pp 161-166.
5. T. L. Brown, S. Swaminathan, B. C. Rao, R. F. Kezar, S. Chandrasekar, W. D. Compton, K. P. Trumble and A. H. King in *Ultrafine Grained Materials III* edited by Y. T. Zhu, T. G. Langdon, R. Z. Valiev, S. L. Semiatin, D. H. Shin and T. C. Lowe, (TMS (The Minerals, Metals and Materials Society, 2004)) pp 167-172.
6. T. Steffens and C. Schwink, *Phil. Mag.* **A56**, 161 (1987).
7. I. R. Harris, I. L. Dillamore, R. E. Smallman and B. E. P. Beeston, *Phil. Mag.* **14**, 325 (1966).
8. P. R. Swann and J. Nutting, *J. Inst. Met.* **50**, 133 (1961).
9. A. Lucci, G. Riontino, M. C. Tabasso, M. Tamanini and G. Venturello, *Acta. Metall.* **26**, 615 (1978).
10. S. Ramalingam, Ph.D. Thesis, University of Illinois, Urbana-Champaign, (1967). Also see M. Ravi Shankar, M. S. Thesis, Purdue University, (2002).
11. M. C. Shaw, *Metal Cutting Principles*, (Oxford University Press, 1984).

Mat. Res. Soc. Symp. Proc. Vol. 821 © 2004 Materials Research Society P5.10

Bimodal Microstructures in Nanocrystalline Al and Al-Mg Alloy Powders Prepared by Cryogenic Ball Milling

F. Zhou[1], Z. Lee[1], E. J. Lavernia[2], S.R. Nutt[1]
[1]Department of Materials Science, University of Southern California, Los Angeles, CA 90089
[2]Department of Chemical Engineering and Materials Science, University of California, Davis, CA 95616

Abstract

The microstructures of nanocrystalline Al and Al-7.6 at% Mg alloy powders processed by cryogenic ball milling (i.e., cryomilling) and subsequent thermal annealing were characterized using transmission electron microscopy. The as-milled samples primarily consisted of equiaxed grains with average sizes of ~ 26 nm. The annealing treatments at elevated temperatures resulted in bimodal grain-size distributions, with sub-micrometer-sized grains embedded in a matrix of nanocrystalline (<50 nm) grains. The bimodal microstructures were formed during the processes of recovery and recrystallization.

Introduction

Considerable interest has recently been drawn in processing ultrafine-grained and/or nanocrystalline (nc) metals and alloys with superior mechanical properties using novel thermo-mechanical methods [1-4]. It has been demonstrated that high tensile ductility can be achieved without losing strength in a number of nc materials with bimodal grain-size distributions (e.g., Ti [3], Cu [4], and Al alloys [5, 6]). A two-step procedure consisting of extremely large plastic deformation (as, for example, via torsion, multiple extrusion or forging) followed by annealing is used to yield the novel microstructures. As most nc metals (with relatively uniform grain size distributions) show little ductility while exhibiting enhanced strength [7], it is expected that this class of inhomogeneous microstructure, with coarse-grains embedded in a matrix of nc grains, will have implications in the development of tough nanostructured metals and alloys for high-performance structural applications [4].

Mechanical alloying [8], and specially cryomilling in a liquid nitrogen medium [9-12], has been used to produce nanostructures in a variety of metals and alloys in reasonable quantities and the resulting powder has been warm/hot consolidated to full/nearly-full density with insignificant grain growth [13]. The method has been particularly successful with aluminum alloys and the presence of fine oxide/nitrides that result from the milling process is the key to maintaining the fine grain size during annealing and consolidation at elevated temperatures [10, 14]. While the control of the average grain size is important to achieve high strength properties, it becomes increasingly interesting to control the grain size distribution for the sake of high ductility. Here we report the bimodal microstructures observed in the cryomilled Al and Al-Mg alloys subjected to annealing at elevated temperatures.

Experimental procedures

Cryogenic ball milling of spray-atomized Al and Al-7.6 at% Mg powders was carried out in liquid nitrogen using a modified Union Process 01-HD attritor (Szegvari) with a stainless steel vial and balls (with 6.4-mm-diameter) at a rate of 180 rpm. The starting powders have an average particle size of 50 μm. The ball-to-powder ratio was 32:1. Prior to milling, ~ 0.2 wt% of stearic acid was added to the powders to inhibit a cold welding process. Details of the cryomilling were reported elsewhere [9]. After 8 h of milling, the powders were removed from the attritor and stored in a glove box at room temperature pending microstructure analysis. The as-milled powders were mostly equiaxed with typical agglomerate diameters of ~ 20-50 μm. Chemical analysis of the milled Al and Al-Mg alloy indicated that nitrogen, oxygen, and carbon were absorbed during cryomilling.

Annealing treatments were performed under a flowing Ar in a Perkin-Elmer DSC-7. The powders were sealed in an Al pan and heated to the desired temperatures at 80 K/min, isothermally held for 1 h, and then rapidly cooled to room temperature. TEM observations were carried out with a Philips CM20 microscope operated at 200 kV. The powder samples were mixed in G-1 epoxy (Gatan Inc.) to form a small disk by curing the epoxy-powder mixture at room temperature. Following epoxy mixing, the samples were mechanically polished, dimpled and ion milled. The ion-milling was carried out in a dual ion-beam miller (Gatan Inc) at liquid nitrogen temperature at 4 keV and a final thinning angle of 11 deg.

Results and Discussions

The microstructures of as-milled powders were not homogeneous in both Al and Al-Mg samples, consisting of two types of representative microstructure. Most frequently observed is a random dispersion of equiaxed grains with a typical diameter scale of 10-30 nm, as shown in Fig. 1a for Al. The neighboring crystals appear to be separated by high angle grain boundaries. This contention is supported by the corresponding diffraction pattern shown in the inset of Fig. 1a. Elongated grains with a length of 100 – 200 nm and a width around 30 nm were also found (Fig.1b), albeit less frequently. The diffraction pattern indicated a strong texture in this area. An area mixed with these two types of microstructures is shown with a low magnification image in Fig. 1c. Grains of larger than 100 nm were rarely observed, and the average grain size obtained by the mean linear intercept method is 28 nm (Fig. 1d), in good consistent with the value obtained from the X-ray diffraction (XRD) measurements. Similar microstructures were also found in the cryomilled Al-Mg powders (see Fig. 2a showing a dispersion of equiaxed grains). Additionally, only fcc-Al diffraction rings were found for both samples, implying that the Mg atoms were essentially dissolved to form a supersaturated solid solution in Al-Mg. This result was also in consistent with the XRD measurements [9, 12, 15].

Annealing of the as-milled samples at elevated temperatures resulted in distinct changes in both the grain sizes and morphologies. Representative microstructures for the Al samples annealed at 250°C for 1 h are shown in Figs. 1e to 1g. A remarkable feature is the appearance of some particularly large-sized grains growing into the matrix of nanometer-sized grains. Large grains, typically ~ 200 nm, are seen in a matrix of small grains with typical sizes of ~ 20 to 50 nm. The microstructure of the small grains appears to be non-recrystallized. The corresponding SAD indicates a ring pattern, as shown in the inset. Although the coarse grains were frequently observed, they account only a small fraction and the majority remains to be nanometer-sized crystallites.

Similar microstructures were also found in annealed Al-Mg samples, as shown in Fig. 2. In the sample annealed at 150°C for 1 h, large grains, typically from 200 to 500 nm, are seen in a matrix of small grains with typical sizes of ~ 20 to 50 nm (see Fig. 2b). The coarse grains

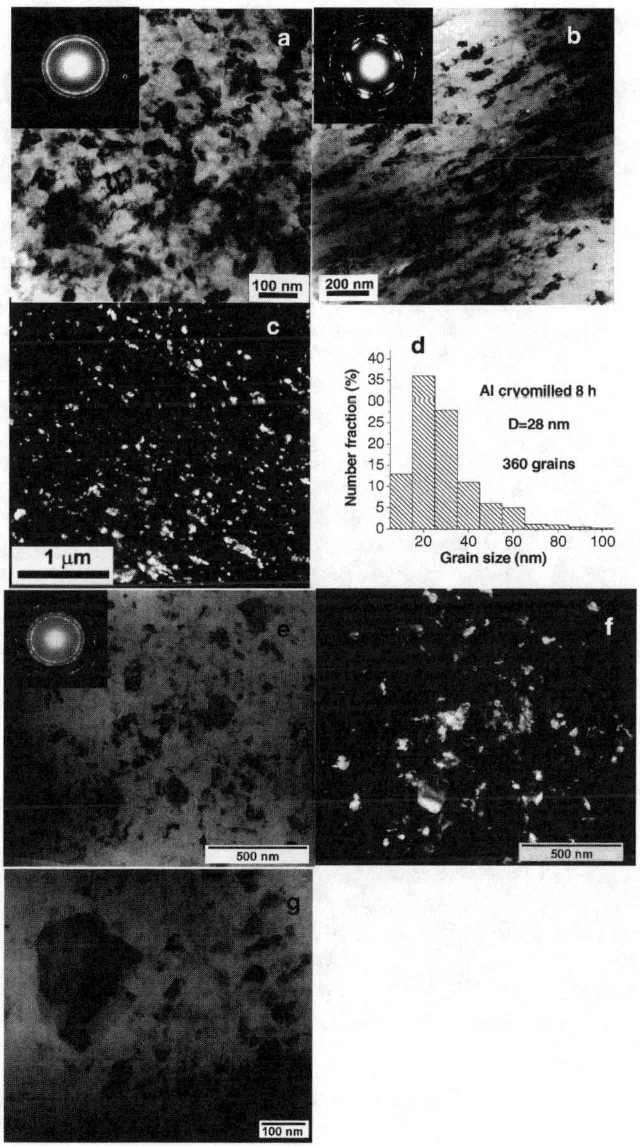

Fig. 1. TEM bright field (BF) and dark field (DF) images of nc Al powders: a) to c) showing the equiaxed grains (BF), elongated grains (BF), and a mixture of both (DF) in as-milled state; d) the grain size distribution in as-milled state; e) BF, f) BF, and g) BF images showing a bimodal microstructure in the sample annealed at 250°C for 1 h.

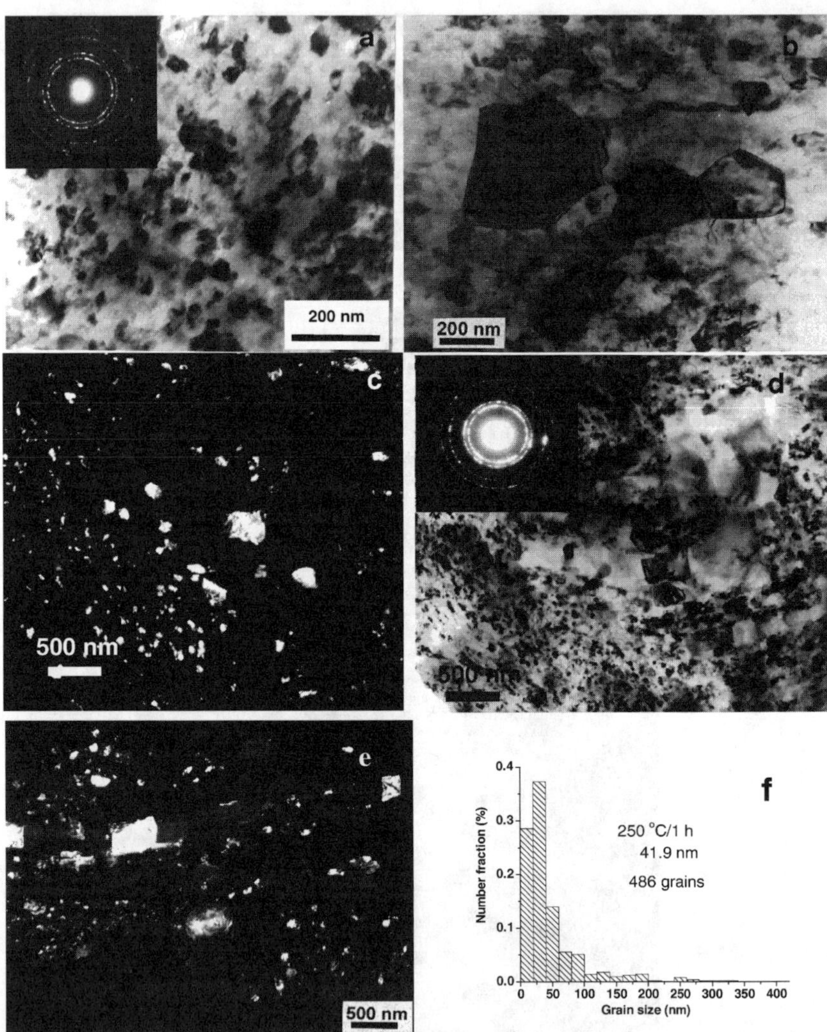

Fig. 2. TEM bright field (BF) and dark field (DF) images of nc Al-Mg powders: a) equiaxed grains in as-milled state (BF); b) a typical coarse grain structure (BF) in the sample annealed at 150ºC for 1 h; c) DF and d) BF images and corresponding SAD pattern of bimodal microstructures in the sample annealed at 250ºC for 1 h; e) DF image showing a mixture of coarse grain bands and ultrafine grains; and f) the grain size distribution.

generally were found to form coarse-grained clusters or zones, where individual large grains with sharp grain boundaries are present. In the sample annealed at 250°C, the duplex grain structure become more distinct, as shown in Figs. 2c and 2d. The corresponding SAD pattern shows a ring pattern indicative of an ultrafine-grained structure, with some randomly distributed extra-strong spots indicative of coarse grains. Large grains in the two samples have the same size range but a larger fraction of coarse grains was formed at the higher temperature annealing. The grain size distributions were obtained from TEM dark-field images. Broad and bimodal-like grain size distributions are clearly seen. A typical grain size distribution is shown in Fig. 2f for the sample annealed at 250°C for 1 h, with the mean grain size of 41.9 nm.

Another type of coarse grained microstructure in the annealed Al-Mg samples is the bands. Shown in Fig. 2e is a typical region for the sample annealed at 250°C, where the coarse grain bands with a width of ~ 0.3 μm were generally adjacent to or among fine grains of 20 to 50 nm. The dislocation-free coarse grains appear to be recrystallized microstructures [16]. In the same region, there are also individual coarse grains.

The non-uniform, bimodal microstructure was formed via recovery and recrystallization during annealing [15]. This is not a surprise, considering the inhomogeneity in the nanostructures of the as-milled sample (in terms of grain size and morphology, deformation defects and grain boundary structures). First, the microstructure in the as-milled powders was not uniform, consistent with the stochastic nature of the milling process, and there are two distinct grain morphologies. Second, manifestations of different deformation modes were also observed in the sample, i.e., a set of high density dislocations ($1.3 \times 10^{17}/m^2$) in the large grains and extensive twins in small grains [17]. Such a difference of defect structures may affect subgrain coarsening during the recovery process, which is driven by the annihilation of defects (dislocations) introduced during deformation and the reduction of boundary areas [16], and thus may lead to a heterogeneous state in microstructures from large- to small-sized grains. Third, as suggested by our previous results [17], both non-equilibrium and equilibrium grain boundaries exist in the as-milled state. All of these inhomogeneities within the as-milled material may inevitably lead to inhomogeneous recovery and recrystallization. In addition, it is apparent that the segregation of Mg solutes and other impurities introduced by cryomilling should also insert an influence on the GB mobility and hence on the growth rate of recrystallizing grains [16]. Although in the present sample the effects of impurities have not been corroborated experimentally, it is reasonable to assume they are effective in determining the inhomogeneity of the annealed microstructures.

It appears that the formation of the coarse grains is due to subgrain rotation and coalescence during recovery, which was originally proposed by Hu [18]. This type of growth is that subgrains might rotate by a boundary diffusional process until adjacent subgrains were of similar orientation. The two subgrains involved would then coalesce into a larger subgrain with little boundary migration, with the driving force arising from a reduction in boundary energies. Moreover, for grain growth in nc fcc metals, it has also been shown by a recent molecular-dynamics simulation that grain rotations play an *equally* important role as GB migration and such rotations lead to the formation of large, elongated grains [19]. In the as-milled Al-Mg alloy nanograins with small angle grain boundaries, which appear to be favorable for the occurrence of the rotation and coalescence, are frequently found [17]. Consequently, we propose that the occurrence of subgrain rotation and coalescence during recovery seems highly feasible to form the large and elongated grains, although there is no evidence to exclude the subgrain boundary migration mechanism, which is usually considered to play a more significant role in recovery and recrystallization.

Compared with the nc Al-Mg alloy, the cryomilled nc Al displays a higher stability against grain coarsening. Under the same annealing condition (e.g., 250ºC for 1 h), coarse grains formed in the Al-Mg sample are larger and denser than those in the nc Al. The difference may be attributed to the addition of Mg that lower the recrystallization temperature and affects the diffusion of Mg in Al, which governs the boundary migration during recovery and recrystallization.

In summary, the cryogenic ball milling produced a heterogeneous microstructure with nanoscale grain sizes in Al and Al-7.6Mg alloy. Annealing the as-milled powders formed non-uniform microstructures with bimodal grain size distributions, which may result from the inhomogeneous nanostructures in the as-milled state. Grain coalescence is highly feasible during the grain coarsening in recovery. It is clear that, by modifying the annealing temperature and time, one can control the grain size and its distribution of the bimodal microstructures in cryomilled powders during consolidation process to optimize the mechanical properties.

Acknowledgments

Financial support for this research was provided by the Office of Naval Research under grants N00014-02-1-1053 and N00014-03-1-0149.

References

1. K.S. Kumar, H. Van Swygenhoven, S. Suresh, Acta Mater **51**, 5743 (2003).
2. S.X. McFadden, R.S. Mishra, R.Z. Valiev, A.P. Zhilyaev, A.K. Mukerjee, Nature **398**, 684 (1999).
3. R.Z. Valiev, I.V. Alexandrov, Y.T. Zhu, T.C. Lowe, J Mater Res **17**, 5 (2000).
4. Y. Wang, M. Chen, F. Zhou, E. Ma, Nature **419**, 912 (2002).
5. Z. Lee, R. Rodriguez, R.W. Hayes, E.J. Lavernia, S.R. Nutt, Metall Mater Trans A **34**, 1473 (2003).
6. D. Witkin, Z. Lee, R. Rodriguez, S.R. Nutt, E.J. Lavernia, Scripta Mater **49**, 297 (2003).
7. C.C. Koch, Scripta Mater **49**, 657 (2003).
8. C.C. Koch, Nanostructured Mater **9**, 13 (1997).
9. F. Zhou, R. Rodriguez, E.J. Lavernia, Mater Sci Forum **386-8**, 409 (2002).
10. J. Lee, F. Zhou, K.H. Chung, N.J. Kim, E.J. Lavernia, Metall Mater Trans **A32**, 3109(2001).
11. F. Zhou, D. Witkin, S.R. Nutt, E.J. Lavernia, Mater Sci Eng A (in press, 2004).
12. F. Zhou, S.R. Nutt, C.C. Bampton, E.J. Lavernia, Metal Mater Trans A **34**, 1985 (2003).
13. B.Q. Han, E.J. Lavernia, F.A. Mohamed, Phil Mag Lett **83**, 89 (2003)
14. F. Zhou, J. Lee, S. Dallek, E.J. Lavernia, J Mater Res **16**, 3451 (2001).
15. F. Zhou, X.Z. Liao, Y.T. Zhu, S. Dallek, E.J. Lavernia, Acta Mater **51**, 2777 (2003).
16. F.J. Humphreys, M. Hatherly. Recrystallization and Related Annealing Phenomena. Oxford, UK: Pergamon, 1996.
17. X.Z. Liao, J.Y. Huang, Y.T. Zhu, F.Zhou, E.J. Lavernia, Phil Mag A **83**, 3065 (2003).
18. H. Hu, In: L. Himmel ed. Recovery and Recrystallization of Metals. New York, NY: Interscience Publishers, 1963. p.311.
19. A.J. Haslam, S.R. Phillpot, D. Wolf, D. Moldovan, H. Gleiter, Mater Sci Eng A **318**, 293 (2001).

Synthesis and Characterization of Fe-Co Nanocrystalline Alloys

Mohammad Mujahid and Jie Zhu
School of Materials Engineering, Nanyang Technological University,
Nanyang Avenue, Singapore 639798.

ABSTRACT

Nanostructured alloys have great potential as soft magnetic materials. In particular, nanocrystalline Fe-Co based alloys are believed to be good candidates for imparting improved magnetic behavior in terms of higher permeability, lower coercivity, reduced hysteresis losses and higher Curie temperatures. In the present work, Fe-50at.%Co alloy powders have been prepared using mechanical alloying (MA) in a planetary ball mill under controlled environment. The particle size and morphology of MA powders was investigated using scanning electron microscopy. The crystal size and internal strain was measured using X-ray diffraction. It has been shown that the crystal size could be reduced down to less than 15 nm in these alloys. Finally, the influence of grain size and internal strain on the magnetic properties has been discussed.

INTRODUCTION

Soft magnetic materials are usually characterized by low coercivity (< $1 kAm^{-1}$) and high permeability. Therefore, a little change in the applied magnetic field can cause large change in the magnetization [1]. Since magnetization involves domain wall movement and domain rotation, grain boundaries are considered as impediments to these processes especially in conventional magnetic materials. It has been found that the magnetic hardness is approximately inversely proportional to the grain size for sizes exceeding 0.1-1μm [2]. However, recent developments in nanostructured soft magnetic materials, coercivity was found to follow a proportional relationship with the grain size for average grain diameters below the ferromagnetic exchange length (L). The underlying reason is related to the reduced effective average magnetocrystalline anisotropy due to many grains being exchange coupled within a volume of L^3. Based on the random anisotropy model, Herzer predicted the well known D^6 law (D, grain size) and extended it to explain lower exponent for the D-power dependence [3,4].

The term "Nanomaterials" refers to those nanostructured bulk materials, nanophase dispersive particles, or thin films, whose characteristic length scale of microstructure ranges from 1-100 nm [5]. Mechanical alloying (MA) has been extensively applied to produce nanostructured bulk materials with a broad range of chemical compositions and crystal structures [6]. In the MA process, continuous collisions between milling media and powders lead to the repeated plastic deformation, fragmentation and re-welding of powder particles [7,8]. The dynamics of milling process has been summarized in a recent report [9]. According to Fecht [10], plastic deformation during milling leads to successive accumulation of dislocations in the shear bands such that the crystals are subsequently disintegrated into subgrains delineated by low-angle grain boundaries.

The size of subgrains can then be continuously reduced with further milling so that the crystallographic orientations of final grains become randomly orientated.

The present work deals with the development of nanocrystalline alloys based on the Fe-Co system via mechanical alloying. The objective of this work was to study the influence of working parameters on microstructures and relate them with magnetic properties.

MATERIALS AND METHODS

High purity Fe (bcc) and Co (hcp) powders with an average particle size of 32 μm and 38 μm, respectively, were employed for MA. Milling was conducted under inert gas atmosphere in a planetary ball mill operated at 300 rpm. Blended powder mixtures of Fe-50at.% Co together with tungsten carbide balls (10 mm diameter) were loaded into tungsten carbide milling jars (250 ml). A ball-to-powder ratio (BPR) of 15:1 and 30:1 was used for various samples. For specimens with BPR of 15:1, the milling time ranged from 0.5 h to 120 h, while for samples with BPR 30:1, the milling time was 30 h and 60 h. Structural characterization was carried out using scanning electron microscopy (SEM) and x-ray diffraction (XRD), whereas the magnetic properties were measured using a vibrating sample magnetometer (VSM).

RESULTS AND DISCUSSION

The data from particle size analysis of as-milled MA powders is shown in figure 1. The particle size with BPR 15:1 first increases to a maximum (about 50 μm) after 13 h of milling, and then decreases gradually (to 3 μm) after milling for 120 h. This agrees with similar results obtained for changes in particle size by other researchers [11]. The increase in particle size is

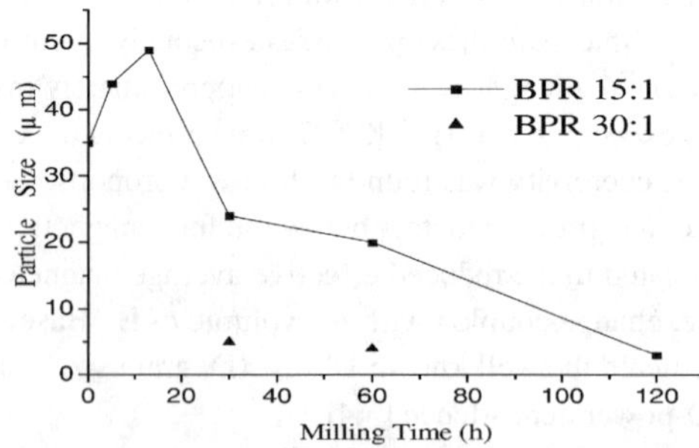

Figure 1. Change in average particle size with milling time.

believed to be due to the plastic deformation (that promotes particle flattening) and cold welding of powder particles during milling. As the milling time increases, the initial "soft" powders become relatively brittle due to the work hardening. This would result in fracturing to become dominant in the subsequent milling process reducing the overall particle size. However, it is thought that saturation in particle size will be reached when a balance between the rate of fracturing and re-welding is reached [8]. The changes in particle size for BPR 30:1 shows that the milling process with higher BPR is more effective in terms of particle size reduction.

X-ray diffraction patterns from milled samples (BPR 15:1) along with a pattern for Fe/Co mechanical mixture are illustrated in figure 2. With the increasing milling time the Co peaks were observed to gradually broaden and eventually disappeared, whereas, the Fe peaks were largely broadened and intensities reduced. Therefore, it was believed that Co atoms gradually diffused into Fe lattice and formed Fe-based solid solution with A2-type bcc crystal structure.

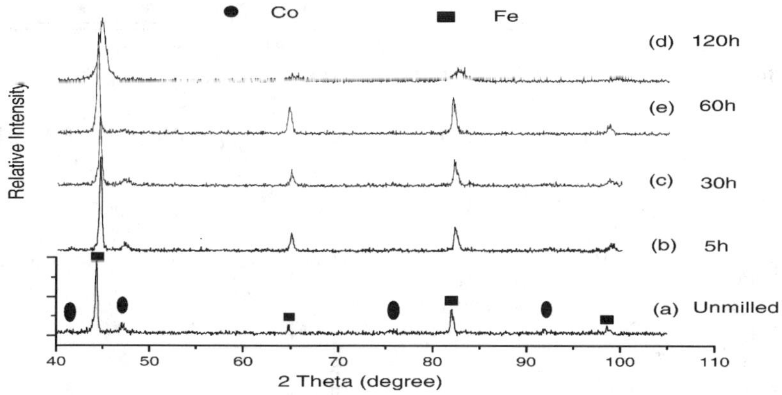

Figure 2. XRD patterns of as-milled powders with different milling time (BPR15:1).

A comparison of XRD patterns for both 15:1 and 30:1 BPR is shown in figure 3. The peaks observed were broadened more for as-milled powders with higher BPR for the same milling time. However, higher BPR may lead to incomplete alloying and composition heterogeneity due to severe agglomeration of as-milled powders. This conclusion was also strengthened by the fact that a small Co peak from (101) planes was revealed in figure 3 (e).

Figure 4 shows the changes of grain size and lattice strain for all the as-milled powder samples. The average grain size and internal strains were calculated from XRD data by using Hall-Williamson method [12] after the instrumental broadening calibration. For BPR 15:1, the grain size was found to be refined quickly to about 150 nm after 5 h of milling, and then continuously reduced to 12 nm after 120 h of milling. For the same samples, the internal strain increased to a maximum value of 0.2% after milling for 60 h and then decreases to 0.03% after 120 h, as shown in figure 4(b). The first rise in the internal strain could be due to the accumulation of dislocations generated by plastic deformation and the increasing grain boundary fraction. The subsequent decrease in internal strain was believed to be due to the disintegration of fine grains to subgrains.

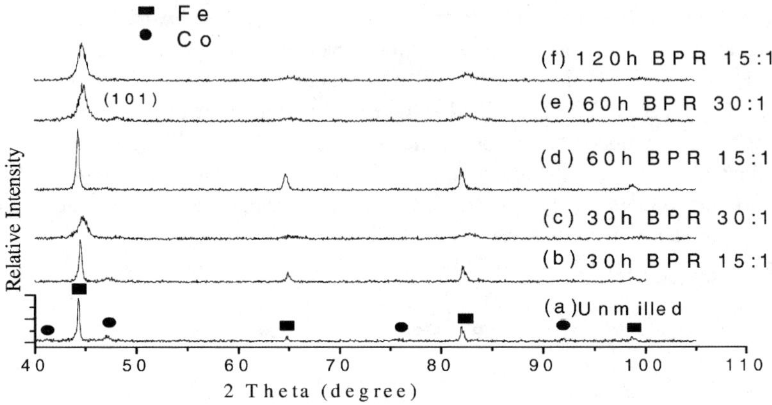

Figure 3. Comparison between XRD patterns of as-milled powders with BPR 15:1 and 30:1.

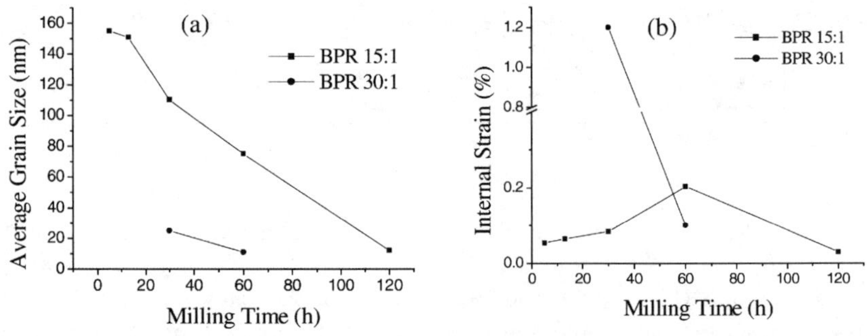

Figure 4. Comparison of grain size reduction (a) and internal strain (b) with milling time between as-milled powders with BPR 15:1 and 30:1.

Grain size reduction and internal strain comparison between BPR 15:1 and 30:1 can also be observed in figure 4. The milling process with higher BPR was observed to be more effective in terms of grain size reduction since the crystal size was reduced to 11 nm after 60 h milling at BPR 30:1 while the same milling time produced the as-milled powders with grain size of 75 nm at BPR 15:1. It was, however, noted that the internal strain was much higher for milling process with higher BPR.

Figure 5 shows the changes in coercivity of as-milled Fe-Co powders at different milling times. For samples with BPR 15:1, the coercivity increased initially with milling time and then

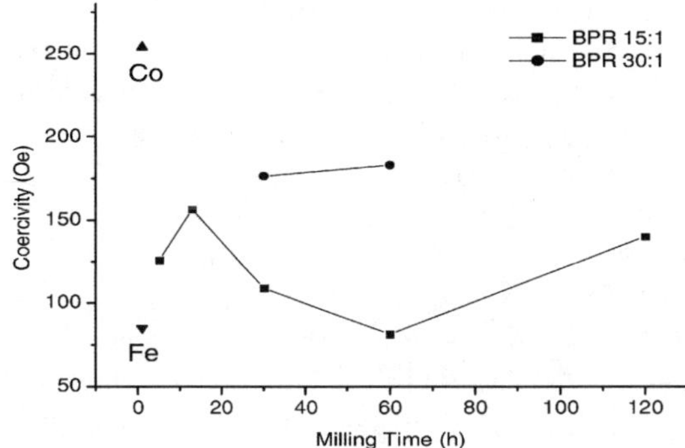

Figure 5. Coercivity changes with milling time for as-milled Fe-Co powders.

decreased to a minimum value for 60 h milling time before rising again for longer milling durations. According to 'strain theory' [1], solids with highly inhomogeneous microstrains provide local energy barriers for the domain walls to overcome, and therefore the coercivity is considered to be proportional to the level of internal strain. However, this proportional relationship was not observed in the present case except for the initial stages of milling. In the intermediate stage, the coercivity was in fact reduced with decreasing grain size (i.e. increasing milling time).

Figure 6 shows that saturation magnetization (M_s) of milled powders (BPR 15:1) increased continuously with the increasing milling time, and reached a maximum value of 240 emu/g (~2.09 T) after 60 h.

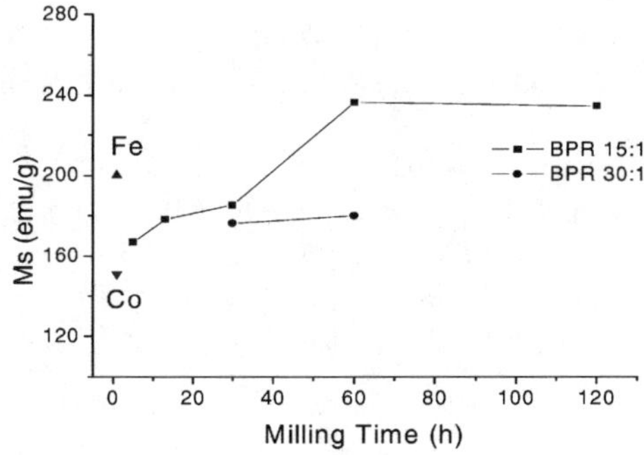

Figure 6. Saturation magnetization of as-milled Fe-Co powders against milling time.

Cobalt, when alloyed with Fe, is known to cause an increase in M_s and Currie temperature [1]. Therefore, the initial increase in M_s could be attributed to the diffusion of Co into Fe to form Fe-Co alloy. After 60 h of milling, little change in M_s was observed suggesting that the alloying was complete. The M_s values for samples milled with BPR 30:1 were observed to be lower than that of BPR 15:1 and showed no significant change with the milling time.

CONCLUSIONS

The Fe-Co alloy produced by MA is believed to have A2-type (bcc) structure. The minimum grain size obtained was 11 nm after 60 h of milling with BPR 30:1. The change in grain size is predicted to be roughly inversely proportional to the milling time. Higher BPR is more effective in particle/grain size reduction, but it also tends to produce alloys with compositional heterogeneities and large internal strain. The minimum coercivity (75 Oe) and maximum saturation magnetization (2.09 T) was obtained for the powders milled for 60h at BPR 15:1.

REFERENCES

1. D. Jiles, *Introduction to Magnetism and Magnetic Materials*, (Chapman and Hall, New York, 1991).
2. M.E. McHenry and D.E. Laughlin, *Acta Mater.* **48**, 223-238 (2000).
3. G. Herzer, *Scripta Metall. Mater.* **33**, 1741-1756 (1995).
4. K. Suzuki, G. Herzer, J.M. Cadogan, *J. of Magnetism and Magnetic Mat.* **177-181**, 949-950 (1998).
5. Zhong Lin Wang, *Characterization of Nanophase Materials*, (Wiley-VCH, New York, 2000).
6. H. J. Fecht, *Scripta Mater.* **44**, 1719-1723 (2001).
7. C.C Koch, *Materials Science and Technology* **Vol. 15**, ed. R.W. Cahn, P. Hassan, E. J. Kramer (VCH, New York, 1991) p193.
8. C. Suryanarayana, *Progress in Mat. Sci.* **46**, 1-184 (2001).
9. P. P. Chattopadhyay, I. Manna, S. Talapatra and S. K. Pabi, *Materials Chemistry and Physics* **68**, 85-94 (2001).
10. H. J. Fecht, *Nanostuctured Mat.* **6**, 33-42 (1995).
11. Y. D. Kim, J. Y. Chung, J. Kim and H. Jeon, *Mat. Sci. Eng.* **A291**, 17-21 (2000).
12. B.D. Cullity and S.R. Stock, *Elements of X-Ray Diffraction* **3rd Ed.**, (Prentice Hall, New Jersey, USA, 2001).

B. Modeling

Mat. Res. Soc. Symp. Proc. Vol. 821 © 2004 Materials Research Society P9.6

The Role Of Vorticity In The Formation Of Tribomaterial During Sliding

Karthikeyan Subramanian, Jian-Hui Wu and David A. Rigney
Dept. of Materials Science and Engineering, The Ohio State University, Columbus, Ohio, U.S.A.

ABSTRACT

The formation of tribomaterial during sliding is well documented for a wide range of materials. Transfer, adhesion, mechanical mixing and formation of nanocrystals are commonly reported, but the mechanisms involved have not been well understood. Recently, molecular dynamics (MD) simulations have been performed to obtain new information on the response of materials to sliding interactions. The results help to explain experimental observations on simple metals such as copper, bulk metallic glasses and multi-component nanocomposite coatings. When the sliding speed is sufficiently high, the strain rate allows vorticity to develop. It is suggested that the resulting eddies are largely responsible for frictional energy dissipation and mechanical mixing in both crystalline and amorphous materials and for the shear mixing of soft embedded particles in nanocomposite coatings. Similarities of flow behavior with that of fluids are noted.

INTRODUCTION

Characterization of worn subsurface material in ductile metals commonly reveals two distinct regions [1] (Figure 1). The first region extends well below the surface and is characterized by large plastic strains and associated substructure. Plastic deformation is a dissipative process and accounts for a significant part of the frictional force. The second region consists of tribomaterial produced by mechanical mixing of components from the contacting bodies and the environment. This tribomaterial commonly consists of nanocrystals. The tribomaterial separates the sliding counterfaces and so the properties of the tribomaterial become important in determining the macroscopic tribological properties. Tribology research that has focused on contact mechanics, geometry of asperities and continuum plasticity has provided very useful insights. However, these approaches have not effectively addressed key issues related to the formation of tribomaterial. Exploration of the mechanisms leading to the formation of tribomaterial, the effects of load and velocity on these mechanisms and transient frictional behavior thus form interesting and important areas of research.

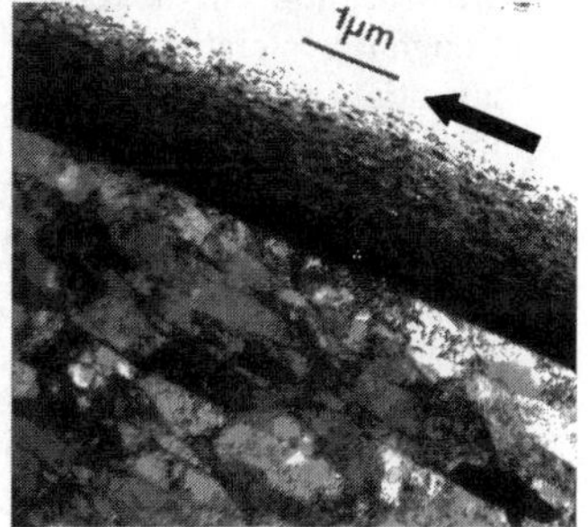

Figure 1. Longitudinal section of OFHC copper block slid against a 440C steel ring shows tribomaterial immediately under the sliding interface and the deformed substructure further below. The tribo-layer has small iron grains mixed in with copper. The sliding direction is shown [1].

The formation of tribomaterial is similar to the production of mechanically mixed material during mechanical alloying [1,2]. Though processes such as mechanical mixing of individual components, material transfer across an interface and adhesion are known to be closely connected with the formation of tribomaterial, the mechanisms are not well understood. Post-test characterization provides information on the chemistry and microstructure of the nanocrystalline layer, but it does not provide information on the processes involved in the formation of tribomaterial. To our advantage, the development of computational tools such as Non-Equilibrium Molecular Dynamics (NEMD) and recent increases in computational speed have made it possible to explore the evolution of the structures generated during sliding. In this paper, we summarize recent results from MD simulations of sliding pairs and discuss the insights gained with respect to mechanical mixing and the formation of tribomaterial. We also examine the flow of material close to the interface and compare it to the flow of fluids.

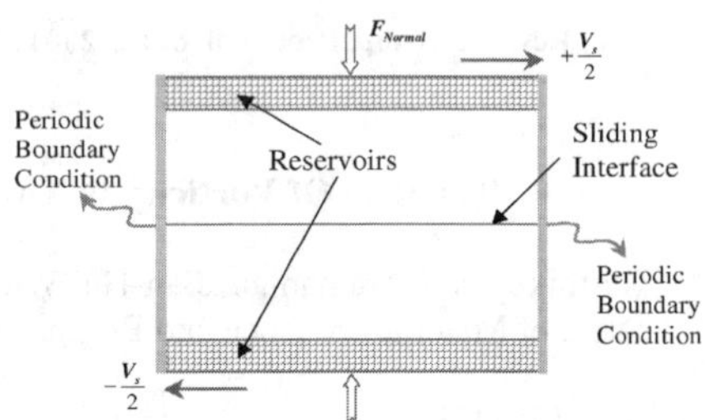

Figure 2. Schematic of the simulation setup.

MD SIMULATIONS

In this paper, we discuss the results of MD simulations done by Hammerberg et al. [2-5], Fu et al. [6,7] and Wu et al [8]. Fu and Wu have simulated 2D sliding between a pair of two-component amorphous blocks. They have chosen simple Lennard-Jones potentials for interatomic interactions [6-8]. Hammerberg et al. have simulated both 2D and 3D systems including Cu/Cu, Cu/Ag, and Al/Ta using Lennard-Jones and embedded atom method (EAM) potentials [2-5]. Figure 2 shows a simple schematic of the simulation setup. It consists of an upper and a lower block that slide against each other with a relative velocity V_s. Two reservoir regions are created in slabs of material furthest from the sliding interface. The composition in the reservoir and non-reservoir regions is the same and the atoms interact with the same potentials. The reservoir region is maintained at a constant low temperature to allow conduction of the heat generated during sliding. The reservoir is also subjected to an external normal force F_n and a tangential force F_t. All atoms in the top block are assigned a starting velocity of $V_s/2$ (and $-V_s/2$ for the bottom block). Subsequently, only the motion of atoms in the reservoirs is imposed – the tangential force is maintained such that the average velocity of a reservoir region is fixed at $\pm V_s/2$. The atoms in the region between the reservoirs are allowed to interact and respond purely through the interatomic potentials. The friction coefficient is calculated as the instantaneous tangential force divided by the normal force. Periodic boundary conditions are used in the x-direction. The simulation cell size varies from ~20,000 atoms for 2D simulations to several million atoms for 3D simulations [3,4]. The force, acceleration, velocity and position of each atom are updated at each time step by allowing atomic interactions. Owing to the length and time scales involved, each time step is of the order of a femtosecond, so simulations over realistic times and sliding velocities would be computationally intensive. Therefore high sliding speeds and short sliding times have been chosen. The friction coefficient reaches a steady state value quickly, so the simulation results are expected to be valid even for longer sliding times.

MD simulation results

The movement of atoms and the evolution of microstructure are conveniently visualized by looking at snapshots at different times. Figure 3 shows such a series for a simulation run by Hammerberg et al. on a copper tribo-pair [2, 3]. The color or gray scale corresponds to the local orientation angle relative to the initial lattice vectors. The snapshots reveal the development of a nanoscale grain structure quite similar to what is observed macroscopically in actual tribolayers.

The velocity of each individual atom can be readily tracked and visualized by using a velocity vector plot. Figure 4 shows one such plot obtained from simulations on an amorphous 2D system close to the interface. It reveals the formation of eddies that involve the coordinated movement of groups of atoms in vortices that are a few nm in diameter. It is suggested that, in crystalline materials, the formation of eddies during flow leads naturally to local differences in orientation, resulting in the formation of nanocrystals.

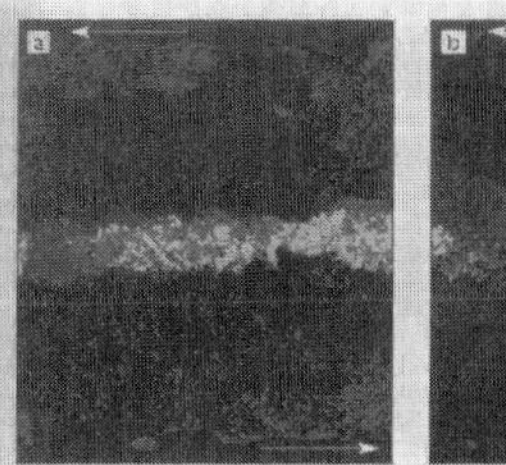

Figure 3. Simulations by Hammerberg et al. suggest evolution of nanoscale grain structure during sliding [3, 4]. The snapshots (a-b) are for copper sliding against copper at times 100 and 200 [2]. Relative velocity is 0.12c; c is the longitudinal sound speed. The time scale uses rescaled units, defined in [3]. Differences in color or gray scale correspond to different crystal orientations.

MD simulations also reveal that sliding leads to mechanical mixing of material between the upper and lower regions [2,6,7] (Figure 5). The mixed layer has also been shown to grow with time in a manner reminiscent of diffusion [6,7]. However, at the small time scales involved in the simulation, this mixing is mechanically driven. Simulations by Wu et al. [8] also show the shear mixing of soft embedded particles in nanocomposite coatings. In light of these observations, it is suggested that the eddies discussed above are responsible for transporting material from one region to another, thus leading to mixing and the subsequent growth of the mixed layer. The nature of mechanical mixing thus appears to be convective.

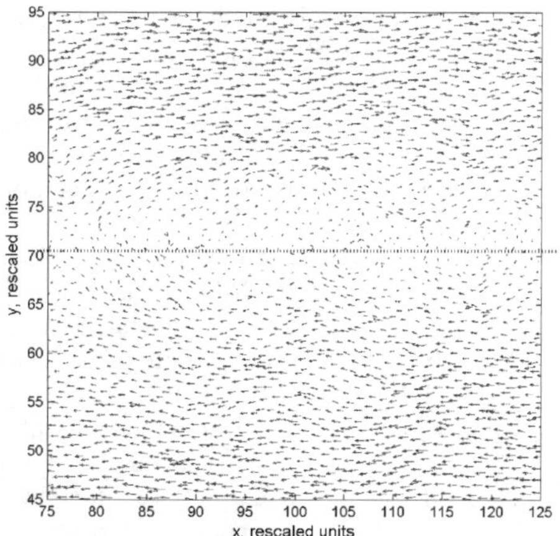

Figure 4. Velocity vector plot near the interface for an amorphous 2D sliding pair after 180 timesteps. Far from the initial interface (dotted line) the movement of atoms appears to be laminar. However, near the interface vortices form that are ~10 atomic lengths in diameter. Sliding velocity V = 1.

DISCUSSION

MD simulations indicate that the flow of material close to the sliding interface is characterized by the formation of eddies that result in intimate mixing and convective growth of the mixed layer. The similarity between this flow and the flow of fluids is easily recognized (Figure 4). The flow observed in the simulations exhibits several classic characteristics of turbulent flow including rapid mixing, vorticity and large velocities [9, 10]. Turbulent flows are always dissipative as is sliding [9, 10]. Dissipation occurs over several length scales. Larger eddies tend to transfer their energy to smaller eddies. External work is required in order to

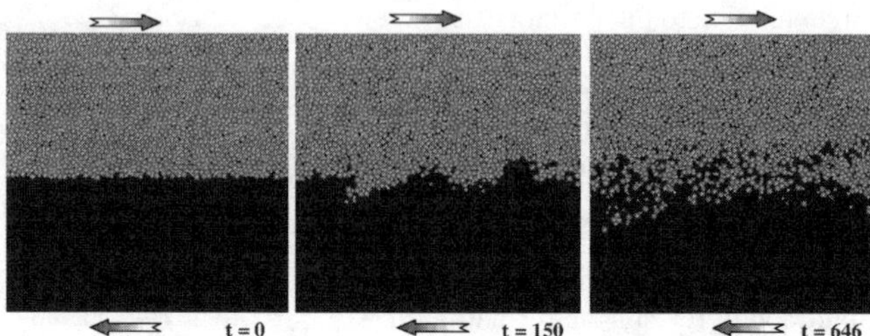

Figure 5. MD simulations of amorphous pairs reveal that sliding leads to mechanical mixing of material between upper and lower regions. Thickness of the mixed layer increases with time [6,7]. Hammerberg et al. have reported similar mixing [3]

maintain turbulence. It has been found that the energy dissipation rate is related to the velocity and length scale of the largest eddies [9]. In the case of the MD simulations, the dissipation rate can be equated to the rate of work done by friction. The friction stress, τ, can be estimated as:

$$\tau \approx \rho \frac{v^3}{V} \tag{1}$$

where ρ is the density of the material, v is the magnitude of the large-eddy velocity maximum and V is the sliding velocity.

In view of the above observations it appears that the velocity profile, mixing kinetics and friction coefficient can be treated by a suitable fluid flow model. The model most appropriate for describing the simulation setup is called the mixing layer or free-shear model. A mixing layer is formed when two parallel streams with different velocities interact, as in Figure 6. The streams do not need to have different properties, such as density and/or viscosity, to generate the layer [10]. The interaction of the two fluids creates a shear layer that changes velocity profiles. The MD simulation setup resembles the mixing layer problem if we envision the top slab and bottom slab to be separate fluids, each with its own velocity. In the MD simulations the top and bottom slabs have equal and opposite velocities, such that the relative velocity, V_s, is twice the velocity of the individual slabs. The free shear flow requires that it be bounded by ambient fluid that is irrotational. This condition is satisfied in the MD simulations by the presence of the reservoirs. A striking characteristic of the free shear flow is that the boundary between the shear

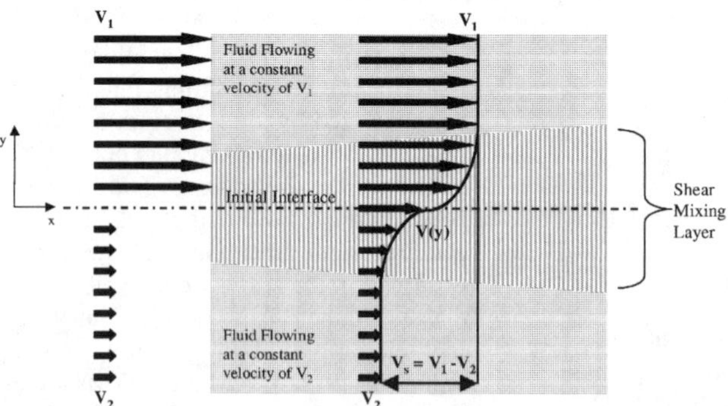

Figure 6. Profile indicating the development of a mixing layer at the interface of two fluids flowing parallel to each other but with different velocities. The relative velocity, V_s, gives rise to a velocity profile $V(y)$ as indicated.

mixing and ambient regions is usually well defined, while the initial interface is quickly lost [10]. The mixing layer typically thickens with time. These features are also observed in MD simulations of sliding. A reasonable model of sliding is expected to predict the velocity profile (the velocity in the x-direction as a function of y, the distance from the initial interface). A hyperbolic tangent function provides a close approximation of the velocity profile for mixing layers [10]. Figure 7 shows that a tanh fit for the velocity profile due to sliding is excellent. The qualitative arguments and velocity profile fit are consistent with a free shear layer model. Furthermore, the mixing layer model suggests that the large-scale eddy velocity v and *V* are related [10], with

$$\left[\frac{\mathrm{v}}{V}\right] = k' \approx 0.24 \tag{2}$$

Eqn 1 and 2 can be combined now to give a value of the friction stress,

$$\tau \approx \rho \frac{\mathrm{v}^3}{V} = \rho \frac{(k'.V)^3}{V} = \rho.(k')^3 V^2 \tag{3}$$

This predicts that the friction stress is proportional to the square of the mean flow velocity. However, it has been shown in MD simulations that the friction force increases to a maximum below a critical velocity and then drops with further increase in velocity. The prediction from Eqn 3 is contrary to observations at high velocities. Preliminary observations suggest that this anomaly can be resolved if we take into account the fact that the mixing layer model is a continuum flow model, while the systems used in MD simulations are driven at velocities close to that of the speed of sound, and the associated time and length scales are correspondingly very small. At high velocities, the largest eddies of size *l* are of the same scale as the smallest eddy possible in the material! A similar breakdown of theory has been observed for fluid mixing layers approaching supersonic speeds [11]. Further investigation into the similarity between high speed sliding and supersonic mixing layers is warranted.

Turbulence is a general characteristic of flow for a wide range of flowing materials. Typically, at relatively low velocities, turbulent flows are almost independent of material properties [9]. At high velocities, dissipation occurs at such small time and length scales that material properties could impose limits on the convective velocity v, and the eddy size *l*. These limits are possibly related to some combination of elastic modulus, density, plasticity and diffusion coefficient. This suggests that while a generic elastic-plastic solid model for flow may be suitable at low speeds, material properties might dictate the high velocity behavior. A systematic study of mixing layers formed during sliding at high velocities and for

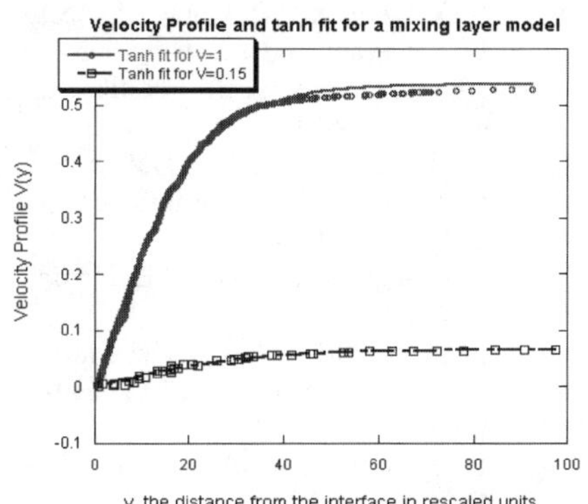

Figure 7. The tanh fit for velocity profiles from MD simulations at two different velocities is excellent, supporting a mixing layer model.

different materials would test this hypothesis. It would also be interesting to investigate the effects of the temperature rise that accompanies sliding. An increase in temperature can be expected to change material properties and to increase the free volume. This should make the flow more localized and progressively easier.

CONCLUSIONS

MD simulations of sliding pairs involving crystalline and amorphous materials reveal the coordinated movement of atom clusters having diameters of a few nanometers. It is suggested that eddies may be present during the formation of nanocrystals. These eddies also seem to be responsible for the mechanical mixing of components across the interface. The presence of eddies suggests that the interfacial processes that occur during sliding are similar to those that occur in fluid flow. Sliding pairs can be described qualitatively by assuming a shear mixing layer. Velocity profiles predicted for a mixing layer match observed velocity profiles in the sliding simulations very well. At high velocities, the sliding scenario is not a continuum problem (as assumed by the mixing layer model), because the length and time scales of eddies are of the same order as the atomic length and time scales. It is proposed that there exist material limits to v and l such that energy cannot be dissipated faster than a critical rate. Such an assumption helps to resolve discrepancies between MD simulations and mixing layer model predictions.

ACKNOWLEDGMENTS

The authors wish to acknowledge Dr. J. Hammerberg at Los Alamos National Labs, Dr. X.Y. Fu at Merck Research Lab. and Dr. M. Falk at the University of Michigan for their ideas and encouragement. This research was sponsored by the National Nuclear Security Administration under the Stewardship Science Academic Alliances program through DOE Research Grant # DE-FG03-03NA00069/A000

REFERENCES

1. D.A. Rigney, L.H. Chen, M.G.S. Naylor and A.R. Rosenfield, Wear **100**, 195 (1984).
2. D.A. Rigney and J.E. Hammerberg, MRS Bulletin **23**(6), 32 (1998).
3. J.E. Hammerberg, B.L. Holian, J. Röder, A.R. Bishop and S.J. Zhou, Physica D **123** (1-4), 330 (1998).
4. J.E.Hammerberg, B.L. Holian and S.J. Zhou, AIP Conf. Proc. **370,** 307 (1996).
5. R.P. Mikulla, J.E. Hammerberg, P.S. Lomdahl and B.L. Holian, Mater. Res. Soc. Proc. **522**, 385 (1998).
6. D.A. Rigney, X.Y. Fu, J.E. Hammerberg, B.L. Holian and M.L. Falk, Scr. Mater. **49**(10), 977 (2003).
7. X.Y. Fu, PhD. Thesis, The Ohio State University, 2001.
8. J.-H Wu, D.A. Rigney, M.L. Falk, J.H. Sanders, A.A. Voevodin and J.S. Zabinski., in press.
9. H. Tennekes and J.L. Lumley, *A First Course in Turb*ulence, 1st ed. (MIT Press, Cambridge, 1972).
10. A.A. Townsend, *The Structure of Turbulent Shear Flow*, 2nd ed., (Cambridge University Press, Cambridge,1976).
11. D. Papamoschou and A. Roshko, J. of Fluid Mech. **197**, 453 (1988).

Mat. Res. Soc. Symp. Proc. Vol. 821 © 2004 Materials Research Society P6.5

Nanoscale Structure and High Velocity Sliding at Cu/Ag Interfaces

James E. Hammerberg, Timothy C. Germann, Brad Lee Holian[1], and Ramon Ravelo[2]
Applied Physics Division, Los Alamos National Laboratory,
Los Alamos, NM 87545
[1]Theoretical Division, Los Alamos National Laboratory,
Los Alamos, NM 87545
[2]Department of Physics and Materials Research Institute, University of Texas-El Paso,
El Paso, TX 79968-0515

ABSTRACT

We present the results of large-scale NonEquilibrium Molecular Dynamics (NEMD) simulations for Cu/Ag interfaces sliding in the velocity regime $0 \leq v \leq 1$ km/sec. System sizes of 2.8×10^6 atoms are considered using Embedded Atom Method (EAM) potentials. Single crystals with (010) interfaces sliding along the ⟨100⟩ direction are considered. We discuss the observed velocity weakening in the tangential force at high velocities, and its connection with the observed dislocation structure and nanostructure that are nucleated during dry sliding.

INTRODUCTION

The frictional properties of smooth Cu/Ag interfaces are interesting from several points of view. Both materials are ductile fcc metals whose deformation properties are well known. They also represent a tribo-pair for which the plastic flow properties are substantially different, but not as dramatically different as for the Ta/Al tribo-pair. Consequently, they may provide a test of ideas inherent in subsurface plasticity theories of the frictional forces between ductile metals in dry sliding [1]. A further reason for studying this interface is that there are now Embedded Atom Method (EAM) atomic interaction potentials that credibly incorporate local defect properties such as defect, stacking fault, and surface energies, as well as thermodynamic properties such as compressibilities, elastic constants, and equations of state [2].

The focus of the present work is the behavior of the frictional force as a function of sliding velocity at fairly large compression. We discuss large-scale NEMD simulations of single crystal Cu and Ag interfaces sliding in the velocity range from 0 – 1 km/s at a pressure of 5.1 GPa, a pressure readily obtainable in impact experiments. The methods, potentials and results for (010) faces sliding along the ⟨100⟩ direction are discussed in the following sections.

METHOD AND POTENTIALS

Method

For the three-dimensional dry sliding simulations we consider here, we use a constant velocity, constant temperature boundary condition for a large simulation cell. A two-dimensional section through the middle of this cell, normal to the Cu/Ag interface and along the ⟨100⟩ sliding direction, is shown schematically in figure 1. It consists of two material regions separated by an interface with an imposed initial roughness. There are also two reservoir regions that are shown as shaded regions. The reservoir regions consist of the same Cu (above) and Ag (below) atoms as

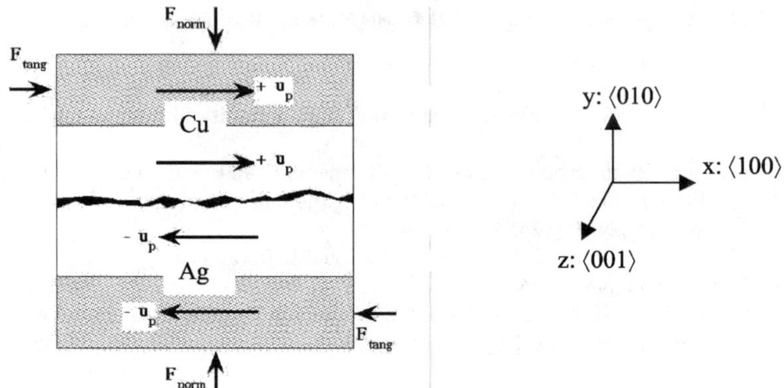

Figure 1. Schematic drawing of the computational cell showing the reservoir regions shaded. F_{norm} is a constant normal (y-direction) load and F_{tang} a time dependent tangential force that enforces constant average tangential velocities $\pm u_p$ in the upper (Cu) and lower (Ag) reservoirs. Boundary conditions in the x and z directions are periodic.

in the contiguous main regions. However, in the reservoir regions, two external forces are applied, uniformly to each reservoir atom. A constant normal force f_{norm} is applied in the normal direction such that the total force per unit area is equal to the initial pressure. A time-dependent tangential force $f_{tang}(t)$ is also applied in the sliding direction such that the average velocity in this direction is equal to $\pm u_p$ in the upper or lower reservoir. The tangential force is then $F(t) = N_r f_{tang}(t)$, where N_r is the number of reservoir atoms. The reservoir is also thermostatted to a constant temperature T [3,4]. The reservoir region typically contains 20 lattice planes, a small fraction of the total number of atoms in the simulation. The initial condition at time $t=t_i$ has all atoms in the upper region moving to the right with velocity $+u_p$ (apart from thermal fluctuations). Similarly, in the lower region all atoms have velocity $-u_p$ at $t=t_i$. For $t>t_i$, Newton's equations of motion are integrated subject only to the reservoir boundary conditions (normal and tangential forces, and thermostatting), periodic boundary conditions (in the x and z directions), and the interatomic potentials.

The frictional force, $F_{tang}(t)$, is a fluctuating quantity. For long times, however, its average value achieves a nonequilibrium steady state limit. This average value, F_{tang}, is the frictional force reported here:

$$F_{tang} \equiv \frac{1}{\Delta t} \int_{t-\Delta t}^{t} F_{tang}(\tau) d\tau . \tag{1}$$

Both upper and lower values of F_{tang} give rise to the same average value since the center of mass remains stationary.

Potentials

For frictional simulations, it is important to have potentials that accurately describe the inhomogeneous structures that occur near the interface and give rise to the dissipation that is the ultimate source of friction. Thus it is necessary to have potentials that accurately reproduce not only elastic constants and phonon spectra, but also defect and stacking fault energies. The EAM potentials have been developed to treat such locally inhomogeneous situations [2]. They are typically short-ranged and are easy to implement in the massively parallel SPaSM [5] NEMD computer code that has been used in our simulations. The Cu and Ag EAM potentials are taken from Sprague et al. [6], who treated Cu growth phenomena on Ag (001) surfaces.

RESULTS

We have considered a system of approximately 2.8×10^6 atoms, at a pressure of P=5.1 GPa, consisting of Cu (upper) and Ag (lower) regions. The reservoir regions each contained approximately 2×10^5 atoms. The Cu and Ag regions were fcc single crystals at densities appropriate to P = 5.1 GPa and T = 300 °K, the temperature to which the reservoirs were thermostatted. The sliding was along the ⟨100⟩ direction for these (010) faces. The overall dimensions of the computational region (x,y,z) were (41,32,33) nm, although in some instances we doubled the dimension normal to the interface (y-direction) to assure ourselves that size was not playing a significant role. We considered two kinds of interfaces, atomically smooth and rough. Both were defined by four sinusoidal peaks in the sliding direction with amplitudes 0.025 nm and 10 nm, respectively. At high sliding velocities, we have found that roughened interfaces lead to similar frictional forces and dynamically formed interfacial microstructure as initially smooth interfaces, but with a more rapid approach to steady state. The samples were allowed to equilibrate and relax for 10 t_0 (sufficient to relax initial strains) at the desired pressure and $u_p = 0$, where $t_0 = 270$ fs is a characteristic time scale (roughly an optical phonon period) and the integration time step was 0.01 t_0. The equilibrated system was then subjected to the initial conditions described in the previous section at $t_i = 10\ t_0$. It then evolved under the influence of the atomic EAM potentials and the boundary conditions described in the previous section for $t > t_i$. The steady state was usually achieved by 500 t_0 and the average tangential force was calculated by the method above. The tangential force per unit area vs. velocity is shown in figure 2; the rougher interface was used for v > 400 m/s.

There is a distinct transition from an increasing velocity dependence of the frictional force to a decrease in the frictional force with velocity at high velocities, i.e. a velocity weakening. This transition occurs at approximately 300 m/s. For reference, the bulk adiabatic sound speeds of Cu and Ag at P=0 are 3.94 km/s and 3.23 km/s, respectively. This change in behavior is related to the development of substructure in the weaker Ag at high velocities. To describe the change in nanostructure we consider three velocities 25 m/s, 300 m/s and 500 m/s. In figure 3 we show for each velocity (1) images of the mid (001) plane which is normal to the interface and parallel to the sliding directions, with lighter and darker vertical lines denoting marker particles which were vertically aligned at $t = t_i$; and (2) a three-dimensional image of a centro-symmetric order parameter, $\eta(R)$, originally proposed by Kelchner et al. [7], which we write as

$$\eta(R) = \frac{1}{2a^2} \sum_{<j>} \left| (R_j - R) + (\hat{R}_j - R) \right|^2 , \tag{2}$$

where the sum is over nearest neighbors, a is the nearest neighbor separation, and $\hat{R}_j$ is the closest neighbor to $R-R_j$. This quantity vanishes for a perfect lattice and is non-zero for dislocation cores and stacking faults.

At low velocities, for smooth interfaces, the dissipation is mainly due to the phonons in the medium (figures 3a,b); the interface remains smooth, although the crosshatch pattern indicates a continual dynamic reconstruction of the Cu/Ag interface. At these low velocities, the imposition of a larger magnitude roughness may change this picture of incommensurate sliding by introducing a more efficient source of dislocations. Beyond a critical velocity, the behavior in the weaker Ag changes for the smooth interface, with a copious generation of dislocations and stacking faults resulting in the formation of nanoscale structure seen in figures 3c and 3d for a velocity of 300 m/s. The Cu workpiece is still responding elastically, with phononic dissipation; the result is a continued increase in the frictional force with velocity. The final figures 3e and 3f show the situation at 500 m/s, beyond the maximum in the tangential force. For these figures we show a simulation that had an initial larger roughness of amplitude 10nm. The nano-asperities are plastically deformed by the Cu and entrained by the harder material with a similar subsurface dislocation and nanostructure developing. At this point a stacking fault has also developed in the Cu. At higher velocities, the structural and physical phenomena are very similar to that discussed above for 300 m/s and 500 m/s. Indeed, the high velocity dependence of the frictional force is well represented by a power law with $F_{tang} \sim v^{-\beta}$ and β=3/4, as shown by the dotted line in figure 2. This form of velocity weakening at high velocities, with a similar value for β, has been observed for other tribo-pairs [4,8,9], and interpreted in terms of the power-law relation between flow stress and plastic strain rate for very high rates of plastic deformation, and the localization of plastic strain [4, 9].

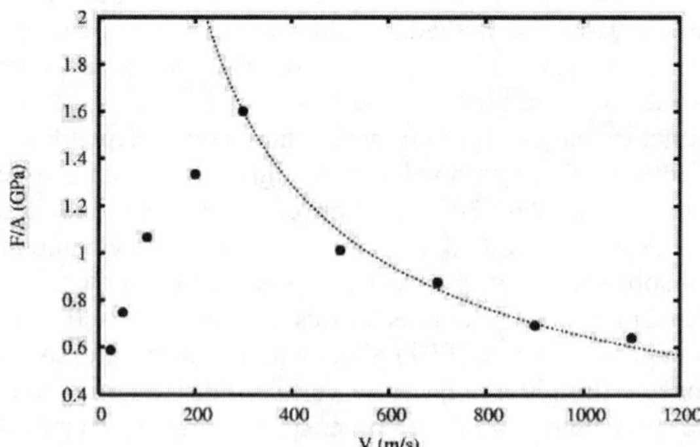

Figure 2. Tangential force per unit area for Cu (010) sliding on Ag (010) in the ⟨100⟩ direction. Symbols represent NEMD simulations, and the dotted line a $v^{-3/4}$ power law velocity weakening at high velocities.

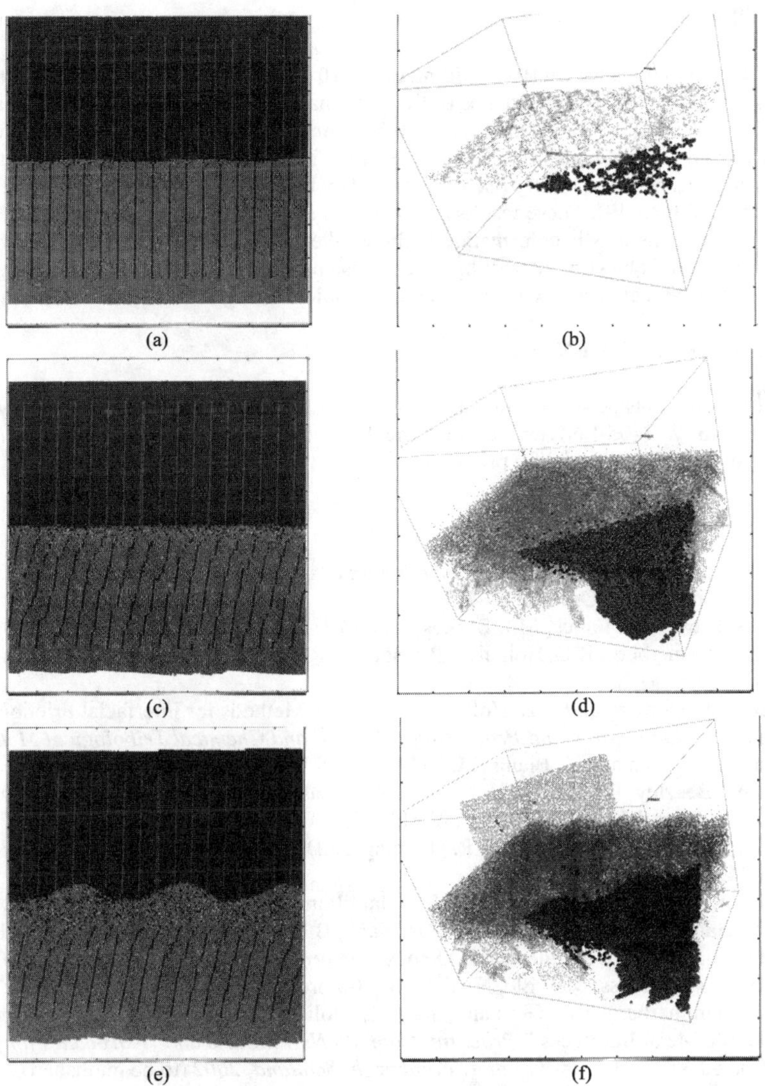

Figure 3. Configurations (left-hand column) and centro-symmetric order parameter (right-hand column) for velocities 25 m/s (a,b); 300 m/s (c,d); and 500 m/s (e,f). Darker regions in the order parameter images are artifacts of the visualization software representing more distant atoms by a single pixel, and nearer atoms by multipixel "spheres". Surface steps in (c) and (e) result from dislocations reaching the reservoirs.

CONCLUSIONS

The simulations for Cu (010) sliding on Ag (010) in the ⟨100⟩ direction exhibit a generic behavior in the velocity dependence of the frictional force (figure 2) which is similar to that in Ta/Al [8], Cu/Cu [3], and Lennard-Jones tribo-pairs [9]. We attribute the velocity weakening of the frictional force at high velocities to a dynamic instability resulting in the formation of nanoscale structure and giving rise to a power law behavior for the high velocity dependence of the frictional force [9]. These results are also consistent with the notion that the frictional force is determined by the plastic deformation in the weaker of the two materials forming the pair, in this case Ag. It would be very interesting to test these predictions both at high velocities in dynamic experiments, and at very low velocities with samples having very smooth surfaces.

ACKNOWLEDGEMENTS

This work was performed under the auspices of the U.S. Department of Energy under contract No. W-7405-ENG-36, through the Advanced Simulation and Computing (ASC) Materials & Physics Modeling program.

REFERENCES

1. F.P. Bowden and D. Tabor, *The Friction and Lubrication of Solids, Parts I and II*, (Clarendon Press, Oxford, 1964).
2. M.S. Daw, S.M. Foiles, M.I. Baskes, *Materials Science Reports* **9**, 251-310 (1993).
3. J.E. Hammerberg, B.L. Holian, J. R Röder, A.R. Bishop, and S.J.Zhou, Physica **D123**, 330-340 (1998).
4. J.E. Hammerberg and B.L. Holian, "Simulation Methods for Interfacial Friction in Solids," *Surface Modification and Processing: Physical and Chemical Tribological Methodologies*, ed. G.E. Totten and H. Liang (M. Dekker, New York, 2004) (in press).
5. D.M. Beazley, P.S. Lomdahl, N. Grønbech-Jensen, R. Giles and P. Tamayo, *Ann. Rev. of Comp. Phys. III*, ed. D. Stauffer (World Scientific, Singapore, 1996), pp.119-175.
6. J.A. Sprague, F. Montalenti, B.P. Uberuaga, J.D. Kress, and A.F. Voter, *Phys. Rev.* **B66,** 205415 (2002).
7. C.L. Kelchner, S.J. Plimpton, and J.C. Hamilton, *Phys. Rev.* **B58**, 11085-11088 (1998).
8. J.E. Hammerberg, R. Ravelo, T.C. Germann, B.L. Holian, and J.D. Kress, "Sliding Friction at Compressed Ta/Al Interfaces," *Shock Compression of Condensed Matter-2003*, ed. J. Forbes (AIP Press, Melville, N. Y., 2004) (in press).
9. J.E. Hammerberg, T.C. Germann, and B.L. Holian, "Modeling High Speed Friction at Ductile Metal Interfaces," *Proc. Int. Conf. on New Models and Hydrocodes for Shock Wave Processes in Condensed Matter, Edinburgh, Scotland, 2002* (to be published).

Part III: Processing—Sintering, Solidification, and Gel Techniques

Mat. Res. Soc. Symp. Proc. Vol. 821 © 2004 Materials Research Society P3.25

Characterization of Carbon Nanotubes/Cu Nanocomposites Processed by Using Nano-sized Cu Powders

Kyung Tae Kim, Kyong Ho Lee, Seung Il Cha, Chan-Bin Mo and Soon Hyung Hong

Department of Materials Science and Engineering,
Korea Advanced Institute of Science and Technology,
373-1 Guseong-dong, Yuseong-gu, Daejeon 305-701, Korea

Address all correspondence to this author
Email: shhong@kaist.ac.kr

ABSTRACT

Carbon nanotubes (CNTs) have attracted remarkable attention as reinforcement for composites owing to their outstanding properties[1-3]. CNT/Cu nanocomposites were fabricated by mixing the nano-sized Cu powders with multi-wall carbon nanotubes and followed by the spark plasma sintering process. The CNT/Cu nanocomposite fabricated from nano-sized Cu powders shows more homogeneous distribution of CNTs in matrix compared to that fabricated from macro-sized Cu powders. The hardness of CNT/Cu nanocomposite fabricated from nano-sized Cu powders increases with increasing the volume fraction of CNTs, while the hardness of that fabricated from macro-sized Cu powders decreases with the addition of CNTs.

INTRODUCTION

Carbon nanotube (CNT) has been considered as an ideal reinforcement for composite materials due to its outstanding properties such as high strength, elastic modulus, aspect ratio and conductivities. In CNT/polymer nanocomposite, the addition of carbon nanotube as reinforcement in polymer matrix improves the strength of matrix by several times[4,5]. However, compared to CNT/polymer nanocomposites, the CNT/metal and CNT/ceramic nanocomposites attract less attention due to their lower mechanical properties than expected. Even though several researchers have fabricated CNTs/metal composite by powder metallurgy process including mixing of CNTs with metal powders and followed by hot pressing[5-8], the CNTs/metal nanocomposites show low relative density ranged 85~95% and, at the same time, the CNTs are severely agglomerated in matrix. As a result, the addition of carbon nanotubes have not been successful to enhance the mechanical properties of matrix materials. In previous researches on CNT/metal nanocomposites, there are two critical problems remained to be solved to produce sound CNT/metal nanocomposites. One is the homogeneous dispersion of CNTs in matrix with avoiding the agglomeration of CNTs. The other is the higher densification of CNT/metal nanocomposite during the consolidation process. Concerning to the homogeneous dispersion of CNTs and matrix, Peigney and Flahaut[9,10] introduced CNT reinforced composites synthesized by the in-situ growth of CNTs in Fe-Al_2O_3 powders. Recently, Mukherjee et al.[11] use nano-size Al_2O_3 powders for CNT/Al_2O_3 nanocomposites, and H. Cong et al.[12] used nano-sized Al powders to fabricate CNT/Al nanocomposites in order to distribute the carbon nanotubes more homogeneously. However, the mechanical properties were not enhanced as expected due to the low density after sintering process.

EXPERIMENTAL PROCEDURES

In this study, CNT/Cu nancomposites were fabricated by powder metallurgy process by mixing of CNTs with Cu powders and followed by spark plasma sintering of the mixed powders. In order to investigate the effect of initial size of Cu powders on microstructure and mechanical properties of CNT/Cu nanocomposite, two different sizes of Cu powders such as 100nm and 200μm, were mixed with the CNTs. The spark plasma sintering process, which is known as a promising process for consolidation of ceramic nanocomposites[11-13], was used for higher densification of the CNT/Cu nanocomposites.

Nano-sized Cu powders, average diameter of 200nm, were fabricated by spray drying process from aqueous solution of Cu nitrate($Cu(NO_3)_2$), and followed by calcination and reduction process. The details on the fabrication process of Cu nanopowders has been described in previous papers[14,15]. The multi-wall carbon nanotubes, produced by chemical vapor deposition process[16] were mixed with Cu powders. The average diameter of the multi-wall carbon nanotubes was measured as 40nm.

Two-stage mixing process was employed for homogeneous mixing of CNTs with nano-sized Cu powders. In the first stage, the CNTs were sonicated in ethanol and then dried to separate the agglomerated CNTs. In the second stage, the nano-sized Cu powders were mixed with the dried CNTs and ball-milled for 24hours by using alumina balls. For the comparison purpose, macro-sized Cu powders with average diameter of 200μm was mixed with CNTs by the same process for nano-sized Cu powders. The volume fractions of CNTs mixed with Cu powders were controlled from 0% to 15%.

The spark plasma sintering(SPS) process was used for densification of the mixture of carbon nanotubes and Cu powders. The sintering mechanism of SPS process is based on plasma generated momentarily in the gaps between powders by electrical discharge during d.c. pulsing. The densification can be promoted by additive effects including spark impact pressure, Joule heating and electrical field diffusion during spark plasma sintering process. The mixture of carbon nanotubes and Cu powders was consolidated into nanocomposites having relative density above 95% at a temperature of 750°C with a holding time of 1 min under an applied pressure of 40MPa and a heating rate of composite powders in SPS process was 50℃/min. The microstructural observation of consolidated CNT/Cu nanocomposites was carried out by optical microscopy and high-resolution scanning electron microscopy (HRSEM). The Vickers microhardness of CNT/Cu nanocomposites was measured under a load of 25g and duration time of 10 seconds.

RESULTS AND DISCUSSION

The microstructure of mixed powders of carbon nanotubes and Cu was shown in Fig. 1. When the nano-sized Cu powders were used, the CNTs were homogeneously intermixed with Cu powders, while, when the macro-sized Cu powders were used, most of the CNTs were agglomerated at the surfaces of Cu powders.

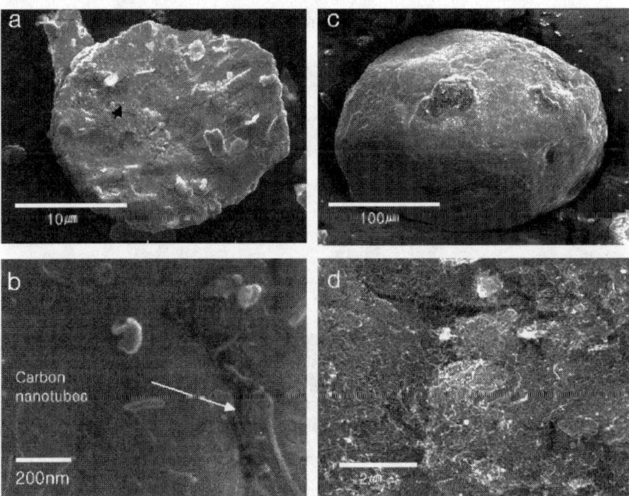

Fig. 1. SEM micrographs of mixture of CNTs and Cu powders after ball milling for 24 hours: (a) Morphology of mixture of CNTs and nano-sized Cu powders; (b) Micrograph showing the distribution of CNTs on nano-sized Cu powders; (c) Morphology of mixture of CNTs and macro-sized Cu powders; (d) Micrograph showing the distribution of CNTs on macro-sized Cu powders.

The relative densities of spark plasma sintered CNT/Cu nanocomposites are measured in a range from 94% to 100%. The measured densities are higher than those reported for CNT reinforced metal matrix nanocomposites, which is ranged from 85% to 95%[5-8], processed by hot pressing technique. The relative densities of CNTs/Cu nanocomposites using macro-size Cu powder were measured from 94 to 100%. Therefore, it can be concluded that the CNT/Cu nanocomposites can be fully densified by SPS process regardless of its initial powder size. Both microstructures show core-shell structure, in which the CNTs are located in shell regions at grain boundaries of Cu and the core regions at grain interior of Cu are devoid of the CNTs. The core-shell microstructure is initiated from the morphologies of mixtures of carbon nanotubes and Cu powders. In mixture powders before sintering, the

CNTs were located just on the surfaces of Cu powders. Even though the two microstructures of sintered CNT/Cu nanocomposites showed similar core-shell structure, the microstructures in shell region and the size of core are quite different by the initial Cu powder size. In case of CNT/Cu nanocomposite using nano-size Cu powders, the shell region consists of CNTs dispersed within the Cu matrix as shown in Fig. 2(c), while in case of using macro-sized Cu powders, the shell consists of CNTs with little Cu matrix phase as shown in Fig. 2(d). The different shell microstructure according to the size of initial Cu size was caused by the difference of CNT dispersion status in CNT/Cu mixed powders.

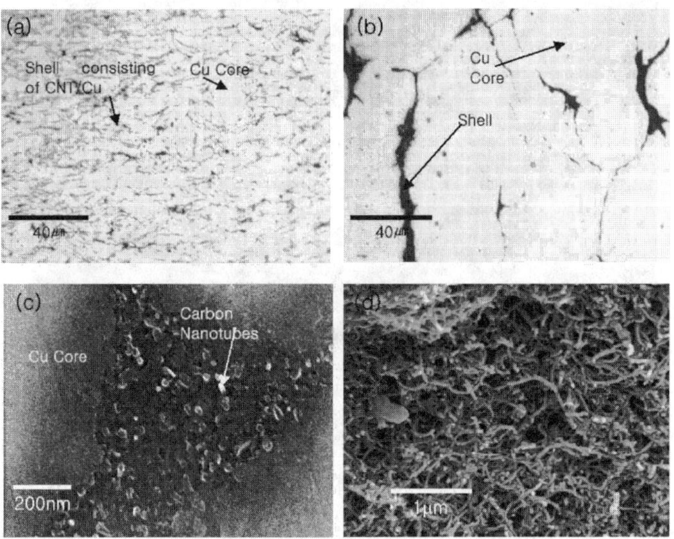

Fig. 2. Optical micrographs showing the overall distribution of CNTs in spark plasma sintered CNT/Cu nanocomposites, (a) fabricated by using the nano-sized Cu powders and, (b) fabricated by using the macro-size Cu powder. SEM micrographs of shell in spark plasma sintered CNTs/Cu nanocomposite, (c) fabricated by using the nano-szie Cu powder and, (d) fabricated by using the macro-size Cu powder.

The Vicker's hardness of the CNT/Cu nanocomposites measured by Vicker's hardness test is shown in Fig.3. A considerable enhancement of hardness is observed in CNT/Cu nanocomposites fabricated using the nano-size Cu powders. The hardness increases with increasing the CNT volume fraction up to 10%, and then slightly decreases at CNT volume fraction of 15% due to the agglomeration of CNTs in shell region. The peak hardness of $102kg/mm^2$ was obtained for the CNTs/Cu nanocomposites, which was approximately 1.5 times higher than that of unreinforced Cu consolidated using the nano-size Cu powders. However, in case of CNT/Cu nanocomposite fabricated using the macro-size Cu powders, the hardness slightly decreases with the addition of CNTs. The reason for the decrease of hardness is considered that the agglomerated CNTs, loosely bonded with Cu matrix, cannot effectively share the external load.

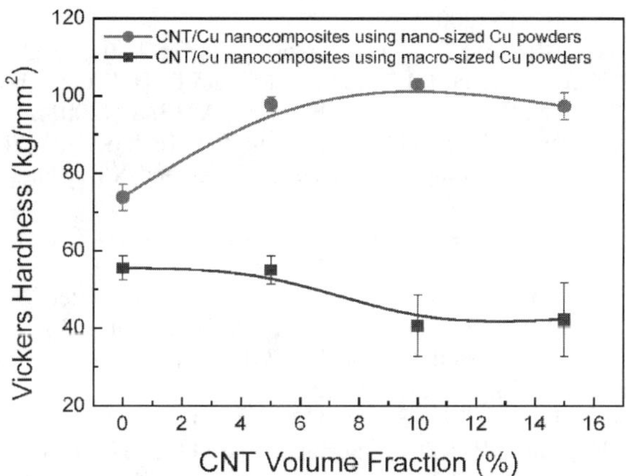

Fig. 3. The variation of Vicker's hardness of CNT/Cu nanocomposites with varying the volume fraction of carbon nanotubes.

Based on these results, it can be shown that the improvement of mechanical properties of CNT reinforced nanocomposite can be obtained only when the external load can be shared by the CNTs through the load transfer from matrix to carbon nanotubes. In order to maximize the load transfer efficiency, the carbon nanotubes must be dispersed homogeneously in matrix with strong bonding at the interfaces with the Cu matrix.

SUMMARY

In summary, CNT/Cu nanocomposites using the nano-sized Cu powders have been successfully fabricated by ball milling and followed by spark plasma sintering process. The hardness of CNTs/Cu nanocomposites enhanced up to 50% according to the addition of CNTs. These results demonstrate that the homogeneous dispersion of CNTs and high densification are the critical problems to be solved to enhance the mechanical properties of CNT/metal nanocomposites.

ACKNOWLEDGEMENT

This research was performed with the financial support of the Center for Nanostructured Materials Technology under the 21st Century Frontier R&D Program of the Ministry of Science and Technology, Korea.

REFERENCES

1. Y. Zhang, N. W. Franklin, R. J. Chen and H. Dai, Chem. Phys. Lett. **331**,35 (2000)
2. K. T. Lau and D. Hui, Carbon **40**, 1597 (2002)
3. H. Dai, Surf. Sci. 500, 218 (2002)
4. E. T. Thostenson, Z. Ren and T. W. Chou, Comp. Sci. Tech. 61, 1899 (2001)
5. C. L. Xu, B. Q. Wei, R. Z. Ma, J. Liang, S. K. Ma and D. H. Wu, Carbon **37**, 855 (1999)
6. S. R. Dong, J. P. Tu, X. B. Zhang, Mater. Sci. Eng. A313, 83 (2001)
7. X. Chen, J. Xia, J. Peng, W. Li and S. Xie, Comp. Sci. Tech 60, 301 (2000)
8. W. X. Chen, J.P. Tu, L. Y. Wang, H. Y. Gan, Z. D. Xu and X.B. Zhang, Carbon 41, 215 (2003)
9. A. Peigney, E. Flahaut, Ch. Laurent, F. Chastel and A. Rousset, Chem. Phys. Lett. **352**, 20 (2002)
10. A. Peigney, E. Flahaut, Ch. Laurent, and A. Rousset, Ceram. Inter. 26, 677 (2000)
11. G.. D. Zhang, J. D. Kuntz, J. Wan and A. K. Mukherjee, Nature Mater. **2**, 38 (2003)
12. R. Zhong, H. Cong and P. Hou, Carbon **41**, CO1-851 (2003)
13. M. Omori, Mater. Sci. Eng. A **287**, 183 (2000)
14. S. H. Hong and B. K. Kim, Mater. Lett. **4337**, 1 (2003)
15. D. W. Lee, G. H. Ha and B. K. Kim, Scripta mater. **44**, 2137 (2001)
16. H. Yokomichi, F. Sakai, M. Ichihara and N. Kishimoto, Physica B: Condensed Matter, 323, 311(2002)

Mat. Res. Soc. Symp. Proc. Vol. 821 © 2004 Materials Research Society P3.20

Fabrication and Characteristics of Zirconia-Alumina Composites with Bilayer Structure

Sung-Jae Je[1], Myung-Won Lee[1], Jae-Ean Lee[1], Yeon-Gil Jung[1], Dong-Soo Park[2], and Ungyu Paik[3]
[1]Department of Ceramics Science and Engineering, Changwon National University Changwon, Kyungnam 641-773, South Korea
[2]Ceramic Materials Group, Korea Institute of machinery and Materials, Changwon, Kyungnam 641-010, South Korea
[3]Department of Ceramic and Engineering, Hanyang University, Seoul 133-791, South Korea

ABSTRACT

Bilayer structures consisting of ZrO_2-3mol% Y_2O_3 (TZ-3Y) and zirconia-alumina composites as inner (substrate) and outer (coating) layers, respectively, are fabricated using gel-casting for the inner layer and dip-coating for the outer layer in aqueous system. The relatively tough TZ-3Y is used as the inner layer for damage absorption due to its mechanical properties. The mixture ratio of alumina/zirconia slips for dip-coating is 1:9, 2:8, and 3:7 as mole ratio. The processing additives for gel casting, such as dispersant, monomer, dimer, and initiator, are adjusted and optimized by measuring viscosity. From which the solid loading of starting material (TZ-3Y) is determined. The optimum amount of dispersant (D-3019; anionic dispersant agent) for TZ-3Y is 0.7 wt%. The slip pH affects the electric double layer in sols, which causes the different rheological behaviors and solid loadings. The castable solid loading of TZ-3Y is 37 vol%, showing a pseudoplastic rheological behavior. The effect of slip type (different mixture ratio) on sintered body is investigated through hardness (Vickers indentation), microstructure (SEM), and strength (4-point bending tests). Strength of sintered bodies after dip-coating into the slips is higher than that before dip-coating, but hardness is not much different among cases. The effects of thickness in the outer layer on damage resistance and mechanical properties of the bilayer structures are discussed extensively.

INTRODUCTION

Zirconia has become a rapid development in recent decades and has been taking an active role in many applications, such as ceramic cutting tools, oxygen sensors, and high temperature fuel cells [1]. However, the application of pure zirconia as bulk is not feasible due to the large volume change in transformation from tetragonal to monoclinic phases. Alumina with flexural strength as high as 700 MPs has been prepared by using pure and fine powder [2]. However, fracture toughness of alumina is very low, typically approximately 3 MPa$\cdot$m$^{1/2}$, which is much lower than those of other structural ceramics such as zirconia and silicon nitride. Recently, coating systems including layered structure ceramics have become interesting due to the development of multifunctional materials for practical applications [3-8]. Also, gel-casting technique is an attractive new ceramic forming process for making high quality and complex-shaped ceramic parts with rapid forming cycle, and good wet and dry strength [5]. Therefore, a new approach to develop and modify ceramic properties is attempted as bilayer structure through gel-casting process. The objective of this study is to design and prepare the microstructure providing optimum mechanical properties in alumina/zirconia bilayer structures – hard outer layer for wear resistance and relatively tough inner layer for damage absorption. The bilayer, consisting of zirconia (TZ-3Y) and zirconia-alumina composites as inner and outer layers respectively, is fabricated by gel-casting for the inner layer and dip-coating for the outer layer in an aqueous system. Here, we describe processing parameters in the gel-casting and fabrication method of the layered structure, and show how the microstructure is controlled.

EXPERIMENT PROCEDURE

Commercial grade ZrO_2-3mol% Y_2O_3 (TZ-3Y, Tosho Corp., Japan) and α-Al_2O_3 (AKP-30, Sumitomo Co., Japan) were used as starting materials for inner and outer layers, respectively. The characteristics of these starting materials are shown in table I. The inner layer was prepared by gel-casting process. The Duramax^{TM} D-3019 (ammonium salt, Rhom and Haas Co., Philadelphia, USA) was used as the dispersant. Acrylamide (AM, H_2C=$CHCONH_2$, M.W = 71.08, 98 %, Duksan Pure Chemical Co., Korea) and N, N´-methylenebisacrylamide (MBAM, $C_7H_{10}N_2O_2$, F.W = 154.17, 96 %, Acros Organic Co., USA) of 20 % of the monomer concentration were added as monomer and dimer (cross-linker), respectively. Initiator of ammonium persulfate (APS, $(NH_4)_2S_2O_8$, F.W = 228.19, $\leq$ 95 %, Duksan Pure Chemical Co., Korea) and catalyst of N, N, N´, N´-tetra-methylethylenediamine (TEMED, $C_6H_{16}N_2$, F.W = 116.21, 99 %, Acros Organic Co., USA) were used for boosting gelation reaction through continuously free radical formation.

To determine the optimum amount of dispersant and solid loading of TZ-3Y powders, the rheological behavior was characterized by a Concentric-Cylinder Viscometer (Model RS75, Haake Inc., Karlsrube, Germany). Dried green bodies were pre-sintered at 900 ºC for 1h for elimination of residual water and organic materials, and then which was dip-coated into slips made by mixing alumina and zirconia powders as 1:9, 2:8, and 3:7 mole ratio after pre-wetting. The green bodies pre-sintered were re-sintered at 1600 ºC for 3h in ambient atmosphere. Scanning Electron Microscopy (SEM, S2400, Hitachi, Japan) was used to observe the fracture surfaces of sintered bodies in the inner and the outer layers. Four-point flexural strength tests were conducted at a crosshead speed of 0.5 mm/min on a comuter-controlled Universal Testing Machine (Instron 5567, INSTRON, UK). Vickers indentations were performed to measure hardness in each region, inner and outer layers, using micro-hardness tester (Mitutoyo, AVK-C2, Japan) at a normal load of 1 N.

RESULTS AND DISCUSION

The rheological behavior of slips is dependent on the amount of dispersant added, due to the polymer electrolyte adsorbed around particles. The adsorbed polymer electrolytes induce the effects of electrostatic and steric hindrance between suspended particles. Figure 1 shows that optimum addition amount of dispersant. The optimum amount of dispersant for TZ-3Y is 0.7 wt%. Excess addition of dispersant increases the slip viscosity. The reason for this is that residual polymer electrolytes of shielding particles form a loop by steric stabilization effects. The optimum range of viscosity for gel-casting is found to be 700 mPa·s at shear rate of 20 s^{-1}, in which the optimum solid loading is determined. Dispersion stability in higher solid loading has a

Table I. Characteristics of 3 mol% Y_2O_3-doped ZrO_2 and alumina powders used in this study

Properties materials	Melting point (℃)	Density (g/cm^3)	Mean particle size (㎛)	Modulus of elasticity (GPa)	Poission's ratio
3 mol% Y_2O_3-ZrO_2	2719	6.05	0.024	186 (20℃)	0.31
AKP-30	2015±15	3.98	0.32	-	-

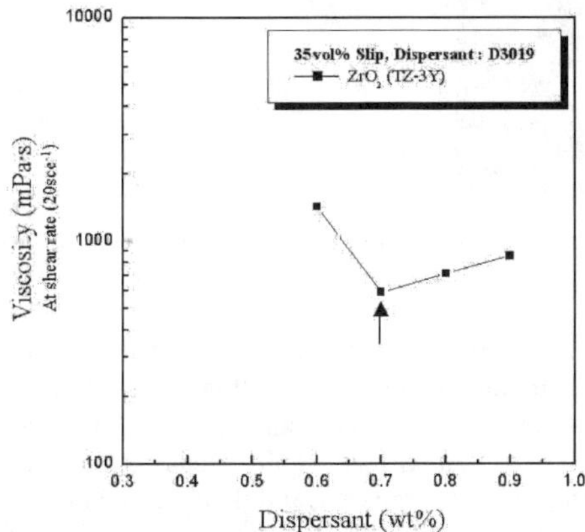

Figure 1. Apparent viscosity with addition amount of dispersant in TZ-3Y slip.

very important role on the quality of the ultimate products, especially in preparation of ceramic green body. The higher solid loading, the lower liquid content of the slip. This leads green body to higher packing factor or higher density, which minimizes the shrinkage of green body. Figure 2 represents the viscosity (figure 2(A)) as a function of shear rate at each volume fraction of the TZ-3Y slips containing 0.7 wt% dispersants, and gelation reaction behavior (figure 2(B)) in the solid loading of 37 vol%. It is found that castable solid loading of TZ-3Y under the presence of 0.7 wt% dispersant is 37 vol% in initial pH condition (pH = 6.5). The TZ-3Y slips show shear-

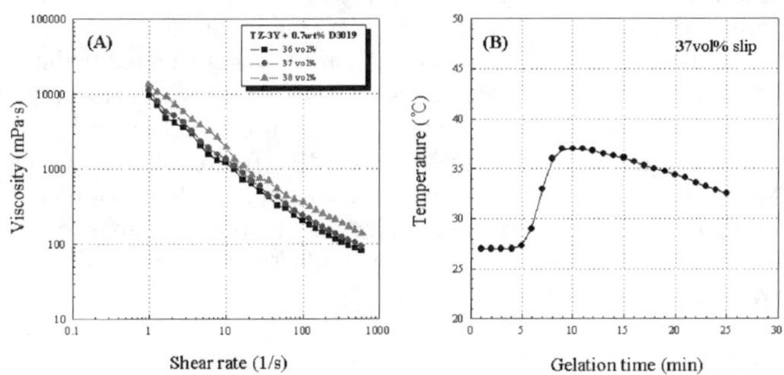

Figure 2. (A) apparent viscosity with volume fraction of TZ-3Y slips and (B) gelation reaction behavior.

Figure 3. Micrographs of fractured surface after sintering at 1600℃ for 3 hours; (A) inner layer and (B), (C), and (D) outer layer.

thinning (pseudo-plastic) behavior with increasing shear rate. Generally, radical polymerization of AM is directly proportional to the amount and the ratio of monomer and initiator and is dependent upon temperature. Therefore, the temperature of slips is sustained at 27 ± 1 °C, and the time for gelation reaction is adjusted within 5 ~ 10 min with considering castability. With above condition, the green bodies of TZ-3Y with pores are prepared, and then zirconia/alumina slips are infiltrated into the porous green bodies prepared at 900 °C for 1h.

Micrographs of fracture surfaces in the inner and the outer layers of the bilayer structure sintered at 1600 °C for 3h are shown in figure 3. Figure 3(A) is microstructure of the inner layer (TZ-3Y), and figure 3(B), (C) and (D) are of the outer layer (zirconia-alumina composites) dip-coated into slips of different mixture ratios of 1:9, 2:8, and 3:7 in alumina and zirconia mole ratio, respectively. Grain size of TZ-3Y in both layers is approximately 1 μm, indicating that grain size of alumina is remained as that of starting powder and is much smaller than that of TZ-3Y. Therefore, layer structures are successfully fabricated using gel-casting and dip-coating methods even in aqueous system. Figure 4 shows EDX analysis of bilayer structures with the inner layer (TZ-3Y) and outer layer (zirconia-alumina composites) dip-coated into slips of different mixture ratio in alumina and zirconia mole ratio. From figure 4, it is found that the thickness of the outer layer in the bilayer is increased with increasing the mole ratio of alumina in slips.

Figure 5 represents strength (figure 5(A)) and hardness (figure 5(B)) of sintered bodies. The strength of sintered body without dip-coating is 156 ± 23 MPa (figure 5A(a)), which is relatively low value compared to the previous reported ones, due to the high porosity in the inner layer. While through dip-coating the strength is increased to 282 ± 43 MPa (figure 5A(b)), and reached about 500 MPa (figure 5A(d)). In accordance, the mole ratio in the slips for dip- coating is important to improve the strength of the layer structure, indicating that the strength is improved as the amount of alumina in the slips increases. However, the hardness of sintered bodies is not different enough between each case, independent of dip-coating.

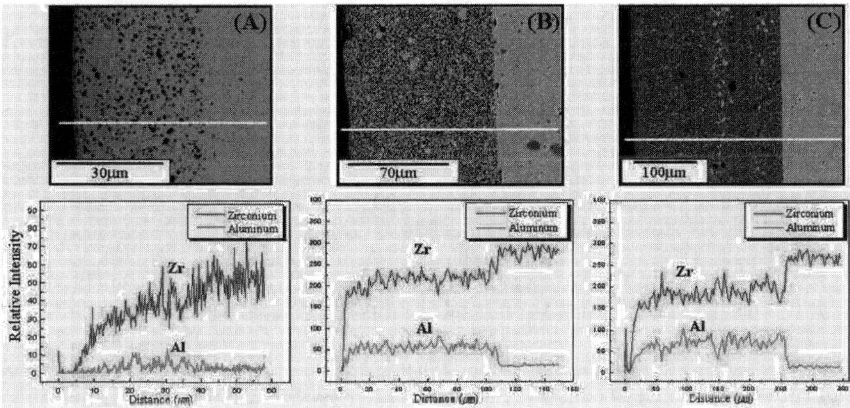

Figure 4. EDX analysis of bilayer structures with the inner layer and the outer layer dip-coated into slips of different mixture ratio in alumina and zirconia mole ratio; (A) 1:9, (B) 2:8, and (C) 3:7.

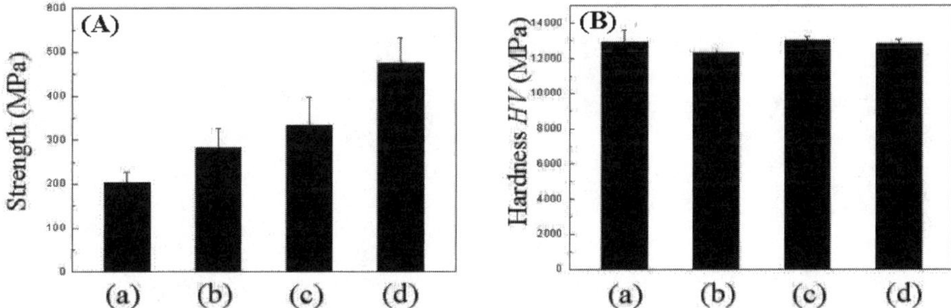

Figure 5. (A) Strength and (B) hardness of sintered bodies using gel-casting; (a) sintered body without dip-coating, (b) sintered body after dip-coating into slip of 1:9 mole ratio, (c) sintered body after dip-coating into slip of 2:8 mole ratio, and (d) sintered body after dip-coating into slip of 3:7 mole ratio.

CONCLUSIONS

The bilayers with ZrO_2-3mol% Y_2O_3 and zirconia-alumina composites as the inner and outer layers, respectively, are successfully fabricated using gel-casting and dip-coating methods in an aqueous system. The optimum amount of dispersant for TZ-3Y is 0.7 wt% and the castable solid loading of TZ-3Y is 37 vol%. The mixture ratio of alumina and zirconia in slips for dip-coating affects the thickness of outer layer, showing a thicker outer layer with increasing amount of alumina in slips. The strength of sintered bodies with dip-coating is higher than that without dip-coating. The mole ratio of alumina in slips acts as an important factor in strength. However, the hardness is not different enough between each case, independent of dip-coating.

ACKNOWLEGEMENTS

This work was financially supported by the Machine Tool Research Center at Changwon National University and by Korea Electric Power Research Institute (KEPRI).

REFERENCES

1. N.Q. Minh, *J. Am. Ceram. Soc.* **76** , 563 (1993).
2. Alford NMcN and Brichall JD, *Nature*, **330**, 51 (1987).
3. J.C. Kinght, T.F. Page and I.M. Hutchings, *Thin Solid Films*, **117**, 177 (1989).
4. S.W. Wuttiphan, B.R. Lawn and N.P. Padture, *J. Am. Ceram. Soc.* **79**, 634 (1996).
5. A.C. Young, O.O. Omatete, M.A. Janney and P.A. Menchhofer, *J. Am. Ceram. Soc.* **74(3)**, 612 (1991).
6. S. Kang, H. Kim, S. Kim, S. Lee and D. Kim, *J. Kor. Ceram. Soc.* **31(5)**, 513 (1994).
7. I.A. Aksay, F.F. Lange, and B.I. Davis *J. Am. Ceram. Soc.* **66(10)**, c190 (1983)
8. K. Kang and K. Han, *J. Kor. Ceram. Soc.* **37(12)**, 1187 (2000)

Mat. Res. Soc. Symp. Proc. Vol. 821 © 2004 Materials Research Society P8.36

Effect of Oxygen and Nitrogen Doping on Mechanical Properties of Silicon Using Nanoindentation

A. Karoui, G. Rozgonyi, and T. Ciszek[1]

Materials Science and Engineering Dept.
North Carolina State University, Raleigh, NC 27695-7916
[1]: National Renewable Energy Laboratory, Golden, CO

ABSTRACT

The effects of oxygen and nitrogen on the mechanical properties of Czochralski (CZ) and float zone silicon have been studied using nano-indentation. Nitrogen free FZ Si exhibited low hardness of 6.49 GPa and elastic modulus of 104 GPa. When doped with $2x10^{15}cm^{-3}$ nitrogen, FZ Si hardness and elastic modulus increased to 8.2 and 182 GPa, respectively. In the near-surface denuded zone of N-doped CZ Si (N-CZ) the hardness correlates well with the O and N profiles. Distinct high hardness points, found in the O- and N-rich subsurface region, were attributed to precipitates. Nano-scratch tests of N-CZ Si confirmed the existence of hard phases, mostly small precipitates, whose density, estimated to be $2x10^{13}$ cm^{-3}, is in the range of previously suggested nuclei density in as-grown N-CZ silicon.

INTRODUCTION

The trend of using large wafers for IC fabrication is challenged by the high thermo-gravitational stresses during high temperature treatments, especially during rapid thermal processing. This heat treatment produces slip at wafer edges and heavy damage at contact points.[1] Relaxation of thermo-gravitational stresses also produces wafer warpage; which is incompatible with micron and submicron IC processing. The tendency of reducing O in modern CZ can be compensated for by using N doping[2,3] at $5x10^{14}cm^{-3}$ or less. Meanwhile, Float Zone (FZ) silicon is the ideal material for high efficiency sliver[4] and high concentration solar cells,[5] but FZ does not withstand handling during solar cell fabrication. Dislocation locking by N in FZ silicon has been investigated.[6,7] Orlov et al, have shown that N-doping causes a decrease of dislocation mobility in CZ Si (suggested as possible mechanisms of hardening) and increases residual stresses.[8] Eremenko et al[9] have studied the interaction between moving dislocation and point defects in silicon. Zozime and Castaing[10] discussed hydrogen effects on dislocation properties. More recently, Yonenaga provided a set of experimental evidences on dislocation-impurity interaction in silicon.[11] Options for toughening FZ Si wafers to improve the mechanical yield of photovoltaic Si wafers without scarifying its high lifetime were proposed based on N doping in conjunction with O control[12,13]

The objective of this work is to measure the mechanical characteristics of N-CZ and N-FZ silicon at submicrometer and nanometer scales. The approach used is nanoindentation and load controlled nano-scratching at room temperature. In prior work we found that N doped CZ silicon (N-CZ) wafers processed with denuded zone thermal cycles have three distinct regions where precipitate density and O and N concentrations profiles vary significantly.[14] These regions are: (i) subsurface, which features a high density of defects, (ii) very low defect density zone (LDZ) instead of the conventional denuded zone (DZ) in CZ Si, and (iii) bulk region.[15] The latter exhibits a high density of precipitates that peaks close to the LDZ and decays farther in the bulk. In previous work, we discussed the structure of defects in these samples, investigated by conventional and high resolution transmission electron microscope (TEM) and STEM in Z-contrast mode. Equilibrium octahedral precipitates (diagonal ~20 to 30 nm) and submicron stacking faults (SF) have been detected in the subsurface. The host crystal in the subsurface zone is almost undistorted around the precipitates and the interfaces is in the {111} plane. In contrast, in the bulk region the precipitate morphology strongly differs from those equilibrium octahedral precipitates.[14,16] In addition, they generate large strain in the matrix. Z-contrast imaging has shown that the precipitates have a N-rich shell. Using electron energy loss spectroscopy (EELS), the N to O content ratio is found to be 2% in the precipitate core and 17% in the shell. The origin of precipitates in the three regions was discussed in relation with point defect dynamics including diffusion and interactions/generation-recombination.[16,17]

Correlation of the local hardness of the shallow oxynitride subsurface layer with the O and N depth distributions and the effect of precipitates on hardness are sought in this study.

EXPERIMENTAL

The samples listed in Table I, were cut for nanoindentation tests from the following N-FZ and N-CZ silicon wafers:

1) dislocated N-free FZ Si intentionally grown with dislocations (referred to as: disl-FZ),
2) two N doped FZ Si wafers that were cut from different positions of the ingot (N-FZ1, N-FZ2), ,
3) one N free CZ Si wafers and one N-CZ Si wafer were cut from ingots grown at the same rate with A and N ambients, respectively. These wafers were heat treated with thermal annealing processes for DZ formation (i.e., Lo-Hi and Hi-Lo-Hi). The Lo-Hi thermal anneal was done at 650°C for 8 h then 1050°C for 16 h and the Hi-Lo-Hi is similar except for the Hi pre-anneal at 1250°C for 1 h. These wafers contain oxygen precipitates; their size distribution and density depth profiles were discussed in Ref. 18. These samples are examined to check if precipitates affect mechanical properties of Si.

Table I: Description of the tested materials. All wafers are [111] oriented.

disl-FZ	N-FZ1	N-FZ2	N-CZ	CZ
Pure FZ Si grown with dislocation. $[O_i]= 7x10^{16}$ cm^{-3} N-free	N-doped FZ Si, (Ga doped). $[O_i]=7x10^{16}$ cm^{-3} $[N] < 2x10^{15}$ cm^{-3}	Similar to N-FZ1, cut from different ingot positions.	N-CZ Si 3" wafer $[O_i]= 1.1x10^{18}cm^{-3}$ $[N]= 10^{15}$ cm^{-3}	CZ Si 3" wafer $[O_i]= 1x10^{18}cm^{-3}$ N-free
As grown	As grown	As grown	Hi-Lo-Hi	Lo-Hi

Hardness and elastic modulus were obtained by nanoindentation, using a Hysitron Triboindenter. This apparatus provides *in-situ* images in scanning probe microscopy mode (SPM) using the indenter tip itself as a probe. A combination of various loads and indenter tips was used in order to remove tip and load effects. Large indentation data sets were produced and average material parameters were deduced. The first measurement series was made on the initial surface of all samples using a peak load of 5 mN and a Berkovitch indenter tip. The load function is trapezoidal and consists of a 5 s loading step, a 3 s holding, and a 5 s unloading. A second series of arrays of indents was made in order to explore the hardness inhomogeneities across the DZ of the Hi-Lo-Hi annealed N-CZ samples. This was made on a finely polished cross-section sample, which is compared to the initial wafer surface. Four indentation grids were performed. Grid 1 was an array of 7 by 7 indents spaced 2 μm, the grid was centered at 150 μm from the sample surface. A Berkovitch indenter tip and peak loads of 5 mN were used. Both Grid 2 and Grid 3 were arrays of 20 indents (4 by 5) spaced 1.5 μm apart, but centered at 5 μm and 150 μm away from the sample surface, respectively. A Berkovitch indenter and 400 μN peak load were used for both grids. Grid 4 consisted of 3 by 10 indents 50 nm apart. The array was centered at just 2 μm from the initial wafer surface. Since Grid 4 was close to the surface, a lower peak load limited to 175 μN, and a sharper indenter tip "Northstar" were used.[19] Even though large grids with up to 49 indents and small steps have been utilized, the mechanical properties of the Si matrix/O-precipitate system in N-CZ sample were tested by controlled nano-scratches. The goal was to intercept inhomogeneities in the hardness and eventually precipitates. These tests were made by applying a normal force to the sample surface, which was ramped from 1980 μN to 2000 μN during the scratching to insure contact continuity. The Northstar tip indenter, was first driven into the sample then slightly withdrawn while controlling the applied force. Fifteen seconds were allowed for indentation and tip stabilization before the scratching begun. A low scratching speed of 0.25 μm/s was used to allow relaxation of the material.

RESULTS

A) Indentation Arrays

The image in Fig.1 is a 5x5 μm^2 *in-situ* scan in SPM mode of an indent made with a Berkovitch tip and 5mN load on the initial surface of N-CZ wafer. The bright side is an elevated ledge and a large dislocation pileup. Cracks are primarily initiated at the corner of the trihedral indent from where they propagate. The elastic modulus and hardness obtained at the initial surface are summarized for the five samples in Table II. In addition, the surface roughness, measured by AFM on $10x10\mu m^2$ areas, is presented. Standard deviations given for the N-CZ sample are relative to 14 indents. Note that the roughness is ten times higher for the N-FZ samples than for N-CZ and CZ samples, which was found to affect the indent penetration phase only. The elastic modulus was calculated from measured reduced modulus, E_r:[20]

$$E_2 = \left(1 - \nu_2^2\right)\left[(1/E_r) - (1 - \nu_1^2)/E_1\right]^{-1} \quad (1)$$

where (1) and (2) indices are relative to the indenter and the tested material.

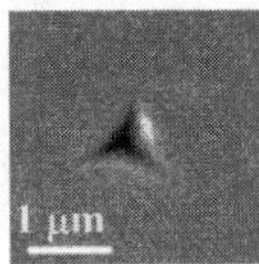

Fig.1: Indent made on the initial surface of N-CZ sample using 5 mN load and Berkovitch tip .

Table II: Summary of indentation data of tests made on initial wafer surfaces. Berkovitch tip and a trapezoidal load function (peak of 5 mN) were used. Averaged parameters are: contact depth <Depth>, reduced modulus <E_r>, elastic modulus <E>, and AFM determined roughness <Ra>.

Samples	<Depth> (nm)	< E_r > (GPa)	< E > (GPa)	<H> (GPa)	<Ra> (nm)
disl-FZ	151.4	101.98	103.78	6.49	163.1
N-FZ1	135.2	164.13	177.61	8.00	110.8
N-FZ2	134.2	167.49	181.87	8.11	103.3
N-CZ (Hi-Lo-Hi)	114.4 ±0.7	173.2 ±0.9	189.2 ±1.1	10.8 ±0.1	10.5 ±0.5
CZ (Lo-Hi)	111.9	182.48	201.23	11.29	12.5

B) Scratch Test

Figure 2 shows a 7.5 µm x 12.5 µm *in-situ* SPM image of a scratch made with a Northstar tip and an enlargement of the starting portion, (the boxed region in Fig.2(a)). The bright spot is a pit dug during the tip insertion in the material, the scratch average depth being 70 nm. Figure 2 (c) provides the variations of the lateral force resulting from the normal force, which was ramped from 1980 to 2000µN. The zoomed area is within the profile segment labeled A, in Fig.2(b).

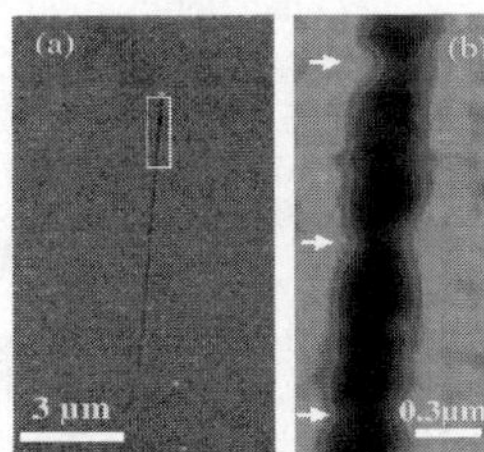

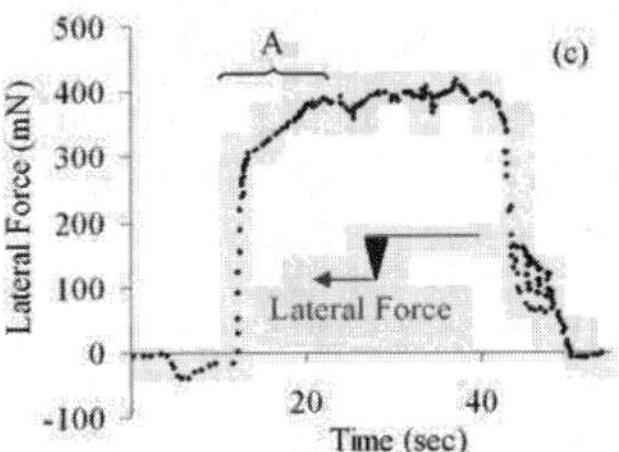

Fig. 2: (a) SPM image of a 10 µm long load controlled scratch on the N-CZ sample, (b) 6x enlargement of the in-boxed segment at the beginning of the scratch, and (c) lateral force during scratching.

DISCUSSION

The FZ samples (i.e., disl-FZ, N-FZ1, and N-FZ2) appeared softer and much rougher than the CZ and N-CZ samples. As expected, N doping of float zone Si increased significantly the hardness (from 6.49 to 8.11 GPa), and reduced the surface roughness (from ~163 to 107 nm). Since the FZ materials underwent the same polishing steps as the CZ wafers, the roughness root mean square height appears strongly correlated with the material mechanical properties. Figure 3 provides a plot of hardness versus contact depth for the four materials (for which averaged values are given in Table II). The chart clearly shows the large dispersion in the dependency hardness-depth with material type. Definitely, the hardness and elastic modulus decrease from CZ, N-CZ, N-FZ to N free FZ. The latter is distinctly softer, while N-FZ Si becomes significantly tougher by N doping. Note that Hi-Lo-Hi cycled N-CZ Si is only slightly softer than Lo-Hi treated CZ. This is likely to be due to the difference in the nature of active species in regard to dislocation locking. In addition, precipitate size and composition are totally different. It was demonstrated that oxygen precipitates in annealed N-CZ Si wafers possess a fair amount of N, particularly at the precipitate interface, where a thick oxynitride shell is observed.[13] For an octahedral-like precipitate of 20x15 nm the shell thickness was 3 nm.

In Fig.3 the hardness appears to vary significantly with the contact depth. The effect of the tip shape is ruled out and the change is attributed to essentially a decrease in the hardness with increasing depths. In low oxygen CZ silicon, N doping increases the hardness to a comparable level as in conventional O-rich CZ, i.e., 13 GPa[21] for [100] and 11.7 GPa for [111] oriented CZ silicon. The latter was measured by AFM nanotribology using a 100 μN load.[22] It is worth noting that the locking effect of N is close to that of O but slightly smaller. This observation was found by molecular dynamics calculations using three body Tersoff potential[23] for the atomic interactions.[24] It was found that O atoms, in bond centered interstitial positions, develop a binding energy (BE) with an edge dislocation of 1.80 eV and a separation shear stress (SSS) of 0.73 eV/Å, while N develops slightly smaller values of 1.66 eV and 0.6 eV/Å, respectively.[24] The ratios of nitrogen to oxygen BE and SSS appear in agreement with the measured ratio of N-FZ to CZ hardness.

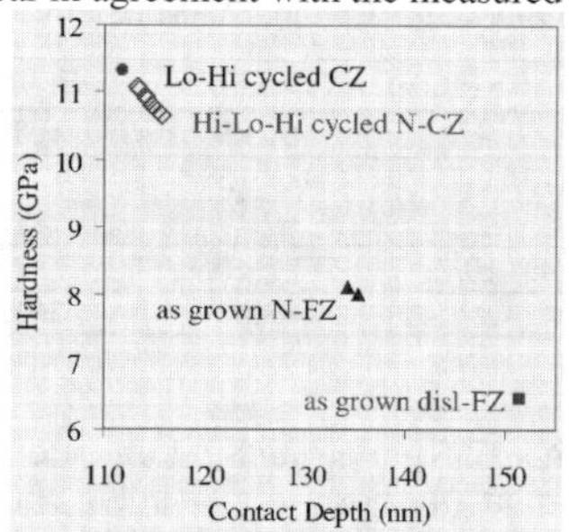

Fig. 3: Hardness vs. contact depth at the initial wafer surface (Berkovitch, 5mN).

Table III summarizes the data obtained from each of the four indentation grid on a cross-section sample from the Lo-Hi cycled N-CZ wafer. The averages of contact depth, modulus, and hardness are given in the DZ and in the bulk along with the standard deviation. Note that:

1. comparing the average hardness (Grid 2 and Grid 3 in Table III) made with the same tip and the same load indicates a softer LDZ than the bulk region 11.5 ±1.0, and 12.3 ±0.9 GPa. This is due to lower O and N concentrations as well as the precipitate density, which attains its lowest value in that region.
2. the hardness of Hi-Lo-Hi treated N-CZ sample done right at the wafer initial surface, which is 10.8±0.1GPa, see Table II, is close to that measured at 2 μm from the wafer surface, 10.7±1.5GPa, see Grid 4 in Table III. The former is obtained using a Berkovitch indenter and a load of 5mN and the latter is done with a Northstar indenter and a load of 175 μN on cross section sample. Thus, it appears that the high load of 5mN probed the subsurface region rather than the surface. Therefore, one can consider that the hardness of the DZ is 10.75 ±0.1 GPa on average, independent of the tip.
3. the high load used for Grid 1 provided the most homogeneous data with a standard deviation equal to 0.25 GPa that is one fourth of that obtained from Grid 2 and 3. The high load and large probed volume, resulted in stabilization of material response, meaning a reduction in sensitivity to heterogeneities.

Figure 4 presents the hardness obtained from Grid 4 which shows the data point distributions along the depth and the lateral positions, i.e., Z- and Y-axis. The origin of Z-axis is taken at the wafer initial surface. One can see that hardness decreases with the depth, while the variation is random along the lateral direction (i.e., Y-axis). This suggests a harder material close to the subsurface, as compared to the center o the LDZ. The dispersion in the Y-direction delineates the hardness fluctuations, most likely due to discrete hard phases including precipitates. Figure 4 (a) shows a large deviation between the three indentation lines that are parallel to the wafer initial surface. The deviation reaches 6 GPa at 1950 nm from the wafer surface with an average hardness of 12.2 GPa versus 10 GPa in LDZ. Because of the continuous character of the deviation, the material toughness changes is likely to result from the dissolved impurities rather than the precipitated ones. However, the dispersed group of high hardness data points is most likely due to precipitates, found to be enriched with 2% N and covered with 3 nm thick oxynitride shell.

Table III: Summary of data from various indentation grids on the cross-section of N-CZ sample. d: distance from wafer surface, B: Berkovitch indenter, N: Northstar indenter, σ: standard deviation

Region	d (μm)	Tip	Load (μN)	Grid	Indents Array Y - Z	Spacing (μm)	<Depth>±σ (nm)	<E_r> ±σ (GPa)	<E> (GPa)	<H> ±σ (GPa)
II	2	N	175	4	10 x 3	0.05	30.0±3.3	157.8±9.8	169.7	10.7±1.5
	5	B	400	2	4 x 5	1.50	21.5±1.5	181.3±9.7	199.7	11.5±1.0
III	150	B	400	3	4 x 5	1.50	20.5±1.2	185.9±7.6	205.7	12.3±0.9
			5000	1	7 x 7	2.00	112.0±1.4	188.0±2.5	208.5	11.3±.25

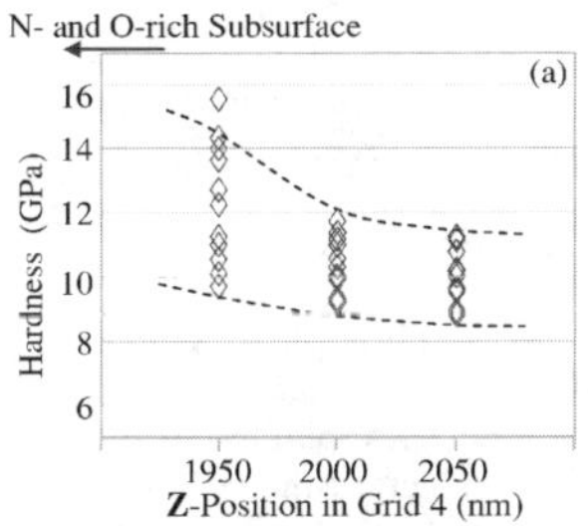

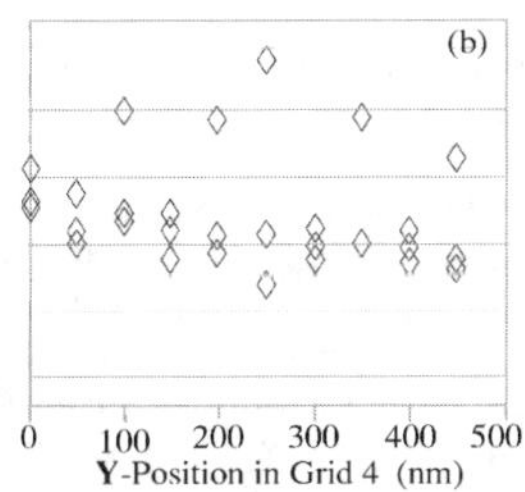

Fig. 4: Averaged hardness along the depth (i.e., Z-axis) and lateral position (i.e., Y-axis). The depth-origin is chosen on the wafer surface and indents are 50 nm apart.

Mechanical properties of the Si matrix/O-precipitate are relevant for the shear of precipitates by dislocations in order to evaluate the locking effects. The zoomed-in area in Fig.2 (c) shows the material heterogeneities originated by harder phases, with probably some instabilities of the tip. The bright zones in each side of the scratch are due to increased height (plastically deformed material full of nano-cracks and dislocation pile-ups). The ellipsoid-like domains are deep, thus pertain to the matrix softer material. The pinch-off spots shown with arrows, see Fig.2 (b), are due to increase in the hardness resulting from either oxynitride precipitates (which have higher hardness ~19 GPa[25] as compared to the silicon matrix that is 12 GPa at most) or just fluctuations of dissolved species [Oi] , $[N_2]$, and [NO]. The scratch zone is examined by micro-Raman scattering spectroscopy after CMP fine polishing to remove the upper part of the damage caused by the ejected debris at the surface. Intensity of line 400cm^{-1}appeared to vary significantly along the scratch independent of the location of the pinch-off spots.[26] This dominating line indicates the existence of a hard phase Si_{II}, which has β-tin structure.[27] The average spacing between pinch-off spots, the scratch width, and the scratch depth are about 0.9 μm, 0.6 μm, and 50 nm, respectively. The average density of the hard phases is then ~2×10^{13} cm^{-3}, which matches the density of N related nuclei proposed in as-grown N-CZ Si.[28] This is also supported by AFM imaging, which provided a density of $1.25\times10^{13} cm^{-3}$ of hard features in the subsurface region. These were remnant at the surface of a cross-sectional stack (from the same material) after an ultra-fine chemical-mechanical polishing.[26]

SUMMARY

Silicon hardness and elastic modulus vary significantly with O and N contents. CZ Si appeared about twice as hard as conventional FZ Si. Doping FZ Si with N increased the hardness significantly. Indentation made at the surface of Hi-Lo-Hi annealed N-CZ wafers, using Berkovitch tip and a load of 5mN, yielded similar hardness and reduced modulus data of subsurface obtained from cross-section sample, made using a Northstar indenter and a load of 175μN. At about 150 μm away from the surface (Grid 3) the N-CZ bulk material appeared harder than the LDZ at 5 μm away from the original surface (Grid 2,), i.e., 12.3 GPa versus 11.5 GPa. The LDZ characterized by a low O and N appears softer (11.5 GPa) than the bulk region (12.3 GPa), both measured with Berkovitch indenter and 400 μN load. Starting from the subsurface, the hardness and reduced modulus decrease with depth due to similar decreases in the N and O concentrations. Distinct high hardness values obtained at the subsurface region were attributed to precipitates. Scratch tests confirmed the existence of harder phases in N-CZ with a density of $\sim 1.5 \times 10^{13}$ cm^{-3}, in agreement with the predicted density of N related nuclei. Nanoindentation showed some effects of precipitates on the matrix mechanical properties in contrast with dissolved O and N accumulated in the near surface.

Acknowledgement: The authors are indebted to Dr Deren Yang from Zhejiang University, China for the N doped CZ silicon and Dr Dehua Yang from Nanomechanics Research Laboratory, Hysitron Inc., Minneapolis, MN, for nanoindentation and nanoscratch tests. This work is supported by the National Renewable Energy Laboratory NREL Grant # AAT-2-31605-05.

References

1. A. Karoui, et al, *Dislocation Control in Silicon Wafers in a Susceptor Based Rapid Thermal Processing Tool* Proc. of *6th International Conference on Advanced Thermal Processing of Semiconductors*, Kyoto, Japan Sep.9-11,`98.
2. D. Yang, D. Que, and K. Sumino, Phys. Stat. Sol. (b) **210**, 295 (1998).
3. K. Nakai, et al, J. Appl. Phys., **89**, 4301 (2001).
4. P.J.Verlinden, A.W.Blakers, K.J. Weber, V. Everett, M. J. Kerr, M. F. Stuckings, D. Gordeev, and M. J Stocks, *Sliver Solar Cells: A New Thin Crystalline Silicon Photovoltaic Technology*, **PVSEC-14**, Jan. 2004.
5. R.B. Bergmann, *Crystalline Si Thin-Film Solar Cells: a Review*, Appl. Phys. A **69**, 187–194 (1999).
6. T. Abe, K. Kikuchi, S. Shira, and S. Muraoka, in "Semiconductor Silicon/1981", H. R. Huff, J. Kriegler and Y. Takeishi, eds., p.54, Softbound Symp. Serv. Electrochem. Soc., Pennington, New Jersey, 1981.
7. K. Sumino, I. Yonenaga, M. Imai and T. Abe, J. Appl. Phys. 54, 5016 (1983).
8. V.I.Orlov, Yu.L.Iunin, M.V.Badylevich, O.Lysytskiy, H.Richter, *Solid State Phenomena*, **95-96**, p 465 (2004)
9. Eremenko, V.G. A.V.Fedorov, *Materials Science Forum*, **196-201**, pt.3, 1219 (1995).
10. Zozime, A; Castaing, *Materials Science & Engineering B*, **42**(1-3), 57 (1996).
11. I. Yonenaga, *Diffusion and Defect Data Part B (Solid State Phenomena)*, **95-96**, 423(2004).
12. A. Karoui, G. A. Rozgonyi, "*Optimization of Silicon Crystal Growth and Wafer Processing for High Efficiency and High Mechanical Yield*", Ann. Rep.1 of Project: NREL, AAT-2-31605-05, Nov. 2002.
13. T. F. Cizek, et al. Solar Energy Materials and Solar Cells **41/42**, 61-70 (1996).
14. A. Karoui, F. Sahtout Karoui, G. Rozgonyi, and D. Yang, Solid State Phenomena, Vols. **82-84**, 69 (2002).
15. A. Karoui, F. Sahtout Karoui, A. Kvit, G. Rozgonyi, and D. Yang, Appl. Phys. Lett. **80**(12), 2114 (2002).
16. A. Kvit, A. Karoui, G. Dusher, and G. Rozgonyi, Appl. Phys. Lett. Appl. Phys. Lett. **84**(11), 1889 (2004)
17. F.Sahtout Karoui, A. Karoui, and G. Rozgonyi, MSM 2000, San Diego, March 27-29, p. 98, (2000).
18. A. Karoui, F. Sahtout Karoui, A. Kvit, G. Rozgonyi, and D. Yang, J. Appl. Phys. **80**(12), 2114 (2004).
19. "Northstar" is a 90 cube corner tip with a radius of 40-60nm.
20. K. L. Johnson, Contact Mechanics, Cambridge University Press, Cambridge, 1985.
21. B. Bhushan, and X. Li, J. Mater. Res., **12**(1), p. 54. (1997).
22. B. Bhushan, *Nanotribology and Nanomechanics of MEMS Devices*, IEEE Micro Electro Mechanical Systems Workshop,SanDiego, California, Feb. 1996, p.91-p.97 (1996).
23. J. Tersoff, Phys. Rev. B, **39**(8), 5566 (1989).
24. A. Karoui, *Atomistic Modeling of Impurity Atmospheres in Silicon and Dislocation Locking Effect*. Symposium (P), MRS Spring Meeting, San Francisco, Apr.12-16, 2004.
25. CRC Materials Science and Engineering Handbook, p.473.
26. A. Karoui, to be published.
27. A. George, in *Properties of Crystalline Silicon,* ed. By R. Hull (INSPEC, London, 1999), p.104.
28. A. Karoui, F.Sahtout Karoui, G.Rozgonyi, M. Hourai, K.Sueoka, J. Electrochem. Soc. **150**(12), G771 (2003

Mat. Res. Soc. Symp. Proc. Vol. 821 © 2004 Materials Research Society P3.17

Synthesis of Zeolite as Ordered Multi-Crystal Arrays using Uniformly Aligned Polyurethane as Templates

Jin Seok Lee, Yun-Jo Lee, Eunju Lee Tae, Yong Soo Park and Kyung Byung Yoon*
Department of Chemistry, Sogang University, Seoul 121-742, Korea

ABSTRACT

Two types of glass plates coated with uniformly aligned polyurethane films were produced by repeating the cycles of the alternative exposures of the glass plates to 1,4-phenylene diisocyanate (PDI) and 2-butyne-1,4-diol (BDO) and to PDI and terephthalic acid bis-(2-hydroxy ethyl) ester (TBE) for 500 times at elevated temperatures. The glass plates coated with uniformly aligned poly-$(PDI\text{-}BDO)_{500}$ (500 represents the number of the cycle) produced monolayers of closely packed 2D arrays of silicalite-1 crystals with the average size of 370 × 200 × 500 nm on the glass plates upon immersion of the glass plates into a dense gel consisting of tetraethylorthosilicate (TEOS), tetrapropylammonium hydroxide (TPA^+OH^-), and water (mole ratio = 7:1.5:330) followed by the hydrothermal reaction at 180 °C for 2 h. The silicalite-1 crystals were aligned with the c-axes perpendicular to the substrate plane. Upon switching the polymer from poly-$(PDI\text{-}BDO)_{500}$ to poly-$(PDI\text{-}TBE)_{500}$ the orientations of the silicalite-1 crystals in the 2D arrays changed from c to a axes perpendicular to the substrate. This report therefore demonstrates that the uniformly aligned polyurethane films serve as the templates for the growth of closely packed multi-crystal arrays of silicalite-1 in uniform orientations and the nature of the polyurethane film affects the resulting orientations of the crystals. We propose that the supramolecularly organized organic-inorganic composites consisting of the hydrolyzed organic products and the seed crystals responsible for the above phenomena.

INTRODUCTION

Great efforts have been directed at applications of zeolites as advanced materials [1] such as imaging and data storage media [2], light energy harvesting devices [3-5], electrode modifiers [6], and hosts for various materials such as semiconductor quantum dots [7], molecular wires [8], chemical sensors [9], luminescence materials [10], nonlinear optical molecules [11,12], and lasing dyes [13]. One of the essential elements to make such efforts fruitful is to acquire the abilities to organize nano- and microcrystalline zeolite hosts into uniformly aligned and closely packed two- (2D) and three-dimensional (3D) multi-crystal arrays on substrates. The uniformly aligned and closely packed monolayers of microcrystals also bear a great potential to be utilized as excellent seeds for the preparation of uniformly aligned continuous zeolite films, which could be employed as molecular sieve membranes [14]. However, despite a large progress made during the last several decades in the synthesis of new zeolites and the related molecular sieves [15], there have been no synthetic methods for production of uniformly aligned zeolite multi-crystal arrays. We have explored a possible post synthetic route to the production of ordered multi-crystal arrays of zeolite on substrates through organization of preformed zeolite crystals with well-defined chemical linkages [16-27]. Although the above route has shown great potential, it has not yet shown the ability to control the orientation of zeolite crystals with respect to the substrate plane. Inspired by biomineralization [28-31] and the phenomenon that polyurethane sponges serve as templates for facile synthesis of large, self-supporting silicalite foams [32], we have recently demonstrated that the polyurethane films with uniformly aligned backbones can serve as the matrices for production of uniformly aligned 2D and possibly 3D arrays of zeolite crystals with uniform orientations and the nature of the polymer determines the orientation [33].

EXPERIMENTAL DETAILS

Materials

Aqueous TPA^+OH^- solutions were prepared from the corresponding solutions of tetrapropylammonium bromide (ACROS) by anion exchange of bromide with hydroxide using a column charged with anion exchange resin (DOWEX MSA-1, Aldrich) pre-treated with sodium hydroxide. The concentration of TPA^+OH^- was adjusted by diluting the as-prepared solution with water. Tetraethylorthosilicate (TEOS) was purchased from Aldrich. The key components for the growth of uniformly aligned polyurethane films were 1,4-phenylenediisocyanate (PDI), 2-butyne-1,4-diol (BDO) and terephthalic acid bis-(2-hydroxy ethyl) ester (TBE). PDI and BDO were purchased from Aldrich and used as received. TBE was synthesized via terephthaloyl chloride from terephthalic acid. Thus, a mixture of terephthalic acid (7 g, 42.1 mmol) and 20 mL of thionyl chloride was refluxed overnight. The solution was cooled to room temperature and the mixture was evaporated to yield terephthaloyl chloride as a white powder (8.3 g). A dichloromethane solution (20 ml) dissolved with terephthaloyl chloride (8.3 g, 40.9 mmol) was added dropwise to a mixture of ethylene glycol (14.1 mL, 250 mmol) and pyridine (200 mL) which was being stirred at 0 °C. After additional stirring for 12 h at room temperature pyridine was evaporated from the reaction mixture, and the residue was pored into water (200 mL). The solid was filtered, and the filtrate was extracted with ethyl acetate. The organic phase was dried, evaporated almost, and filtered to afford 8.0 g of trephtalic acid bis-(2-hydroxy ethyl) ester as a white powder. Yield. 75 %. H^1 NMR (300 MHz, $CDCl_3$) 8.19(4H, s), 4.45(4H, t, J = 5.0 Hz), 4.17(2H, t, J = 6.0 Hz), 3.92(4H, dt, J = 3.9, 5.7 Hz) [34]. Cover glasses (18 x 18 mm^2, Marienfeld) were cleaned by dipping into a piranha solution (H_2SO_4:H_2O_2 = 7:3) for 1 h. The glasses were then washed with copious amounts of distilled deionized water, with dilute aqueous ammonia solution, and finally with fresh water. The washed glasses were dried by a gentle stream of nitrogen. The water contact angles of the glass plates were usually less than 5°. Silicon wafers were cleaned similarly.

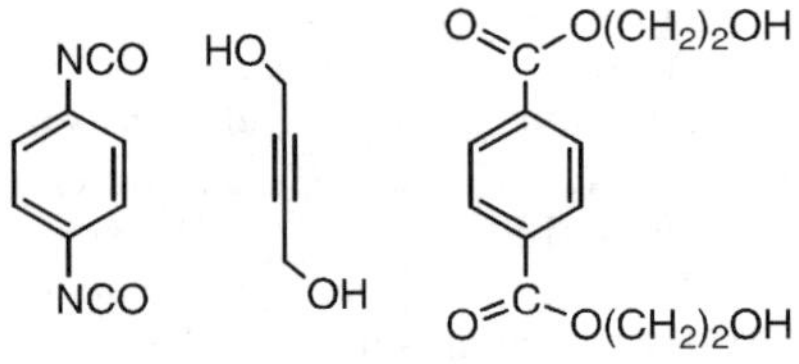

NCO HO $O(CH_2)_2OH$ NCO OH $O(CH_2)_2OH$

PDI BDO TBE

Preparation of $(PDI/BDO)_x$/Si, $(PDI/BDO)_x$/G, $(PDI/TBE)_x$/Si, and $(PDI/TBE)_x$/G (x = 100, 500, and 1000)

$(PDI/BDO)_x$/Si and $(PDI/BDO)_x$/G were prepared by repeating the desired number of cycles of alternative introduction of the PDI and BDO vapors into a vacuum-tight stainless steel chamber containing a Teflon support mounting 20 pieces of silicon wafers or glass plates. The stainless steel chamber was connected to vacuum (10^{-4} Torr) and to two glass flasks containing PDI and BDO, respectively, through individual solenoid valves. The chamber was placed in an oven whose temperature was maintained at 120-125 °C. The chamber was initially kept under vacuum and each vapor was sequentially fed into the chamber by opening the corresponding solenoid valve for 60 s while keeping the other two valves closed. After each introduction of a vapor into the chamber, all the three valves were kept closed for 60 s to allow reaction between the introduced vapor and the glass or polymer film already produced on the glass. After each reaction period, the non-reacted, excess vapor was removed from the chamber by opening the vacuum valve for 120 s. $(PDI/TBE)_x/Si$ and $(PDI/TBE)_x/G$ were similarly prepared by replacing BDO with TBE. The oven temperature for preparation of $(PDI/TBE)_x/G$ was 180 °C. The period of PDI introduction into the chamber was reduced to 0.2 s at 180 °C. The pressure of each vaper inside the reaction chamber was ~4.7 Torr.

Growth of Multicrystal Arrays on glass

A piece of $[(PDI/BDO)_{500}/G]$ was placed at the bottom of a Teflon cylinder (internal diameter = 4.2 cm, height = 7.6 cm) placed in a stainless steel autoclave. The cylinder was subsequently charged with 40 mL of a synthesis gel consisting of TEOS, TPA^+OH^-, and water (in the molar ratio of 7:1.5:330), and hydrothermal synthesis of silicalite-1 was carried out at 180 °C for 2 h. After rapid cooling of the reactor to room temperature, the glass plates were removed from the reactor, and gently shook in fresh water. The washed glass plate was then placed in an oven preheated at 120 °C for 30 min. The glass plate was then cooled to room temperature, and briefly sonicated for ~10 s in water to remove non-reacted gel particles and random aggregates of zeolite crystals deposited from the bulk on the multicrystal arrays.

Characterization

FT-IR spectra were measured on a Jasco FT-IR 620 spectrometer. The polarized grazing angle FT-IR spectra were abtained from the above spectrometer using a Spectra-Tech grazing angle accessory (model 0014-051) equipped with a plane polarizer. The scanning electron microscope (SEM) images were measured on a Hitachi S-4300 FE-SEM. EDX data were obtained from a Horiba EMAX ENERGY EX-220. X-ray diffraction (XRD) patterns were measured on a Rigaku diffractometer D/MAX-1C with Cu K_α radiation by $2\theta/\theta$ scanning with the speed of $1^\circ\,min^{-1}$.

RESULTS AND DISCUSSIONS

The diffuse reflectance Fourier-Transform infrared (FT-IR) spectra of the polymer films grown on silicon wafers [$(PDI/BDO)_x/Si$, x = 100, 500, or 1,000] (figure 1A) showed characteristic N-H, C-H (aliphatic), C=O, and C-O stretching bands of polyurethane at 3318, 2959, 1735, and 1107 cm^{-1}, respectively [35]. The intensities of the stretching bands linearly increased as shown in the inset of figure 1A for N-H and C=O bands, indicating that the thickness of the polymer increases with increasing the number of the cycle. Consistent with the FT-IR data, the thicknesses of the cross sections measured by SEM images were ~95, ~360, and ~610 nm.

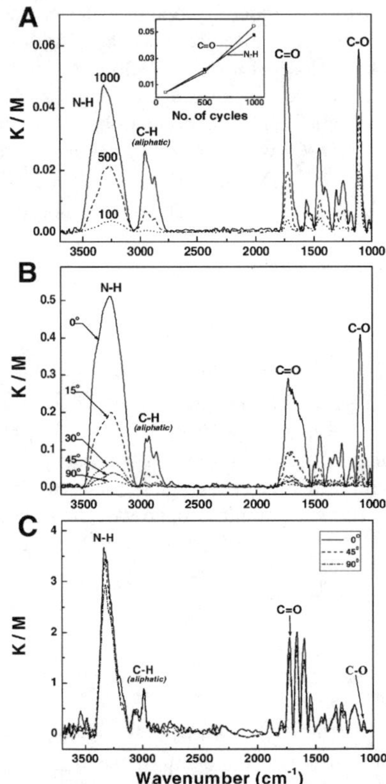

Figure 1. Diffuse reflectance FT-IR spectra of PDI/BDO polyurethane films grown on silicon wafer $(PDI/BDO)_x/Si$ (x = 100, 500, and 1,000) by repeating the cycle of alternate exposures of the substrate into the vapors of PDI and BDO (A), the plane-polarized grazing angle (80° with respect to surface normal) FT-IR spectra of $(PDI/BDO)_{500}/Si$ (B), and the plane-polarized grazing angle of randomly oriented PDI-BDO polymer fibers supported on a silicon wafer, which was prepared by spin-coating the wafer with the slurry of randomly oriented PDI-BDO polymer fibers in ethyl acetate (C).

The plane-polarized grazing angle FT-IR spectra further revealed that the intensities of N-H, C-H (aliphatic), C=O, and C-O stretching bands decrease as the dihedral angle between the plane of polarization and the silicon wafer increases from 0° (p-polarization) to 90° (s-polarization) (figure 1B), indicating that N-H, C-H (aliphatic), C=O, and C-O bonds are oriented parallel to the substrate plane. For comparison, we also prepared silicon wafers coated with randomly oriented poly-(PDI-BDO) fibers [r-(PDI-BDO)/Si] by spin-coating the wafers with the slurry of poly-(PDI-BDO) fibers in ethyl acetate. In contrast to $(PDI/BDO)_x/Si$, the FT-IR intensities of r-(PDI-BDO)/Si remained nearly invariant regardless of the dihedral angle (figure 1C). Furthermore, the stretching bands of the horizontally oriented N-H and C=O bonds from $(PDI/BDO)_x/Si$ were about two times broader [full width at half maximum (fwhm), N-H; 285 and C=O; 78 cm^{-1}] than those of r-PDI-BDO/Si (fwhm, N-H; 124 and C=O; 38 cm^{-1}) (figure 2), indicating that hydrogen bonding occurs between the H atom in N-H and the O atom in C=O in the case of the stepwise grown polyurethane films. It is therefore concluded that most of the PDI/BDO polymer chains in $(PDI/BDO)_x/Si$ were uniformly aligned on substrates in such a manner depicted in figure 3.

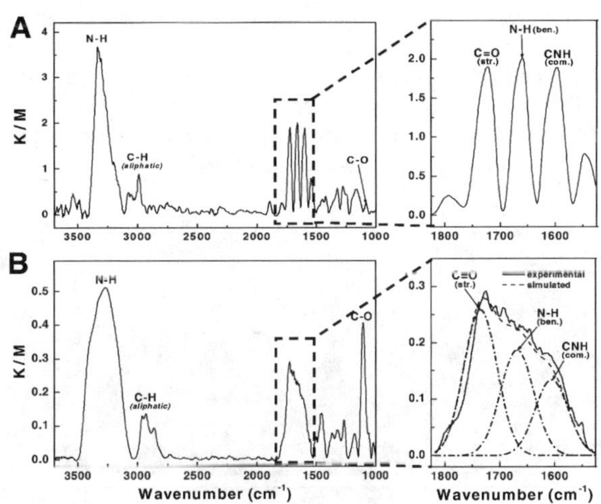

Figure 2. Comparison of the bandwidths of the carbonyl stretching bands in the plane-polarized grazing-angle (80°) FT-IR spectra from the randomly oriented (A) and the stepwise grown (B) PDI-BDO polyurethane polymers supported on silicon wafers. The broad band in the region of 1500-1800 cm^{-1} in (B) was deconvoluted into three bands as indicated.

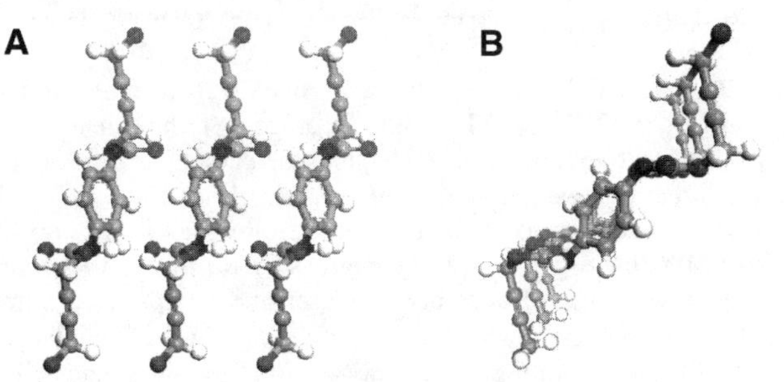

Figure 3. Proposed structure and interchain interaction of PDI/BDO polyurethane polymer on silicon wafer and glass, which were created based on the structure and the packing of the polymer obtained by energy minimization using the Forcite module from Accelrys and the FT-IR data shown in Fig. 1. (A) depicts the hydrogen bonding between the H atom on N and the O atom in C=O unit and (B) shows the horizontal orientation of N-H, C-H, C=O, and C-O bonds.

As the silicon wafers rapidly corroded in our highly basic gels, we instead immersed glass plates similarly coated with uniformly aligned poly-(PDI-BDO) [$(PDI/BDO)_{500}$/G] into a dense gel, and the hydrothermal reaction was carried out (see Experimental Details). The SEM images of the glass plates revealed that they were indeed covered with closely packed 2D arrays of silicalite-1 crystals with the average size of 370 × 200 × 500 nm (figure 4A). Interestingly, the SEM images of the cross sections (figure 4B) revealed that a great majority of the silicalite-1 crystals are aligned with the c-axes perpendicular to the substrate plane despite the fact that the crystals are narrow and round along the c-axes. Accordingly, they are closely packed with the flat (1 0 0) and (0 1 0) faces contacting each other. In strong contrast, all the silicalite-1 crystals grown on the bare glass plates were oriented with the b-axes perpendicular to the substrate (figure 4C) despite the fact that they were in the same gel.

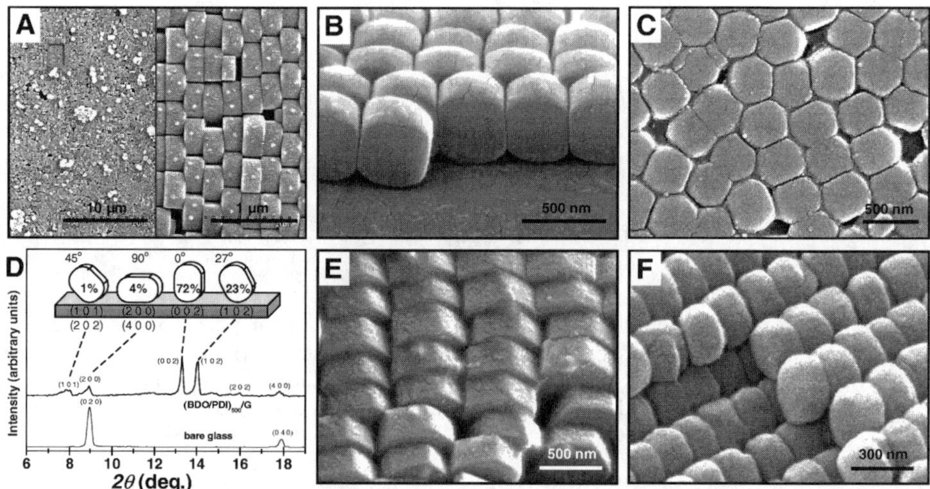

Figure 4. SEM images showing a 2D array of closely packed, uniformly aligned silicalite-1 crystals grown on $(PDI/BDO)_{500}$/G in high coverage (A), the orientation of the crystals with their c-axes perpendicular to the glass plane (B), and the crystals grown on a bare glass plate from the same gel with the b-axes perpendicular to the glass plane (C). The corresponding x-ray diffraction patterns are compared in (D) and the drawings in the top of the panel indicates the orientations of silicalite-1 crystals, their degrees of tilting, and the corresponding percentages. (E) shows the SEM image of an array of larger and sharp-edged silicalite-1 crystals obtained from the gels with the composition of $TEOS:TPA^{+}OH^{-}:H_2O$ = 7:1:330. (F) shows the SEM image of a double layer multi-crystal array of smaller and round silicalite-1 crystals obtained from a gel with the composition of 4:2:330. The zeolite crystals in (E) and (F) are also aligned with the c-axes perpendicular to the glass plane.

The multi-crystal arrays gave the x-ray diffraction peaks at 8.00°, 8.90°, 13.30°, 14.00°, 16.00°, and 17.75° in the 2θ region between 6° and 19°, with the relative intensities of 14, 25,

100, 83, 8, and 12, respectively. The above diffraction peaks are assigned to the reflections arising from (1 0 1), (2 0 0), (0 0 2), (1 0 2), (2 0 2), and (4 0 0) planes, respectively, that are oriented perpendicular to the glass plane [36]. The corresponding crystal orientations are illustrated in figure 4D. To compare the above result with the relative reflection intensities of the randomly oriented powders we simulated relative intensities of the (1 0 1), (2 0 0), (0 0 2), and (1 0 2) reflections of the randomly oriented powders by taking the reflection multiplicities and structure factors into account. They are 68.4, 31.3, 6.8, and 17.4, respectively. By comparing the two sets of data, the calculated contribution of each orientation was 1, 4, 72, and 23%, respectively (figure 4D). Thus, 72% of the crystals were oriented with the c-axis perpendicular to the plane, and the remaining 28% were tilted by 27° (23%), 45° (1%), and 90° (4%) from the upright orientation. We found that the tilting of the crystals occurs during the removal of the sedimented, randomly aggregated crystals from the underlying multi-crystal arrays.

The fact that most (95%) of the crystals are oriented with the c-axes either perpendicular to the glass plane or slightly (by 27°) tilted from the upright orientation demonstrates that the $(PDI/BDO)_{300}$ serves as the matrix for the growth of ordered multi-crystal arrays of silicalite-1 that are uniformly oriented with the c-axes perpendicular to the substrate plane. It is noteworthy that the corresponding x-ray diffraction peaks recorded from the silicalite-1 crystals grown on bare glass plates were 9.00° and 17.95° within the same 2θ region (figure 4D, bottom), consistent with their b-axes being oriented perpendicular to the substrate planes.

The crystal shapes and sizes readily varied by slightly changing the relative amount of TPA^+OH^- in the gel. For instance, larger (550 × 200 × 900 nm) crystals with sharp edges appeared upon slightly decreasing the basicity of the gel (7:1:330) (figure 4E). The sharp-edged crystals also oriented with the c-axes perpendicular to the plane despite the fact that such an orientation leads to minimal contacts between the crystals and the substrates. On the other hand, the crystals became smaller and round (250 × 170 × 300 nm) (figure 4F), upon slightly increasing the basicity of the gel (4:2:330). The x-ray diffraction pattern also revealed that most of the round crystals were aligned with the c-axes perpendicular to the substrate plane. Interestingly, there were some areas covered with the second layers of silicalite-1 crystals in the same orientation as shown in figure 4F, indicating that production of even 3D arrays of uniformly aligned silicalite-1 crystals is also possible by optimization of the condition. It appeared to us that the second layers more easily fell off the first layer than did the first layer from the glass substrate during our experimental procedure. Control experiments showed that at least 300 layers of PDI/BDO unit are necessary for the production of self-assembled monolayers of c-oriented multicrystal arrays and more than 500 layers of PDI/BDO unit are necessary for the production of self-assembled multilayers of c-oriented multicrystal arrays. It should also be noted that the glass plates coated with randomly oriented poly-(PDI-BDO) fibers gave only randomly aggregated crystals (figure 5), emphasizing that the uniform orientation of polyurethane film is crucial for growth of ordered multi-crystal arrays.

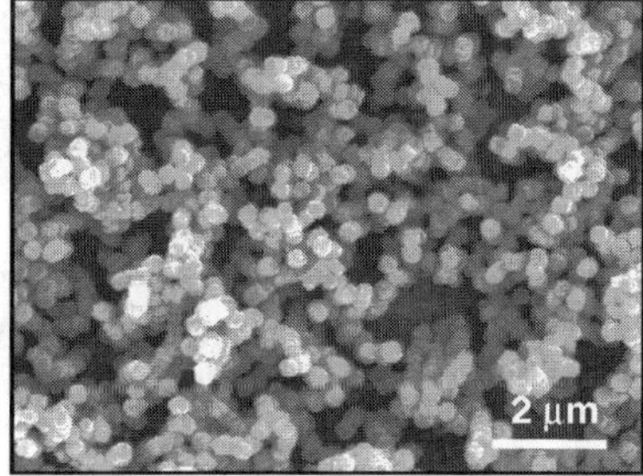

Figure 5. SEM image of randomly aggregated silicalite-1 crystals grown from the randomly oriented PDI/BDO polyurethane fibers supported on glass.

The plane-polarized grazing-angle FT-IR spectra of the $(PDI/TBE)_{500}$ polyurethane films grown on silicon wafers $[(PDI/TBE)_{500}/Si]$ (figure 6) also showed that they are uniformly aligned. The glass plates coated with $(PDI/TBE)_{500}$ $[(PDI/TBE)_{500}/G]$ gave rise to a remarkable increase in size of the 'perfect domain' where all the three principal axes of each crystal were uniformly aligned as shown in figure 7A. Interestingly, the x-ray diffraction peaks of the arrays appeared at 8.90°, 17.75°, 26.70°, 36.0°, and 45.30° while the silicalite-1 crystals grown on bare glass gave at 9.00°, 17.95°, 27.00°, 36.2°, and 45.65° as compared in figure 7B, demonstrating that all the crystals were aligned with the a-axes perpendicular to the glass plate. Thus, the above result shows that the orientation of the crystals in the multi-crystal arrays can also be changed by varying the nature of the polymers.

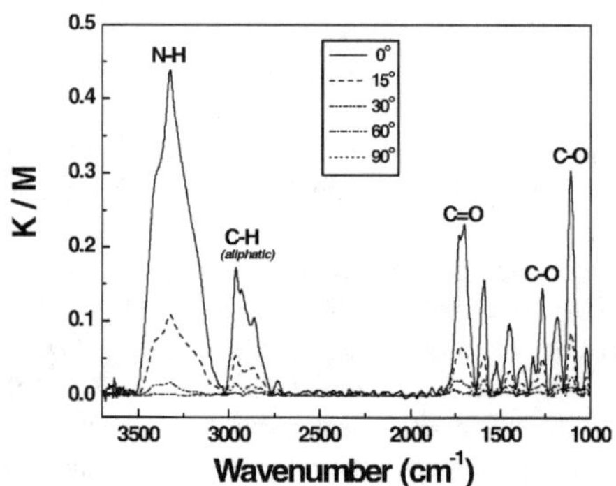

Figure 6. Diffuse-reflectance plane-polarized grazing-angle (80°) FT-IR spectra of $(PDI/TBE)_{500}/Si$ at different dihedral angles between the plane of polarization and the substrate plane. The angle dependence shows that the polymer backbones are uniformly aligned.

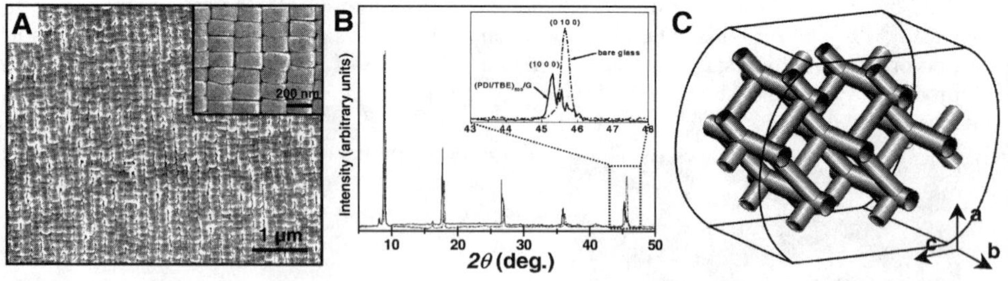

Figure 7. SEM images showing a 2D array of silicalite-1 crystals grown on $(PDI/TBE)_{500}/G$ (A) and the corresponding x-ray diffraction pattern showing that the crystals are aligned with the a-axes perpendicular to glass plane (B). (C) shows the orientations of the straight and sinusoidal channels in a silicalite-1 crystal.

The polyurethane thin films completely disappeared from the glass plates after reaction for 2 h. The films also readily disappeared even in an aqueous, TEOS-free TPA^+OH^- solution whose TPA^+OH^- concentration was adjusted to be the same with the synthesis gel after keeping at 180 °C for 2 h. It was also found that the addition of TPA^+ ion into the synthesis gel was essential for the formation of the silicalite-1 arrays. Dilution of the gel (7:6.6:3300) was detrimental to the formation of multi-crystal arrays, indicating that the gel should be dense and viscous. Based on the above results, and by taking into account the finding of Martens and co-workers that the synthesis gel readily produces TPA^+-incorporating nanoslabs (seeds) during a few hours of aging even at room temperature [37,38], we propose that the following series of events occur sequentially.

The uniformly aligned polyurethane chains are first hydrolyzed by TPA^+OH^- at the reaction temperature. The hydrolyzed products (such as phenylenediamine and BDO or phenylenediamine, terephthalic acid, and ethylene glycol) retain uniform alignment under the dense and viscous synthesis gel. The TPA^+-incorporating nanoslabs in the synthesis gel penetrate into the uniformly aligned organic layer with either the a- or c-axes pointing perpendicular to the substrate depending on the nature of hydrolyzed products as illustrated in figure 8. The nanoslabs and the surrounding aligned organic products form an ordered supramolecularly organized organic-inorganic composite. The incarcerated nanoslabs undergo cross linking between them accompanying slow release of the hydrolyzed organic fragments into the mother liquor leading to growth of ordered silicalite-1 multi-crystal arrays with specific orientations.

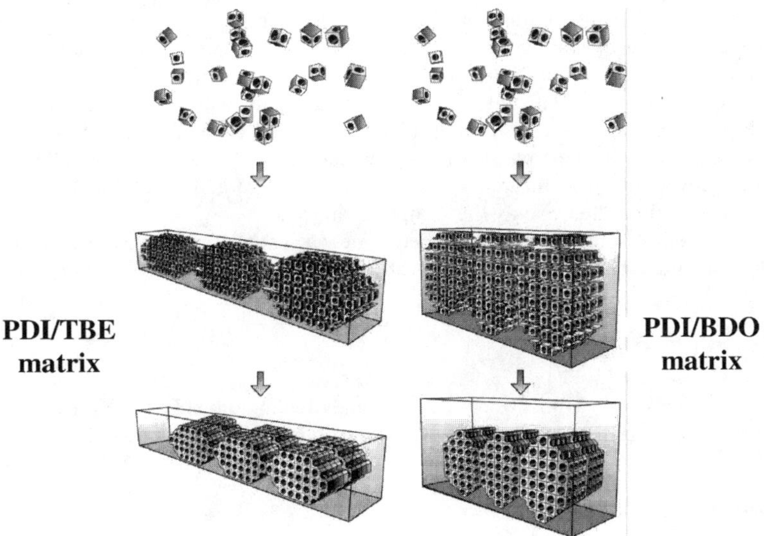

Figure 8. Schematic representations showing the mechanism of the formation of 2D arrays of silicalite crystals with uniform orienations depending on the nature of the hydrolyzed organic fragments from the uniformly aligned polyurethane films by forming supramolecular assemblies of organic fragments-silicalite nanoslab composites.

In support of the above formulation, we observed that the polymer layers transformed into much thicker films (by ~10 times) consisting of carbon, oxygen, and silicon in the average atomic ratio of 42:38:20 after reaction for 1 h 20 min, indicating that the layers are indeed composed of hydrolyzed polyurethane fragments and silicon species (nanoslabs).

It has been well known that the organic species in the synthesis gels (such as TPA^+) serve as 'structure directors' for creation of nanopores in certain shapes, sizes, and networks in zeolites and the related molecular sieves [37,38]. Based on our results we also propose that the organic species incorporated within the synthesis gels can also play the role of 'orientation directors' for the resulting crystals by formation of supramolecularly organized organic-inorganic composites. The reason for each glass plate was covered with small perfect domains rather than with a single perfect domain may be ascribed to the imperfect alignment of the polymer chains arising from the unevenness of glass surfaces, defect formation during sequential exposures to the component vapors, and evolution of CO_2 during hydrolysis of polyurethane film.

Nevertheless, it is certain that we now can orient silicalite-1 crystals on glass with any principal axis in the upright orientation. In fact, the growth of silicalite-1 and the related ZSM-5 crystals on glass with their b-axes perpendicular to the substrate planes has also been observed by others [39]. Caro et al. showed that electric field (2 $kVcm^{-1}$) can align very long ZSM-5 crystals (> 200 μm in the c-direction) with the c-axes perpendicular to the substrate plane [40]. However, the above physical alignment method was not effective for small crystals and in particular for aluminum-free silicalite crystals. Furthermore, orientation of either ZSM-5 or silicalite-1 crystals with their a-axes perpendicular to the substrate planes has never been achieved by any means. Bein and the co-workers showed oriented growth of zincophosphate and $AlPO_4$-5 crystals on the self-assembled phosphate monolayers on gold [41,42]. However, the monolayer-directed crystallization leads to the formation of randomly scattered crystals rather than closely packed multi-crystal arrays.

The orientation of silicalite-1 crystals on substrates with the a- or c-axes perpendicular to the substrate planes also means orientation of the straight channels parallel to the substrate plane and the sinusoidal channels parallel or perpendicular to the substrate plane. This aspect will be useful for novel optical applications of the arrays after dye-inclusion [43,44], because molecules longer than ~1 nm can enter only the straight channels.

We believe our finding will open a new avenue of ordered 2D and 3D multi-crystal arrays of zeolite, which will trigger application of zeolites for innovative materials and devices.

ACKNOWLEDGMENTS

We thank the Ministry of Science and Technology of the Korean Government for supporting this work through the Creative Research Initiatives programs.

REFERENCES

1. G. A. Ozin, A. Kuperman, A. Stein, *Angew. Chem. Int. Ed. Engl. Adv. Mater.* **28**, 359 (1989).
2. G. A. Ozin, A. Stein, G. D. Stucky, J. P. Godber, *J. Inclusion Phenom.* **6**, 379 (1990).
3. M. Borja, P. K. Dutta, *Nature* **362**, 43 (1993).
4. M. Sykora, J. R. Kincaid, *Nature* **387**, 162 (1997).

5. Y. I. Kim, S. W. Keller, J. S. Krueger, E. H. Yonemoto, G. B. Saupe, T. E. Mallouk, *J. Phys. Chem. B* **101**, 2491 (1997).
6. D. R. Rolison, C. A. Bessel, *Acc. Chem. Res.* **33**, 737 (2000).
7. N. Herron, Y. Wang, M. M. Eddy, G. D. Stucky, D. E. Cox, K. Moller, T. Bein, *J. Am. Chem. Soc.* **111**, 530 (1989).
8. T. Bein, P. Enzel, *Angew. Chem. Int. Ed. Engl.* **12**, 1737 (1989).
9. G. Grubert, M. Stockenhuber, O. P. Tkachenko, M. Wark, *Chem. Mater.* **14**, 2458 (2002).
10. G. Calzaferri, M. Pauchard, H. Maas, S. Huber, A. Khatyr, T. Schaafsma, *J. Mater. Chem.* **12**, 1 (2002).
11. S. D. Cox, T. E. Gier, G. D. Stucky, J. Bierlein, *J. Am. Chem. Soc.* **110**, 2986 (1988).
12. H. S. Kim, S. M. Lee, K. Ha, C. S. Jung, Y.-J. Lee, Y. S, Chun, D. S, Kim, B. K. Rhee, K. B. Yoon, *J. Am. Chem. Soc.* **126**, 673 (2004).
13. U. Vietze, O. Krauß, F. Laeri, G. Ihlein, F. Schüth, B. Limburg, M. Abraham, *Phys. Rev. Lett.* **81**, 4628 (1998).
14. Z. Lai, G. Bonilla, I. Diaz, J. G. Nery, K. Sujaoti, M. A. Amat, E. Kokkoli, O. Terasaki, R. W. Thompson, M. Tsapatsis, D. G. Vlachos, *Science* **300**, 456 (2003).
15. M. E. Davis, *Nature* **417**, 813 (2002).
16. A. Kulak, Y.-J. Lee, Y. S. Park, K. B. Yoon, *Angew. Chem. Int. Ed.* **39**, 950 (2000).
17. S. Y. Choi, Y.-J. Lee, Y. S. Park, K. Ha, K. B. Yoon, *J. Am. Chem. Soc.* **122**, 5201 (2000).
18. G. S. Lee, Y.-J. Lee, K. Ha, K. B. Yoon, *Tetrahedron* **56**, 6965 (2000).
19. G. S. Lee, Y.-J. Lee, K. B. Yoon, *J. Am. Chem. Soc.* **123**, 9769 (2001).
20. Y. S. Chun, K. Ha, Y.-J. Lee, J. S. Lee, H. S. Kim, Y. S. Park, K. B. Yoon, *Chem. Comm.* **17**, 1846 (2002).
21. K. Ha, Y.-J. Lee, H. J. Lee, K. B. Yoon, *Adv. Mater.* **12**, 1114 (2000).
22. K. Ha, Y.-J. Lee, D.-Y. Jung, J. H. Lee, K .B. Yoon, *Adv. Mater.* **12**, 1614 (2000).
23. K. Ha, Y.-J. Lee, Y. S. Chun, Y. S. Park, G. S. Lee, K. B. Yoon, *Adv. Mater.* **13**, 594 (2001).
24. J. S. Park, G. S. Lee, Y.-J. Lee, Y. S. Park, K. B. Yoon, *J. Am. Chem. Soc.* **124**, 13366 (2002).
25. J. S. Park, Y.-J. Lee, K. B. Yoon, *J. Am. Chem. Soc.* **126**, 1934 (2004).
26. A. Kulak, Y. S. Park, Y.-J. Lee, Y. S. Chun, K. Ha, and K. B. Yoon, *J. Am. Chem. Soc.* **122**, 9308 (2000).
27. G. S. Lee, Y.-J. Lee, K. Ha, K. B. Yoon, *Adv. Mater.* **13**, 1491 (2001).
28. S. Mann, *Nature* **365**, 499 (1993).
29. E. Dujardin, S. Mann, *Adv. Eng. Mater.* **4**, 461 (2002).
30. L. Addadi, S. Weiner, *Angew. Chem. Int. Ed. Engl.* **31**, 153 (1992).
31. S. Weiner, L. Addadi, *J. Mater. Chem.* **7**, 689 (1997).
32. Y.-J. Lee, J. S. Lee, Y. S. Park, K. B. Yoon, *Adv. Mater.* **13**, 1259 (2001).
33. J. S. Lee, Y.-J. Lee, E. L. Tae, Y. S. Park, K. B. Yoon, *Science* **301**, 818 (2003).
34. M. Yamaguchi, T. Takata, T. Endo, *J. Org. Chem.* **55**, 1490 (1990).
35. D. L. Pavia, G. M. Lampman, G. S. Kriz, *Introduction to Spectroscopy* (Saunders College Publishing, Orlando, 1996).
36. M. M. J. Treacy, J. B. Higgins, *Collection of Simulated XRD Powder Patterns for Zeolites* (Elsevier, Amsterdam, 2001).
37 C. E. A. Kirschhock, R. Ravishankar, P. A. Jacobs, J. A. Martens, *J. Phys. Chem.* **103**, 11021 (1999).

38. C. E. A. Kirschhock, S. P. B. Kremer, P. J. Grobet, P. A. Jacobs, J. A. Martens, *J. Phys. Chem.* **106**, 4897 (2002).
39. J. H. Koegler, H. van Bekkum, J. C. Jansen, *Zeolites* **19**, 262 (1997).
40. J. Caro *et al., Adv. Mater.* **4**, 273 (1992).
41. S. Feng, T. Bein, *Nature* **368**, 834 (1994).
42. S. Feng, T. Bein, *Science* **265**, 1839 (1994).
43. G. Calzaferri *et al., J. Mater. Chem.* **12**, 1 (2002).
44. U. Vietze *et al., Phys. Rev. Lett.* **81,** 4628 (1998).

Part IV: Internal Stress and Strengthening Mechanisms

A. Experiments

Mat. Res. Soc. Symp. Proc. Vol. 821 © 2004 Materials Research Society P1.6

A New Synchrotron-based Technique for Measuring Stresses in Ultrathin Metallic Films

Jochen Böhm[1], Patric Gruber[1], Ralph Spolenak[2], Alexander Wanner[3], and Eduard Arzt[1,2]
[1]Institut für Metallkunde, Universität Stuttgart, Heisenbergstr. 3, D-70569 Stuttgart, Germany
[2]Max Planck Institute for Metals Research, Heisenbergstr. 3, D-70569 Stuttgart, Germany
[3]Institut für Werkstoffkunde I, Universität Karlsruhe (TH), D-76128 Karlsruhe, Germany

ABSTRACT

A novel synchrotron-based X-ray diffraction technique is presented by which it is possible to characterize the evolution of stresses in polycrystalline metallic films thinner than 100 nm. The film under investigation is deposited on a flexible polyimide substrate which is subjected to a uniaxial tensile test. The evolutions of longitudinal and transverse elastic strains in the film are monitored simultaneously by means of a high-resolution area detector. The strain resolution is better than 10^{-4}. The samples are typically subjected to 6 % total strain and subsequently unloaded. First experiments carried out on Au films in the thickness range between 20 nm and 1000 nm show the usefulness and the power of this new technique.

INTRODUCTION

Most investigations of size effects on the mechanical properties of thin metallic films are based on wafer-curvature experiments involving thermal cycling. In order to better understand the fundamental mechanisms, it is favorable to decouple the changes in temperature and stress. A well-established isothermal technique to characterize the flow behavior of a crystalline thin film is to perform *in situ* tensile tests in an X-ray diffractometer [1]. The film of interest is deposited on a dog bone-shaped substrate. Film and substrate are strained simultaneously while the stress in the thin film is monitored by X-ray measurements. Polyimide has been shown to serve well as substrate material [2,3], because this amorphous polymer is compliant, can be deformed homogeneously up to very large strains (>5%), and is available with excellent surface quality. The standard X-ray technique applied so far to measure the stress evolution in the metal film is the "$\sin^2\psi$ method", an adaptation of which to thin films has been described extensively elsewhere [4-6]. This X-ray *in situ* tensile testing technique has been shown to work well for metal films with thickness in the micrometer range. Serious experimental problems arise, however, for films considerably thinner than 1 μm: The X-ray counting times required per data point increase to an impractical level as the gauge volume is decreased and the signal-to-noise ratio of the diffraction peaks becomes unfavorable, given that the amorphous substrate yields a considerable scattering background. Materials characterization by X-ray techniques is currently being revolutionized by the combined availability of three recently developed components: (1) 3rd generation synchrotron sources providing high-flux, tunable X-rays, (2) large-area X-ray detectors exhibiting low noise, high sensitivity and spatial resolution, and (3) storage and processing capabilities for large amounts of data created by such detectors. We have developed an improved experimental procedure for *in situ* X-ray tensile testing of thin crystalline films, taking advantage of all three developments. For in-depth treatment of the theoretical background and a more detailed description of the experimental procedures we refer to a more extensive methodology paper published recently [7]. In the present paper, we will outline the method and present first results on sputter-deposited gold (Au) films in the thickness range from 20 nm to 1000 nm.

EXPERIMENTAL DETAILS

Specimen preparation: Dog bone-shaped tensile specimens with gauge section 20 x 6 mm^2 were cut out of 125 μm thick polyimide foil. The specimen surface was cleaned on one side by argon ion bombardment and a polycrystalline gold (Au) film was deposited under ultrahigh vacuum (10^{-9} mbar) conditions by magnetron sputtering. The film thickness was varied from specimen to specimen. Fine tungsten (W) powder was dispersed in vacuum grease, a thin layer of which was applied on the reverse side of the polyimide substrate as a calibration substance for the X-ray measurements.

The typical microstructure of the polycrystalline Au is shown in Fig. 1a for a 500 nm thick film. The grains exhibit a columnar structure, the grain diameter is considerably smaller than the film thickness, and a pronounced <111> fiber texture is observed (Fig. 1b).

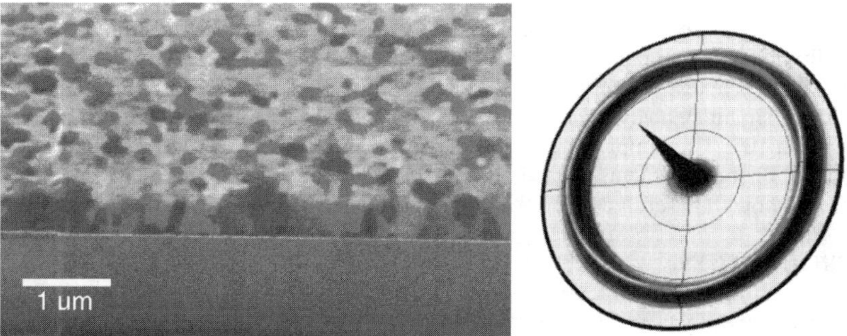

Fig. 1. (a) Focused ion beam image of a 500 nm thick gold film on a 125 μm thick polyimide substrate. The Au film was not annealed after sputter-deposition and exhibits a columnar grain structure. (b) (111) pole figure of a 160 nm thick Au film deposited in the same way, showing a strong <111> fiber texture.

Tensile testing and X-ray diffraction: *In situ* tensile tests were performed at the Max Planck Institute Surface Diffraction Beamline located at the synchrotron radiation source ANKA (Karlsruhe, Germany) where monochromatic X-rays tunable in the wavelength range from 0.6 to 2.1 Å (corresponding to photon energies ranging from 20 to 6 keV) are available. A miniature, screw-driven microtensile tester (Kammrath & Weiss GmbH, Dortmund, Germany) was mounted on the sample stage of a 2+3 circle diffractometer. Diffraction patterns were recorded using a CCD area detector, the X-ray sensitive screen of which has a diameter of 165 mm (2048 x 2048 pixels). The spot size was typically 1 x 1 mm^2, the exposure time was 30 s, and the time required for readout and storage was 3 s per image. Two different detector positions were used for (a) characterizing the initial residual stress produced in the thin film during fabrication and for (b) monitoring the stress evolution in the film during deformation (plane stress):

(a) The equi-biaxial residual stress existing in the as-deposited state was characterized by measuring the interplanar spacing of (111)-planes exhibiting different inclination angles with respect to the film plane ($\sin^2\psi$ method). Due to the texture, strong (111)-peak intensities can be found only at inclination angles near $\psi_1 = 0°$ and $\psi_2 = 70.53°$ (Fig. 2a). In order to eliminate effects of unintentional specimen movement it is essential to detect the Au (111) ring as well as at least two rings of the calibration substance (W) with a single CCD frame. The tunability of the X-ray source is crucial in this context because only by adjusting the wavelength all experimental constraints (e.g., shadowing of the tensile tester and the need for measuring the majority texture component, compare Fig2b) can be accounted for. In the present study, optimum results were obtained for $\lambda = 1.7735$ Å, resulting in Bragg angles of $2\theta = 44.34°$ for the Au (111) reflection and of $2\theta = 46.76°$ and 68.28° for the W (110) and W (200) reflections, respectively. The X-ray detector was positioned $L = 220$ mm away from the specimen center, inclined by an angle of $\delta = 50°$ from the primary X-ray beam.

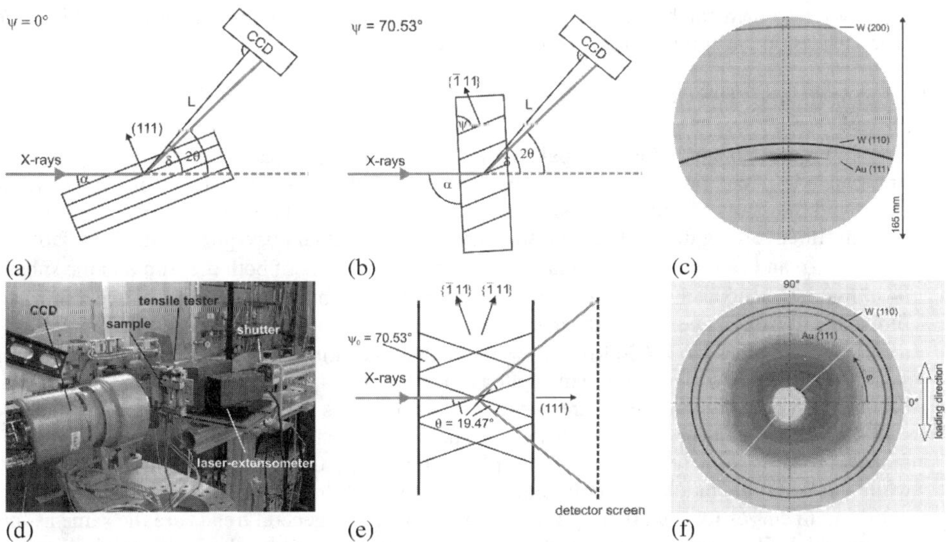

Figure 2 (a)-(b): Diffraction geometries for the "$\sin^2\psi$ method": (a) $\psi = 0°$, (b) $\psi = 70.53°$; (c): Diffraction pattern showing partial Debye-Scherrer rings of a 40 nm Au film and W powder (calibration substance). This pattern was recorded using the following parameters: $\lambda = 1.7735$ Å, $\alpha = 22.67°$, $\delta = 50°$, exposure time: 30 s. (d): Experimental set-up for *in situ* experiments using the "$\sin^2\varphi$ method"; (e): Diffraction from { $\bar{1}11$ } planes in a <111> fiber-textured film ("$\sin^2\varphi$ method"); (f): Complete Debye-Scherrer rings recorded in normal-incidence geometry used for "$\sin^2\varphi$" measurements at $\lambda = 1.5577$ Å, $L = 83$ mm, exposure time: 30 s.

(b) For measuring the evolution of non-equibiaxial stress changes in the course of *in situ* tensile tests, the area detector was positioned "on axis" on the downstream side of the specimen, thus capable of recording the complete Au (111) Debye-Scherrer ring. Due to the pronounced <111> fiber texture, an intense (111) Debye-Scherrer ring is obtained at $\psi_2 = 70.53°$. As depicted in Fig. 2e, this corresponds to a Bragg angle of $\theta = 19.47°$. In order to obtain the Au (111) reflection at this Bragg angle, the incident X-ray beam must be tuned to a wavelength of $\lambda = 1.57$ Å which corresponds to a photon energy of 7.9 keV. In order to fit the Au (111) and the

W (110) ring on the detector screen, the sample-to-detector distance L had to be reduced to 83 mm. For the tensile test, the displacement of the crossheads was increased in steps of 20 to 100 μm. During X-ray exposure, the crosshead position was kept constant. The sample was strained up to a maximum total strain of about 7 % and subsequently unloaded. The experimental arrangement is shown in Figure 2d. The tensile testing device was equipped with a 250 N load cell. A laser extensometer (Fiedler Optoelektronik GmbH, Germany) was used for contact-free measurement of the total strain of the specimen gauge section.

Data analysis: The lattice strains were extracted from the diffraction patterns by fitting ellipses to the diffraction rings [8] and determining the semi-axes of these ellipses. The lattice strains were converted into stresses using the generalized Hooke's law for <111>-textured Au polycrystals, which can be derived from the single-crystal elastic constants of this material. For more details on the computational procedures we refer to Ref. [7].

RESULTS AND DISCUSSION

In Figures 3a and b the principal stresses σ_1 (macroscopic loading direction) and σ_2 (transverse in-plane direction) are plotted against the total longitudinal strain for two of the Au films with thickness 500 nm and 20 nm, respectively. These plots show that the stress states in the thin film evolving during tensile testing are neither uniaxial nor equibiaxial. The relative values of σ_1 and σ_2 depend on the elastic and plastic properties of both the film and the substrate.

Figure 3a shows the evolution of σ_1 and σ_2 in the 500 nm thick film. At the beginning the film is nearly stress-free. After linear elastic straining and some strain hardening the stresses reach plateau values (σ_1 = 600 MPa, σ_2 = 380 MPa), indicating steady state plastic deformation. Transverse stresses depend on the mismatch of the Poisson's ratios of the film and the substrate. The fact that here the transverse stresses are tensile indicates that the in-plane Poisson's ratio of <111>-textured Au is greater than that of the substrate in both the elastic and plastic regimes. Indeed, it can be shown that the elastic in-plane Poisson's ratio of <111> textured Au is 0.57 [7], while the Poisson's ratio of the polyimide foil is only 0.34. In Figure 3b, the corresponding stress-strain curves for the 20 nm thick film are shown. The general trends are the same as for the 500 nm thick film. However, lower plateau stresses are observed for the 20 nm thick film.

In order to compare the results to yield stresses from other kinds of tests (e.g., true uniaxial tensile tests or wafer-curvature measurements) or to compare the films of the present study to each other, it is useful to consider the von Mises equivalent stress, which can be computed from σ_1 and σ_2 assuming $\sigma_3 = 0$. We define the yield stress as the von Mises stress observed at 0.5% longitudinal plastic strain. This quantity is plotted against the film thickness in Figure 4. It is evident that the yield stress is nearly constant in the thickness range from 100 nm to 1000 nm. Here the grain diameters are smaller than the film thickness (see Figure 1) and approximately the same for all films. Thus, the strength is governed by the grain size and not by the film thickness. Following this argument it would be expected that the thinnest films of the present study (20 nm and 40 nm) are stronger than the other films because here the thickness is smaller than the grain diameter. However, the opposite behavior is observed in the experiments: The yield stresses of the 20 nm or 40 nm thick films are only about half of that of the thicker films and compare well to the result obtained by Leung et al. [9] from wafer-curvature experiments on 40 nm thick Au. A possible explanation for this unexpected trend is that for these ultra-thin films the thickness variations due to grain boundary grooving or due to non-uniform film deposition are in the range of the film thickness. For films strained via the substrate, the average stress developing in a non-uniform film is generally smaller than the homogeneous stress in a corresponding uniform film. In addition, effects from constrained diffusional creep (model by Gao *et al.* [10,11]) may become relevant for the ultra-thin films of this study. The diffusion lengths scale both with grain size and film thickness and thus diffusive processes are expected to have a stronger influence at smaller film thicknesses. It was shown that for free-standing polycrystalline Au films the strain rate sensitivity of the yield stress increases with decreasing grain size [12] which is supporting evidence for this argument.

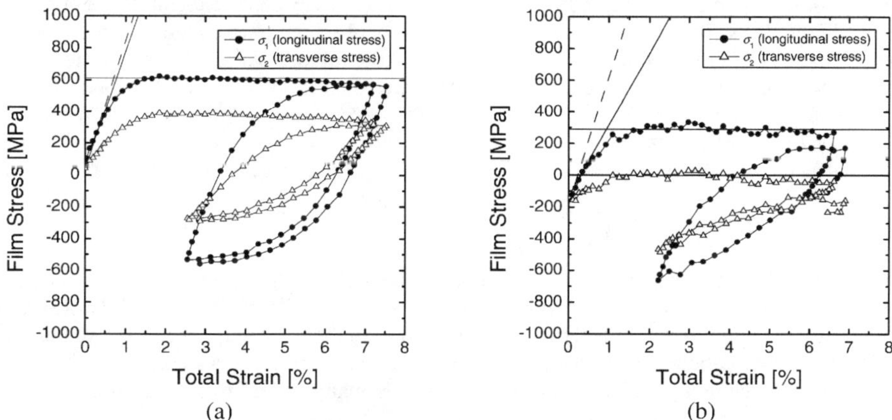

Figure 3: Stress vs. total strain for thin Au films: (a) 500 nm (b) 20 nm. The stress values were calculated from the lattice strains measured by the "$\sin^2\varphi$ method". The initial residual stress state was measured by the "$\sin^2\psi$ method".

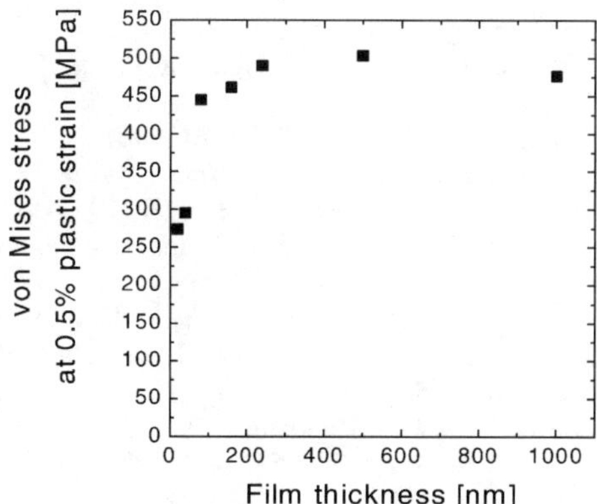

Figure 4. Yield stress vs. film thickness for Au films in the thickness range between 20 nm and 1000 nm.

CONCLUSIONS

In this paper we have described an X-ray diffraction technique for the measurement of mechanical properties of thin metallic films in a thickness range down to 20 nm. The sample preparation is simple and compatible with a wide range of crystalline materials. A strain resolution of better than 10^{-4} is achieved at measurement times of 30 s per load step. Longitudinal and transverse lattice strains are obtained simultaneously. Measurements on sputter-deposited Au films in a wide thickness range showed two regimes of mechanical behavior. In the thickness range from 100 nm to 1000 nm the yield behavior does not depend on the film thickness because all films exhibit the same columnar grain structure, where the grain diameter is smaller than the film thickness. Below 100 nm the strength drops significantly. At present it is not possible to unambiguously determine the origin of this stress drop. Film thickness variations that reduce the measured stresses or diffusional processes that become more dominant at small film thicknesses are plausible explanations. Further studies that will analyze the surface morphology of the films as well as time dependent measurements are underway.

ACKNOWLEDGMENTS

The *in situ* experiments were carried out at the MPI-MF beamline at ANKA, Karlsruhe Research Center, Germany. We thank the beamline team (A. Stierle, R. Waigel, N. Kaspar and A. Rühm) for their excellent technical support. We also thank T. Wagner and G. Adam of the MPI Thin Film Laboratory for sample preparation. Part of this work was supported by the Deutsche Forschungsgemeinschaft (DFG), Bonn, Germany under Project No. SFB381/A1.

REFERENCES

1. A. Kretschmann, W.-M. Kuschke, S. P. Baker, and E. Arzt, *Mater. Res. Soc. Symp. Proc.* **436**, 59-64 (1997).
2. M. Hommel and O. Kraft, *Acta Mater.* **49**, 3935 (2001).
3. P.O. Renault PO, P. Villain, C. Coupeau, P. Goudeau P, and K.F. Badawi , *Thin Solid Films* **242**, 267 (2003)
4. P.A. Flinn and C. Chiang, *J. Appl. Phys.* **67**, 2927 (1990).
5. P.R. Besser, B. Brennan, and J.C. Bravman, *J. Mater. Res.* **9**, 13 (1994).
6. W.M. Kuschke and E. Arzt, *Appl. Phys. Lett.* **64**, 1097 (1994).
7. J. Böhm, P. Gruber, R. Spolenak, A. Stierle, A. Wanner, and E. Arzt, *Rev. Sci. Instrum.* **75**, 1110 (2004).
8. A. Wanner and D. C. Dunand, *Metall. Mater. Trans. A* **31A**, 2949 (2000).
9. O.S. Leung, A. Munkholm, S. Brennan, and W.D. Nix, *J. Appl. Phys.* **88**, 1389 (2000).
10. H. Gao, L. Zhang, W.D. Nix, C.V. Thompson, and E. Arzt, *Acta Materialia* **47**, 2865 (1999).
11. D. Weiss, H. Gao, and E. Arzt, *Acta Materialia* **49**, 2395 (2001).
12. R.D. Emery, G.L. Povirk, *Acta Materialia* **51**, 2079 (2003).

Mat. Res. Soc. Symp. Proc. Vol. 821 © 2004 Materials Research Society P4.5/N4.5

Slip Transfer across Hetero-Interfaces in two-phase Titanium Aluminum Intermetallics

Jörg M.K. Wiezorek[1], Xiao-Dong Zhang[2], Hamish. L. Fraser[3]
[1] Department of Materials Science and Engineering, University of Pittsburgh, Pittsburgh, PA;
[2] Center for Materials Science, Bose Corporation, Framingham, MA;
[3] Department of Materials Science & Engineering, The Ohio State University, Columbus, OH;

ABSTRACT

The role of slip transfer processes across the heterophase interfaces in two-phase TiAl intermetallics has been studied. Polysynthetically twinned crystals of TiAl (PST-TiAl) have been used as model systems for individual grains in technologically relevant polycrystalline lamellar TiAl alloys. Compressive plastic loads have been applied for orientations of the lamellar interfaces parallel and perpendicular to the loading directions to produce hard mode slip activity. Transmission electron microscopy has been used to determine the active deformation modes in the constituent phases and to study the slip transfer across heterophase interfaces. The results are discussed with respect to the mechanical behavior of PST-TiAl and lamellar TiAl alloys, which is of relevance to in-service performance and metallurgical processing operations.

INTRODUCTION

Two-phase TiAl based intermetallics with slightly Ti-rich composition, e.g. Ti-48at.%Al, with microstructures containing large volume fractions of morphologically lamellar grains are promising candidate materials for applications in advanced gas-turbine-engines [1]. However, the widespread use of these alloys is hindered by the limited tensile elongation to failure and problems with cracking during metallurgical processing by forging for instance. Hence, it is important to develop a detailed understanding of the deformation behavior of the constituent phases, γ-TiAl and α_2-Ti_3Al, and the interplay between them during plastic straining. The basic deformation accommodating processes in the two phases, i.e. relevant dislocation glide and mechanical twinning, are understood reasonably well [1, 2-8]. However, the synergistic interplay between the deformation processes active in the γ- and α_2-phases in lamellar TiAl requires further attention, since for general plasticity of the technologically important polycrystalline materials both phases must deform compatibly across the heterophase interfaces.

The lamellar grains consist mainly of lamellae of the ordered tetragonal γ-TiAl phase with a small volume fraction of lamellae of the ordered hexagonal α_2-Ti_3Al phase, which obey a crystallographic orientation relationship, such that $[1\text{-}10]\gamma//[11\text{-}20]\alpha_2$ and $(111)\gamma//(0001)\alpha_2$ [2]. For the γ-lamellae six orientation variants are related by multiple 60°-rotations about the common lamellar interface normal [111] and coexist with the α_2-lamellae, which form crystallographically equivalent heterophase interfaces with each of the γ-phase variant lamellae [2]. The anisotropic mechanical properties associated with the lamellar grains are very well documented in the literature and have been studied from a fundamental point of view using directionally solidified material comprising only a single lamellar grain, so-called polysynthetically twinned TiAl (PST-TiAl) [1, 2]. The yield strength of PST-TiAl is very high for loading normal to the lamellar interface plane, still quite high for loading parallel to the interface plane, and much lower for intermediate orientations of the loading direction relative to the lamellar interface plane [2]. Hence, "soft" and "hard" loading orientations and deformation modes exist for the lamellar grains.

With reference to the majority phase two fundamentally different *hard* loading conditions are distinguished, namely loading normal to the lamellar interfaces, (111) (i.e. along [111], N-orientation [2]), and loading parallel to the interfaces, e.g. along [-110] and [-1-12] (i.e. orientations A1 and A2 [2]). *Hard* deformation modes (octahedral plane dislocation glide and mechanical twinning) are associated with slip vector components normal to the lamellar interface, while *soft* deformation modes involve slip vectors contained in the lamellar interface plane. The large differences in yield strength and ductility of PST-TiAl for the *soft* and *hard* orientations have been attributed to a Hall-Petch type effect associated with the much shorter slip distances available for *hard* deformation systems than for *soft* deformation systems [2, 9, 10]. However, the *hard* orientation deformation behavior observed for PST-TiAl with the significant differences in yield strength for the A-type and the N-orientations is less well understood than the *soft* orientation behavior. Interestingly, the deformation modes active in the γ-lamellae for A- and N-type *hard* orientation loading are essentially the same [e.g. 2, 11-13]. They involve octahedral plane deformation systems, namely slip of both ordinary dislocations, e.g. Burgers vector **b**=1/2[110], and superdislocations, e.g. **b**=[011], as well as *hard* translamellar twins associated with slip vectors of the type **b**=3x1/6[1-12] [2, 11-13]. Furthermore, deformation can occur compatibly between the six different γ-variant lamellae for the different *hard* loading conditions and is achieved by direct slip transfer for which numerous suitable boundary reactions have been identified [12-19]. Thus, it appears that an explanation for the differences between the A-type and N-orientation *hard* mode deformation characteristics has to be related to either changes in the deformation behavior of the minority α_2-phase lamellae or to the behavior of the lamellar heterophase (γ/α_2)-interfaces or a combination of both. Deformation of technologically relevant polycrystalline lamellar TiAl alloys always involves some grains experiencing *hard* mode loading and thus the minority phase must also deform. Hence, it is important to understand the role of the α_2-phase and the heterophase interfaces for the *hard* orientation deformation behavior of lamellar TiAl. Recognition of this fact is reflected in a series of recent reports related specifically to the study of deformation processes involving the heterophase interfaces in lamellar TiAl alloys using transmission electron microscopy (TEM) [11, 12, 14, 15, 20-22].

This paper presents results of an experimental study of the transfer of slip across the lamellar heterophase interfaces in TiAl with the aim to elucidate the origins for the characteristic *hard* orientation deformation. Hence, mechanical test of PST-TiAl in *hard* mode loading conditions are complemented with detailed transmission electron microscopy (TEM) studies to determine the active deformation modes for both the γ- and α_2-phase lamellae with a specific focus on slip transfer processes across the (γ/α_2)-interfaces.

EXPERIMENTAL PROCEDURE

Rods of PST-TiAl 12.5 mm in diameter and up to 175 mm in length have been obtained from the Materials Directorate, Wright Laboratories, Wright-Patterson Air Force Base and have been grown using an optical float zone technique akin to that in [2]. The composition of this PST-

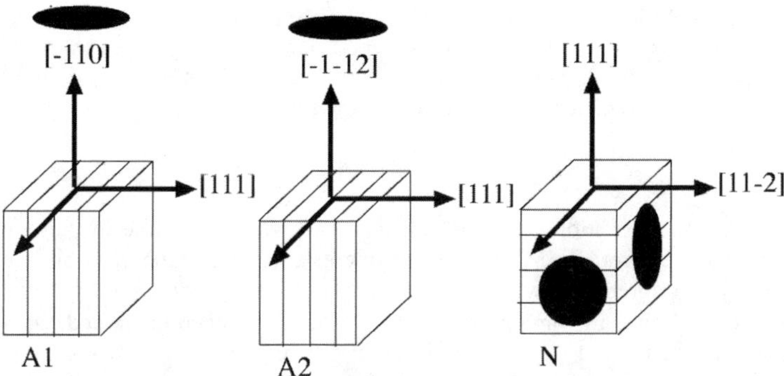

FIG_01: Schematic of the compression coupons for the three different hard orientations. The black shapes indicate TEM foils that have been extracted from the coupons.

material has been determined by wet-chemical analysis as Ti-48.1at.%Al with impurity levels of 610 wt.ppm oxygen, 35 wt.ppm nitrogen, and less than 100 wt.ppm carbon. After crystallographic alignment using a combination of Laue Backscatter X-ray diffraction and TEM rectangular compression coupons 6.0 mm tall with square cross sections (4.0x4.0) mm^2 have been cut with a precision diamond saw. Three types of compression coupons have been obtained for hard orientation loading parallel to the interface plane along [-110] (A1-orientation) and [-1-12] (A2-orientation), and for loading normal to the interface plane along [111] (N-orientation) as depicted schematically in FIG_01. The coupons have been deformed at room temperature at a constant rate of about $6x10^{-4}s^{-1}$ to total linear strains between 0.01 and 0.07. Thin disks have been sectioned both parallel and perpendicular to the macroscopic loading axes from the coupons with side-faces parallel to the (111), (-110) and the (-1-12) planes as indicated in FIG_01. Thin foils for the TEM investigation have been obtained by twin-jet electropolishing with an electrolyte of 6vol.% perchloric acid, 34vol.% butoxyethanol in methanol at a temperature of about -35°C and ion-beam milling with a liquid nitrogen cooled stage. Operating at 200 kV a Philips CM200 TEM has been used during this study for defect identification.

RESULTS AND DISCUSSION

Compression tests

Two compression tests have been performed for each of the three hard orientations (FIG_01). The compressive yield stress at 0.002 plastic strain, $\sigma_{0.2}$, and the total linear plastic strain, ε, determined from these tests are listed in table 2. On average the yield strength for the A-type

Table 1: Some pertinent data obtained from the compression tests.

	A1	A1	A2	A2	N	N
$\sigma_{0.2}$ [MPA]	533	512	513	541	847	913
ε [%]	7	1	7	3	1	3

orientation loaded PST-TiAl has been determined as ≈525MPa. For the N-orientation geometry yield stresses almost twice as high as those for the A-type orientations have been recorded. These mechanical characteristics of the PST-TiAl are consistent with previous reports and confirm the anisotropic behavior for the two different types of *hard* orientations [2, 8].

TEM observations

The deformation systems, dislocation glide and mechanical twinning, active in the constituent phases have been determined from TEM tilting experiments using the (**g.b**)-invisibility criterion (not shown for brevity). In agreement with previous reports [11-14], *hard* modes of dislocation glide and mechanical twinning in the majority phase γ-lamellae accommodated most of the plastic strain for each of the three different *hard* orientations (e.g. FIG_02). Dislocation glide in the α_2-phase lamellae was generally only observed in the vicinity of hard-mode mechanical twins piled-up against the lamellar heterophase interfaces (FIG_02). The results of the TEM tilting experiments for the strain accommodating deformation systems in the various types of γ-variant lamellae are summarized in Table 2. Generally, the observed deformation systems in the γ-phase are those that experience a large resolved shear stress of sufficient magnitude to exceed the relevant critical resolved shear stresses. Some of the notable exceptions to this general observation (Table 2) may be attributed to hard mode mechanical twin related slip transfer processes across the relevant lamellar (γ/γ)-interfaces. The slip transfer across these lamellar homophase interfaces has been studied in detail in a number of earlier works and generally only obtains in the presence of local pile-up stresses because the processes are often energetically up-hill [12-19]. Slip transfer commonly involves dislocations reacting via a boundary reaction, $\mathbf{b_1}$=

Table 2: Deformation systems active in the γ-lamellae. Slip systems and Schmid factors are shown. Less abundant systems are shown in curly brackets, { }. The roman numerals I - III refer to the three variants related by multiple 120°-rotations about the common [111]. Subscripts M and T refer to the matrix and twin variant [2].

	A1	A2	N
I_M	[10-1](1-11), 0.41 [0-11](-111), 0.41	[1-1-2](1-11), 0.31 [110](1-11), 0.27 { [112](11-1), 0.31 }	[-11-2](-111), 0.16 [1-1-2](1-11), 0.16 [-1-10](-111), 0.28
I_T	as I_M	as I_M	as I_M
II_M	[1-1-2](1-11), 0.24 [110](1-11), 0.41 { [-101](1-11), 0.41 }	[1-10](11-1), 0.41 { [10-1](1-11) 0.41 }	as I_M
II_T	as II_M	[1-10](11-1), 0.41 { [10-1](1-11). 0.41 }	as I_M
III_M	[-11-2](-111), 0.24 [110](-111), 0.41 { [0-11](-111), 0.41 }	[1-10](11-1), 0.41 { [0-11](-111), 0.41 }	as I_M
III_T	as III_M	[0-11](-111), 0.41 { [1-10](11-1), 0.41 }	as I_M

b_2 = Δ**b** [23, 24], and tend to dissipate the hard slip components normal to the lamellar (γ/γ)-interfaces, leading to improved compatibility of deformation in lamellar grains [12-14, 25].

The deformation modes active in the α_2-lamellae during compression in the A1- and A2-orientations were identical and involved only slip of <**a**>-dislocations, **b**=1/3<11-20>, on prism, {1-100}, or first-order pyramidal planes, {2-201} (FIG_03). Despite significant Schmid factors for some of the <**c**>-component pyramidal plane slip systems, e.g. <2**c**+**a**>{11-21}, **c**-component dislocation have not been observed after A-type compression (FIG_02, FIG_03) [11, 22]. The absence of dislocations with a Burgers vector component parallel to the [**c**]-direction can be attributed to the very high $\tau_{CRSS}\approx$900MPa measured for such pyramidal slip in off-stoichiometric Ti_3Al [6, 11, 22], which significantly exceeds the maximum stress applied during the A-type compression tests (σ_{max}<650MPa). For Ti_3Al of similar composition as the α_2-lamellae in PST-TiAl prism plane slip of <**a**>-dislocations has been reported to occur at $\tau_{CRSS}\approx$100MPa [6]. Interestingly, the <**a**>-dislocation slip systems were associated with high Schmid factors ($0.38\leq m\leq 0.46$) and the average yield stress measured for the A-type tests (Table 1) appears to be sufficient for their general activation. However, <**a**>-slip has been observed only in the vicinity of the intersections of *hard*-mode deformation twins in the γ-phase with the heterophase interfaces (FIG_02b, FIG_03). This indicates that the local stress concentrations associated with the deformation twin pile-ups significantly influenced the <**a**>-slip behavior in A-type orientation compressed PST-TiAl. Indeed, detailed analyses of the deformation substructures in α_2- and adjacent γ-lamellae in A-type orientation compressed PST-TiAl reported previously showed that <**a**>-dislocations are generated by both the activation of sources in the lamellar interfaces and direct slip transfer across these interfaces involving a boundary reaction [11, 12, 14, 22]. Both of these processes require external driving forces, which are available due to the macroscopically applied stress and also locally as a result of the considerable pile-up stresses [11,12]. The observed <**a**>-slip in the α_2-lamellae can only accommodate strains associated with components contained in the lamellar interface plane. This is sufficient for a macroscopic shape change of the PST-TiAl coupons involving thickening only along directions contained in (111)

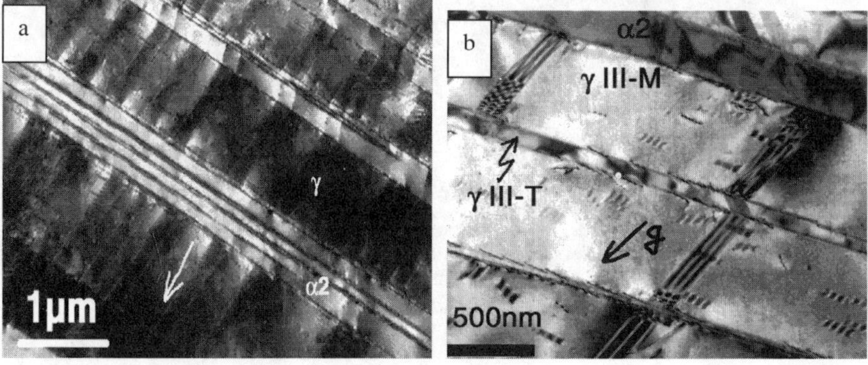

FIG_02: a) Example of deformation sub-structure after A-type hard loading. Twinning and dislocation glide in the γ-lamellae, **g**=111. In b) slip transfer across lamellar boundaries after A1-compression, small plastic strain, **g**=1-11. Arrows mark **g**-vectors.

but not normal to it, which is consistent with experimental observations for A-type loading [25]. However, it may be noted that the strains associated with the *hard* twinning, e.g. [-11-2](-111) (Table 2), and the *hard* ordinary dislocation slip systems, e.g. [110](-111) (Table 2), operative for A-type compression in some γ-lamellae cannot be accommodated entirely by the <**a**>-dislocation slip in the α_2-lamellae [11]. The resulting slip incompatibilities and local stress concentrations must be dissipated to prevent the initiation of microcracks and premature failure of the PST-TiAl. Possible micro-mechanism facilitating this include slip transfer processes across the (γ/α_2)-interface and are discussed further in *the next section* [11, 20-22]).

After N-orientation compression of the PST-TiAl the α_2-lamellae exhibited inhomogeneous slip of <**c**+**a**/2>-superpartial dislocations, **b**=1/6<11-26>, on first- and second-order pyramidal planes, {2-201} and {11-21} respectively, while <**a**>-dislocation slip has not been detected (FIG_04, FIG_05). This is not surprising as no stress is resolved onto the latter slip systems, while the <11-26>{2-201} and <11-26>{11-21} slip systems experienced high Schmid factors, 0.40 and 0.45 respectively. The pyramidal slip of the <**c**+**a**/2>-dislocations occurred inhomogeneously in slip bands, which are associated with *hard*-mode deformation twins impinging on the lamellar interfaces (FIG_04). These observations are in concert with previous studies [11, 12, 20-22]. The applied stress required for general activation of pyramidal slip systems of approximately τ_{CRSS}~(900/0.45)MPa =2000MPa significantly exceeds the maximum stress of about 1000MPa applied during N-orientation loading to the larger strain (Table 1). The macroscopically applied stress was insufficient for the activation of the experimentally observed <**c**+**a**/2>-slip systems, implying that additional local stress concentrations assisted activation of these pyramidal slip systems. Alternatively, unidentified but fairly abundant sources in or at the lamellar (γ/α_2)-interfaces, which would permit the activity of these slip systems at lower stresses than in monolithic Ti_3Al, for which the shortage of sources might present an additional problem for slip activity, may be responsible for the pyramidal slip. After compression to 0.01 strain in the N-orientation the α_2-lamellae remained essentially free of defects and thus practically

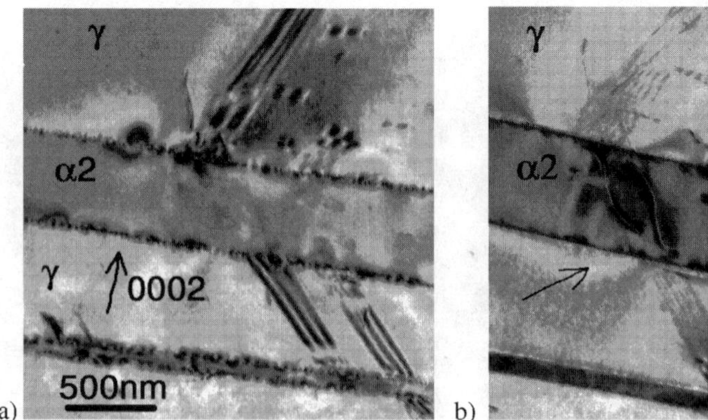

FIG_03: Hard A-loading and twin transfer across heterophase interface producing <a>-slip on {10-10} and {20-21}. a) g=0002 and b) g=2-201, arrows mark **g**-vectors.

undeformed (FIG_05). For the small strain, 0.01, only very few and very short segments of <**c**+**a**/2>-dislocations on both types of pyramidal planes, {11-21} and {2-201}, have been observed to bow out into the α_2-lamellae at or near deformation twins piled-up at the heterophase interface (FIG_05). The slip planes of the dislocations near A and B in FIG_05 have been identified as follows. The two hook-shaped defects near B exhibited a nearly perfectly straight morphology when viewed along a **BD**≈[-1100], whereas at least the defect above the label A appeared to have significant curvature in this foil geometry (FIG_05). When viewed along a **BD**≈[-2110] essentially the reverse is the case (FIG_05). This contrast behavior suggests that the defect near B be contained in a plane that is edge-on for **BD**≈[-1100], e.g. (11-2±1). Conversely, the defect near the label A appears to be contained in (0-22±1). These <**c**+**a**/2>-dislocations may have been generated by the local pile-up stresses of the translamellar deformation twins, T1 and T2, impinging on the (γ/α_2)-interface, involving possibly a direct slip transfer mechanism near A [11]. However, the systems <**c**+**a**/2>{11-21} cannot be generated by direct slip transfer of the twinning shear involving a boundary reaction, whereas <**c**+**a**/2>{2-201} can, albeit with great difficulty (next *section* [11, 20-22]). Hence, it appears that after compression to 0.01 strain the local stress associated with the piled-up *hard* twins were sufficient to activate sources for <**c**+**a**/2>-dislocations available at the (γ/α_2)-interface. While these sources have not been identified yet they may include the typical growth ledges in the interfaces as well as lattice dislocations trapped in the interface [11]. However, at the small strain, 0.01, stress concentrations appeared to be insufficient to drive the <**c**+**a**/2>-dislocations across the full width of the α_2-lamellae, which has been observed in the form of slip bands for larger strain, 0.03 (FIG_04) [11]. At the larger strain the **c**-component pyramidal slip clearly contributes significantly to strain accommodation. The local stress concentrations from the deformation twin pile-ups increase with increasing strain and locally appear to reach levels sufficient to facilitate pyramidal glide of

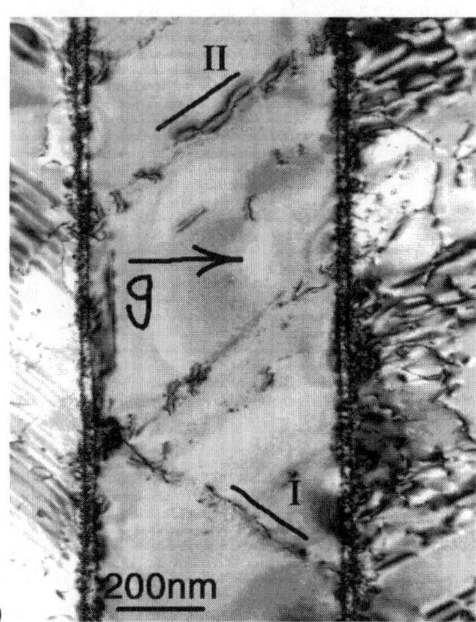

a)

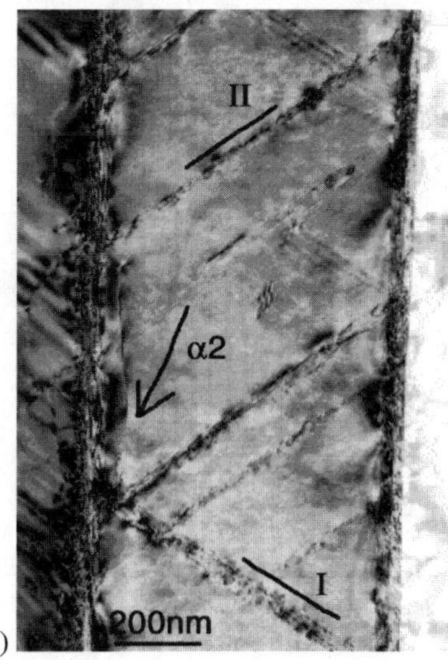

b)

FIG_04: Hard N-loading to 3% strain and twin transfer across heterophase interface producing slip bands of <c+a/2>-type dislocation slip on (1-211), (I), and (2-1-1-1), (II). a) g=0002 and b) g=2-201, both for BD~[11-20]. Arrows mark **g**-vectors.

<c+**a**/2>-dislocations across α_2-lamellae in N-orientation compressed PST-TiAl. Strain in excess of 0.01 appears to be necessary to produce these critical pile-up stresses for α_2-lamellae significantly wider than 100nm in the material studied here. Interestingly, after tensile tests at room temperature <c+**a**/2>-slip systems have not been observed in lamellar TiAl [11, 22]. This is attributed to the difficulty in reaching stress levels sufficient for the activation of such slip systems prior to fracture and is consistent with the essentially vanishing tensile strain to failure for N-oriented PST-TiAl [2, 13, 25].

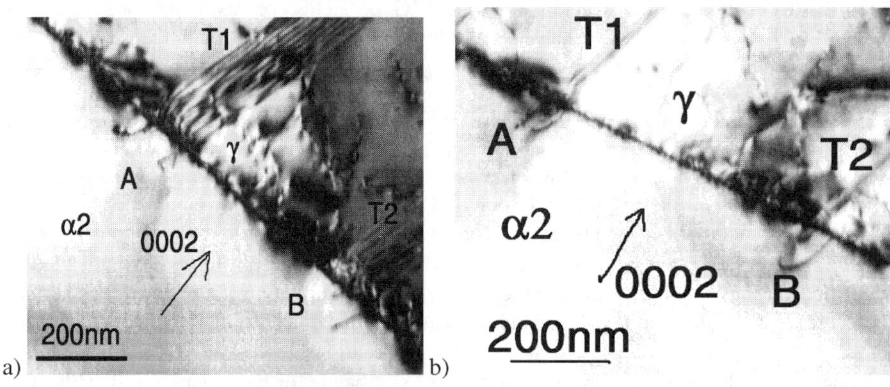

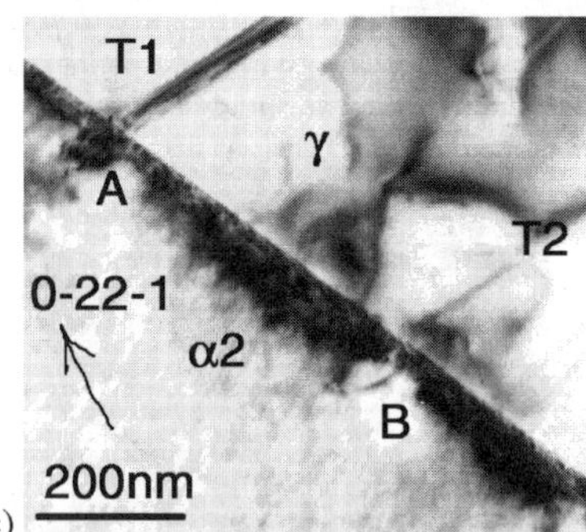

FIG_05: Hard N-loading to only 1% strain and twin transfer across heterophase interface producing very short segments of <c+a/2>-type dislocation slip on two types of pyramidal planes, i.e. (11-2±1) at B and (0-22±1) at A. a) g=0002, BD~[-1100], b) g=0002, BD~[-2110] and c) g=0-22-1, BD~[-2110]. These c-component dislocations are generated by stress concentrations associated with the hard twins T1 and T2 in the γ-lamellae, potentially by direct slip transfer at B and by local source activation at A.

Heterophase Interface Slip Transfer in PST-TiAl

Examples of post-mortem TEM observations of slip transfer processes across the lamellar heterophase interfaces in the A-type and N-orientation compressed PST-TiAl involving *hard*-mode twinning and dislocation shears have been presented in the previous section. For reasonably easy slip transfer across interfaces some important requirements have been identified previously [23, 24]:

a) Sufficient amount of resolved shear stress of the correct sense acting on the slip systems involved in the slip transfer process.
b) Minimal angle of rotation, φ, of the interfacial traces of the two relevant slip planes. Ideally, $\varphi=0$, the two relevant slip planes have a common interfacial trace.
c) Conservation of Burgers vector in the slip transfer reaction, i.e. $\Sigma \mathbf{b_i} = \mathbf{b_1} + \mathbf{b_2} + \Delta\mathbf{b} = 0$, where $\mathbf{b_1}$ and $\mathbf{b_2}$ are the Burgers vectors in the two phases and $\Delta\mathbf{b}$ is the residual Burgers vector content due to the difference in magnitude and direction of $\mathbf{b_1}$ and $\mathbf{b_2}$.
d) Reactions with smaller $\Delta\mathbf{b}$ are favored and in general more likely.

All six possible heterophase interfaces in PST-TiAl are crystallographically equivalent. The close-packed directions in the two phases match almost perfectly in the lamellar interface and the *hard*-mode octahedral planes, {11-1}, and the prism and first-order pyramidal planes, {h-h0l} with l=0, ±1,±2 etc., in the γ- and α_2-lamellae share interfacial traces, $\varphi=0$. The interfacial traces of the second-order pyramidal planes, {11-21}, and hard {11-1}, are rotated by $\varphi=30°$.

The heterophase interface slip transfer processes observed experimentally have been associated with pile-ups of either twinning superpartials or superdislocations against the respective lamellar boundary (FIG_02 to 05) [11, 12, 14, 20-22]. The local stress concentrations associated with the pile-ups provide driving forces required for the slip transfer processes. The localized deformation in the α_2-lamellae is due to the higher τ_{CRSS} values for the slip systems in Ti_3Al as compared to those typical of TiA. Most of the dislocation reactions that can be envisaged for *hard*-mode slip transfer across the heterophase interface in PST-TiAl are unfavorable when only line energy is considered. However, the presence of applied stresses can compensate for this elastic energy penalty, especially if dislocation reactions enable the relaxation of incompatibility stress. Possible processes for the direct transfer of γ-phase twins across the lamellar-interface in A-type orientation compressed PST-TiAl have been considered in earlier studies [11, 12, 14, 20-22]. The following type of reactions can describe the twin transfer as depicted in the example of FIG_03, generating <**a**>-dislocations in the α_2-lamella:

$$3x1/6[-11-2](-111) \Rightarrow [01-1](-111) + 1/2[-1-10](-111) \quad (1a)$$

followed by
$$[01-1](-111) \Rightarrow 1/3[2-1-10](0h\text{-}hX)\text{, where } X=0,\pm1,\pm2,\ldots \quad (1b)$$

This type of reaction may produce prism and pyramidal <**a**>-slip and has been confirmed to occur by detailed TEM experiments reported previously [11, 22]. Numerous other possible reactions for the transfer of twinning shears across the (γ/α_2)-interface can be envisaged [15] but have not been observed experimentally here. However, the type of reaction described in (1a, b) satisfies nearly all the criteria (a-d) outlined above and is well suited to increase the plastic compatibility between the α_2- and γ-lamellae during A-type loading [11, 22]. However, notably <**a**>-dislocation slip is not compatible with the hard slip component in of the γ-phase twins. Incompatibility stresses arise and must be dissipated. Interestingly, the twinning system and the superdislocation slip systems involved in reaction (1a,b) have reasonably high Schmid factors of the same sign, whereas the significant Schmid factor for the ordinary dislocation slip system is of opposite sign (Table 1 and reaction (1a), [11, 22]). The decomposition (1a) produces a soft superdislocation in the γ-phase that is compatibly transferred to <**a**>-dislocation slip in the α_2-

lamella (reaction (1b)) and a hard ordinary dislocation, which accounts for the remaining hard slip component normal to the interface and moves back into the γ-lamella under the applied stress. Thus, the incompatibility stresses associated with the *hard*-mode twinning systems, such as [-11-2](-111), are effectively reduced and the plastic compatibility between the phases increases. Furthermore, other hard deformation systems in the γ-variants, e.g. [110](-111) (Table 1), have not been observed frequently. It is tempting to propose that the infrequent hard ordinary dislocation slip systems detected by TEM may actually be by-products of reactions as described in (1a, b), rather than a result of general slip activity. Thus, slip transfer reactions of the type described in (1a,b) are very important during A-type loading of PST-TiAl and prevent premature failure.

Transfer of <**a**>-dislocations from α_2-lamellae across the (γ/α_2)-interface has also been observed (FIG_03) [11, 12, 22] and produces soft superdislocations with **b**=±[10-1] and [0-11] or soft ordinary dislocations with **b**=±1/2[1-10] in the γ-phase. Such transfer processes are expected to be relatively easy since the Burgers vectors involved are almost exactly parallel and in magnitude near integer multiple equivalents to the <**a**/2>-superpartial dislocation shears, **b**=1/6<11-20>, in the α_2-phase [11, 15]. Furthermore, the octahedral slip planes {1-11} in γ-lamellae and appropriate prism and pyramidal, {h-h0X}, slip planes in the α_2-lamellae have common interfacial traces, and for the A-type orientation loading conditions exhibit high Schmid factors [11]. These slip transfer processes may proceed according to reaction (2):

$$<01\text{-}1>\{-111\} \Longleftrightarrow 1/3<2\text{-}1\text{-}10>\{0h\text{-}hX\},\ X=0.\ \pm 1, \pm 2, \ldots \qquad (2a)$$

or

$$2x1/2<1\text{-}10>\{-11\text{-}1\} \Longleftrightarrow 1/3<\text{-}1\text{-}120>\{0h\text{-}hX\},\ X=0.\ \pm 1, \pm 2, \ldots \qquad (2b)$$

These transfer processes require small driving forces, which are provided externally and by pile-up stresses. Thus, for the A-type loading hard twinning and dislocation systems active in the γ-lamellae are predicted to transfer frequently across the (γ/α_2)-interfaces into <**a**>-dislocation slip, which easily produces soft dislocations in γ, affording good compatibility and ductility.

Possible heterophase interface slip transfer mechanisms active during N-orientation loading of the PST-TiAl have to enable the generation of <**c**+**a**/2>-dislocations in the minority phase [11, 21, 22]. Our observations indicate that stress concentrations ahead of the *hard*-mode deformation twin pile-ups are sufficient and suitable to produce local slip activity of <**c**+**a**/2>-dislocations on two types of pyramidal planes, {11-21} and {-2201}. However, realistic slip transfer processes can only be envisaged for the latter type of pyramidal slip systems, <**c**+**a**/2>{-2201}[11]. The former slip systems, <**c**+**a**/2>{11-21}, do not share a common trace with the hard {1-11} planes in the γ-phase. A large reorientation of the dislocation lines incipient at the lamellar interface by an angle of $\varphi \approx 30°$ would be required for direct slip transfer [11]. This is cumbersome and constitutes a major obstacle for slip transfer involving <**c**+**a**/2>{11-21} [11]. However, slip transfer processes producing <**c**+**a**/2>{-2201} can be described by reactions such as:

$$3x1/6[-112](-11\text{-}1) \Rightarrow 1/6[-2116](0\text{-}221) + 1/6[-110\text{-}3](0\text{-}221) \qquad (3)$$

Here, $\Delta\mathbf{b}_{res}$ =1/6[-110-3] is a residual Burgers vector of the type ±1/6<1-10X>, with 0≤X≤3, and under certain circumstances is contained in the interface [11, 21]. These twinning shear transfer processes involve fairly large residual Burgers vectors and must also overcome the high τ_{CRSS} for the pyramidal slip activity in the α_2-lamellae. Both of these factors contribute to the large energy barrier for this type of slip transfer process. In the N-orientation loading condition the local stresses required for the slip activity of the <**c**+**a**/2>-systems appear to be very high and may be reached for compression, consistent with the observation of this deformation mode in the PST-TiAl samples here [11, 21]. During tensile tests of PST-TiAl, for which near-zero elongation to failure results at applied stresses well below the high τ_{CRSS} for pyramidal slip in the α_2-lamellae, the stress concentrations fracture the lamellar grains prior to slip transfer suitable for dissipation of the incompatibility stresses [2-4, 6]. Previous studies showed that compressive and tensile loading along the **c**-axis of the α_2-phase unit cell leads to preferred **c**-component slip on {11-21} and {2-201} type planes, respectively [7, 13]. This change in preferred <**c**+**a**/2>-slip systems may be related to the fine structure of the dislocation cores during glide under these different loading conditions. Finally, the residual Burgers vector content, $\Delta\mathbf{b}$ or $\Delta\mathbf{b}_{res}$, produced by the various possible heterophase slip transfer processes under hard orientation loading conditions may alter the interfacial structure. Under suitable thermo-mechanical loading conditions these effects of the slip transfer processes can promote an acceleration of the decomposition or destabilization of the lamellar microstructures in TiAl [26].

CONCLUSIONS

Defect-interface interactions in room temperature deformed PST-TiAl have been studied by TEM and the importance of the deformation behavior of the α_2-phase lamellae and the characteristics of the slip transfer across the heterophase interfaces for the mechanical properties and behavior of lamellar TiAl has been demonstrated. Some of the main conclusion can be summarized as follows:

A) The differences in the heterophase interface slip transfer behavior for N- and A-type hard orientation loading are suitable to rationalize the observed deformation behavior (tension vs. compression) of lamellar TiAl.

B) The heterophase interface slip transfer processes observed experimentally generally adhere to the continuum elasticity and geometry based 'rules' for slip transfer across general boundaries accomplished by a boundary reaction [23, 24]. Effects from anisotropic elastic properties of the phases appear to be second-order effects. A common trace of the slip planes involved in the heterophase interface slip transfer constitutes a primary prerequisite for relatively easy slip transfer. Conversely, in the absence of common interfacial traces the direct slip transfer boundary reactions are likely to be impeded (e.g. **FIG_05**).

C) The core structure of the <**c**+**a**/2>-dislocations may influence the choice of preferred type of pyramidal slip plane and thus affect possible slip transfer under tension versus compression.

D) Narrower α_2-lamellae require smaller pile-up stresses to drive <**c**+**a**/2>-slip across them. Narrower γ-lamellae require larger stresses to deform (Hall-Petch effect). It may be proposed that a finer scale of lamellae with a narrow width distribution would lead to better plastic compatibility between the phases and thus should lead to strength and ductility improvement in lamellar TiAl.

ACKNOWLEDGEMENTS

Support for this work by a grant from the National Science Foundation with Dr. Bruce MacDonald as program manager and from the Office of Research, University of Pittsburgh are gratefully acknowledged. The authors also thank Dr. Dennis Dimiduk, Materials Directorate, Wright Laboratories, Wright-Patterson Air Force Base, for provision of the PST-TiAl rods.

REFERENCES

[1] Y. W. Kim, Acta metall. **42**, 1221 (1992).
[2] T. Fujiwara et al., Phil. Mag. A **61**, 591 (1990).
[3] Y. Minonishi and M.H. Yoo, Phil. Mag. Lett. **61**, 203 (1990).
[4] Y. Umakoshi et al., Acta metall mater. **41**, 1144 (1993).
[5] H. Inui, Y. Toda and M. Yamaguchi, Phil. Mag A **67**, 1315 (1993).
[6] H. Inui, Y. Toda, Y Shirai and M. Yamaguchi, Phil. Mag A **69**, 1161 (1994).
[7] M. Legros, Y. Minoninshi and D. Caillard Phil. Mag. A. **76**, 1013 (1997).
[8] Y. Umakoshi and T. Nakano, Acta metall. mater. **41**, 1155 (1993).
[9] P.M. Hazzledine and B.K.Kad, Mat. Sci. Eng A **95**, 340 (1995).
[10] B.K.Kad, P.M. Hazzledine and H.L. Fraser, MRS Symp Proc **288**, 495 (1993).
[11] J.M.K. Wiezorek, X.D. Zhang, W.A.T. Clark and H.L. Fraser, Phil. Mag. A **78**, 217 (1998).
[12] J.M.K. Wiezorek, X.D. Zhang and H.L. Fraser, Structural Intermetallics II, (TMS, Warrendale, 1997) p. 195.
[13] K. Kishida, H. Inui and M. Yamaguchi, Phil Mag. A **78**, 1 (1998).
[14] J.M.K. Wiezorek, X.D. Zhang and H.L. Fraser, MRS Symp Proc **578**, 383 (2000).
[15] D. Hu and M.H. Loretto, Intermetallics **7**, 1299 (1999).
[16] M.A. Gibson and C.T Forwood, Phil. Mag A **80**, 2747 (2000).
[17] M.A. Gibson and C.T Forwood, Phil. Mag A **82**, 1381 (2002).
[18] S. Zghal and A. Couret, Phil. Mag A **81**, 365 (2001).
[19] S. Zghal and A. Couret, Phil. Mag A **81**, 345 (2001).
[20] J.M.K. Wiezorek, , A. Godfrey, X.D. Zhang, D. Hu, M.H. Loretto and H.L. Fraser, Scripta mater. **38**, 811 (1998).
[21] A. Godfrey, D. Hu and M.H. Loretto, Phil. Mag. A 77, 287 (1998).
[22] M.A. Gibson and C.T Forwood, Phil. Mag A **80**, 2785 (2000).
[23] Z. Shen, R.H. Wagoner and W.A.T. Clark, Acta metall. mater. **36**, 3231 (1988)
[24] W.A.T. Clark, R.H. Wagoner, Z.Y. Shen, T.C. Lee, I.M. Robertson and H.K. Birnbaum, Scripta metall. mater. **26**, 203 (1992).
[25] K. Kishida, J. Yoshikawa, H. Inui and M. Yamaguchi, Acta mater. **47**, 3405 (1999).
[26] A. Gupta and J.M.K. Wiezorek, Intermetallics **11**, 598 (2003).

Mat. Res. Soc. Symp. Proc. Vol. 821 © 2004 Materials Research Society P4.9/N4.9

The Mechanical Properties in the Vicinity of Grain Boundaries in Ultrafine-Grained and Polycrystalline Materials Studied by Nanoindentations

E. Schweitzer, K. Durst, D. Amberger and M. Göken
Department of Materials Science & Engineering, University Erlangen-Nuremberg
91058 Erlangen, Martensstrasse 5, Germany

ABSTRACT

The strength of structural materials strongly depends on the structure and properties of grain boundaries. Interfaces usually act as barriers to dislocation motion and therefore strengthen materials with decreasing grain size, quantitatively described by the well-known Hall-Petch relation. However, interfaces in nanocrystalline materials are often covered with impurities or second phases, which may influence the mechanical properties. With nanoindentation testing it is now possible to probe the strength of interfaces like grain boundaries directly on a nanometer scale. Therefore this method was used to investigate the properties in the vicinity of grain boundaries in polycrystalline materials with conventional grain size and in ultrafine-grained metals prepared by equal channel angular pressing (ECAP), where no impurities are introduced during processing. Measurements on an austenitic steel clearly show a decreasing hardness close to the interface opposite to the general expected behavior of strengthening. In this case segregation effects strongly influence the mechanical properties near the boundaries. The nanoindentation investigations on ultrafine-grained Al and Cu show a strong strain rate sensitivity. Inelastic effects are also found between unloading-loading segments during indentations.

INTRODUCTION

The strength of structural materials depends strongly on the structure and properties of grain boundaries. Interfaces in crystalline solids usually act as barriers to dislocation motion and therefore strengthen materials with decreasing grain size, quantitatively described by the well-known Hall-Petch relation. In addition, segregation effects and the nucleation of precipitates are more likely to occur at grain boundaries. Although the width of a grain boundary extends only over a few interatomic distances, the effect arising from the presence of a boundary may extend over a much greater distance (1-100 μm). Grain boundaries have also an important influence on corrosion cracking of austenitic stainless steels used as tubing materials.

One of the easiest and perhaps most promising approaches to studying the influence of interfaces is to measure their mechanical properties and in particular plastic deformation. And indeed the hardness in the vicinity of grain boundaries has been studied intensively in early work by Westbrook et al. [1, 2]. They used Vickers hardness measurements to study hardness variation at grain boundaries in different metallic and intermetallic materials and found in many cases a strong distance dependence of the hardness. However, the measurements on intermetallic materials strongly depend on the chemical composition. Thus a significant decrease of hardness with increasing distance from the boundary was only found for off-stoichiometric compositions, where large quantities of constitutional defects are present. The defects concentration at the boundary then probably leads to an increasing hardness. With nanoindentations, hardness and modulus of elasticity measurements are now possible on a local scale of say 100 nm. Such measurements on grain boundaries in NiAl only show a weak influence of grain boundaries on the hardness, due to the interaction of the plastic zone generated under the indenter with the grain boundaries [3].

In this work nanoindentation measurements on austenitic steel 1.4541 and on ultrafine-grained Cu and Al samples were performed. Although the investigated 1.4541 steel is Ti stabilized, sensitization occurs in a temperature range of around 650°C, which consists of carbide precipitation at grain boundaries, making the materials susceptible to intergranular corrosion. Nanohardness profiles were measured across grain boundaries with a maximum load of 0.4 mN. Segregation and purity issues might be of relevance also for the deformation behavior of ultrafine-grained metals (UFG) prepared by severe plastic deformation or for other nanocrystalline materials.

EXPERIMENTAL DETAILS

In this work grain boundaries in X6CrNiTi 18-10 austenitic steel (1.4541 or AISI Type 321) have been evaluated. Specimens had been investigated after three different heat treatments. Heat treatments at a temperature of 650°C lead to a sensitization, during which carbides are precipitated at grain boundaries. The specimen surface was either electrolytically etched using a 10% oxalic acid or conventionally etched with a grain boundary etchant based upon a mixture of hydrochloric and nitric acid in ethanol. The measurements on ultrafine grained materials were performed on Al and Cu samples of technical purity 99.5% processed with equal channel angular pressing (ECAP).

The hardness measurements on ultrafine-grained metals were performed with the Nanoindenter XP (Nanoinstruments, Knoxville, TN). The measurements on individual grain boundaries require a very high lateral resolution, which is achieved by a combination of an indenter from Hysitron (Minneapolis, MN) with an atomic force microscope (multimode AFM from Digital Instruments). Diamond tips with a Berkovich type geometry were used for all measurements. At very low indentation loads a discontinuity is found in the load-displacements curves, which marks the transition from elastic to plastic deformation [4]. The tip-shape function was determined with the Oliver and Pharr method [5] by reference indentations in fused quartz.

RESULTS

Grain boundaries in austenitic steels

Nanoindentation measurements were performed only on large angle grain boundaries in an austenitic steel. The grain boundary type has been determined with EBSD measurements. Figure 1 shows a hardness profile close to a grain boundary after sensitization heat treatment at 650 °C. A significant decrease of the hardness at the grain boundary is found after heat treatment, whereas a more or less constant hardness is found for a sample without heat treatment, not shown here. The hardness decreases from around 4.5 GPa in the grain interior to around 3 GPa close to the grain boundary. This influence is probably caused by a segregation effect. The heat treatment leads to a sensitization, which means carbides precipitate at the boundaries. These carbides are clearly visible after etching with oxalic acid, but not in figure 1 where a

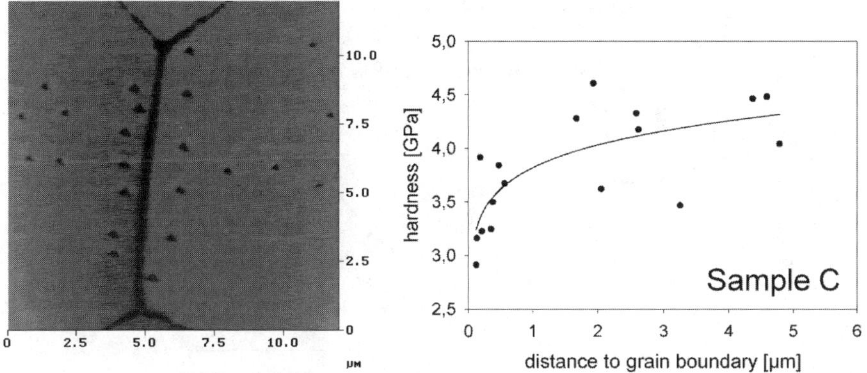

Figure 1. AFM image of a grain boundary in an austenitic steel. The results from the nanoindentations visible in the AFM image are shown in the hardness plot.

conventional etchant was used. The oxalic etching leads to a relatively high surface roughness, not suitable for nanoindentation measurements. The grain boundary softening therefore is due to a depletion of carbide forming elements at the interface. Since many different grain boundaries show the same results, possible influences from an inclination of the boundary to the surface normal could be neglected. More results and a more refined discussion will be given elsewhere [6].

Nanoindentations on ultrafine-grained metals

To evaluate influences from grain boundaries in materials with high interface density, nanoindentation measurements were performed on ultrafine-grained materials and in comparison on the same materials with conventional grain size. Such measurements were performed on Cu and Al, where the ultrafine-grained microstructure was produced with the ECAP-process.

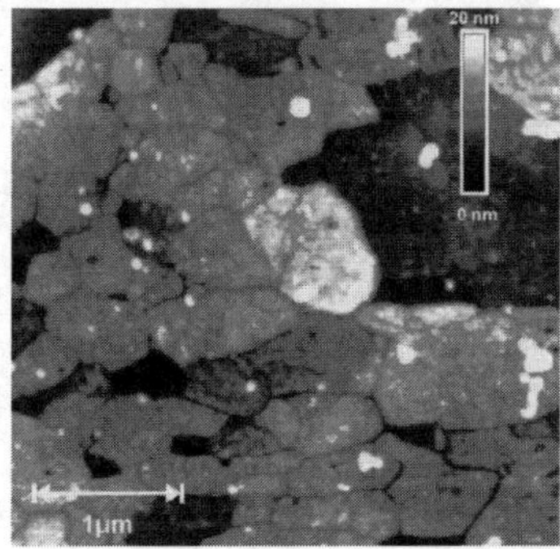

Figure 2. The microstructure of ultra-fine grained Cu as imaged with an AFM after electropolishing.

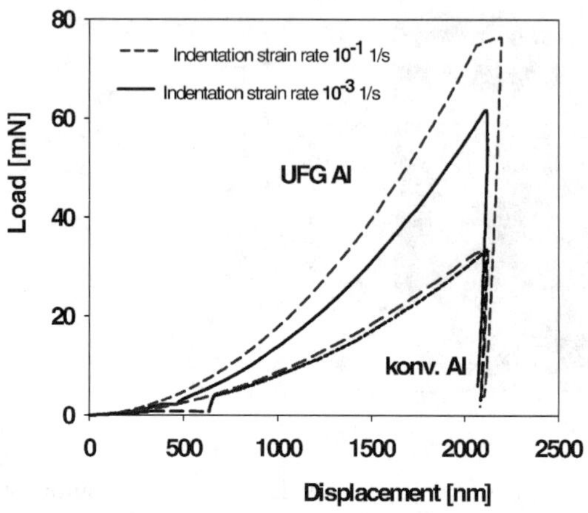

Figure 3. Load-displacements curves on UFG Al and Al with conventional grain size at different indentation strain rates.

Figure 2 shows the typical microstructure of an UFG-Cu specimen after electropolishing. The AFM-image clearly shows the microstructure with a grain size of some hundred nanometer. The grain size of the Al sample as determined by a TEM analysis was in the range of 850 nm, whereas the Cu samples showed a grain size of ca. 250 nm.

Nanoindentation measurements show a large influence of the indentation strain rate, which is given by $\dot{h}/h$, where h is the indentation depth. Figure 3 shows typical load-displacement curves obtained with the Nanoindenter XP on UFG-Al and in comparison on Al with conventional grain size at two different indentation strain rates. The indents were performed under strain control,

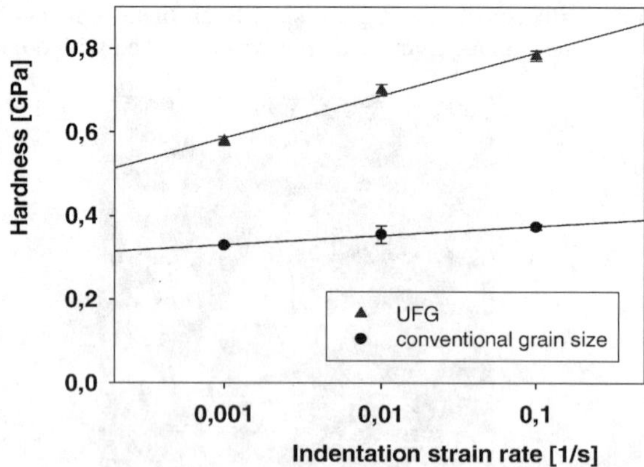

Figure 4. Hardness of UFG Al and Al with conventional grain size at different strain rates. Each data point represents an average over 4-10 indents.

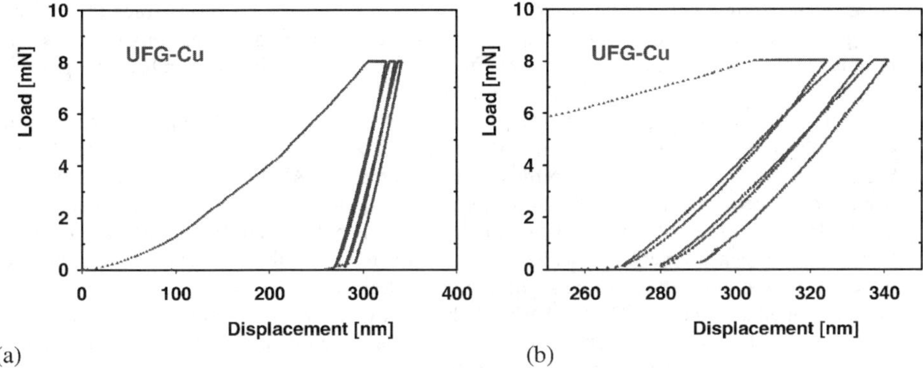

Figure 5. (a) Load displacements curve for UFG Cu with three loading-unloading cycles. The magnified part of the load-displacement curves (b) clearly shows an inelastic deformation behavior.

where the load increases until a given indentation depth is reached. A higher maximum load therefore reflects a higher hardness.

The hardness as calculated with the Oliver-Pharr method from the load-displacement curves clearly shows a grain size influence. The hardness of UFG Al is around double as high compared to materials with conventional grain size. Furthermore the UFG hardness shows a significant strain rate influence in hardness testing, whereas materials with conventional grain size only show a weak influence from the strain rate. The strain rate was varied between 10^{-3} and $1\ s^{-1}$. However, the Nanoindenter XP which was used for these measurements is only specified for strain rates between 10^{-2} and $10^{-1}\ s^{-1}$. At strain rates higher than $0.1\ s^{-1}$ no reliable data could be obtained, therefore these data were not included in figure 4. From measurements at a strain rate of $10^{-3}\ s^{-1}$ reasonable results are only expected if thermal drift is very low, since such indentations last more than 1 hour. However, these measurements performed at a thermal drift rate below 0.05 nm/s and with drift correction do follow the same trend as at faster strain rates.

An inelastic loading-unloading deformation behavior of UFG metals is also found in simple load controlled indentations if loading-unloading is repeated at least twice. This is exemplary shown in figure 5 for UFG-Cu. After the first loading and unloading cycle loading to the same maximum load is repeated twice. The unloading and loading cycle, however, do not match, which would have been expected for ideal elastic deformation behavior. A significant inelastic deformation is obvious, which is not observed for similar materials with conventional grain size. However, a similar behavior has also been found in tensile unloading-reloading tests, where the microyielding behavior of ultrafine-grained Cu and Al was studied [7].

CONCLUSIONS

In this paper the hardening or softening influence of grain boundaries was investigated. Measurements on individual interfaces such as grain boundaries in many cases only show a very weak influence on the hardness. This has been shown for example on stoichiometric intermetallic NiAl bicrystals [3]. In austenitic steels on the other hand a significant softening effect was found

close to large angle grain boundaries, which is due to segregation effects. Since this effect leads to a grain boundary hardness which is nearly reduced by a factor of two this effect is far more pronounced than the small hardening effect caused by the interaction of the plastic zone generated under the indenter with the interface.
Since the segregation effect is very pronounced in this austenitic steel, similar effects might also play a role in ultrafine-grained metals, which became a very interesting class of metals recently. Some first nanoindentation measurements on UFG Cu and Al show large inelastic deformation behavior. In previous nanoindentation studies [8, 9] on nanocrystalline and microcrystalline materials a strain rate influence during indentations was only found for nanocrystalline materials. However, the measurements presented here clearly show that this influence is also prevalent on materials with grain sizes of some hundred nanometers. The effect is especially significant at the lowest applied strain rate of 10^{-3}, whereas Schwaiger et al. [8] investigated the materials at strain rates above 10^{-2}. Significant rate sensitivity has also been found in tension tests on UFG Al processed with accumulative roll bonding [10]. The inelastic deformation regime is also found in repeated unloading-loading cycles during indentations. Probably this inelastic deformation behavior could be understood by a thermally activated grain boundary relaxation mechanism. The prevalent high internal stresses in these ultrafine-grained materials contribute to the unusual deformation behavior at the small grain size. More measurements have to be performed on these materials and especially experiments close to boundaries and in the grain interior will shed more light on these effects.

ACKNOWLEDGMENTS

The UFG-Al has been prepared in the laboratory of T. G. Langdon, Los Angeles, CA and the UFG-Cu sample has been supplied by R. Valiev, Ufa, Russia. These materials were prepared for a joint research project on the fatigue mechanism for the group in Erlangen (M. Kautz, H.W. Hoeppel and H. Mughrabi). Support by this group is gratefully acknowledged. The authors furthermore thank U. Ilg EnBW AG and R. Kilian Framatome ANP GmbH, Germany for supplying the steel specimens.

REFERENCES

1. J. H. Westbrook and D.L. Wood, Journal of the Institute of Metals **91**, 174 (1962-63)
2. M. Braunovic in: The science of hardness testing and its research applications, ed. Westbrook, J.H. and Conrad, H., American Society for Metals, Ohio, p. 329 (1973)
3. M. Göken, M. Kempf, M. Bordenet and H. Vehoff, Surface and Interface Analysis **27**, 302 (1999)
4. M. Göken and M. Kempf, Z. Metallkde. 92, 1061 (2001)
5. W.C. Oliver and G.M. Pharr, J. Mater. Res. **7**, 1564 (1992)
6. E. Schweitzer, M. Göken, Z. Metallkde. in press (2004)
7. H. W. Höppel, C. Xu, M. Kautz, N. Barta-Schreiber, T. G. Langdon, H. Mughrabi, in: Nanomaterials by Severe Plastic Deformation, edited by M. Zehetbauer and R. Valiev, (Wiley-VCH, Weinheim, 2004) pp 677-683
8. R. Schwaiger, B. Moser, M. Dao, N. Chollacoop, and S. Suresh, Acta Mater. **51**, 5159 (2003)
9. K. S. Kumar, H. van Swygenhoven and S. Suresh, Acta Mater. **51**, 5743 (2003)
10. H. W. Höppel, J. May and M. Göken, Adv. Eng. Mater. 6, 219 (2004)

Mat. Res. Soc. Symp. Proc. Vol. 821 © 2004 Materials Research Society P8.8

Stress Distribution in Ultra Thin SiO_2 Film/Si Substrate System Measured by a Low Level Birefringence Detection Technique

X. H. Liu*, H. J. Peng*, S.P. Wong*[1], Shounan Zhao**,
*Department. of Electronic Engineering and Materials Science and Technology Research Centre, The Chinese University of Hong Kong, Hong Kong, China
** Department of Applied Physics, South China University of Technology, Guangzhou, China

ABSTRACT

In this work, a low level birefringence detection system was employed to study the stress distribution in Si substrate induced by thermally grown ultra-thin SiO_2 film. According to traditional bi-metallic strip theory, it is expected that the stress should show a linear dependence on depth with the zero stress plane located at the position of two third of the thickness of the substrate from the SiO_2/Si interface. The linear dependence of stress on depth in accordance with the bi-metallic strip theory was observed only in part of the substrate. For the region below the SiO_2/Si interface extending to a depth of about 1/5 of the thickness of the substrate, the magnitude of the stress was found to be significantly smaller than expected. The position of the zero stress plane was found to depend on the thickness of the SiO_2 film and the oxidation conditions. The zero stress plane seemed to move towards the bottom of the Si substrate as the thickness of the SiO_2 film became thinner and no zero stress plane was observed in the Si substrate when the SiO_2 films became sufficiently thin.

INTRODUCTION

The oxidation of silicon is an important industrial process, which has been studied intensively for tens of years. Though several models have been proposed and successfully fit the case of thick oxidation, there are still some arguments on the initial oxidation or ultra-thin oxidation. For example, “initial enhanced oxidation" and "pattern dependent oxidation" phenomena cannot be explained by the conventional thermal oxidation theory satisfactorily [1]. One of the difficulties in addressing this issue is the lack of tools to characterize the thin oxidation process.

Stress measurement in the oxide layer and silicon substrate is a popular method helping us to understand the oxidation mechanism. Some stress measurement techniques have been used to characterize the stress in the silicon substrate and oxide film. Curvature method is widely adopted in the film stress measurement [2]. But this method could not suit the need of the ultra thin oxidation stress characterization due to its stress sensitivity limitation. Another popularly used method is the infrared photoelastic (PE) technique which is non-destructive [3]. The traditional full-field imaging method determines the stress state using a plane or a circular polariscope. However, its sensitivity is quite low since it provides phase information by counting

[1] Contacting Author: S. P. Wong

the fringes which are proportional to the magnitude of the birefringence. Thus, the resulting minimum measurable phase retardation is in general a fraction of 180° or π.

A new low-level birefringence detection (LLBD) system has been developed in our group through which very small internal stress of semiconductors could be measured [4]. The birefringence measurement sensitivity of this system was found to be about 0.03°, corresponding to a stress level of about 10^5 dyne/cm^2 in silicon sample for a small measurement optical path length of 0.5 mm. In this work, this system was employed to study the stress distribution in silicon substrate induced by thermally grown ultra-thin oxide layer.

BACKGROUND THEORY

When a uniform oxide layer exists on one side of a silicon substrate, stresses will result leading to the bending of the structure due to the difference in the thermal expansion coefficients. Supposing that the silicon substrate is thin enough, bi-metallic strip theory can be applied to model the stress in the substrate [2]. In this theory the stress in the substrate perpendicular to its surface could be negligible. Neglecting the edge effects, the stress dependence on the depth in the substrate parallel to the surface is:

$$\sigma_s(z) = \frac{E_s}{1-\upsilon_s}\frac{2h+3z}{3R_0} \qquad (1)$$

where E_s, ν_s, z, and h are the Young's modulus, Poisson's ratio, depth and the thickness of the substrate, and R_0 is the radius of curvature of the bending. It is expected by the bi-metallic strip theory that the stress should show a linear dependence on depth with the zero stress plane located at the position of two third of the thickness of the substrate from the SiO_2/Si interface.

Based on the LLBD system we developed, we designed a scan system with a sample moving stage to measure the stress distribution in the silicon substrate. In the LLBD system, a laser beam with a wavelength of 1152 nm passes in succession through a polarizer, a photoelastic modulator (PEM), the first quarter-wave plate, the stressed sample, the second quarter wave plate, a polarization analyzer, and then incident to a detector. The principle of LLBD is as follows: after a polarized light passes through the PEM and the first quarter-wave plate, the electric vector components of it along the fast and slow axes of the PEM are modulated to sinusoidal signals at a frequency given by the PEM. Since the phase difference of the two components is π/2, the total intensity of the light beam is a constant. After passing through the stressed sample, due to the PE effect, an additional retardation is introduced. Therefore the total intensity of the light passing through the second quarter wave plate and the analyzer, is not constant anymore but an AC signal consisting of multiple frequencies and intensities, dependent on the retardation and its orientation with respect to the fast axis of the PEM. Using a lock-in amplifier, the intensities of the DC, first harmonic and second harmonic signals are measured, from which the retardation (δ) and orientation (ρ) can be calculated. According to the PE theory, we have

$$\delta = \frac{2\pi}{\lambda}Cd(\sigma_1 - \sigma_2) \qquad (2)$$

where σ_1 and σ_2 are the secondary principal stresses in the plane normal to the incident direction, *C* is the relative stress-optical coefficient, *d* is the thickness of the sample along the propagation direction, and λ is the wavelength of the light source. From this equation, the difference of the secondary principal stresses can be calculated. The direction of the secondary principal stresses is decided by the orientation with respect to the reference coordination system.

EXPERIMENTAL DETAILS

To measure the stress distribution in the silicon substrate, a slit and a moving stage was added to the LLBD system. The detail of the arrangement is shown in Fig. 1. After the light beam passing through the sample, only the light at the position of the slit (width 10 μm) can go through it. Hence the stress distribution can be measured.

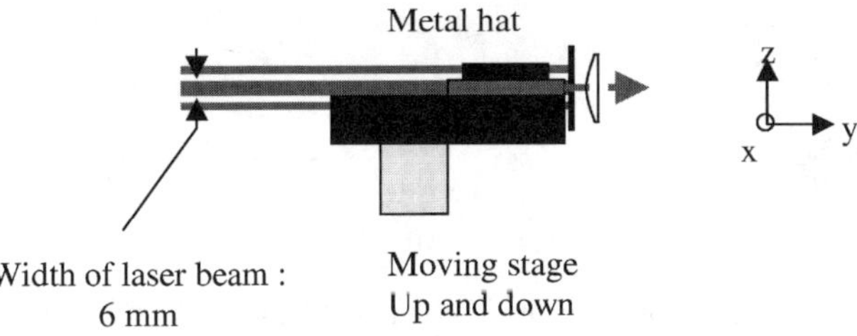

Fig. 1 Schematic of the sample block

In this work, single-sided polished p-type (100) silicon wafers with a resistivity of 65~135 Ω-cm and a thickness of about 400 or 500 μm were used. The oxide layers on these wafers were prepared using two different oxidation processes: dry thermal oxidation and electron-beam (E-beam) deposition process. The dry thermal oxidation was performed at 1100°C with an oxygen flow of 3 lit./min. The E-beam depositing temperature was 220 °C. Samples with different oxide thicknesses were prepared for both the dry thermal oxidation and the E-beam process, depending on the processing time. Ellipsometry was used to characterize the thickness of the oxide layer. After the oxide layers were prepared, the wafers were cut into pieces with a size of about 2 cm x 1 cm. The (010) planes of the silicon substrates were polished where the light beam is incident on and coming out. The dry thermal oxidation samples were dipped into HF with the polished side covered by photoresist so that only the oxide layer on the polished side was left after the etching. Before the stress distribution measurement was carried out, the samples were stored in an electronic dry cabinet with a humidity of 49% at room temperature for several days, helping to relax the stress of the sample induced by the polish process.

Using the LLBD system and the sample stage described above, the distribution of stress in the cross-sectional plane of the sample were measured with a moving step of 10 μm. The slit was put at the center of the sample and moved along the direction perpendicular to the surface. The depth resolution is decided by the slit width of about 10 μm.

RESULTS AND DISCUSSION

The variation of oxide layer thickness with the preparation time is plotted in Fig. 2. For dry thermal oxidation (Fig. 2a), the thickness of the sample with an oxidation time of 500 seconds deviates from the "Deal-Grove" fitting [5] indicating that it is under the initial enhanced oxidation regime. The data for other dry thermal oxidation samples fit the model quite well. For E-beam oxidation (Fig. 2b), it is clear that the growth rate is a constant.

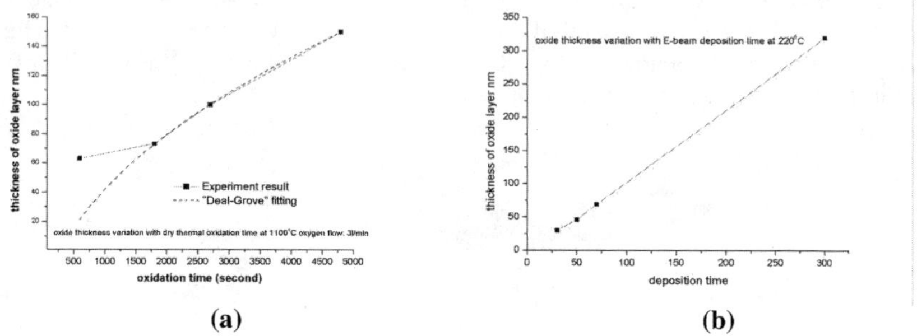

(a) (b)

Fig. 2 Variation of oxide layer thickness with processing time. (a) Dry thermal oxidation at 1100°C with an oxygen flow rate of 3 lit./min. (b) E-beam deposition at 220°C.

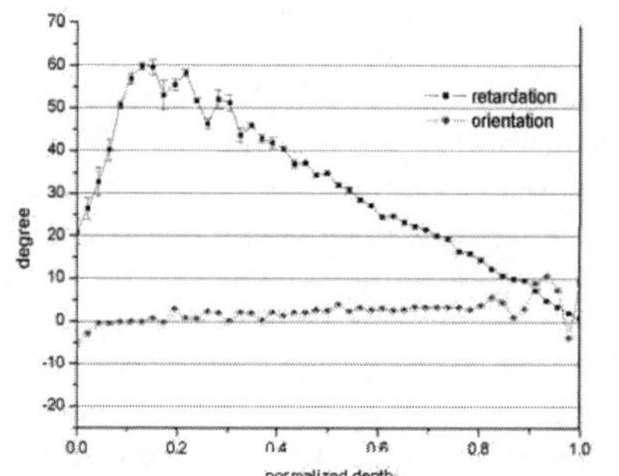

Fig. 3 Measured retardation and orientation results for a dry thermal oxidation sample with an oxidation time of 4800s. The normalized depth is measured from the SiO_2/Si interface to the bottom of the substrate.

The measured retardation and orientation of a typical sample are plotted in Fig. 3. The retardation and orientation were calculated using the method we discussed above. The orientation data for this sample are all close to 0 degree indicating that the directions of the secondary principal stresses are basically along the x and z-axes. Under the bi-axial stress assumption the stress component perpendicular to the sample surface (z-axis) is negligible compared to the component parallel to the surface (x-axis). Thus we have

$$\sigma_x \approx \sigma_x - \sigma_z = \sigma_1 - \sigma_2 \qquad (3)$$

Substitute Eq. 3 into Eq. 2, we have

$$\sigma_x \approx \frac{\delta\lambda}{2\pi C d(\sigma_1 - \sigma_2)} \tag{4}$$

from which the horizontal stress component at the point can be estimated directly.

The stresses parallel to the surface direction versus normalized depth for different samples are shown in Fig. 4. For all the dry thermal oxidation samples, two different regions, labeled as region A and B, respectively, can be identified from their stress distribution (Fig. 4a). Region A starts from the interface to a position of about 1/5 of the thickness of the substrate, where the stress arrives the maximum point. The other part of the substrate is region B, where a linear dependence of the stress on the depth is found and a non-zero minimum stress appears at the bottom of the substrate. On the other hand, for the E-beam oxidation samples, three regions, labeled as region A, B, and C, respectively, can be identified (Fig. 4b). Region A also starts from the interface to a position of about 1/5 of the thickness of the substrate. Next it comes the linear region B, and then the region C near the bottom where there is an increase in stress. The presence of Region C in the stress distribution of the E-beam oxidation samples which does not show up in the dry thermal oxidation samples could be attributed to stress effects induced by the bottom surface roughness of the single-sided polished silicon wafer. For dry thermal oxidation samples, the rough bottom surface was greatly smoothed by the process of thermal oxidation and etching.

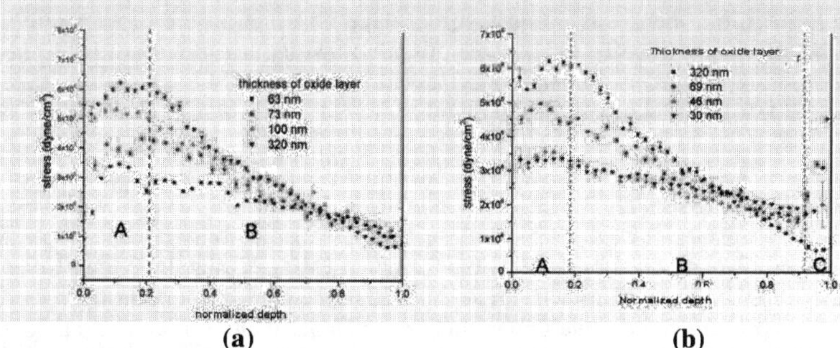

(a) **(b)**

Fig. 4 Stress distribution in the Si substrate under oxide layers for samples with various oxide layer thickness prepared by (a) dry thermal oxidation and (b) E-beam deposition.

According to the bi-metallic strip theory, it is expected that the stress should show a linear dependence on depth with the zero stress plane located at the position of two third of the thickness of the substrate from the SiO_2/Si interface. The linear dependence on depth was verified only for part of the depth as shown by our measurement results. In region A for both the dry thermal oxidation and E-beam oxidation samples, the magnitude of the stresses is significantly smaller than expected at the interface. There must be some physical mechanisms leading to the relaxation of stresses near the interface.

In the bi-metallic strip theory, the mechanical properties (Young's modulus, Poisson's ratio, etc.) of the substrate are assumed to be homogeneous. In reality, during the oxidation process, significant volume expansion near the oxide/silicon interface will occur and stress will generate in both the silicon dioxide layer and the silicon substrate. As the expansion of the layer is free in the direction perpendicular to the surface of the silica film but strained parallel to it, a tensional stress is induced close to the interface in the silicon substrate. To relax the stress, point defects would be emitted or captured by dislocation near the interface in the silicon lattice. Meanwhile the kick-out self-interstitials of the silicon atoms would also cause mechanical inhomogeneity in the silicon substrate [6-8]. All these effects make the properties of the silicon substrate at the region close to the interface different from the bulk.

It was reported that the grinding of one side of the silicon wafer could also lead to a linear stress distribution depending on the depth in the substrate. In our experiment, it means that our single-sided polished wafer will have a tensional stress on both the oxidation side and the grinding side. Superposing of the stresses originating from the two opposite sides we have a stress distribution with a reduced slope. This could be a possible reason why there was no zero stress plane observed in our measurement results.

CONCLUSIONS

In this study, a LLBD system was used to measure the stress distribution in the silicon substrates covered with oxide layers prepared using either dry thermal oxidation or E-beam deposition. The linear dependence of stress on depth in accordance with the bi-metallic strip theory was observed only in part of the substrate. The magnitude of the stress in the region near the oxide/silicon interface deviates significantly from the prediction of the be-metallic strip theory. In addition, there was no zero stress plane observed in the Si substrate when the SiO_2 film was sufficiently thin. To explain the observed results, it is believed that a theory taken into consideration of the inhomogeneous mechanical properties of the silicon substrate in the region near the oxide/silicon interface caused by microscopic stress relaxation processes is necessary.

ACKNOWLEDGMENTS

This work is partially supported by the Research Grants Council of Hong Kong SAR (ref. no. CUHK4370/02E). The authors wish to acknowledge the research support from Dr. N. Ke and Dr. W. Y. Cheung of the Solid State Laboratory of the Department of Electronic Engineering at the Chinese University of Hong Kong.

REFERENCES

1. A Athanassouli, Th Ganetsos, F Klose and S Messoloras, Semiconductor Science and Technology, 17, 65 (2002).
2. Bernard Leroy, Philosophical Magazine B, 55, 159 (1987).
3. H. Liang, Y. Pan, S. Zhao, G. Qin and K.K. Chin, J. Appl. Phys, 71, 2863 (1992).
4. H.J. Peng, S.P. Wong, Y. W. Lai, X. H. Liu, H. P. Ho, and Shounan Zhao, Rev. Sci. Instr. 74, 4745 (2003).
5. B.E. Deal and A.S. Grove, J. Appl. Phys. 36, 3770 (1965).
6. H. Kageshima and K. Shiraishi, Phys. Rev. Lett. 81, 5936 (1998).
7. S.M. Hu, Appl. Phys. Lett. 27, 165 (1975).
8. T.Y. Tan and U. Gosele, Appl. Phys. A 37, 1 (1985).

Mat. Res. Soc. Symp. Proc. Vol. 821 © 2004 Materials Research Society P2.3

Namomechanical Imaging and Nanoscale Elastic Modulus Measurements of SnO_2 Nanobelts

Y. Zheng and R. E. Geer
College of Nanoscale Science and Engineering, University at Albany, State University of New York, Albany NY 12203

ABSTRACT

The relative surface contact stiffness of SnO_2 nanobelts has been investigated via ultrasonic force microscopy (UFM). The nanobelt crystal structure, as determined via transmission electron microscopy, was indexed to the tetragonal rutile structure (with lattice constants identical to those of bulk SnO_2) as reported previously. The atomic Sn:O composition of the nanobelts studied was at or near 1:2. Topographic imaging studies revealed the nanobelt surface to be atomically flat with the exception of surface nanodots, assumed to be local SnO_2 crystallites. Preliminary reduced modulus measurements were carried out via differential UFM on both the flat and nanodot regions of the nanobelt. Using the underlying Si substrate as a calibration standard the SnO_2 modulus was estimated at 157±12 GPa, significantly lower than corresponding bulk values for any of the observed crystal orientations. We speculate this discrepancy is due in part to a combination of the aspherical probe tip and unknown adhesive properties of nanobelt. An intrinsic reduction of the SnO_2 nanobelt modulus cannot be ruled out.

INTRODUCTION

Semiconducting oxide nanobelts have been the focus of considerable experimental activity for applications in both nanoelectronics and nanosensors, in addition to their importance in elucidating the properties of self-assembling inorganic nanostructures[1-3]. Recently, nanomechanical analyses of individual SnO_2 nanobelts have been reported[4]. Nanoindentation was utilized to investigate the inelastic mechanical response of nanobelts including hardness and the load profile for fracture propagation. Such information is critical for predictive reliability analyses of nanoelectronic or nanosensor devices incorporating such nanostructures. However, a nondestructive approach is preferred to enable eventual in situ mechanical analysis of nanobelt-containing device structures. Pursuant to these goals nanomechanical, structural, and compositional characterization of individual SnO_2 nanobelts has been undertaken. The nanomechanical characterization consists of nanomechanical imaging based on ultrasonic force microscopy (UFM) and quantitative reduced modulus extraction via differential-UFM utilizing Si as a calibration standard. Structure characterization employed atomic force microscopy (AFM) for measurements of nanobelt surface topography and overall dimension, and transmission electron microscopy (TEM) for determination of nanobelt crystal structure and orientation. Auger electron spectroscopy (AES) was used for elemental compositional analysis.

Elemental compositional analysis via AES confirmed the expected 2:1 ratio of O:Sn. Electron diffraction confirmed a single crystal rutile structure in agreement with prior work[1,5]. However, in the present work separate, single-crystal nanobelts exhibited a variety of crystal orientations (relative to the nanobelt axis), in contrast to published reports. Topographic imaging revealed atomically smooth nanobelt surfaces with the exception of nano-dot regions presumed to be local SnO_2 crystallites. Nanomechanical imaging via UFM similarly revealed a uniform mechanical response across the surface of the nanobelts with the exception of the nanodots. The contrast of the latter was dominated by tip-surface contact area effects. Quantitative extraction of reduced modulus from both smooth and nanodot regions via differential UFM (using Si for calibration) yielded a value of 157±12 GPa, significantly lower than corresponding bulk values. This disagreement is unexpected considering the high degree of crystalline and compositional similarity between the nanobelts and bulk SnO_2. We speculate this disagreement is due, in large part, to a combination of differential-UFM experimental factors, although an inherent reduction in the reduced modulus of the nanobelt compared to bulk SnO_2 cannot be ruled out.

EXPERIMENTAL

SnO_2 nanobelts were synthesized following the procedure outlined in Ref. 6. Tin powder was placed in a crucible within a quartz tube furnace. The reaction chamber was heated to 800°C under an Ar/O_2 flow and maintained at that temperature for two hours followed by a slow cooling ramp to room temperature. Deposited material is removed from alumina substrates adjacent to the crucible in the oven and dispersed in a solvent (acetone or ethanol) for solvent-based deposition on Si (100) substrates. Auger electron spectroscopy (AES) was carried out using a PHI 600 Scanning Auger Multiprobe spectrometer with a 5keV primary electron beam. Relative elemental composition can be determined to an accuracy of 2%. AES depth-profiling employed a 4keV Ar-ion sputter beam. The AES probe beam diameter is approximately 1 μm. Consequently, relatively wide nanobelts were used for compositional analysis. Transmission electron diffraction data were acquired with a JEOL 2010 HR-TEM. Nanobelts were dispersed on standard C-coated Cu TEM grids for diffraction.

UFM imaging scans were acquired using an ultrasonic force microscope configured from a commercial scanning probe microscope (JEOL SPM4200) employing crystalline Si_3N_4 cantilever contact tips. In elastic imaging mode a super-resonant ultrasonic vibration (f = 2.2MHz) is applied to the sample via a piezoelectric crystal. In this dynamic regime the cantilever oscillation amplitude is inertially damped resulting in a periodic deformation of the sample surface by the Si_3N_4 tip. Modulation of the sample ultrasonic waveform (half duty-cycle triangular waveform at 2.7 kHz) results in a periodic cantilever deflection signal varying nonlinearly with ultrasonic vibration amplitude. This tip deflection is quasi-static on the order of the ultrasonic vibration frequency. The periodic cantilever deflection is measured via lockin amplification[7-8]. The amplitude of the phase-locked deflection signal is proportional to the elastic/adhesive response of the sample surface to the immobile silicon nitride tip. Neglecting adhesive interactions, the resultant image contrast is proportional to the local contact stiffness.

Differential UFM extracts the reduced Young's modulus via determination of the onset of scanning probe tip deflection as a function of applied tip force[9-11]. This threshold is calculated for relative applied tip forces ranging from 0nN to 25nN (calibrated for a given cantilever spring constant). The ratio of the force differential to the threshold amplitude differential is assumed equal to the surface contact stiffness[9]:

$$S \equiv \frac{\partial F_{Tip}}{\partial A_T}. \tag{1}$$

F_{Tip} denotes the applied tip force and A_T denotes the threshold amplitude for the onset of quasi-static tip deflection. Neglecting adhesive effects (Hertzian model), the contact stiffness for a spherical indenter on a planar surface can be related to the reduced modulus as follows[12]:

$$S_{Hertz} = \sqrt[3]{6\bar{F}RE_r^2}\,. \quad (2)$$

$\bar{F}$ denotes the average tip/sample force over the interval used to determine S. E_r is defined as the reduced modulus[12]. For purposes of the extraction of E_r we utilize a Si substrate as a calibration standard. The Hertzian contact stiffness is measured similarly as for the nanobelt and the reduced modulus ratio is determined. The contact stiffness ratio is approximately independent of the average tip/sample force if the adhesive portion of the force does not vary substantially between the nanobelt and the Si calibration standard and the same applied tip forces are used for both.

RESULTS AND DISCUSSION

The elemental composition, determined from AES, of an individual 850nm-wide nanobelt is shown Fig. 1. The O:Sn ratio is approximately 2:1. The Si signal is due to the slight illumination of the Si substrate from the 1 μm diameter electron probe. AES depth profiling of this and other nanobelts was undertaken to investigate compositional uniformity. To within the experimental resolution of AES the O:Sn ratio was constant throughout each nanobelt investigated.

Investigation of the crystalline structure of the nanobelts was undertaken via electron diffraction. Previous studies of similar materials indicated a rutile structure as for bulk SnO_2 with corresponding lattice constants[1,5]. The nanobelts investigated here exhibited a single crystal rutile structure with measured lattice constants a=b=4.73 Å and c=3.19Å which agree precisely with bulk SnO_2 values. In contrast to published reports we observed multiple nanobelt crystalline orientations. Specifically, four distinct orientations were observed for the various SnO_2 nanobelts studied: $[10\bar{1}]$, $[1\bar{3}1]$, $[\bar{1}13]$, $[10\bar{3}]$. The orientations are defined by the surface normal direction of the nanobelt in a real space rutile lattice. The first orientation is consistent with a [101] growth direction and (010) side facets as first reported by Pan et al.[1]

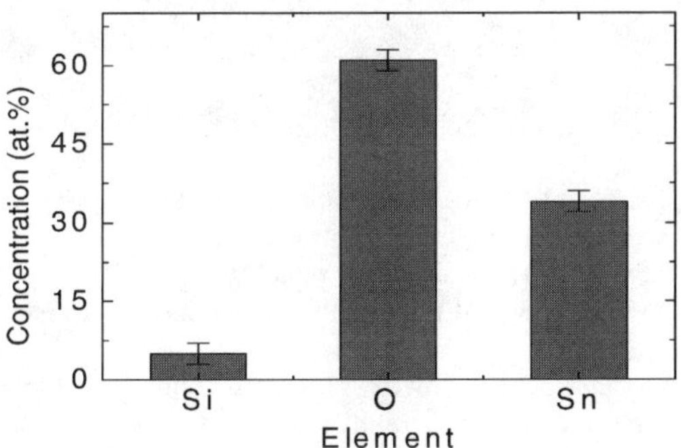

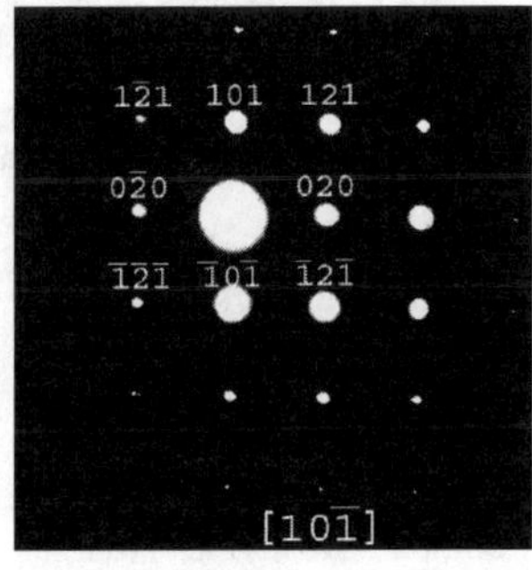

Figure 1. (Left) Elemental composition of a SnO_2 nanobelt as determined from AES. The O:Sn ratio is nearly 2:1. The Si is attributed to a substrate contribution. (Right) Electron diffraction pattern from a SnO_2 nanobelt. The peak indexes are as shown. In real space coordinates the normal of the nanobelt is parallel to the $[10\bar{1}]$ axis.

The latter three have not, to the best of our knowledge, been reported. This has important implications for the expected values for the reduced Young's modulus along the surface normal direction (discussed below). Figure 1 also shows an indexed diffraction pattern from a $[10\bar{1}]$ SnO_2 nanobelt. Corresponding figures for the other observed orientations are omitted for brevity.

Topographic (AFM) and nanomechanical (UFM) imaging data were acquired for a variety of nanobelts. Figure 2a shows topographic detail of a single nanobelt with a rectangular cross section measuring 196nm x 1,255nm. The nearly 1:6 aspect ratio is similar to previous reports regarding SnO_2 nanobelts[1,4-6]. The cross-section is uniform along the long axis of the nanobelt. The top surface is atomically smooth, possessing an RMS roughness of 0.18±0.02nm. Closer inspection of the top surface of the nanobelt reveals a collection of discrete topographic features. A higher resolution image of two of these features is shown in Fig. 2b. Typical length, width, and height dimensions are 90nm, 50nm, and 5nm, respectively. The stability of the features during topographic scanning and their uniform orientation with respect to the nanobelt axis implies they are instrinsic features of the nanobelt, presumably nucleated SnO_x domains. Figures 2c and 2d display UFM image scans acquired simultaneoulsy with the topographic data. The UFM image of the nanobelt is, excluding the nanodot regions, highly uniform. The latter features possess a reduced contrast in the UFM scan of Fig. 2d.

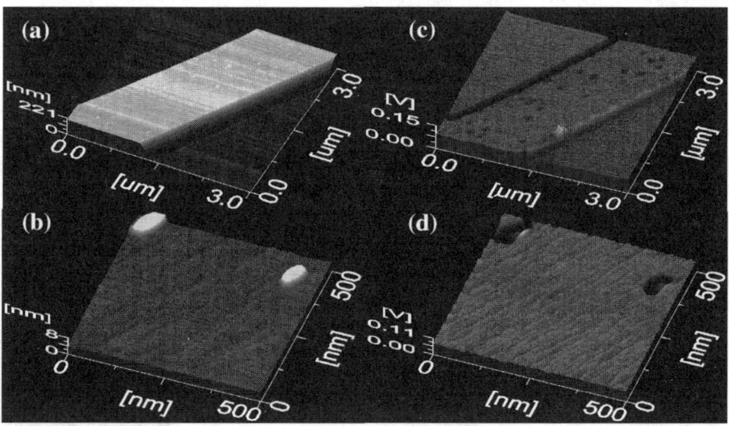

Figure 2. (a) 3μm x 3 μm opographic scan of a SnO_2 nanobelt. The small, high-contrast regions on the nanobelt surface correspond to nanodot features. (b) 500nm x 500nm topographic scan detailing two nanodots. (c) UFM image of the area shown in panel (a). The only significant contrast occurs for the nanodots. (d) UFM image of the nanobelt and nanodots in panel (b).

Typically, lower UFM contrast is associated with a lower value of the nanodot reduced modulus compared with the flat areas of the nanobelt. This is actually not the case for the data in Fig. 2d and is discussed in greater detail below.

Differential UFM was carried out on SnO_2 nanobelts following the methodology outlined in the experimental section and documented in the literature[9-11]. These data are shown in Fig. 3. As previously noted the variation of ultrasonic threshold amplitude for quasi-static tip deflection is plotted as a function of the relative tip/sample contact force. The probe area for each of the plots in Fig. 3 was set at 10nm x 10nm. The leftmost plot was acquired on the Si substrate at a distance of less than 1 μm from the nanobelt.

As evidenced by the two plots the variation of force with threshold amplitude is approximately linear. The Hertzian expression leading to Eq. 2 predicts a power-law dependence of the average force on indentation with an exponent of 1.5. Considering the relatively small variation in the threshold amplitude accessed during the modulation sweep (< 0.35nm) and the errors inherent in determining A_T we expect the experimentally measured slope to be approximately linear. From the data in Fig. 3 the contact stiffness ratio of the nanobelt to Si is 0.91±0.08. This corresponds to a reduced modulus ratio of 0.87±0.07. Taking the modulus of (100) Si to be 179 GPa[13] and the Poisson's ratio to be 0.22[14] we estimate the reduced Young's modulus of the Si substrate to be 188 GPa. Equation 3 yields a value for the reduced SnO_2 modulus of 164±13 GPa. Assuming a Poisson's ratio for SnO_2 of 0.2 the Young's modulus of the nanobelt is 157±12 GPa.

Differential UFM measurements of the nanodot features shown in Fig. 2 were also carried out. The contact stiffness ratio of the nanodot region compared to that of a neighboring flat region was measured to be 1.05±0.15. Consequently, we conclude that there is no resolvable difference in reduced modulus between the nanodot and flat portion of the nanobelt.

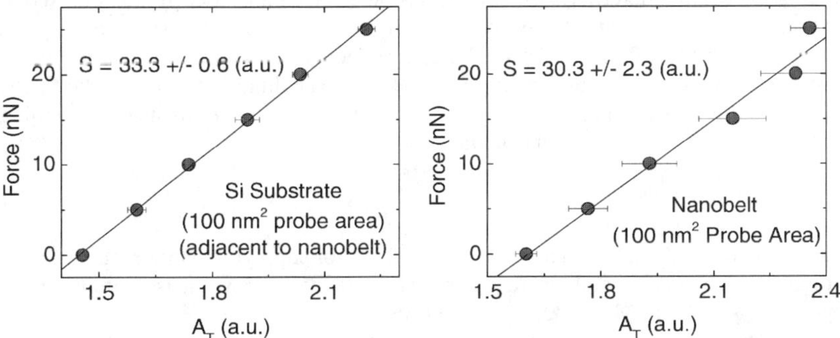

Figure 3. Force versus threshold amplitude plots to experimentally determined the surface contact stiffness from the Si substrate (left) and a smooth region of a SnO_2 nanobelt (right).

In light of this result we conclude the significant UFM contrast associated with the nanodots compared to the flat portion of the nanobelt (Fig. 2d) arises wholly from changes in tip/sample contact area during the scanning. The convex curvature of the nanodot results in local reductions in the tip/sample contact area that increase the local tip pressure. This represents an effective decrease in local contact stiffness that increases the threshold amplitude and reduces the UFM contrast. Similar effects have been pointed out in the literature[7].

The SnO_2 reduced modulus calculated from the data in Fig. 3 is unexpectedly low for bulk SnO_2. A single crystal of rutile SnO_2 is described by 6 independent elastic constants. Using the measured compliances determined for bulk SnO_2[15] the calculated values of the Young's modulus for the nanobelt crystal orientations observed here ranged from 181 GPa for $\lfloor 1\bar{3}1 \rfloor$ nanobelts to 453 GPa for $\lfloor 10\bar{3} \rfloor$ nanobelts. The $\lfloor 10\bar{1} \rfloor$ orientation referred to in the literature would possess a Young's modulus of 185 GPa if it behaved as a bulk sample. It is clear from the data presented above that the measured modulus of the nanobelts is significantly lower than that of a corresponding bulk sample of the same crystallographic orientation. Considering the structural similarity of the nanobelts to bulk samples this disagreement is unexpected. This is complicated by the lack of crystallographic information on specific nanobelts for which differential UFM measurements were undertaken. Of significant concern is the difficulty in using differential UFM on high-modulus materials. Although its accuracy has been verified for relatively 'soft' materials including silicon oxide glasses[11] ($E < 100$ GPa) the reproducibility for differential UFM in harder materials including Si and Al_2O_3 has proved elusive[9]. Most of the challenges relate to reproducible tip/surface contact areas. Significant wear of aspheric asperities of the tip invalidate the use of the Hertzian assumptions. These same issues may be critical for the nanobelts studied here. In addition, the inability to quantify variations in the adhesive properties of the nanobelt/tip interface may be a critical factor in accurate modulus calibration. Finally, a reduction in the Young's modulus of the nanobelts compared to bulk SnO_2 cannot, at this time, be ruled out.

CONCLUSIONS

The relative surface contact stiffness of SnO_2 nanobelts has been investigated via UFM. The nanobelt crystal structure determined via transmission electron microscopy was indexed to the tetragonal rutile structure as reported previously. The atomic Sn:O composition of the nanobelts studied was at or near 1:2. Preliminary local measurements of the SnO_2 modulus were carried out via differential UFM. Using the underlying Si substrate as a calibration standard the SnO_2 indentation modulus was estimated at 157 ± 12 GPa, significantly lower than corresponding bulk values for any of the observed crystal orientations.

ACKNOWLEDGMENTS

The authors gratefully acknowledge Prof. P. Ajayan for supplying nanobelt materials, Dr. R. Moore for assistance in AES, and Dr. K. Dunn for assistance in TEM measurements. This work was supported by the NSF, MARCO, and NYSTAR.

REFERENCES

1. Zheng Wei Pan, Zu Rong Dai, Zhong Lin Wang, *Science* **291**, 1947-49 (2001).
2. E. Comini, G. Faglia, G.Sberveg, Z.Pan and Z.L.Wang, *Appl. Phys. Lett.* **81**, 1869-71 (2002).
3. M.S.Arnold, P. Avouris; W. P. Zheng, and Z.L. Wang, *J. Phys. Chem. B* **107**, 659-63 (2003).
4. Scott X. Mao, Minhua Zhao, and Zhong Lin Wang, *Appl. Phys. Lett.* **83**, 993-995 (2003).
5. Jun Zhang, Feihong Jiang and Lide Zhang, *J. Phys. D: Appl. Phys.* **36,** L21-L24 (2003).
6. S. H. Sun, G. W. Meng, Y.W. Wang, T. Gao, M. G. Zhang, Y. T. Tian, X. S. Peng, and L. D. Zhang, *Appl. Phys. A, Mater. Sci. Process.* **A76**, 287-289 (2003).
7. O. V. Kolosov, M.R. Castell, C. D. Marsh, G. A. D. Briggs, T. I. Kamins, and R. S. Williams, *Phys. Rev. Lett.* **81**, 1046-49 (1998).
8. G. Shekhawat, O. Kolosov, G.A.D. Briggs, R.E. Geer, *J. Appl. Phys.*, **91**, 4549-54 (2002).
9. F. Dinelli, S. Biswas, G.A.D. Briggs, and O. Kolosov, Phys. Rev. B **61**, 13995-14006 (2000).
10. L. Muthuswami and R. E. Geer, *Appl. Phys. Lett.*, accepted for publication.
11. L. Muthuswami, E. S. Moyer, Z. Li, and R. E. Geer, *Proceedings of the IEEE 2002 International Interconnect Technology Conference* (Cat. No. 02EX519C), 239-41 (2002).
12. A. C. Fischer-Cripps, *Nanoindentation*, Springer, New York, p. 3 (2002).
13. Bharat Bhushan and Xiaodong Li, *J. Mater. Res.* **12** 54-63 (1997).
14. J. Dolbow and M. Gosz, *Mechanics of Materials*, **23.** 311-321 (1996).
15. E. Chang and E. K. Graham, *J. Geophys. Res.* **80**, 2595-2599 (1975).

Mat. Res. Soc. Symp. Proc. Vol. 821 © 2004 Materials Research Society P8.7

Investigation of Strain Accommodation upon Phase Transformation of Small Inclusions in Aluminum

L. H. Zhang[1], E. Johnson[2], U. Dahmen[1]
[1] National Center for Electron Microscopy, Lawrence Berkeley National Laboratory, Berkeley, CA94720, USA
[2] Nano Science Center, Niels Bohr Institute, University of Copenhagen, and Department of Materials Research, RISØ National Laboratory, Denmark.

ABSTRACT

The evolution of elastic strain caused by melting and solidification of small inclusions in aluminum was investigated by in-situ transmission electron microscopy. The appearance and subsequent decay of elastic strain during phase transformation of inclusions around 100nm in size were observed directly, and the decay rate was determined as a function of temperature. The mechanism of strain accommodation was studied by determining the activation energy of the process using alloy composition and inclusion size to control the transformation temperature.

INTRODUCTION

Small inclusions embedded in a solid matrix are subject to crystallographic and elastic constraints from the matrix [1,2,3,4,5]. These constraints can have a significant effect on thermodynamic and kinetic behavior of materials. Several such effects have been demonstrated for lead inclusions in aluminum. For example, it was found that residual elastic strain energy was responsible for magic-size behavior of solid Pb inclusions in Al [6], or that the anisotropy of interfacial energy can cause particles to be frozen in non-equilibrium [7,8], or that premelting – the initiation of melting below the bulk melting point - is controlled by the crystallographic orientation of an inclusion [9]. When inclusions melt or solidify, the shape or volume change leads to an elastic strain, which can be observed in a transmission electron microscope under two-beam dynamical diffraction conditions [10,11]. Our previous work showed the appearance and later decay of strain around a small $Pb_{72}Cd_{28}$ inclusion in an Al matrix during solidification of the particle [1]. It was hypothesized that the strain decay was due to vacancy transport, similar to that which operates during annealing of vacancy defects such as dislocation loops or voids. The study of defect shrinkage kinetics has proven to yield accurate measurements of self-diffusion coefficients, stacking fault and surface energies, and binding energies of impurity atoms [12].

However, to date the accommodation of elastic strain by vacancy transport has not been investigated in detail. In the present work we have examined the evolution of this strain systematically for a range of temperatures and sizes during in-situ heating and cooling experiments. By varying alloy composition and inclusion size we were able to control the transformation temperature and thus study the mechanism of strain accommodation by determining the activation energy of the process.

EXPERIMENTAL PROCEDURE

Alloys with nominal composition $Al_{99.5}Pb_{0.5}$, $Al_{99.5}(Pb_{72}Cd_{28})_{0.5}$, $Al_{99.5}(Pb_{50}Cd_{50})_{0.5}$, and $Al_{99.3}(Pb_{50}In_{50})_{0.7}$ were prepared from high purity elements (99.999% for Al, Pb, Cd and In) by rapid solidification, and TEM specimens were prepared by jet polishing in a solution of HNO_3: methanol = 1:3 at a temperature of -40 to -50°C. In-situ transmission electron microscopy (TEM) observations were performed using a Gatan double-tilt heating holder in a JEOL 3010 electron microscope equipped with a video system for recording dynamic behavior. The experiments were carried out at operating voltages of 200 and 300 kV, maintaining precise two-beam imaging conditions with g_{111} and deviation parameter s=0. Interior inclusions far from free surfaces, dislocations or grain boundaries were chosen because of the role of these defects in strain accommodation [13] and to minimize image artifacts caused by surface relaxation. After in-situ experiments, still images were captured to measure the magnitude and decay time of the strain.

To eliminate effect of thermal strain during phase transformation our experiments were carried out at slow heating rates, typically on the order of 0.1°C/s. Thus, the measured strain lobes are caused only by the phase transformation and not by differential thermal expansion.

RESULTS

To control the temperature range of the phase transformation, we varied the alloy composition of the inclusions from pure Pb to $Pb_{72}Cd_{28}$, $Pb_{50}Cd_{50}$, and $Pb_{50}In_{50}$. In this way, a wide range of transformation temperatures from 183°C to 354°C became accessible. Table 1 lists the composition of inclusions used in our experiments, along with their transformation temperatures estimated from the phase diagram, and the experimentally observed temperature ranges of phase transformation. It is apparent from the table that most of the inclusions exhibit a substantial hysteresis between melting and solidification and a strong size dependence of the transformation temperature. The range of particle sizes examined in this work is shown in a histogram in figure 1. Altogether more than 300 individual observations were made. As seen in figure 1, most of the observations were made on inclusions between 80 and 120nm in size.

Strain contrast upon melting was observed only for pure Pb inclusions, whereas during solidification, all inclusions displayed observable strain lobes. This is because melting of pure Pb

Table 1 Temperature range for melting and solidification of inclusions with different compositions and sizes. Tm: melting temperature, Ts: solidification temperature. The range of transformation temperatures for a given alloy composition is due to its size dependence.

Inclusion	Tm (phase diagram)	Tm (experiment)	Ts (experiment)
Pb	327°C	339°C - 354°C	300°C - 307°C
$Pb_{50}Cd_{50}$	248°C	230°C - 260°C	188°C - 220°C
$Pb_{72}Cd_{28}$	248°C	240°C - 250°C	210°C - 220°C
$Pb_{50}In_{50}$	210°C	220°C - 280°C	183°C - 250°C

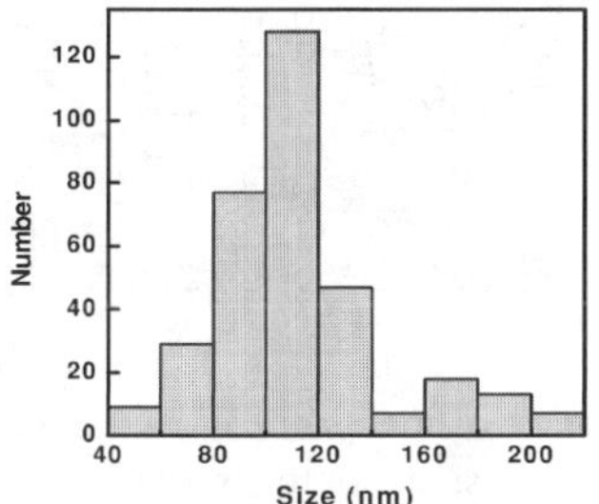

Figure 1 Histogram showing size distribution of inclusions examined in this work.

inclusions is much faster (a fraction of a second) than that of alloy inclusions, which usually proceeds gradually [1].

The evolution of this strain contrast was recorded at video rates. Figure 2 shows typical diffraction contrast images of a $Pb_{50}Cd_{50}$ inclusion before and after its solidification. This experiment was performed at 200 kV under g_{111} two-beam dynamical conditions, close to the<011> zone axis of matrix, where four of the major facets of the cuboctahedral shape of this particle are viewed edge-on. These imaging conditions are sensitive to the distortion of the {111} planes parallel to one pair of facets. Strong strain lobes are clearly apparent in (b) immediately after solidification at 188°C. This strain contrast is due to the 3.6% volume contraction of the $Pb_{50}Cd_{50}$ inclusion during solidification, which results in elastic strain in the matrix and inclusion. The elastic strain in the Al matrix decreased with time, as seen from figure 2(a) to (f), and disappeared completely in 95 seconds.

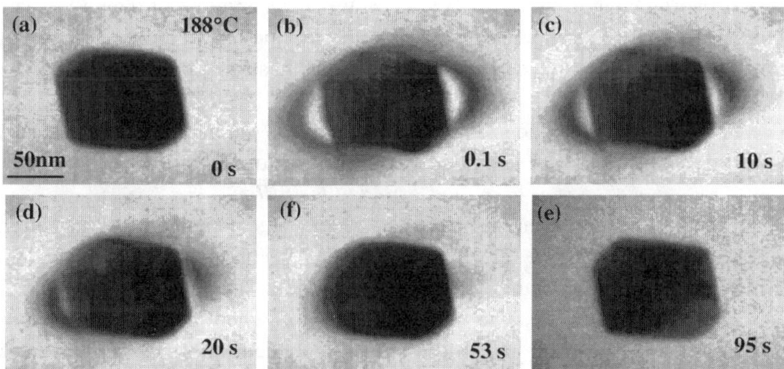

Figure 2 TEM micrographs before and after solidification of a $Pb_{50}Cd_{50}$ inclusion. The strain constrast appeared when the inclusion solidified and decayed completely over the course of 95 seconds.

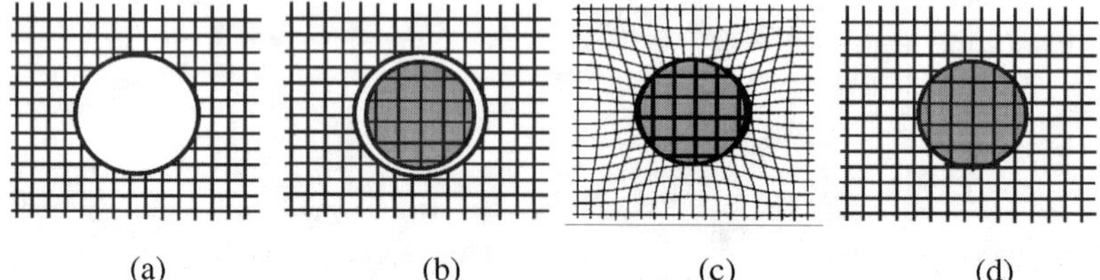

Figure 3 Schematic diagram showing solidification of inclusion. When the particle is liquid, the matrix is strain-free (a). As the inclusion solidifies, it contracts (b) and (c), causing elastic deformation of the matrix, which is relieved by lattice vacancies (d).

The same process was observed and recorded on video tape for many inclusions with different compositions and over a range of sizes. By choosing appropriate alloy compositions and particle sizes, a large temperature range from 183°C to 354°C was accessible in our experiments.

Inclusions undergo a volume change upon phase transformation. The volume difference at the solidification temperature of Pb, $Pb_{50}Cd_{50}$, $Pb_{72}Cd_{28}$ and $Pb_{50}In_{50}$ inclusions is 3.3%, 3.6%, 3.5% and 3.1%, respectively. When an inclusion solidifies, it contracts and distorts the Al lattice around it. A schematic of this process is shown in figure 3. The lattice distortion caused by the phase transformation (figure 3c) gives rise to strain contrast around the inclusion as observed experimentally in figure 2.

We expect that the elastic strain surrounding the inclusion is accommodated by vacancy diffusion in the Al matrix. Generally, the change in local vacancy concentration in the vicinity of a defect is given by [9]

$$c_1 = c_0 \, exp\{(dF / dn)/ kT\} \tag{1}$$

where dF/dn is the change in defect free energy per vacancy emitted and c_0 is the equilibrium vacancy concentration at temperature T. For an elastic inclusion, the free energy of the defect can be approximated by Eshelby's expression for the elastic strain energy of a spherical inclusion

$$E_{el} = 4\mu\varepsilon^2 V \tag{2}$$

where ε is the strain, μ is the shear modulus, and V is the particle volume. Due to the presence of this elastic strain, we thus obtain a local equilibrium concentration

$$c_1 = c_0 \, exp\{8\mu\Omega\varepsilon / 3kT\} \tag{3}$$

where Ω is the atomic volume. Vacancies that are generated at the interface set up a concentration gradient from c_1 at the particle to c_0 in the matrix. As a result, vacancies diffuse away from the particle to be annihilated at vacancy sinks such as grain boundaries or dislocations (see schematic in figure 3). The strain disappears when the vacancy concentration c_1 at inclusion/matrix interface is equal to the vacancy concentration c_0 in the perfect lattice. For a spherical diffusion geometry and diffusion-limited kinetics, it can be shown that the decay time t_d is roughly inversely proportional to the self-diffusion coefficient

$$D_S \propto 1/t_d \tag{4}$$

Here, t_d is the time for the strain to be reduced to fixed fraction of its initial value,

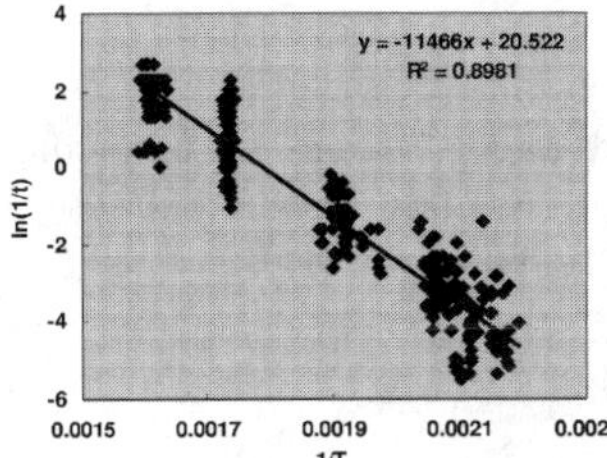

Figure 4 Plot of $ln(1/t_d)$ *versus* $1/T$ for more than 300 measurements of the strain decay time t_d for particles of different sizes and compositions. A linear least squares fit to these measurements gives an activation energy of Q=0.99±0.02 eV.

corresponding to the experimentally observed decay time. The self-diffusion coefficient has the form

$$D_s = D_0 \, exp(-Q / kT) \tag{5}$$

Thus, an Arrhenius plot of $ln(1/t_d)$ *versus* $1/T$ is expected to display a slope Q/k proportional to the activation energy for self-diffusion. Such a plot is shown in figure 4. A linear fit gives a slope corresponding to Q= 0.99±0.02 eV. The data exhibits a fair amount of scatter, which is thought to arise from variations in particle size, local composition and the possible effect of undetected sources or sinks of vacancies (such as out-of-contrast dislocations or voids). The scatter is significantly reduced when only a subset of data spanning a narrow size range is analyzed. However, the slope remains unchanged and significantly lower than expected for a process controlled by self-diffusion alone.

DISCUSSION

Our analysis follows the well-established procedure for the annealing kinetics of vacancy defects such as voids [14] or vacancy loops [15,16,17,18] in Al. In those experiments, the same method was shown to give accurate data on the self-diffusion coefficient because the data are derived from observation of individual defects rather than average quantities. Using this method, the activation energy for self-diffusion was found to be 1.19eV for high-purity Al [15]. This value is deemed to be the most accurate of all available experimental measurements (1.19eV [15], 1.26eV [17] and 1.31eV [14]). The activation energy of 0.99±0.02eV found from the current experiments is significantly lower than these values. It is likely that this is due to electron irradiation effects during observation. A preliminary check revealed no significant difference between experiments conducted at 200kV and at 300kV. However, this is not surprising because Al is subject to electron beam displacement damage at voltages higher than 170kV [19]. The effect of irradiation is known to be less pronounced above a critical temperature (260°C for aluminum) [20] where the thermal vacancy concentration matches that due to electron irradiation. Thus we expect the kinetics to be affected more strongly at lower temperatures, effectively reducing the temperature dependence and hence the measured activation energy of the process. To eliminate this effect, experiments at operating voltages below the displacement threshold for Al are currently underway.

CONCLUSIONS

We have investigated the mechanism and rate of strain accommodation during phase transformation of small inclusions by in-situ TEM. The elastic strain that appeared spontaneously upon melting or solidification of Pb or Pb alloy inclusions in Al was found to decay with time. Measurements of decay times as a function of temperature yielded an activation energy of 0.99±0.02 eV. We conclude that the mechanism of strain accommodation is vacancy diffusion, enhanced by the electron beam.

ACKNOWLEDGEMENTS

This work was supported by the Director, Office of Basic Energy Sciences, Materials Science Division, US Department of Energy, under contract DE-AC3-76SF00098 and the Danish Natural Sciences Research Council. We are grateful to P. Ochin and his group (CNRS-CECM) for preparing rapidly solidified materials.

REFERENCES

1 S. Hagège and U. Dahmen. Phil. Mag. Lett. 74, 259-266(1996).
2 M. Rühle and W.M. Kriven. Ber. Bunsenges. Phys. Chem. 87, 222-228(1983).
3 W. Mader and M. Rühle. Proceedings of the Institute of Physics Electron Microscopy and Analysis Group Conference (London, 1983) pp.385-388.
4 R. Goswami, K. Chattopadhyay and P.L. Ryder. Acta mater. 46, 4257-4271(1998).
5 G.L. Allen, W.W. Gile, and W.A. Jesser. Acta Metall. 28, 1695-1701(1980).
6 U. Dahmen, S. Q. Xiao, S. Paciornik, E. Johnson and A. Johansen. Phy. Rev. Lett., 78, 471-474(1997).
7 W. W. Mullins and G. S. Rohrer. J. Am. Ceram. Soc. 83, 214-16(2000).
8 N. Combe, P. Hensen and A. Pimpinelli. Phy. Rev. Lett., 85, 110-113(2000).
9 U. Dahmen, S. Hagege, F. Faudot, T. Radetic and E. Johnson. Philo. Mag., in press (2004).
10 M. F. Ashby and L.M. Brown. Phil. Mag. 8, 1083 (1963).
11 M. F. Ashby and L.M. Brown. Phil. Mag. 8, 1649 (1963).
12 R. E. Smallman. *Modern Physical Metallurgy*. Butterworths (1985).
13 A. K. Malhotra and D. C. Van Aken. Phil. Mag. A. 71, 949-964(1995).
14 T. E.Volin and R. W. Balluffi. Phys. Stat. Solidi. 25, 163-173(1968)
15 R. L. Peck and K. H. Westmacott. Metal Science. 9, 283-288(1975).
16 R. L. Peck and K. H. Westmacott. Metal Science Journal. 5, 155-159 (1971).
17 P. S. Dobson, P.J. Goodhew, and R.E. Smallman. Phil. Mag. 16, 9-22(1967).
18 R. E. Smallman and K. H. Westmacott. Mater. Sci. & Eng. 9, 249-272(1972).
19 J. J. Hren, J. I. Goldstein and D.C. Joy. *Introduction to Analytical Electron Microscopy* (New York, 1979). Vol.9, pp.447.
20 E. P. Butler and K.F. Hale. *Dynamical Experiments in the Electron Microscope* in *Practical Methods in Electron Microscopy* (1981, Amsterdam), 9, pp.18-22; R.H. Brickness and J.W. Edington. *Acta Metall.* 25, 447-458(1977).

Mat. Res. Soc. Symp. Proc. Vol. 821 © 2004 Materials Research Society P3.10

Computer Simulated 3D Virtual Reality for Dynamical Modeling and Calculations of Carbon-Based Composite Materials

Maksim V. Kireitseu[1†], David Hui[1], Liya Bochkaryova[2†], Sergey Eremeev[3†] and Igor Nedavniy[3]
[1]Composite Nano/Materials Research Center
University of New Orleans,
New Orleans, LA 70148-2220, USA; E-mail: indmash@yandex.ru
[2]United Institute of Informatics Problems NAS of Belarus
[3]Institute of Strength Physics and Materials Science
Siberian Branch of Russian Academy of Sciences
† - these authors have made an equal contribution to the research paper.

ABSTRACT

The principal goal of the present paper is to demonstrate an application of modern software engineering tools for modeling virtual reality and molecular dynamics of novel nanocomposites. The main technical components of presented system are 1) software and nanoengineering tools for modeling of virtual reality, molecular dynamic and 3D video images of novel diamonds-containing nanocomposites and 2) Problem tracking system to be used during modeling of virtual reality. For a realistic simulation of the stability behavior of the reinforced material, the nonlinear intramolecular inter-actions between neighboring atoms have to be taken into account. A comparison shows the buckling sensitivity of different geometries. In order to reduce computational costs, it is necessary to develop suited homogenization techniques, so that shell elements can be applied.

INTRODUCTION

Since the discovery of multi-walled carbon-based materials (nanotubes and diamonds) (MWCNT), NASA centers and researchers worldwide have engaged in fundamental studies of this novel material and have investigated the potential of its applied engineering and technological applications, including carbon nanotube and carbon nanoparticles (diamonds) based composites. The adequate description and numerical simulation of features of deformation and fracture of materials at various kind of loading are of great importance to the problem of design of advanced reinforced materials.

Many of the computer calculations and simulations of composite materials carried out in computational chemistry require powerful computational resources and developments in computer science and information analysis. The software for vector and parallel computing has had great influence on material science over the last 15 years because of the hardware development of processors. It seems likely that the continued increase in capabilities of computer hardware and new software will have great impacts on the nano industry and molecular modeling.

The use of molecular dynamics in modeling of materials is one of the great applications and successes. It has been important both for material crystallography and for establishing the structures of new carbon-reinforced composite materials. However, since the X-ray experimental data in crystallography of material are insufficient to study the structure at atomic level, researchers need to use new model and approach to fill in the gaps.

The XPLOR program [1], developed by Brunger and Karplus, is by far the most popular program for combined molecular dynamics/X-ray refinement and provides a good example of the general methodology. In it, one refines the initial structure using a function that contains both the difference between calculated and observed reflections and the molecular mechanical energy function. By using molecular dynamics and high temperature, one can move the structure out of local minimum and, by changing the weights of the molecular mechanical energy and the fit to experimental reflections, guide the system to a structure that has good stereochemistry and bonding and accurately reproduces the experimental reflections.

Modeling is even more important in deriving macromolecular structures by NMR, where the number of experimental geometric data is significantly less than in crystallography [2]. Thus, using various technologies to guide the determination of the three-dimensional structure, such as distance geometry and molecular dynamics, plays a large role in NMR structure determination of macromolecules.

Improvement in computer graphics has been probably the single largest contributor to a wide use of molecularly based modeling. Although some specific history is mentioned below in the chapter on molecular modeling in the pharmaceutical industry, the ability to visualize the drug target (when its structure was known) or a conformational family of active and inactive analogs was essential in the involvement of nonmodeling experts (i.e., the synthetic chemists) in the modeling. Computer graphics came of age when one could use depth cueing and stereo in color and thus highlight important interactions [3]. It has continued to develop and allow ever more beautiful and imaginative representations of important processes of chemical and biophysical interest [4].

In following sections, we first establish the bases of concept to simulate the deformation of carbon-nanotube reinforced composite materials and then demonstrate the approach by a few computational examples.

STRUCTURAL MECHANIC APPROACH

The potential energy of molecular systems like carbon nanotubes can be calculated with the force field method. We apply the DREIDING approach which is described by [5], among others. Neglecting inversion and non-bonded interactions, the potential energy of this generic force field is of the form

$$E = E_B + E_A + E_T \tag{1}$$

Fig. 1 illustrates the bond stretch energies. The distance between two atoms is R_e=1.39 Å at the equilibrium state. For the bond order n=1.5, the bond energies are given to k_e=1050 kcal/(mol Å²) and D_e=105 kcal/mol.

The bond between two atoms I and J can be described by the harmonic potential

$$E_B = \tfrac{1}{2} k_e \left(R - R_e\right)^2 \tag{2}$$

which leads to the spring constant

$$c_B = \frac{d^2 E_B}{dR^2} = k_e \tag{3}$$

or by the Morse function

$$E_B = D_e \left[e^{-\alpha n(R-R_e)} - 1 \right]^2 \tag{4}$$

which yields the nonlinear spring coefficient

$$c_B = \frac{d^2 E_B}{dR^2} = 4D_e (\alpha n)^2 \left[e^{-2\alpha n(R-R_e)} - \tfrac{1}{2} e^{-\alpha n(R-R_e)} \right] \quad (5).$$

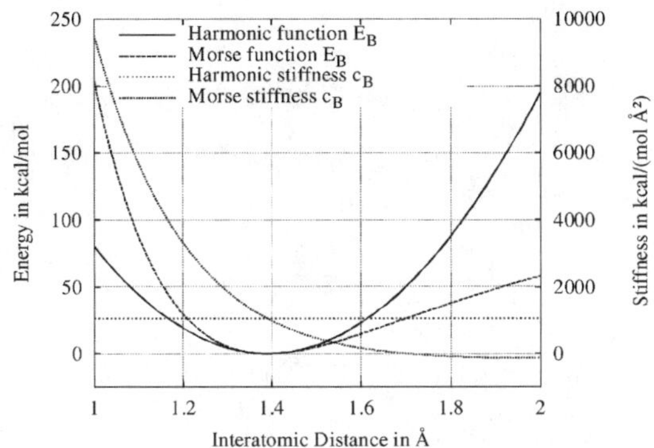

Figure 1. Bond stretch energies

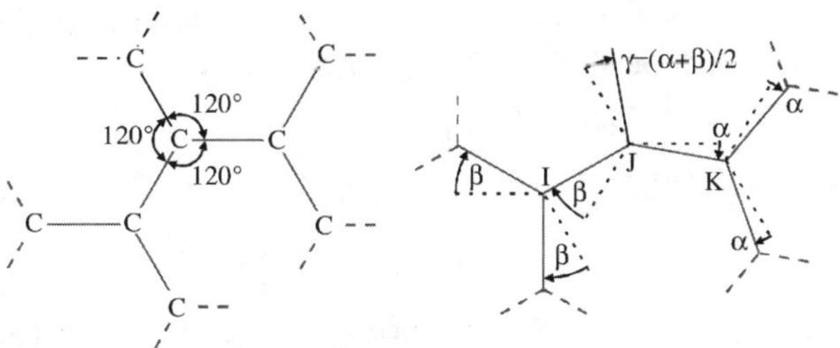

a.) Undeformed configuration b.) Deformed configuration

Figure 2. Bond angles

The second term of DREIDING is determined as

$$E_A = E_{IJK} = \tfrac{1}{2} K_{IJK} \left[\theta_{IJK} - \theta_J^0 \right]^2 \quad (6)$$

describes the variation of the potential energy caused by change of bond angles, see Fig. 3. This energy function leads to the spring constant:

$$c_A = \frac{d^2 E_A}{d\theta_{IJK}^2} = K_{IJK} \quad (7)$$

Alternatively, DREIDING provides a cosine form

$$E_A = E_{IJK} = \tfrac{1}{2} C_{IJK} \left[\cos\theta_{IJK} - \cos\theta_J^0 \right]^2 \quad (8)$$

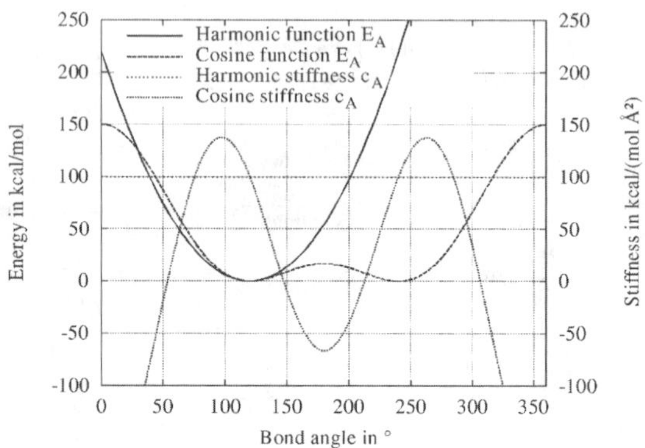

Figure 3. Bond angle energies

with the nonlinear spring coefficient

$$c_A = \frac{d^2 E_A}{d\theta_{IJK}^2} = C_{IJK}\left[\cos\theta_J^0 \cos\theta_{IJK} - \cos\left(2\theta_{IJK}\right)\right] \tag{9}$$

With θ_j^0=120° and K_{IJK}=100 kcal/(mol rad²), we obtain the bond angle energies given in Fig. 4. The torsion energy depends on the dihedral angle φ between the planes *IJK* and *JKL*.

$$E_T = E_{IJKL} = \tfrac{1}{2} V_{JK}\left\{1 - \cos\left[n_{JK}\left(\varphi - \varphi_{JK}^0\right)\right]\right\} \tag{10}$$

The torsional spring constant is

$$c_T = \frac{d^2 E_T}{d\varphi^2} = \tfrac{1}{2} V_{JK} n_{JK}^2 \cos\left[n_{JK}\left(\varphi - \varphi_{JK}^0\right)\right] \tag{11}$$

The equilibrium angle is φ_{jk}^0=180°, the periodicity n_{JK}=2 and the torsion energy V_{JK}=25 kcal/(mol rad²).

It has to be considered that the bond angle energy $E_A = E_{IJK}$ only depends on the angle between the bonds *IJ* and *JK*. As illustrated in Fig. 2b, the adjacent angles are equal to α and β if E_{IJK} is the only energy. Frame elements (Fig. 2a) force the adjacent bonds to rotate, but due to their bending resistance, these angles are unequal to α and β. Since this could only be prevented by additional constraints, we prefer a finite element model using spring elements (Fig. 4b).

In order to be able to compute large structures containing several nanotubes, the number of degrees of freedom of the overall finite element model has to be reduced. Shell elements could be used for that purpose [6, 9-11]. Fig. 5 shows the differences between these finite element models. Because 3D computer simulations based on reasonable physical models cannot only highlight the molecular features of nanomaterials for theoreticians but also provide guidance and interpretations for experimentalists. It is still an ongoing and challenging process to identify effective and efficient computational methods with respect to specific nanomaterials.

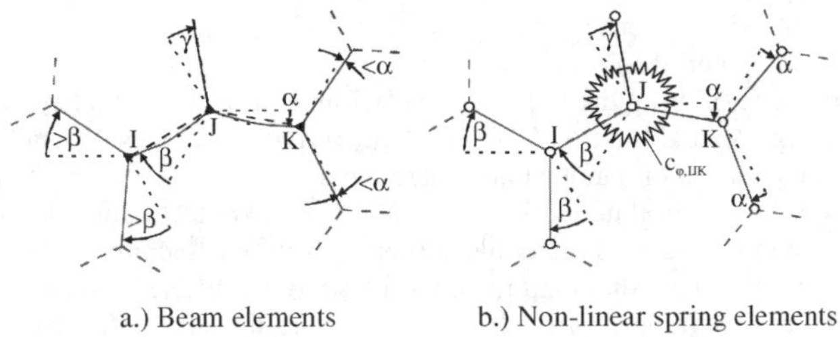

a.) Beam elements b.) Non-linear spring elements

Figure 4. Simple and improved beam models

However, it's true that carbon-based nanomaterials and nanocomposites have been the subject of enormous interest, but development of nanotechnology has been spurred by refinement of IT engineering tools to virtually design the nanoworld, such as more sophisticated information technologies, software and computers to be used for nanoengineering [5]. The principal objective of the paper was to demonstrate an application of modern software engineering tools for modeling virtual reality and molecular dynamics of novel nanocomposites.

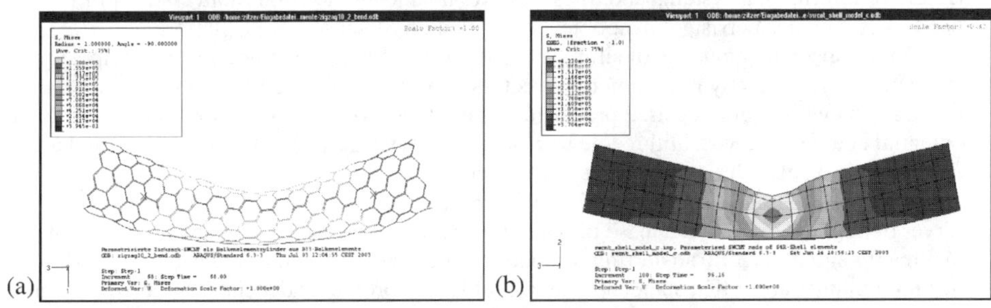

Figure 5. Beam element model (a) and Shell element model (b)

COMPUTATIONAL AND ANALYTICAL TOOLS FOR 3D MODELING

Distance geometry and computer graphics are other areas of scientific development for modeling of reinforced composite materials. The transfer of this esoteric mathematical subject to applied molecular systems was done and used by Crippen, Havel [14]. Crippen has continued to develop it for modeling of macromolecule interactions [13].

When the structure of composite material is complex with many clusters and hard inclusions, it is essential to use computational methods that can accurately predict the structure of entire composite material and a single cluster for an entire database of computational results. The programs DOCK [15] and FLEXX [16] enable one to screen databases of the order of 100,000 clusters and screen for those most likely to bind to a target macromolecule and virtually the composite material. The other nanosimulation software is the program AUTODOCK [17], which carries out more extensive simulation of a single cluster of material using a Monte-Carlo-based method.

Methods to carry out simulations are being developed [18] and should have a major impact in structure-based nanomaterials design.

Our joint group is exploring the use of spatially immersive virtual reality systems and modern IT technologies (e.g., Visual Studio Microsoft.Net, Active X and JAVA2 applets) for interactive modeling and visualization of nanotechnology relevant systems. A computational scheme and software, which utilizes neural networks and/or Microsoft.Net technique, was developed to predict properties of nano-structured materials and optimization and control of nano-devices.

We studied ultra dispersed diamond nanoparticles having different size and shape. The nanoparticles could be pyramid, sphere-like and fullerene-like shape. Its size might be up to 100 nm and higher, but usually their diameter is between 5 and 15 nm. Computer-vision-assisted techniques enable semi-automatic intermittent 3-D graphic model update to match the simulated virtual reality of nanomaterials with their actual video images. Key features of proposed technology are as follows. Developed software and (info/analytical tools) IT tools for modeling of virtual reality of nanostructures and nanocomposites differs from others either on the market or under development as follows.

1. Developed a computational scheme and software utilizes neural networks and/or Microsoft.Net technique that were developed just a year ago by Microsoft Inc. as an advanced analogue of C++ or C Sharp programming languages. Now it differs by developed three-level architecture of information systems and tools for virtual nanoengineering. Main concept of the developed tools includes persistent nanoobjects modeled on Miscosoft.NET platform.

2. Developed an unique set of simple algorithms to encode the structure and composition of the nanomaterial directly into numerical vectors neural network modules can correlate these numeric inputs with a set of desired properties. The interactive system employs a reactive diamond potential coupled to a virtual three-degree-of-freedom hap tic (force-feedback) arm and a real-time graphic display through Internet or intranet.

3. The software and nanoengineering tools to be announced are based on three key parts: server of applications and database of nanoobjects, tools for engineering and modeling of 3D virtual reality and tools for transferring through internet-based solutions and cable networks, utilities for administration. The nanoengineering tools also provide independence between logical structure of modeled nanomaterials and its physical core. It made universal engineering logic for various databases, software shells and interfaces.

While Virtual Nanoengineering a user can easily attach his hand to an atom or molecule and maneuver it in 3D three-dimensional space. 3D pictures (fig. 6, 7) of nano simulation show respectively, for example, crystallize lattice of aluminum matrix in fig. 6a, diamond nanoparticles of chosen shape in fig. 6b (pyramid-like, sphere or fullerene-like shapes) and nanocomposites represented as aluminum matrix with two introduced pyramid-like diamond nanoparticles (green-atoms in fig. 7). The positions of all atoms in the atomic lattices are also calculated and stored in data file. The data are used for further calculations of mechanical properties by the number of techniques (FEM, FEM, MD etc).

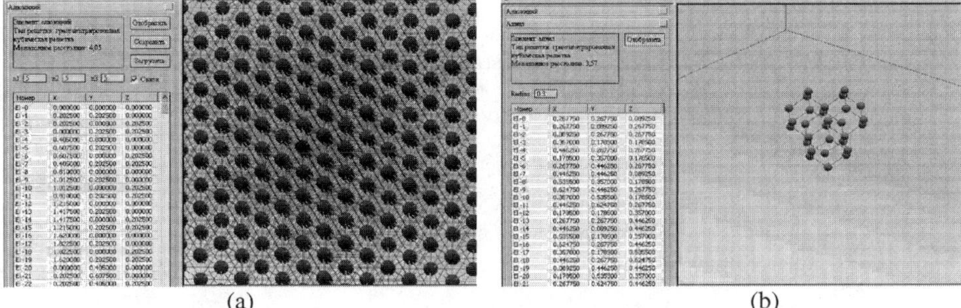

(a) (b)

Figure 6. Atomic lattice of Aluminum matrix and diamond nanoparticle

Valuable solution and tool for modeling of nanostructures provide maximum performance and quality of obtained nanomodels. The reactive potential and mechanical properties are calculated from the resultant forces and motions of all atoms and nanoparticles, including changes in bonding topology. The forces on the atoms being manipulated are continuously fed back to the user through the haptic interface, while the newly calculated positions of all atoms and nanoparticles are continuously updated on the 3D graphics display.

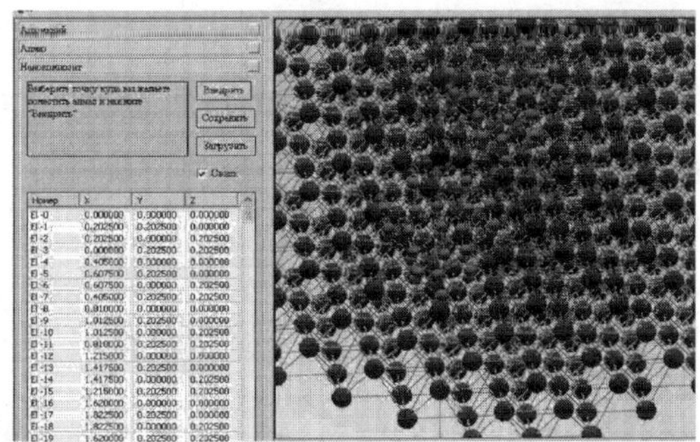

Figure 7. Reinforced Nanocomposite: Aluminum matrix – diamond nanoparticles

We also believe that nanotechnology CAD environments and SIDs offer significant advantages relative to traditional displays or head mounted displays for collaboratively designing and exploring engineered nanostructures. Virtual database of various nanostructures and their computer-based calculations will introduce outstanding possibilities to researchers in this field.

ACKNOWLEDGEMENTS

The research works are being supported by the WELCH foundation scholarship, the American Vacuum Society – Dr. F.R. Shepherd, administrator.

REFERENCES

1. Brunger, A.T., J. Kuriyan, and M. Karplus. Science, vol. 235, (1987), p. 458
2. Case, D., J. Wiley, and Chichester. NMR refinement. Encyclopedia of Computational Chemistry. P.v.R. Schleyer (ed.). (1998), p. 1866.
3. Langridge, R., T.E. Ferrin, I.D. Kuntz, and M.L. Connolly. Science, 211, (1981), p. 611
4. Ferrin, T.E., and T.E. Klein. Computer graphics and molecular modeling. Encyclopedia of Computational Chemistry. P.v.R. Schleyer (ed.). J. Wiley. Chichester. (1998), p. 463.
5. Dresselhaus M. S. and Avouris P. Applied Physics, vol. 80, (2001), pp. 1-9.
6. Govindjee S. and Sackman J. L. Solid State Communications, vol. 110, (1999), pp. 227-230.
7. Lau K.T., Li H.L., Lim D.S. and Hui D. Annals of European Academy off Science Journal, 1, (2003), pp. 314-330.
8. Mayo S.L., Olafson B. D. and Goddard W. A. (III). J. Phys. Chem., vol. 94, (1990), pp. 8897-8909.
9. Qian D., Wagner G. J., Liu W. K., Yu M. F. and Ruoff R. S. Handbook of Nanoscience, Engineering and Technology, (2003), pp. 19.1-19.63.
10. Superfine R., Falvo M., Taylor R.M. (II) and Washburn S. Handbook of Nanoscience, Engineering and Technology, (2003), pp. 13.1-13.20.
11. Yakobson B.I. and Avouris, P. Applied Physics, vol. 80, (2001), pp. 287-327.
12. Nadai A., Theory of flow and fracture of solids. Vol. 1, 2nd ed., McGraw-Hill, New York (1950)
13. Crippen, G. *J. Comp. Chem.,* 1999, vol. 20. p. 1577
14. Crippen, G.M., and T.F. Havel. *Distance Geometry and Molecular Conformations*. Research Studies Press, Wiley, New York, 1988.
15. Zou, X.Q., Y.X. Sun, and I.D. Kuntz. *J. Amer. Chem. Soc.,* 1999, vol. 121, p. 8033
16. Kramer, B., G. Metz, M. Rarey, and T. Lengauer. 1999. *Med. Chem. Res.* 9. 463
17. Rao, M.S., and A.J. Olson. 1999. *Proteins.Structure, Function and Genetics* .34. 173
18. Liu, T.J.A. Ewing, Y.X. Sun, I.D. Kuntz, and J.A. Ellman. *Chemistry and Biology,* 1997, vol. 4, p. 297

NOMENCLATURE

E_A E_B E_T - the potential energy of this generic force field; E_{IJK} - the bond angle energy; IJ and JK - the bonds; α and β - the adjacent angles; C_A, C_T - the spring constants; ε - deformation rate [m/s]; τ - time of relaxation [s]; σ - stress [Pa]; $\sigma(t)$ - function of stress rating at applied load [Pa]; D – diameter [m].

Mat. Res. Soc. Symp. Proc. Vol. 821 © 2004 Materials Research Society P4.8/N4.8

Twin Boundaries and Stacking Faults in Monazite (Monoclinic $LaPO_4$)

Randall S. Hay
Air Force Research Laboratory
Materials and Manufacturing Directorate
WPAFB, OH 45433-6533, U.S.A.

ABSTRACT

Monazite ($LaPO_4$) was indented at room temperature. Deformation twin boundaries and stacking faults were characterized by high resolution transmission electron microscopy. Kinked deformation twins were also characterized and analyzed. Three types of stacking faults associated with climb-dissociated partial dislocations were observed. Two were found on twin boundaries, and a third in the lattice. Formation mechanisms are discussed. The superimposition of stacking faults along twin boundaries during deformation twinning and the glide of climb-dissociated partial dislocations allowed by stacking fault migration are discussed. The possible relationship between the formation mechanisms for these defects and the low-temperature recrystallization and self-annealing of defects in monazite is considered.

INTRODUCTION

Monazite ($LaPO_4$) is monoclinic with lattice parameters a = 0.683 nm, b = 0.706 nm, c = 0.648 nm, and β = 103.2°.(*1*) It is refractory (mp 2072°C), relatively soft (5 GPa), machinable, and stable with many common oxides.(*2-5*) Monazite is a member of a large structural family that includes xenotime, scheelite, fergusonite, vandates, and other refractory crystal structures.(*2, 6*) The mechanical properties of monazite promote distributive damage mechanisms in ceramic fiber-matrix composites (CMCs).(*2, 7-9*) Deformation during pullout or push in of fibers in CMCs is localized in monazite coatings, without observable affects on the surrounding fiber or matrix.(*7, 10*) These deformation mechanisms include cleavage fracture, dislocation slip, and twinning, and they are critical to the mechanical functionality of the coating. Wear of monazite during room-temperature fiber push-out forms fine-grained microstructures that suggest recrystallization.(*10*) The large lattice parameters, low symmetry, and relative complexity of the monazite crystal structure make it useful for testing models for deformation mechanisms and interface structures developed for simpler materials with higher symmetry. For these reasons the fundamental deformation mechanisms of monazite and similar materials are of great interest.

Monazite deformation mechanisms are the subject of recent studies. Five deformation twin modes, including an unusual kinked deformation twin, slip systems associated with deformation twinning, and climb-dissociated dislocations were identified.(*11-15*) This paper reviews some unusual characteristics of twin boundaries and stacking faults introduced during deformation of this refractory, low-symmetry material. Formation mechanisms and the significance to mechanical behavior are discussed.

EXPERIMENTS

Polycrystalline monazite (5 – 20 μm grain size) was indented at room temperature with a 3 mm spherical tungsten carbide indenter using a 20 kg load.(*5, 13*) TEM samples of regions underneath the indents were prepared as previously described.(*11, 16, 17*) TEM observations were done in a Phillips CM200 FEG operating at 200 kV. TEM negatives were digitized at 1200 dpi and measurements were made off the digital images at actual pixel resolution.

The large lattice parameters and the uncharacterized but probably significant elastic anisotropy of monazite do not lend themselves to conventional TEM dislocation imaging with **g·b** analysis.(*13, 14*) Instead, defects were analyzed from Burgers vector circuits in TEM lattice images.(*18-21*) Details of the image analysis methods are given elsewhere.(*13, 14*) We use the $P2_1/n$ space group convention for monazite, as discussed previously.(*13*)

RESULTS AND DISCUSSION

Deformation twins were observed on (100), (001), (120), ($12\bar{2}$), and '(483)' planes in monazite deformed at room temperature.(*13*) The '( )' notation denotes a type II twin with the closest rational approximation to an irrational twin plane. The twin modes on these planes were identified, and the required atom shuffles were calculated.(*13*) The modes for the (100), (001), and (120) twins have the smallest atom shuffles of all possible twin modes in monazite, and the other two twin modes have shuffles nearly as small. In all cases the shuffles were less than 0.12 nm, which is less than ½ the La-O bond distance, so as expected diffusion is not required during deformation twinning at room temperature. The (100) twin was by far the most common. Twin abundance correlated with low-energy inferred for interface structures and with low shear modulus inferred from crystallographic considerations for the twin plane and shear direction.(*13*)

Atom shuffling restores atoms to their correct positions in the crystal lattice after shear during deformation twinning. In some cases the shortest possible shuffles require a virtual translation of atoms in the twin by a fraction of a lattice vector, called an R-vector, which superimposes a stacking fault on the twin boundary.(*13, 22-24*) This was shown by calculation to be the case for the (001) twin in monazite, which requires an R-vector of ½[010].(*13*) Some high resolution TEM (HRTEM) observations are consistent with this structure. An example is shown in figure 1, where an (001) twin is imaged along the [110] zone, and a lattice shift consistent with the projection of ½[010] on this zone is evident. A schematic of the boundary structure from the highlighted area of the micrograph is shown in the lower half of the figure. The projection of the shuffle directions for the La atoms (gray) and PO_4 tetrahedra (black) are shown in the schematic. More thorough descriptions of the structure and observations along other imaging zones are given elsewhere.(*13*)

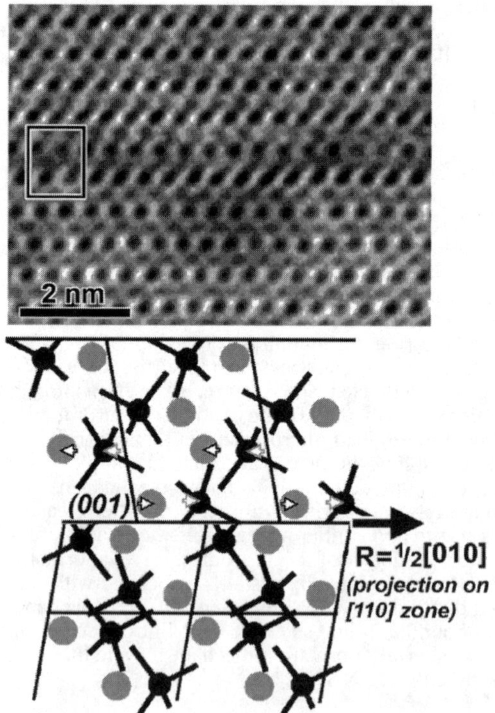

Figure 1. HRTEM micrograph of a (001) twin boundary ([110] zone) and a schematic of the boundary structure showing atom shuffles and the superimposition of a stacking fault by **R**=½[010]

Stacking faults may also be superimposed on monazite twin boundaries by climb-dissociation of lattice dislocations that impinge twin boundaries. This is unexpected because a diffusion mechanism allowing climb-dissociation is not expected at room temperature in a refractory material like monazite. Climb-dissociation was inferred for [100] dislocations along (100) twin boundaries and for [010] dislocations along (120) twin boundaries.(*15*) A HRTEM micrograph showing Burgers circuit closure failure of a defect along a (120) twin boundary of one (120) lattice plane that is consistent with a ½[010] dislocation is shown in figure 2.

Climb-dissociation was also observed for [101] lattice dislocations.(*15*) A HRTEM micrograph of a ½[101] dislocation and the associated stacking fault as imaged along the [01$\bar{1}$] zone is shown in the lower half of figure 3. A schematic diagram of the stacking fault structure in the same zone is shown in the upper half of figure 3. For migration this stacking fault requires only small shuffles (0.11 nm) of the La atoms and PO_4 tetrahedra, and 26° rotation of the PO_4 tetrahedra.(*15*) The projections of the required shuffles and rotations are shown in the schematic by arrows (Fig. 3). The shuffle magnitudes are no larger than those required for some of the deformation twins observed in monazite at room temperature. Room temperature migration of the stacking fault between the climb-dissociated ½[101] partials allows such partials to be glissile and makes them a viable room-temperature deformation mechanism.(*15*) However, the formation of such climb-dissociated structures at room temperature is still problematic. Similar structures and mechanisms have been proposed for climb-dissociated dislocations in spinel at 400°C.(*25*)

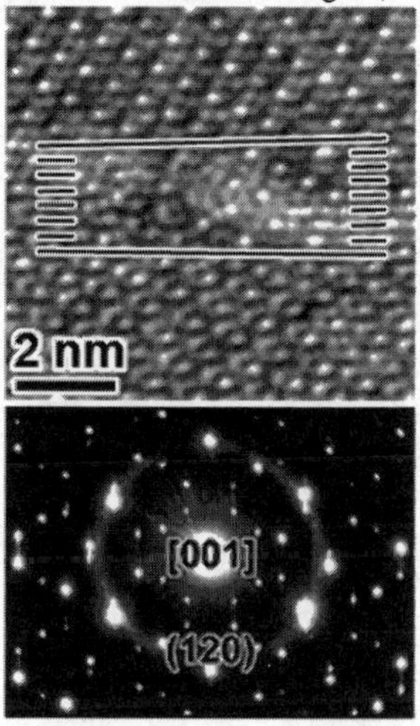

Figure 2. HRTEM micrograph of Burgers circuit closure failure around a defect along a (120) twin boundary. The closure failure is consistent with a ½[010] dislocation.

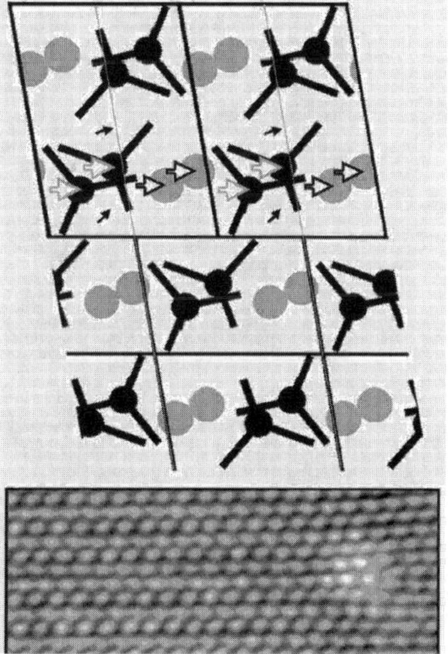

Figure 3. HRTEM micrograph ([01$\bar{1}$] zone) of a climb-dissociated ½[101] dislocation and stacking fault. The schematic shows the stacking fault structure ([01$\bar{1}$] zone), and the shuffles and PO_4 rotations required for migration.

Some (120) twins were kinked on '(483)'. TEM micrographs of kinks imaged along the [2$\bar{1}$0] and [001] zones and a schematic of the kinked twin structure are shown in figures 4b, 4c, and 4a, respectively. The (120) and '(483)' planes are close in orientation, as are the respective shear directions (η_1) of '[4$\bar{2}$1]' and [2$\bar{1}$0]. Strain compatibility requires that formation and growth of the '(483)' kinks be accompanied by extensive plastic deformation in the small volume of the kink. Dislocations diagnostic of such deformation are not evident in the micrographs (Fig. 4).(*12*) However, some (120) twins had highly irregular boundary morphology that may be relict from kinks and possible resorption of the twin by the matrix through boundary migration, which is unexpected in monazite at room temperature (Fig. 4a). It is not clear what causes kinking; some (120) twin lamellae are kinked while adjacent ones are not, and others have the irregular morphology shown in figure 4a.

Deformation mechanisms in monazite such as twinning and stacking fault migration require atom shuffles, but the shuffles are smaller than ½ the La-O bond length and should not require diffusion. However, the climb-dissociated dislocations observed along both twin boundaries

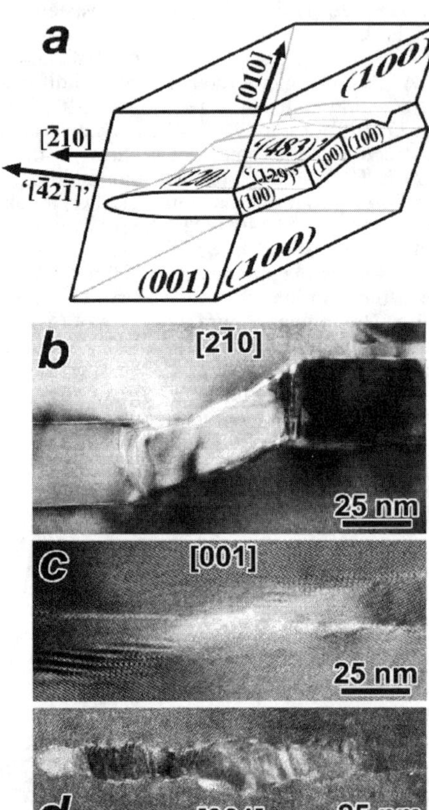

Figure 4. a) Schematic of a (120) kinked twin. b) TEM micrograph of a (120) twin kinked along '(483)' ([21̄0] zone). c) TEM micrograph of a (120) twin kinked along '(483)' ([001] zone). d) TEM micrograph of irregular (120) boundary morphology suggestive of possible lamellae resorption ([001] zone).

and in the lattice require diffusion to form.(*26*) Some unusual structures observed along (120) twin boundaries are also difficult to explain without diffusion.

At least two other phenomenon observed in monazite are possibly consistent with low temperature mass transport by diffusion. Relative to other minerals such as zircon, monazite recovers easily from radiation-induced amorphization at low temperatures – this is frequently referred to as the "self-annealing" characteristic of radiation damage in monazite.(*27-30*) In monazite that is deformed at near-room temperature during fiber push out, the most heavily deformed regions apparently recrystallize.(*10*)

SUMMARY AND CONCLUSIONS

Stacking faults may be superimposed on monazite twin boundaries formed during room temperature deformation. For (001) twins, their formation is intrinsic to the small atom shuffles necessary during formation of the twin. They can also be formed by climb-dissociation of [100] and [010] lattice dislocations that impinge (100) and (120) twin boundaries, respectively.

Climb-dissociation of [101] lattice dislocations to ½[101] forms a stacking fault that requires only small atom shuffles to migrate. Such climb-dissociated dislocations might therefore be glissile. Unusual kinks in (120) twins are a type II twin with an irrational '(483)' twin boundary plane. Their formation requires intense plastic deformation in the kinked twin.

Climb-dissociation of dislocations and possible resorption of (120) twin lamellae suggest a diffusion mechanism must operate in monazite (mp 2072°C) at room-temperature. This is unexpected, but possibly consistent with observations of self-annealing of radiation damage in monazite and with the room temperature recrystallization of monazite.

ACKNOWLEDGEMENTS

J. B. Davis supplied the indented monazite. D. B. Marshall made many valuable suggestions in the course of this work.

REFERENCES

1. D. F. Mullica, W. O. Milligan, D. A. Grossie, G. W. Beall, L. A. Boatner, *Inorg. Chim. Acta* **95**, 231-236 (1984).
2. P. E. D. Morgan, D. B. Marshall, *J. Am. Ceram. Soc.* **78**, 1553-63 (1995).
3. D. B. Marshall, P. E. D. Morgan, R. M. Housley, J. T. Cheung, *J. Am. Ceram. Soc.* **81**, 951-956 (1998).
4. Y. Hikichi, T. Nomura, *J. Am. Ceram. Soc.* **70**, C252-C253 (1987).
5. J. B. Davis, D. B. Marshall, R. M. Housley, P. E. D. Morgan, *J. Am. Ceram. Soc.* **81**, 2169-2175 (1998).
6. O. Muller, R. Roy, *The Major Ternary Structural Families*, Crystal Chemistry of Non-Metallic Materials (Springer-Verlag, New York, ed. 1st, 1974).
7. K. A. Keller *et al.*, *J. Am. Ceram. Soc.* **86**, 325-332 (2003).
8. P.-Y. Lee, M. Imai, T. Yano, *J. Ceram. Soc. Japan* **112**, 29-34 (2004).
9. C. Kaya, E. G. Butler, A. Selcuk, A. R. Boccaccini, M. H. Lewis, *J. Eur. Ceram. Soc.* **22**, 2333-2342 (2002).
10. J. B. Davis, R. S. Hay, D. B. Marshall, P. E. D. Morgan, A. Sayir, *J. Am. Ceram. Soc.* **86**, 305-316 (2003).
11. R. S. Hay, *Ceram. Eng. Sci. Proc.* **21**, 203-218 (2000).
12. R. S. Hay, *Acta Mater.* **51**, 5255-5262 (2003).
13. R. S. Hay, D. B. Marshall, *Acta Mater.* **51**, 5235-5254 (2003).
14. R. S. Hay, *Philos. Mag.* (accepted).
15. R. S. Hay, *J. Am. Ceram. Soc.* (accepted).
16. R. S. Hay, J. R. Welch, M. K. Cinibulk, *Thin Solid Films* **308-309**, 389-392 (1997).
17. M. K. Cinibulk, J. R. Welch, R. S. Hay, *J. Am. Ceram. Soc.* **79**, 2481-2484 (1996).
18. G. R. Anstis, J. L. Hutchinson, in *Dislocations in Solids* F. R. N. Nabarro, Ed. (Elsevier Science Publishers, 1992), vol. 9, pp. 1-55.
19. D. L. Medlin, C. B. Carter, J. E. Angelo, M. J. Mills, *Philos. Mag. A* **75**, 733-747 (1997).
20. J. M. Howe, U. Dahmen, R. Gronsky, *Philos. Mag.* **A56**, 31-51 (1987).
21. J. R. Parsons, C. W. Hoelke, *J. Appl. Phys.* **40**, 866-872 (1969).
22. J. Starkey, in *Deformation Twinning* R. E. Reed-Hill, J. P. Hirth, H. C. Rogers, Eds. (American Institute of Mining, Metallurgical, and Petroleum Engineers, New York, NY, 1964) pp. 177-191.
23. D. B. Marshall, A. C. McLaren, *Phys. Chem Minerals* **1**, 351-370 (1977).
24. W. A. Wooster, *Mineral. Mag.* **46**, 265-268 (1982).
25. P. Veyssiere, C. B. Carter, *Philos. Mag. Lett.* **57**, 211-220 (1988).
26. T. E. Mitchell, W. T. Donlon, K. P. D. Lagerlof, A. H. Heuer, in *Deformation of Ceramic Materials II* R. E. Tressler, R. C. Bradt, Eds. (Plenum Press, New York, 1984), vol. 18, pp. 125-139.
27. A. Meldrum, L. A. Boatner, R. C. Ewing, *Mineral. Mag.* **64**, 185-194 (2000).
28. T. M. Harrison, E. J. Catlos, J.-M. Montel, in *Phosphates: Geochemical, Geobiological, and Materials Importance* M. J. Kohn, J. Rakovan, J. M. Hughes, Eds. (Mineralogical Society of America, Washington, DC, 2002), vol. 48, pp. 523-552.
29. A. J. W. Gleadow, D. X. Belton, B. P. Kohn, R. W. Brown, in *Phosphates - Geochemical, Geobiological, and Materials Importance* M. J. Kohn, J. Rakovan, J. M. Hughes, Eds. (Mineralogical Society of America, Wahington, DC, 2002), vol. 48, pp. 579-630.
30. R. C. Ewing, L. Wang, in *Phosphates - Geochemical, Geobiological, and Materials Importance* M. J. Kohn, J. Rakovan, J. M. Hughes, Eds. (Mineralogical Society of America, Wahington, DC, 2002), vol. 48, pp. 673-699.

B. Modeling

Mat. Res. Soc. Symp. Proc. Vol. 821 © 2004 Materials Research Society P8.17

Atomistic Simulation of Dislocation Interactions with a Σ = 5 (210) Grain Boundary during Nanoindentation of Ni

Ho Jang[1] and Diana Farkas[2]
[1]Department of Materials Science and Engineering, Korea University, Seoul, 136-701, Korea
[2]Department of Materials Science and Engineering, Virginia Polytechnic Institute, Blacksburg, VA 24061, USA

ABSTRACT.

Molecular dynamics simulations of nanoindentation were performed using embedded atom potentials. The indentation was simulated on a Ni substrate using a diamond-like spherical indenter. In this study, we focused on the interaction of a Σ=5 (210) grain boundary with the dislocations that were nucleated and expanded as loops under the indenter during nano-indentation. The results showed that dislocation loops were nucleated beneath the surface and propagated on multiple {111} slip planes. These dislocations impinged into the grain boundary in the course of nanoindentation. The lattice dislocations changed the atom configuration at the boundary region as they merged into the grain boundary and at later stages they transmitted across the grain boundary. The results also showed that the presence of a grain boundary affected the indenting speed and dislocation motion during nanoindentation, with the grain boundary retarding the indentation process.

INTRODUCTION

Understanding of material deformation during nanoindentation has attracted much attention due to its technological and fundamental interests. A large experimental and theoretical effort has been devoted during the last several decades to the understanding of the deformation process in the course of nanoindentation on an atomic scale [1-3]. In experiments, advances in nanoindentation instruments have enabled precise measurements of the incipient plastic deformation including displacement bursts, formation of surface hillocks, and unusual yielding phenomena due to surface effects [4-7]. Atomistic simulation techniques also have successfully progressed to achieve atomic scale understanding of experimentally observed deformations. In particular, the simulation techniques have enabled detailed analysis of dislocation nucleation and defect interactions during nanoindentation, comprising homogeneous nucleation of lattice dislocations under the nanoindenter [8-9], the effect of surface roughness [10], emission of prismatic dislocation loops [11], and the role of grain boundaries in nano-sized polycrystals during deformation [12].

In this paper, we focus on the dislocation interaction with a Σ = 5 {210} grain boundary parallel to the indentation surface during nanoindentation of Ni. The indentation on Ni films with and without a grain boundary was analyzed in detail to study the effect of the grain boundary on the indentation speed and dislocation motion. We particularly analyzed the detailed mechanism of transmission of the dislocations across the grain boundary and the effects on the grain boundary structure itself.

SIMULATION METHODS

A single crystal and a bicrystal Ni with a $\Sigma = 5$ (210) grain boundary were constructed for the indentation simulations. The single crystal Ni contained 60480 Ni atoms and the bicrystal contained 60500 Ni atoms with the simulation cell size of 93Å x 94Å x 94Å. A spherical diamond-like indenter (4.9 nm in diameter) was used in this simulation. A free standing thin film geometry was prepared by eliminating the periodic boundary condition along the indentation direction while maintaining it for two other orthogonal directions. Molecular dynamics simulations were carried out at 100K using an embedded atom potential for Ni [13]. Frictionless indentation, having no chemical interaction between C and Ni atoms on the traction free surface [8], was simulated by using a repulsive pair potential between C and the Ni. The repulsive potential was prepared by replacing the pair potential term in the embedded atom potential with a strong positive pair potential $V(r) = Ar^{-n}$, where A was the force constant and n to adjust the potential energy as a function of atomic distance r. In this work, A was chosen as 10 nN/Å^2 and n as 3. Carbon atoms in the indenter have a diamond cubic structure and have been fixed during the simulation to simulate an infinitely strong indenter. Nanoindentation simulation was carried out under a load-controlled mode at a constant external force of 115 nN and carried out for 16 ps.

RESULTS AND DISCUSSION

Nanoindentation simulations on a bicrystal film containing a $\Sigma = 5$ (210) symmetric tilt boundary was performed using a spherical indenter of 4.9 nm in diameter. Figure 1 shows the simulation block and a detail of the grain boundary region showing the structural units typical of the $\Sigma = 5$ (210) grain boundary. Indentation was performed on the (210) free surface, which was parallel to the grain boundary plane located in the middle of the film. Figure 2 shows a series of simulation results as a function of time. The indentation results obtained from a single crystal block of the same dimensions and with the orientation of the top grain of the bicrystal are also included for comparison (figures 2 a and b). In this figure, the atoms with potential energies corresponding to the perfect lattice are not shown, in order to better visualize the

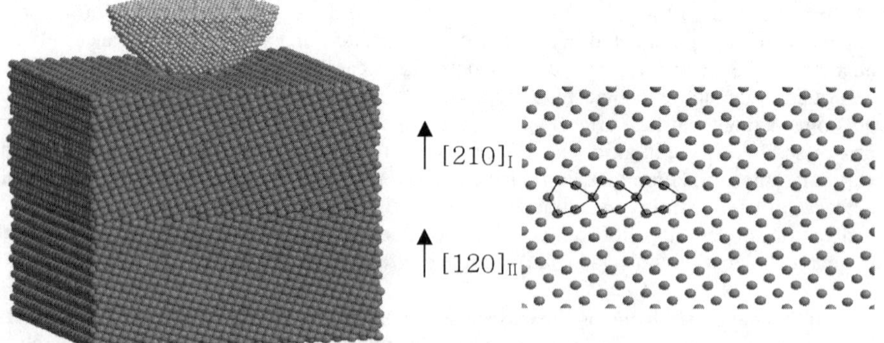

Figure 1. The geometry of the simulation block with a structural unit showing a periodic atom arrangement at the $\Sigma = 5$ (210) symmetric tilt boundary. Indentation was performed on the {210} free surface, which was parallel to the grain boundary plane located in the middle of the film. The grain boundary normal is [210] in crystal I and [120] in crystal II.

dislocation activity. From a relatively early stage of indentation, the initial configurations of the emitted dislocations under the indenter were affected by the presence of the grain boundary.

The dislocations are nucleated from a position close to the indented surface but not exactly at the surface. As the propagation of the dislocation loops proceeds, they form a parallelogram shaped prism surrounded by {111} slip planes [11] and the two pairs of parallel dislocations in the prism extend during indentation until they interact with the grain boundary or escape through the free surface in the bottom of a film. In the case of the film without a grain boundary, a leading partial dislocation reaches the bottom free surface first and a trailing partial completes the interaction, forming a surface step in the size of a Burgers vector. Similar dislocation motions are also observed in the film containing a $\Sigma = 5$ (210) grain boundary. In this case,

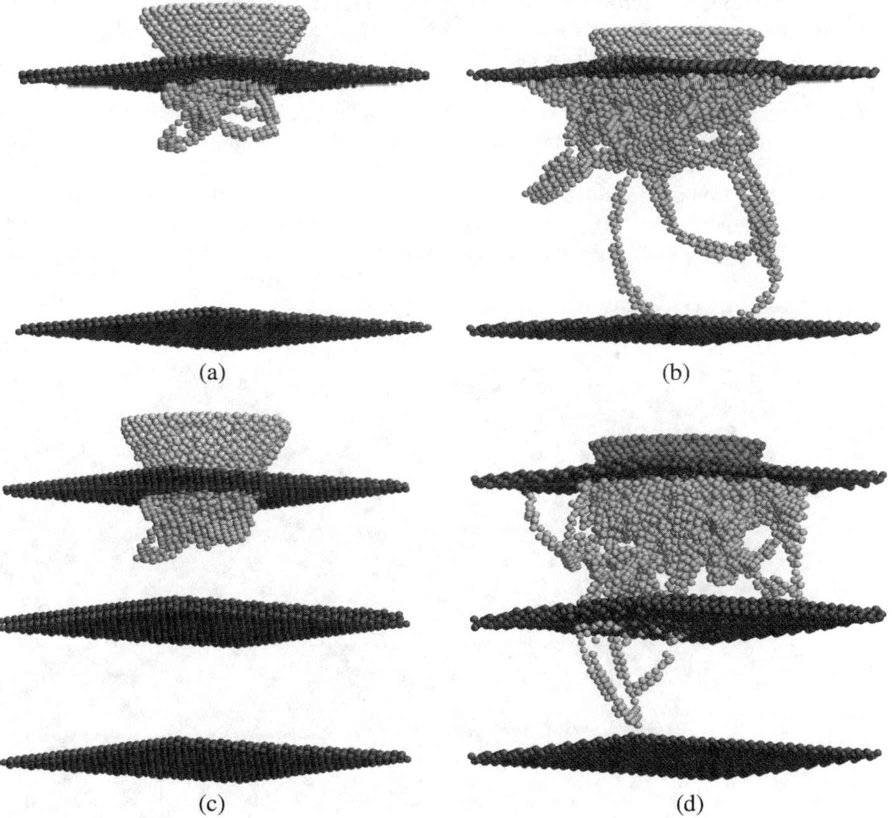

Figure 2. Dislocations produced during nanoindentation on a Ni single crystal and a Ni bicrystal containing a $\Sigma = 5$ (210) tilt grain boundary. The Ni atoms in these figures represent the atoms having less than 12 nearest neighbor atoms (imperfect atoms). (a) and (c) correspond to the atomic structures after 3 ps of MD calculations and 13 ps for (b) and (d).

however, the dislocations interact with the grain boundary and further propagation is delayed until new dislocations appear on the other side of the grain boundary (figure 2(d)). When a dislocation transmits across the grain boundary, the trailing partials are not observed. A similar result has been reported in the case of the deformation simulation of polycrystals containing random boundaries [12]. The absence of trailing partial dislocations is ascribed to the structural relaxation at the grain boundary, avoiding the immediate emission of the second partial. However, in the case of polycrystal deformation, the transmission of a dislocation across the randomly oriented grain boundaries was not observed during nanoindentation [12,14,15].

Figure 3 shows an example of the interaction of dislocations with the $\Sigma = 5$ (210) grain boundary during the propagation of the dislocation. The figure shows two snapshots as a dislocation merges into the grain boundary, disturbing the periodicity of the grain boundary structure, and transmits across the grain boundary. Figure 3 (a) shows the snapshot just before the dislocation propagates into the second grain and figure 3 (b) shows the configuration immediately after the dislocation is emitted into the second grain. This suggests that the grain boundary can provide the resistance against dislocation propagation and slows down the plastic deformation, supporting the well-known Hall-Petch relationship for grain size strengthening [16]. A salient feature observed in this process is that the process is accompanied by the creation of a step in the grain boundary.

The interaction of the lattice dislocation with the symmetrical tilt boundary can be analyzed considering Burgers vectors of lattice dislocations and grain boundary dislocations. Since the [210] direction in crystal I is equivalent to [120] in crystal II in the $\Sigma = 5$ (210) grain boundary, the rotation matrix R between the two grain is given by; $R = \begin{bmatrix} 0.8 & 0.6 & 0 \\ -0.6 & 0.8 & 0 \\ 0 & 0 & 1 \end{bmatrix}$

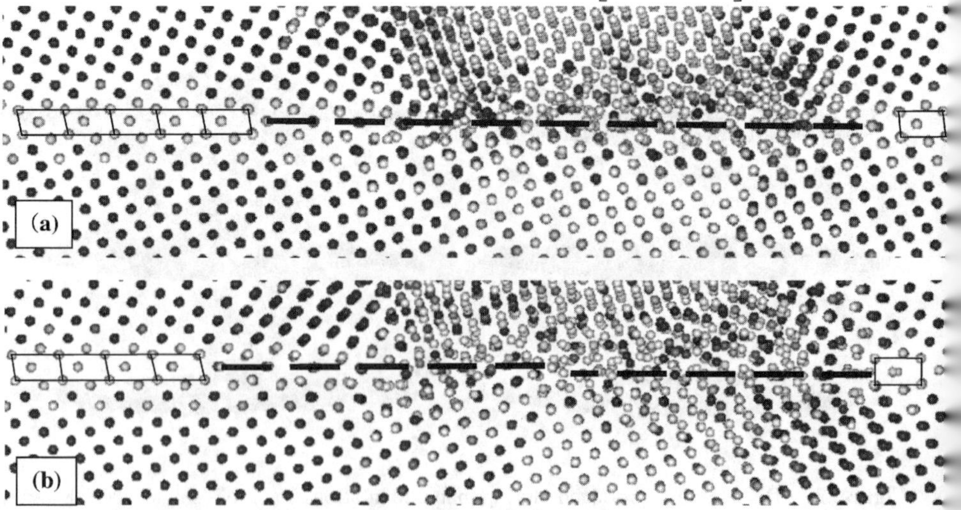

(a)

(b)

Figure 3. The interaction of lattice dislocations with a Σ=5 {210} grain boundary during the transmission of a dislocation across the grain boundary. The shading of the atomic symbols corresponds to local stress states.

The perfect dislocation in the crystal I (top) dissociates into Shockley partials, following the reaction; $\frac{1}{2}[\bar{1}\,\bar{1}\,0]_I \rightarrow \frac{1}{6}[\bar{1}\,2\,\bar{1}]_I + \frac{1}{6}[\bar{2}\,\bar{1}\,1]_I$. The two partial dislocations with the Burgers vectors, $\frac{1}{6}[\bar{1}\,2\,\bar{1}]_I$ and $\frac{1}{6}[\bar{2}\,\bar{1}\,1]_I$ in the crystal I are equivalent to $\frac{1}{30}[2\,\bar{1}\,\bar{1}\,\bar{5}]_{II}$ and $\frac{1}{6}[\bar{1}\,\bar{2}\,1]_{II}$ in the crystal II, respectively. The first dislocation with Burgers vector $\frac{1}{30}[2\,\bar{1}\,\bar{1}\,\bar{5}]_{II}$ may dissociate following the dislocation reaction: $\frac{1}{30}[2\,\bar{1}\,\bar{1}\,\bar{5}]_{II} \rightarrow \frac{1}{6}[\bar{2}\,\bar{1}\,\bar{1}]_{II} + \frac{1}{5}[2\,\bar{1}\,0]_{I,II}$. As a result, the dislocation in grain I produces a lattice dislocation in grain II and a grain boundary dislocation with a DSC (displacement shift complete) vector $\frac{1}{5}[2\,\bar{1}\,0]_{I,II}$ [17,18]. The presence of DSC dislocations in the grain boundary is known to cause a motion of the grain boundary parallel to itself. Indeed, figure 3 (b) shows the creation of a step in the Σ=5 {210} grain boundary during the transmission of the lattice dislocation across the grain boundary.

To examine the effect of grain boundary on the indentation speed, the penetration distance of the indenter was plotted as a function of the MD simulation time (figure 4). It shows that the penetration rate, which is a slope of the curve, is slower when the film contains a grain boundary, suggesting more resistance to the plastic deformation than the single crystal without a grain boundary. The penetration speed decreases as indentation proceeds when the grain boundary is present. Inflection in the curve in the time period 2 ps to 3 ps indicates incipient plastic deformation with dislocation nucleation and propagation

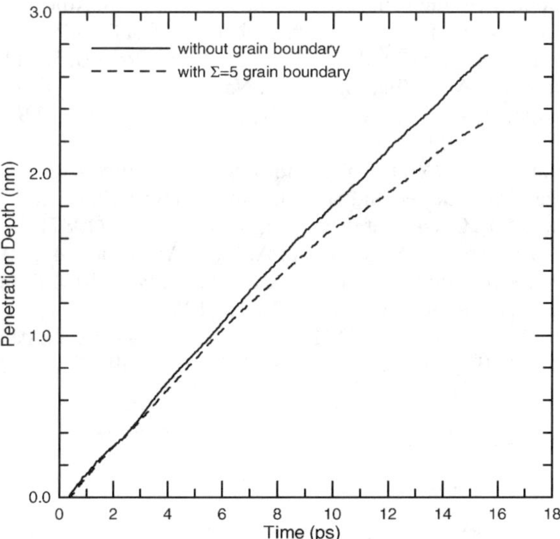

Figure 4. Penetration depth as a function of MD simulation time for films with and without the grain boundary.

CONCLUSIONS

Dislocation generation during nanoindentation and their interaction with a Σ=5 {210} grain boundary were studied by molecular dynamics simulation using embedded atom potentials. Dislocations were nucleated near the free surface beneath the indenter and they produced a parallelogram shaped prism surrounded by {111} slip planes. Transmission of a dislocation across the grain boundary was observed during indentation simulation. Burgers vector analysis suggested that the lattice dislocation in the grain I merged into grain boundary and it was dissociated into another lattice dislocation in the grain II and a grain boundary dislocation, introducing a step in the grain boundary. The indentation on the film with a grain boundary was slower than that without a grain boundary. The slower indentation was ascribed to the interaction of dislocations with the grain boundary.

REFERENCES

[1] N. Gane and F.P. Bowden, J. Appl. Phys. 39,1432 (1968).
[2] N.R. Moody, W.W. Gerberich, N. Burnham, and S.P. Baker (ed.), Fundamentals of Nanoindentation and Nanotribology, MRS Proc. 522 (1998).
[3] D.E. Kramer, K.B. Yoder, W.W. Gerberich, Phil. Mag. A, 81, 2033 (2001).
[4] S.G. Corcoran, R.J Colton, E.T. Lilleodden, and W.W. Gerberich, Phy. Rev. B, 55, R16 057 (1997).
[5] S. Suresh, T.-G. Nieh, and B.W. Choi, Scripta Mater. 41, 951 (1999).
[6] J.D. Kiely and J.E. Houston, Phys. Rev. B, 57, 12 588 (1998).
[7] J. de la Figuera, M.A. Gonzalez, R. Garcia-Martinez, J.M. Rojo, O.S. Herman, V.L. Vazquez de Parga, and R. Miranda, Phys. Rev. B, 58, 1169 (1998).
[8] C. Kelchner, S.J. Plimpton, and J.C. Hamilton, Phys. Rev. B, 58, 11 058 (1998).
[9] A. Gouldstone, K.J. Van Vliet, S. Suresh, Nature, 411, 656 (2001).
[10] J.A. Zimmerman, C.L. Kelchner, P.A. Klein, J.C. Hamilton, and S.M. Foiles, Phys. Rev. Lett., 87, 165507 (2001).
[11] J. Li, K.J. Van Vliet, T. Zhu, S Yip, and S. Suresh, Nature, 418, 307 (2002).
[12] D. Feichtinger, P.M. Derlet, and H. Van Swygenhoven, Phys. Rev. B 67, 024113 (2003).
[13] A.F. Voter and S.P. Chen, Mat. Res. Soc. Symp. 82. 175 (1987).
[14] H. Van Swygenhoven, M. Spaczer, and A. Caro, Acta mater. 47, 10, 317 (1999).
[15] H. Van swygnhoven, P.M. Derlet, and A Hasnaoui, Phys. Rev. B, 66 024101 (2002).
[16] N.J. Petch, J. Iron and Steel Institute, 174, 25 (1953).
[17] H. Grimmer, W. Bollmann and D.H. Warringron, Acta Crystall. A 30, 197 (1974).
[18] R.W. Balluffi, A. Brokman, and A. H. King, Acta Metall. 30, 1453 (1982).

Mat. Res. Soc. Symp. Proc. Vol. 821 © 2004 Materials Research Society P6.7

Atomistic Simulations of Dislocation-Interface Interactions in Thin Films

R. W. Leger and Y.-L. Shen
Department of Mechanical Engineering, University of New Mexico
Albuquerque, NM 87131, U.S.A.

ABSTRACT

Atomistic simulations are carried out to study the effect of atomic sliding capability at the interface between a plastically deforming film and a stiff substrate. Molecular statics modeling is utilized to corroborate the overall film response and the nano-scale defect mechanisms. A free-sliding interface is shown to be able to cause "reflection" of oncoming dislocations and enhance film plasticity. A rigidly bonded interface, on the other hand, is seen to resist approaching dislocations. Partial sliding results in a transitional behavior between the two extremes, as revealed in our parametric analysis. The sliding capability of interface atoms is also seen to dictate the overall film response.

INTRODUCTION

This study focuses on small-scale plasticity and the interaction of dislocations with film-substrate interfaces. Our recent study has aimed at providing an atomistic picture on the interface-mediated plasticity in thin metal films [1]. Through two-dimensional (2D) molecular statics simulations it was shown that the elimination of a free surface facilitated by a stiff substrate enhances the yield strength of the film by restricting dislocation activities. In particular, when atoms at the interface have the capability of unrestricted tangential slide, an oncoming dislocation tends to be "reflected" by the interface by way of a dislocation reaction at the interface. On the other hand, when the interfacial atoms are not allowed to slide, the oncoming dislocation tends to be pinned near the interface and becomes immobile. The "no-slide" case leads to much greater strengthening than the "free-slide" case. In a follow-up work [2], the effects of controlled partial sliding along the interface were examined. When the interface atoms are endowed with certain capabilities of tangential slide, the film displays a yield behavior in between those of a perfectly bonded film and a freely sliding film. A maximum allowable sliding distance of 1% of the current interatomic spacing along the interface results in a film response only slightly different from the unrestricted free-sliding case. With a sliding capability greater than about 0.075%, the interface is seen to have sufficient flexibility to interact with the oncoming dislocation, forcing it to reflect from the interface and to slip toward the free surface of the film. While the above studies provided insight into the dislocation-interface interaction in thin crystals, the influence caused by the sample size is still unclear. In the present work, we extend the analyses to feature an approximately 50% increase in film thickness compared to the previous studies [1,2], with the purpose of investigating both the qualitative and quantitative features affected by the problem size.

APPROACH

Figure 1 shows the model system, which is a close-packed planar crystal (containing 2276 atoms) with one of its close-packed directions parallel to the tensile loading (x) axis. To assist in triggering the onset of crystal plasticity, an artificial defect, in the form of a self-interstitial, is introduced in the model and allowed to equilibrate with its surrounding atoms before the loading steps commence. This technique enables ductile elastic-plastic behavior and the underlying atomistic processes to be simulated with the simple pair potential adopted here [3]. The Morse interatomic potential is used with the parameters determined by fitting to experimental data of the equilibrium lattice parameter, cohesive energy and bulk modulus of copper featuring near-neighbor interactions [4]. The molecular statics simulation of tensile stretching is carried out by prescribing a small displacement in the x direction on the right-hand boundary atoms at each step while fixing the x positions of the left-hand boundary atoms. In response to each loading step the atomic particles are allowed to iteratively reach their new equilibrium positions. The overall tensile load is calculated by summing the x-component force along the boundary atoms where displacements are prescribed. The bottom boundary atoms are conceived to be the interface layer with the substrate, with their motion in y-direction prohibited. No atoms are allowed to go below the interface. Note that this study focuses on the parametric analysis of interface characteristics with *controlled* interface behavior (described below), so there is no attempt to model a true bi-layer system.

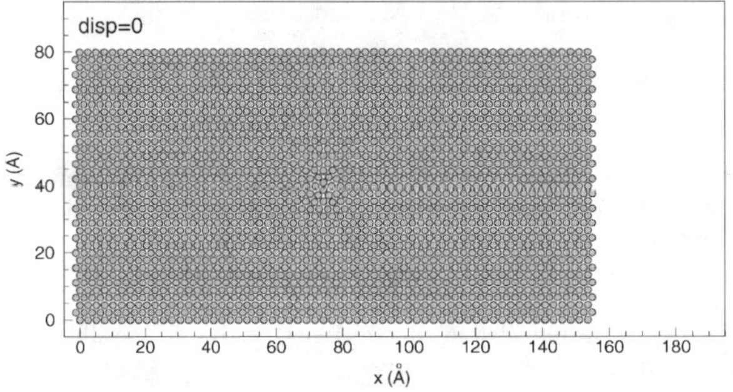

Fig. 1 Initial atomic configuration used for simulating the thin film (the substrate material is not specifically included in the model). A self-interstitial is arbitrarily placed at near $x = 75$ Å and $y = 40$ Å and allowed to equilibrate with its surroundings before the loading simulation starts.

The interface sliding capability is facilitated by restricting the x-direction movement of interface atoms to various extents. In the extreme case of no slide, the x-component displacements of interface atoms are made proportional to the prescribed boundary displacement, simulating perfect bonding with the substrate which controls the macroscopic deformation. Different extents of sliding are controlled by allowing certain maximum x-component displacements of interface atoms. The maximum allowable displacement along the interface, applied individually on interface atoms, is expressed as

$$u_{x,\max} = k \cdot r_{\text{int}}, \tag{1}$$

where r_{int} represents the spacing between adjacent atoms along the interface at the beginning of the *current* loading increment (controlled by the prescribed boundary displacement), and k is the sliding parameter. The parameter k is henceforth used to designate the extent of sliding allowed in the model. (It was found that, at $k > 0.01$, the film response is already close to the free-slide case where no restriction on the x-component displacements is imposed. This is quantitatively similar to the smaller-sample calculations with the total x- and y-spans approximately 50% smaller than those in the present case [2].) The no-slide case is represented by $k = 0$. Note in Eq. (1) that $u_{x,\max}$ is individually set for the interface atoms in the calculation. The actual u_x for each interface atom can vary (and is determined by the analysis itself) so the nonuniform nature of interfacial sliding is accounted for.

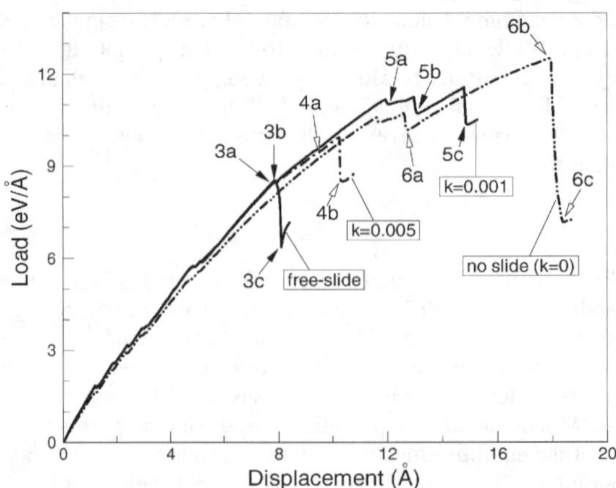

Fig. 2 Overall tensile load-displacement curves obtained from the simulation. The interface sliding capabilities are denoted by the constant k. Select atomic snapshots corresponding to different stages of deformation are shown in Figs. 3-6 as labeled.

RESULTS AND DISCUSSION

Figure 2 shows the overall load-displacement curves of the films with various interface sliding capabilities. The curves are shown only up to their first significant load drop (chosen to be greater than 10% of the peak load), which signifies gross plastic yielding with at least one slip step created at the free surface (see below). There is a general trend that, as interface sliding becomes more difficult (decreasing k), the major load drop becomes more delayed. The symbols such as 3a, 3b etc. labeled along the curves in Fig. 2 indicate the figure numbers below, where the atomic snapshots during the deformation histories are shown.

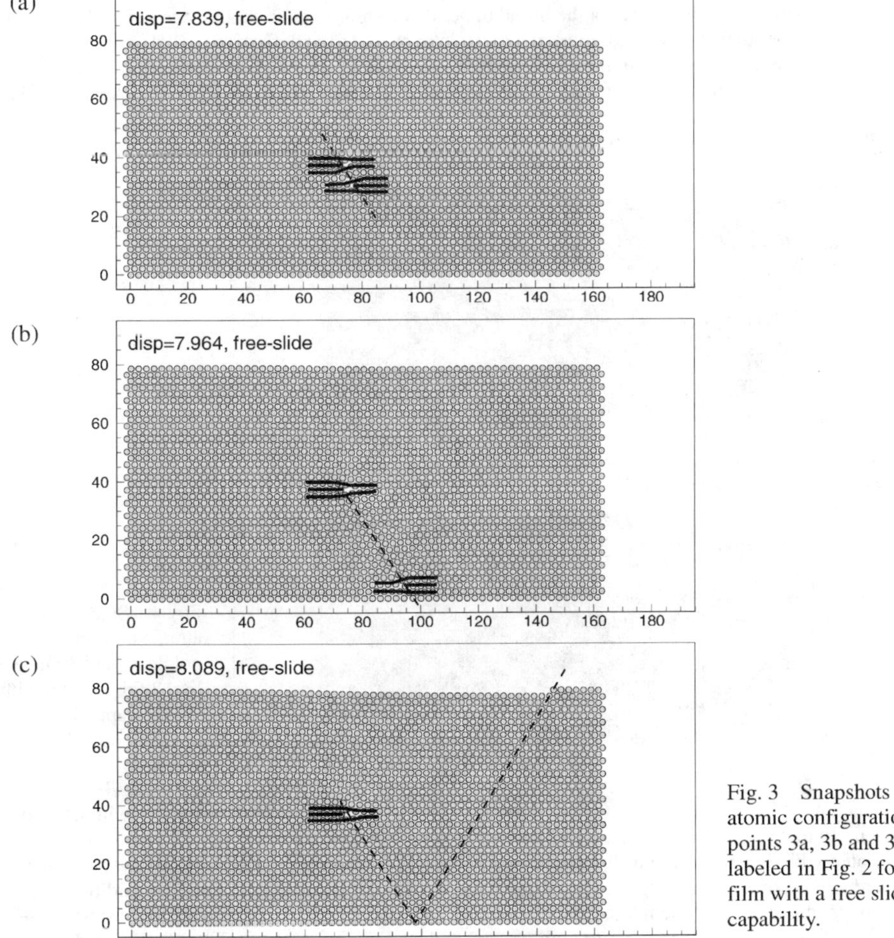

Fig. 3 Snapshots of atomic configurations at points 3a, 3b and 3c labeled in Fig. 2 for the film with a free sliding capability.

The snapshots presented in Figs. 3-6 are largely self-explanatory, so only minimal descriptions are given here for conserving space. In these figures dark lines connecting some atoms are used to highlight the dislocations in an elementary manner. The dashed lines indicate their slip planes. In Fig. 3(a), the initial point defect has evolved into a pair of dislocations in response to the macroscopic deformation. The lower dislocation has slipped to near the bottom interface in Fig. 3(b) and has been "reflected" by the interface and slipped out of the crystal to create a slip step in Fig. 3(c). This process correlates with the large load drop observed in Fig. 2. The "reflected" dislocation, having a different Burgers vector from before the reflection, is in fact the outcome of dislocation reaction at the free-sliding interface as discussed in reference [1]. A similar behavior can be seen in Figs. 4(a) and 4(b). A notable difference is that, with a sliding limit imposed, the ability of the interface atoms to accommodate the oncoming dislocation is reduced so the dislocation-interface reaction is delayed (Fig. 2). There is also a small void generated at the initial upper dislocation in Fig. 4, apparently due to the longer-range effect caused by the interface constraint. In Figs. 3 and 4 the initial upper dislocation stays immobile even after the large load reduction. In fact they will start to move at a much later stage of the deformation (not shown here). This is a major difference compared to the thinner films considered in the previous study [2]. The nature of the dislocation-interface interaction, however, appears similar in both the thinner [2] and thicker (Figs. 3 and 4) films.

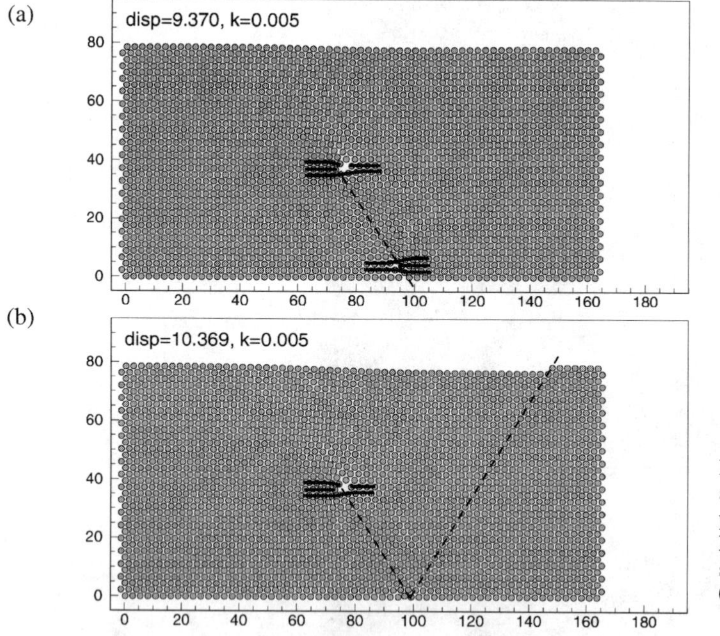

Fig. 4 Snapshots of atomic configurations at points 4a and 4b labeled in Fig. 2 for the film with a sliding capability k = 0.005.

When the sliding capability is further restricted (Figs. 5 and 6), a fundamental change in behavior can be seen. In the case of $k = 0.001$, the load reduction occurs in three successions (5a, 5b and 5c), each associated with the creation of a slip step. The dislocation reaction at the interface and the subsequent upward slip becomes more sluggish, resulting in the last load drop among the three. In addition, the reaction is imperfect because it left behind a local disturbance in atomic packing clearly visible at the interface (Fig. 5(c)). Further, a larger void was created near the initial upper dislocation site. In Fig. 6 where no interfacial sliding is allowed, the dislocation-interface reaction is further delayed. In fact, it occurs *after* a major internal damage in the crystal is formed. The oncoming dislocation stays pinned near the interface until very late in the deformation process. The effect of the "reflection" process on the overall load-displacement response is masked by the ductile fracture of the film. Final fracture occurs as a result of slip-induced void growth toward the interface side (not the free surface side) of the film.

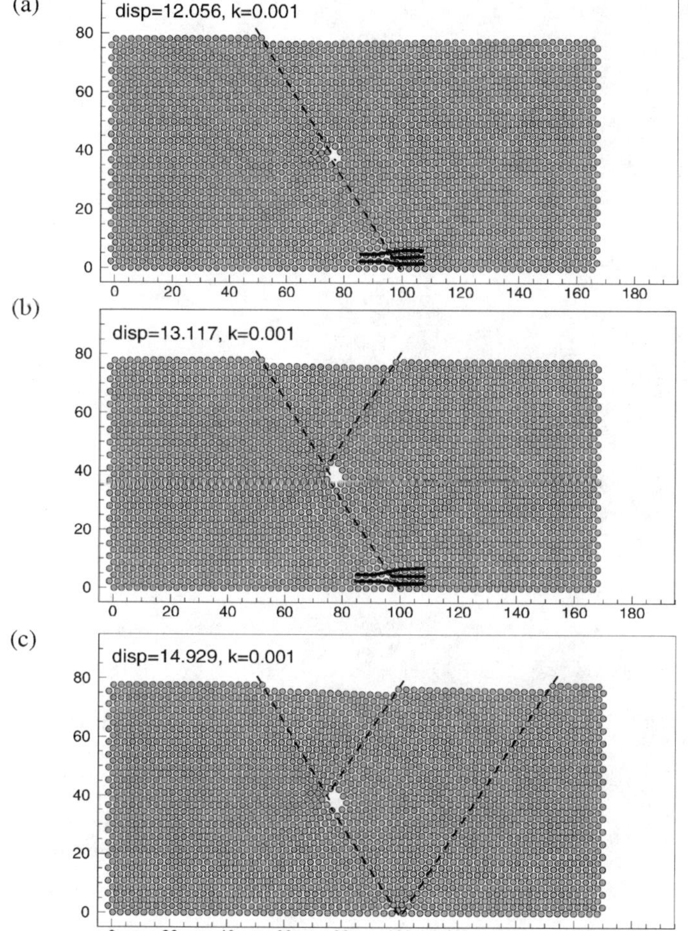

Fig. 5 Snapshots of atomic configurations at points 5a, 5b and 5c labeled in Fig. 2 for the film with a sliding capability $k = 0.001$.

CONCLUDING REMARKS

A comparison between the present study and previous work [1,2] illustrates that the modeling sample size does not affect the simulated dislocation-interface interaction in any fundamental way. However, other defect mechanisms and damage evolution in the crystal may be significantly modified. Although the behavior of interface sliding at the atomistic level in actual metal films is currently not well understood [5], our analyses help to provide some quantitative insight. The sliding parameter k used here to specify the maximum allowable sliding distance corresponds to exceedingly small local displacements. Nevertheless, the very small-scale sliding operation is still able to generate a very wide range of film response, implying the great sensitivity of thin-film plasticity to the atomistic characteristics at the interface.

(a)

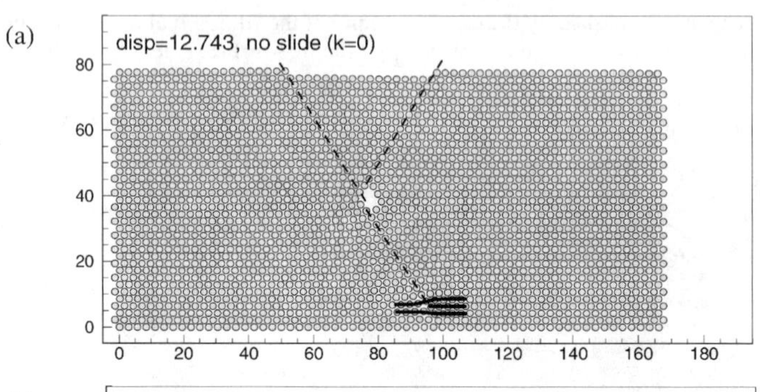

(b)

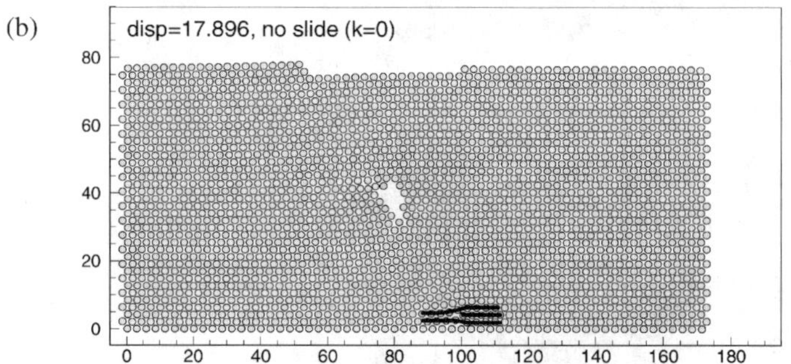

(c)

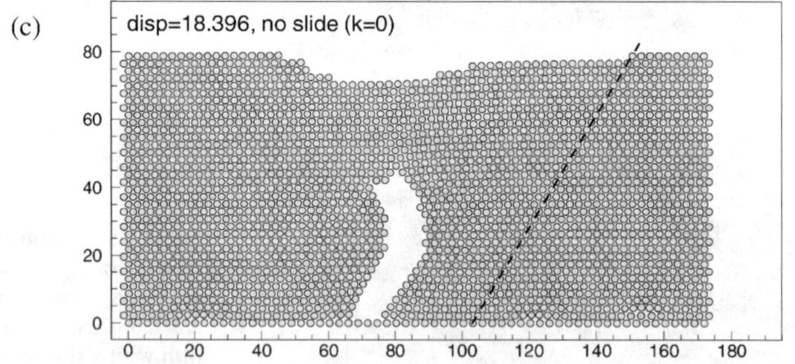

Fig. 6 Snapshots of atomic configurations at points 6a, 6b and 6c labeled in Fig. 2 for the film with no sliding capability, $k = 0$.

REFERENCES

1. Y.-L. Shen, *J. Mater. Res.* **18**, 2281 (2003).
2. E. S. Ege and Y.-L. Shen, in *Thin Films – Stresses and Mechanical Properties X*, Mater. Res. Soc. Symp. Proceedings, Vol. 795, U8.8 (2004).
3. Y.-L. Shen, *J. Mater. Res.* **19**, 973 (2004).
4. R. Phillips, *Crystals, Defects and Microstructures – Modeling Across Scales*, Cambridge University Press, 2001, p. 206.
5. L. B. Freund and S. Suresh, *Thin Film Materials*, Cambridge University Press, 2004, p.534.

Mat. Res. Soc. Symp. Proc. Vol. 821 © 2004 Materials Research Society P8.40

Critical Conditions for Dislocation Nucleation from Surface Steps

Chengzhi Li and Guanshui Xu
Department of Mechanical Engineering
University of California at Riverside, Riverside, CA 92521

Abstract

The critical conditions for dislocation nucleation from the surface steps of various geometries are analyzed based on the Peierls-Nabarro dislocation model. By modeling a surface step as part of a three-dimensional crack surface, the half space problem is transferred into an equivalent three dimensional crack problem in an infinite medium. The profiles of embryonic dislocations, corresponding to the relative displacements between the two adjacent atomic layers along slip planes, are then rigorously solved through the variational boundary integral method. The critical conditions for dislocation nucleation are determined by solving the stress dependent activation energies required to activate embryonic dislocations from their stable to unstable saddle point configurations. For a given slip plane, the effects of step geometry such as the step height and inclined angle on dislocation nucleation are analyzed in detail. The results show that the atomic scale steps may reduce the critical stress required for dislocation nucleation from the surface by several factors. Compared to previous analyses of this type of problem based on continuum elastic dislocation theory, the presented analysis eliminates the uncertain core cutoff parameter by allowing for the existence of an extended dislocation core as the embryonic dislocation evolves. Because of the serious limitation of direct atomic simulation for this type of problem, the presented methodology of incorporating atomic information into continuum approach appears to be particularly noteworthy for providing insights of energetics of the atomic processes involved in dislocation nucleation.

Introduction

The critical conditions for dislocation nucleation from the stressed surface have been of considerable interest for understanding a wide variety scientific and engineering problems such as deformation mechanisms of nanoscale contact and control of dislocation density in strained heteroepitaxial thin films in microelectronic devices. Previous analyses have been focused on critical conditions for dislocation nucleation from the perfectly flat surface using either atomic simulation, the combined atomistic and continuum model such as the Peierls-Nabarro dislocation model, or continuum elastic dislocation theory. Many researches pointed out that the stress concentration caused by surface heterogeneities such as steps and microcracks should significantly facilitate dislocation nucleation from the stressed surface [1-4]. The effect of an atomic scale surface crack on dislocation nucleation has been previously studied based on the Peierls and Nabarro dislocation model [4]. The effects of the surface steps on dislocation nucleation, however, have not been addressed in the literature in detail.

We present an analysis of this problem based on the Peierls-Nabarro dislocation model [5, 6]. By modeling a surface step as part of a three-dimensional crack surface, we transfer the half space problem into an equivalent three dimensional crack problem in the infinite medium. The profiles of embryonic dislocations, corresponding to the relative displacements between the two adjacent atomic layers along slip planes, are then rigorously solved through the variational boundary integral method. The critical conditions for dislocation nucleation are determined by solving the stress dependent activation energies required to activate embryonic dislocations from their stable to unstable saddle point configurations. For a given slip plane, the effects of step geometry such as the step height and inclined angle on dislocation nucleation are analyzed in detail.

Configuration for Dislocation Nucleation from a Step

Dislocation nucleation from a surface step may occur in a wide variety of configurations dependin on the geometry of the surface step, crystallographic orientation, and loading condition. We limit ou attention to a few relatively simple but important nucleation configurations. By comparing the critica conditions for dislocation nucleation in these configurations with those from otherwise perfectly fla surfaces, the goal is to understand the effects of the surface steps on dislocation nucleation.

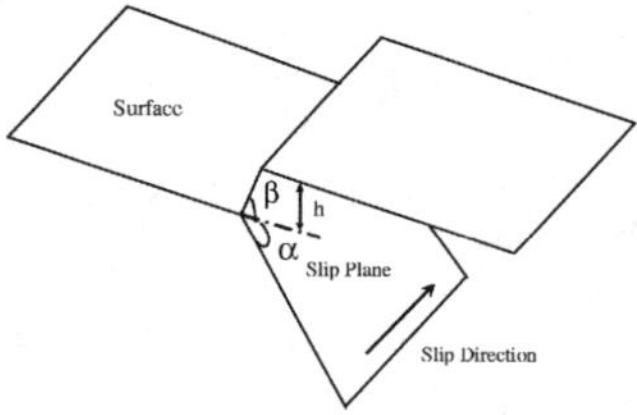

Fig. 1. The geometry of a surface step

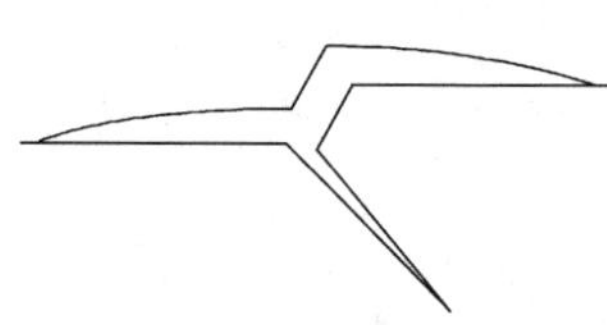

Fig. 2. The surface step treated as part of a three dimensional crack system.

Consider a general step configuration as shown in Fig. 1. The angle between the surface and sli plane on which dislocation nucleation occurs is denoted by α . The angle between the surface and th surface of the step is denoted by β . The height of the step is denoted by h. The slip direction of th slip plane is assumed to be parallel to the surface. The driving force for nucleation of a scre dislocation is the remote simple uniform shear stress acting parallel to the slip direction. We recogniz that this configuration is somewhat uncommon, and we select it for its simplicity and expect it t capture the essential features of the influence of the step on dislocation nucleation in more realisti configurations. The crystal is further assumed to be isotropic characterized by the shear modulus μ an Poisson's ratio $\nu = 0.3$, although the formulation presented in this paper also applies to the gener anisotropic crystals.

Variational Boundary Integral Formulation of the Peierls and Nabarro Dislocation Model

The problem is analyzed based on the framework of the Peierls and Nabarro dislocation model. I this model, the profile of a dislocation along a slip plane, corresponding to the relative displaceme between the two adjacent atomic layers along the slip plane, is modeled as a continuous distribution infinitesimal dislocations. Between the adjacent atomic layers is assumed an interatomic layer potenti along the slip direction, while the surrounding crystal is treated as a linear elastic solid. The profile the dislocation is then solved by the variational boundary integral method [7, 8], which had bee generally developed for solving displacement discontinuities of surfaces in infinite elastic solids. T include the surface effect, we proposed a straightforward approach in which the surface with the step modeled as part of a very large crack embedded in the infinite elastic solid. As the size of this crack selected to be much larger than any relevant size of the dislocation and step configuration, the proble can then be solved in an equivalent system that is essentially composed of the slip plane and the crac embedded in the infinite solid. These developments have been described in detail in the literature ar are only briefly summarized here [7-9].

Let $\Delta(\mathbf{x})$ denote the displacement along the slip plane and the opening displacement of the crack (i.e. the displacement of the surface). The total energy $\Pi[\Delta(\mathbf{x})]$ of the system can be expressed by

$$\Pi[\Delta(\mathbf{x})] = W[\Delta(\mathbf{x})] + V[\Delta(\mathbf{x})] - P[\Delta(\mathbf{x})] \tag{1}$$

where W is the elastic strain energy, V is the interatomic layer potential energy, and P is the external work of the applied force. Xu has shown the elastic strain energy for the general anisotropic solid is given by [10]

$$W[\Delta(\mathbf{x})] = \frac{1}{8\pi^2}\int_S\int_S \frac{1}{R}\int_0^{2\pi} \mathbf{e}_i \cdot \left\{ \begin{array}{l} \left[(\mathbf{n}\times\nabla\Delta_i)_1 \times \mathbf{z}, (\mathbf{n}\times\nabla\Delta_j)_2 \times \mathbf{z}\right] \\ -\left[(\mathbf{n}\times\nabla\Delta_i)_1 \times \mathbf{z}, \mathbf{z}\right]\cdot(\mathbf{z},\mathbf{z})\cdot\left[\mathbf{z}, (\mathbf{n}\times\nabla\Delta_j)_2 \times \mathbf{z}\right] \end{array} \right\} \cdot \mathbf{e}_j d\varphi dS_1 dS_2 \tag{2}$$

where S represents the slip plane and crack surface; $(.)_1$ and $(.)_2$ denote two different points on the surface; R is the distance between these two points; $\mathbf{e}_i$ (i=1,2,3) are Cartesian basis vectors; **n** is the unit normal vector on the surface S; **z** represents any unit vector in the plane perpendicular to **R** (vector between the two points), and φ is the polar angle of **z** within the plane. The components of the second-rank tensor in the notation $(\mathbf{a},\mathbf{b})$ are defined as $(\mathbf{a},\mathbf{b})_{jk} = a_i c_{ijkl} b_l$, where c_{ijkl} are elastic constants. The interatomic layer potential energy is given by

$$V[\Delta(\mathbf{x})] = \int_S \Phi[\Delta(\mathbf{x})]dS \tag{3}$$

where $\Phi[\Delta(\mathbf{x})]$ is the interatomic layer potential energy defined per unit area of the slip plane, which can be modeled by considering two blocks of crystals sliding uniformly across the slip plane. Under simple share, by assuming that the sliding is restricted in the Burgers vector direction, the shear resistance versus displacement may be approximated as the sinusoidal relation. The more general relation can be obtained based on atomic calculations of the energy variation as two blocks of crystals slide uniformly across the slip plane [11, 12]. The sinusoidal relation is adopted in the following study for its simplicity. The effects of other type relations on the energetics of dislocation nucleation have been documented in the previous studies [9]. Finally, the external work of traction **t** applied on the surface and the slip plane is

$$P[\Delta(\mathbf{x})] = \int_S \mathbf{t}\cdot\Delta(\mathbf{x})dS \tag{5}$$

where **t** can be related to the remote stress by invoking the principle of superposition.

The minimization of the potential energy $\Pi[\Delta(\mathbf{x})]$ with respect to the displacement $\Delta(\mathbf{x})$ leads to nonlinear integral equations. The finite element method with the six-node triangular elements is used to discretize the slip plane and the crack. The resulted nonlinear equations are solved by the Newton-Raphson iteration. The saddle-point configurations of dislocations are solved by recourse to displacement control, which is, in essence, the characteristic of a system with equality constraints enforced by Lagrange multipliers. For the flexibility of the method to treat both straight and curved dislocations uniformly, the periodic condition is enforced along the slip direction. The influence of the interaction between the curved dislocations can be justifiably minimized by increasing the period. Again, the details of these techniques are referred to Xu and Ortiz [7], and Xu, et al. [8].

Nucleation of a Straight Dislocation versus a Curved Dislocation

Consider the configuration as shown in Fig. 1. As the shear stress increases from zero, the stre concentration on the slip plane induces a stable slip until the shear stress reaches a critical value τ Because of the periodic interlayer potential, an incipient dislocation can then be formed and becom unstable as the stress decreases from this critical point. Thus, there are two solutions associated wi the same level of stress below this critical value, one is stable and the other is unstable. The ener difference between these two configurations is the activation energy E_{act} required for thermal assisted dislocation nucleation.

Under uniform shear loading in the slip direction, a straight screw dislocation may instantaneously nucleated when the stress reaches the critical value τ_c. When the stress is below t critical value, the more realistic scenario is the process of nucleation of a curved dislocation which c be viewed as a consequence of a localized outward protrusion of slip into an unstable saddle poi configuration from the stable configuration of a straight embryonic dislocation. The computation f nucleation of the curve dislocation is obviously more demanding than nucleation of the otherwi straight dislocation. We first compare the results between these two cases. The objective is to illustra the effects of the step on dislocation nucleation can be adequately interpreted through the calculation energetics of nucleation of straight dislocations.

Without loss of generality, we compare the effect of the step height on dislocation nucleation in particular configuration of $\alpha = 30^\circ$ and $\beta = 90^\circ$. Fig. 3 shows the typical mesh used in the analys Fig. 4 shows the contour plot of the typical saddle point configuration of an embryonic dislocation the slip plane. The period L used in the analysis is 20b, where b is the magnitude of the Burgers vect The stress dependent activation energies for nucleation of a straight or curved dislocation for vario step heights are plotted in Fig. 5 and 6, respectively. The trends of activation energy are similar wh the critical shear stresses for instantaneously nucleation of dislocations are the same. The stresses normalized by the maximum stress of the sinusoidal shear resistance between the atomic laye Interestingly, the results show that the atomic scale step may rapidly reduce the critical shear stress more clearly indicated in Fig. 7. As the height of the step increases, the critical stress is gradua leveled out and becomes less sensitive to the height of the step. We must cautiously note that continuum assumption may provide inaccurate results for the atomic scale steps. Direct atomis simulation may be used to further investigate this issue.

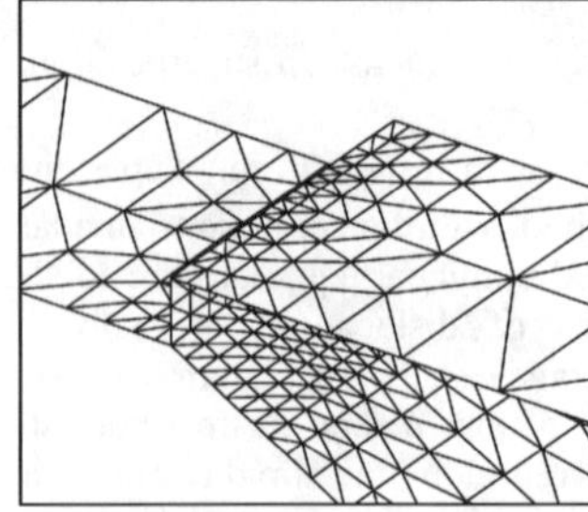

Fig. 3.Typical mesh used in simulation

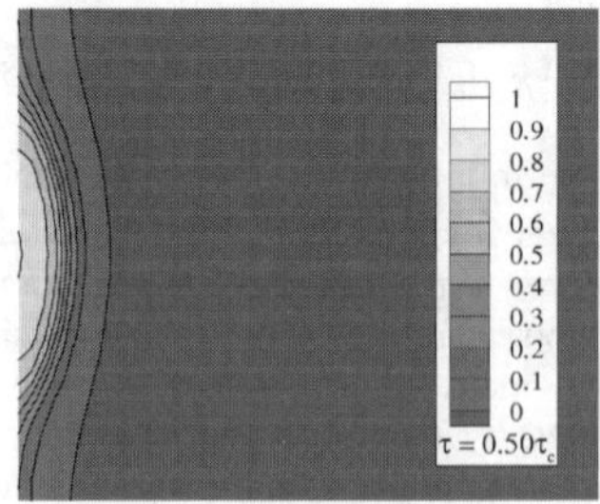

Fig.4. Contour plot of the displacement along the slip plane.

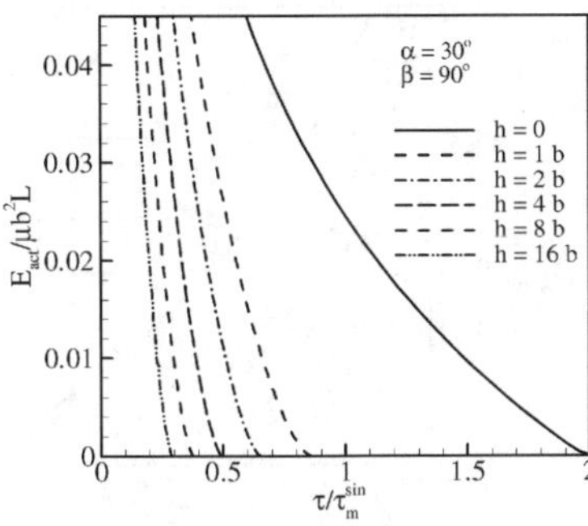

Fig.5. Stress-dependent activation energy for nucleation of a straight dislocation for various step heights.

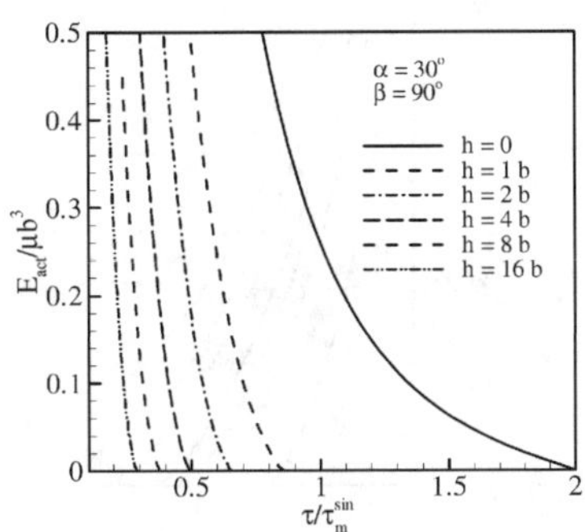

Fig. 6. Stress-dependent activation energy for nucleation of a curved dislocation for various step heights.

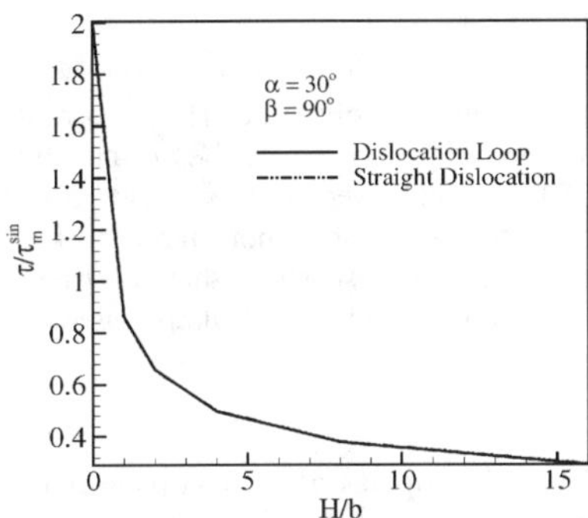

Fig.7. Critical stress for nucleation of a straight dislocation as a function of step height

The Effects of Other Step Geometry on Dislocation Nucleation

Since the activation energies for nucleation of straight and curved dislocations have the similar trend, we will illustrate the effects of the step geometry on dislocation nucleation by only conducting analyses of dislocation nucleation for straight dislocations.

Fig. 8 shows the effect of the slip plane orientation on dislocation nucleation in the configuration of the step height 4b and $\beta = 90^\circ$. The results show that the surface tends to facilitate dislocation nucleation, and that dislocations appear to be more likely nucleated under the step.

Fig. 9 shows the effect of the step surface orientation on dislocation nucleation in the configuration of the step height 4b and $\alpha = 45^\circ$. Expectedly, the step which is more vertical is more likely to facilitate dislocation nucleation than the step which is less vertical. This is obviously due to the larger stress concentration at the foot of the step which is more vertical.

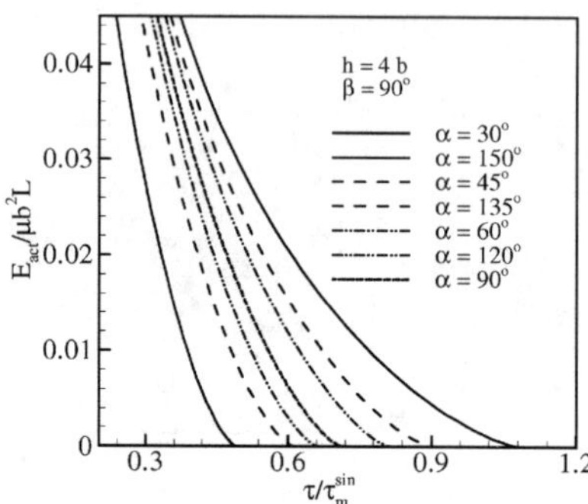

Fig. 8. Stress-dependent activation energy for various slip plane orientations.

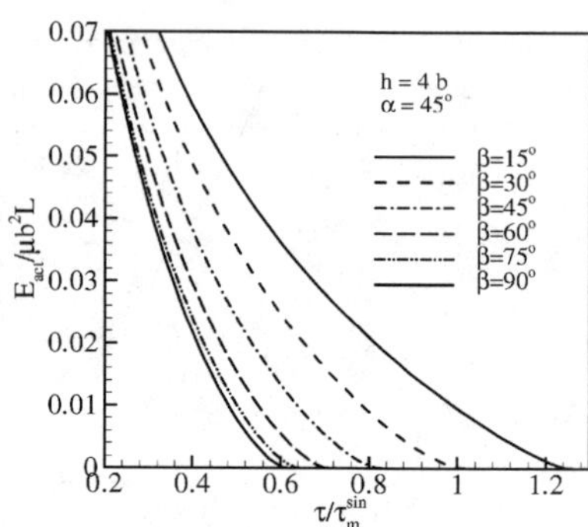

Fig. 9. Stress-dependent activation energy for various step orientations.

5. Summary and conclusions

We have analyzed the energetics of dislocation nucleation from the surface steps based on tl variational boundary integral formulation of the Peierls-Nabarro dislocation model. The results indica that the atomic scale step may significantly reduce the critical stress for dislocation nucleation from th surface by several factors. The critical stress is less sensitive to the height of the large step than tl small step. The step which is more vertical is more likely to facilitate dislocation nucleation than tl step which is less vertical. The results also show that the free surface tends to facilitate dislocatic nucleation, and that dislocations appear to be more likely nucleated under the step.

Acknowledges

We gratefully acknowledge the support of the National Science Foundation through the CAREE (0103257) and NIRT (0134446) awards.

References

1. Gao, H., Ozkan, C. S., Nix, W. D., Zimmerman, J. A. and Freund, L. B., *Phil. Mag. A*, **79**, 349-3 (1999).
2. Wagner, G., *Physica Status Solidi A*, **173**, 385-403 (1999).
3. Brochard, S., Beauchamp, P., Grilhe, J., *Phil. Mag. A*, **80**, 503-524 (2000).
4. Xu, G., *Phil. Mag. A*, 82, 3177-3185 (2002).
5. Peierls, R. E., *Proc. Phys. Soc.*, **52**, 34-7 (1940).
6. Nabarro, F. R. N., *Proc. Phys. Soc.*, **59**, 256-72 (1947).
7. Xu, G. and M. Ortiz, *Int. J. Num. Meth. Engng.*, **36**, 3675-701 (1993).
8. Xu, G., A. S. Argon and M. Ortiz, *Phil. Mag. A*, **72**, 415-51 (1995).
9. Xu, G., Zhang, C., *J. Mech. Phys. Solids*, **51**, 1371(2003)
10. Xu, G. *J. Appl. Mech.*, **67**, 403-408 (2000).
11. Christian, J. W. and V. Vitek, *Rep. Prog. Phys.*, **33**, 307-411 (1970).
12. Schoeck, G., *Phil. Mag. A*, **81**, 1161-1176 (2001).

Mat. Res. Soc. Symp. Proc. Vol. 821 © 2004 Materials Research Society P1.2

CONSTRAINED GRAIN BOUNDARY DIFFUSION IN THIN COPPER FILMS

Markus J. Buehler[1], Alexander Hartmaier[1] and Huajian Gao[1]
[1]Max Planck Institute for Metals Research, 70569 Stuttgart, Germany

ABSTRACT

In a recent study of diffusional creep in polycrystalline thin films deposited on substrates, we have discovered a new class of defects called the grain boundary diffusion wedges (Gao *et al.*, Acta Mat. **47**, pp. 2865-2878, 1999). These diffusion wedges are formed by stress driven mass transport between the free surface of the film and the grain boundaries during the process of substrate-constrained grain boundary diffusion. The mathematical modeling involves solution of integro-differential equations representing a strong coupling between elasticity and diffusion. The solution can be decomposed into diffusional eigenmodes reminiscent of crack-like opening displacement along the grain boundary which leads to a singular stress field at the root of the grain boundary. We find that the theoretical analysis successfully explains the difference between the mechanical behaviors of passivated and unpassivated copper films during thermal cycling on a silicon substrate. An important implication of our theoretical analysis is that dislocations with Burgers vector parallel to the interface can be nucleated at the root of the grain boundary. This is a new dislocation mechanism in thin films which contrasts to the well known Mathews-Freund-Nix mechanism of threading dislocation propagation. Recent TEM experiments at the Max Planck Institute for Metals Research have shown that, while threading dislocations dominate in passivated metal films, parallel glide dislocations begin to dominate in unpassivated copper films with thickness below 400 nm. This is consistent with our theoretical predictions. We have developed large scale molecular dynamics simulations of grain boundary diffusion wedges to clarify the nucleation mechanisms of parallel glide in thin films. Such atomic scale simulations of thin film diffusion not only show results which are consistent with both continuum theoretical and experimental studies, but also revealed the atomic processes of dislocation nucleation, climb, glide and storage in grain boundaries. The study should have far reaching implications for modeling deformation and diffusion in micro- and nanostructured materials.

INTRODUCTION

The study of the mechanical properties of materials at nano- and sub-micrometer scales is motivated by continuing miniaturization of engineering and electronic components, development of nanostructured materials, thin film technology and surface science. When the material volume is lowered, characteristic dimensions are reduced and material properties often deviate from the behavior of bulk materials. Small-scale materials are defined as materials where at least one dimension is reduced. For instance, thin films bond to substrates are a relevant example of materials in small dimensions since the film thickness h_f is small compared to the planar extension of the film and the thickness of the substrate.

Thin films deposited on substrates have become an increasingly active area of research in the last decades. This can partly be attributed to the fact that these materials are becoming critically important in today's technologies, whereas changes in material behavior due to the effects of surfaces, interfaces and geometrical constraints are not completely understood. The focus of this paper is on mechanical properties of ultra thin sub-micron copper films on substrates. We show that in such materials, important effects of the film surface and grain boundaries are observed and that the constraint by the film-substrate interface governs the mechanical behavior.

Different inelastic deformation mechanisms are known to relax the internal and external stresses in a thin film. Experiment shows that for films of thicknesses between approximately 2 and 0.5 μm, the flow stress increases in inverse proportion to the film thickness [1-3], thus

$$\sigma_Y \propto \frac{1}{h_f}. \quad (1)$$

Originally, it was proposed that the strengthening of thin films results from the energetic effort associated with dislocation motion [4-6]. More recent theoretical and experimental work [7-11] indicates that the strength of thin metal films often results from a lack of active dislocation sources.

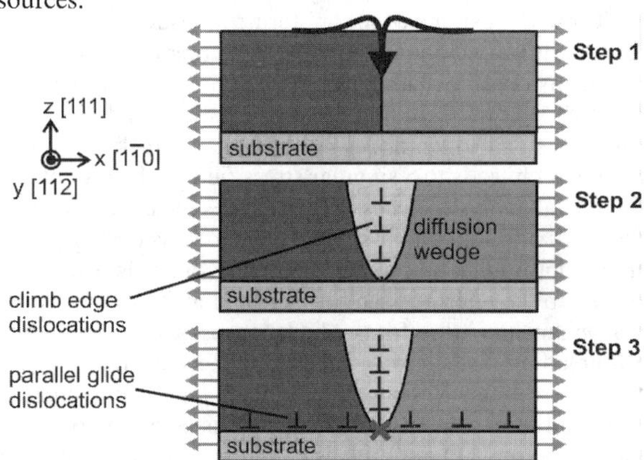

Figure 1: Continuum mechanics model of constrained diffusional creep [13]. Diffusional creep is modeled by climb of edge dislocations in the grain boundary. A singular, crack-like stress field develops in the long-time limit leading to very large resolved shear stress on glide planes parallel to the film surface, causing nucleation of parallel glide dislocations.

In copper films, the regime where plastic relaxation is limited by dislocation nucleation and carried by glide of threading dislocations reaches down to film thicknesses of about $h_f \approx 400$ nm [10]. For yet thinner films experiments reveal a film-thickness-independent flow stress [8, 10, 12]. This observation contradicts the existing theories of plasticity in thin films. *In-situ* transmission electron microscopy observations of the deformation of such ultra thin films reveal dislocation motion parallel to the film-substrate interface [8, 10, 12]. This glide mechanism is unexpected, because in the global biaxial stress field there is no resolved shear stress on parallel glide planes. This indicates that there must be a mechanism involving long-range internal stresses that decay only slowly on the length scale of the film thickness. For sufficiently thin films these internal stresses have a pronounced effect on the mechanical behavior. It has been proposed that constrained diffusional creep may be the origin of this novel deformation mechanism [13]. In the model of constrained diffusional creep, it is assumed that through climb of edge dislocations along the grain boundary, material is transported from the surface toward the substrate, leading to a crack-like stress field with large resolved shear stress on glide planes parallel to the film surface. This explains the occurrence of parallel glide dislocations. The model is schematically summarized in Figure 1. Despite progresses in atomistic and continuum modeling [13-18], deformation by parallel glide dislocation motion in ultra thin films is not well understood as of today. In this paper, we propose atomistic and continuum studies to investigate the behavior of such thin films below 400 nm.

The paper is divided into two main parts. In the first part, we report continuum modeling of thermal cycling experiments of ultra thin copper films, similar to the studies reported in [14]. In contrast to earlier work, here we describe a continuum mechanics model [13, 15] incorporating the concept of a threshold stress for diffusion initiation. The assumption of the threshold stress is motivated by theoretical considerations [16, 17], in particular mesoscopic discrete dislocation simulations [18], as well as experimental results [10, 12]. The model is then used to calculate the stress in the thin film during thermal cycling experiments, and several numerical examples are provided showing qualitative agreement of the continuum model with the experimental results. In the second part, we focus on fully atomistic modeling of constrained diffusional creep in thin copper films. Unlike in previous studies of quasi-two-dimensional simulations [16], here we describe simulation results of a polycrystalline model. We discuss the dependence of the grain boundary structure on the diffusivities and show that nucleation of parallel glide dislocations occurs from those grain boundaries with highest diffusivities leading to fastest grain boundary traction relaxation, in agreement with the predictions by theory [13, 15]. The simulation results close the theory-experiment-simulation linkage.

MODELING OF THERMAL CYCLING EXPERIMENTS

It is known that mechanical stresses in thin films are often significantly higher than those in the bulk form of the same material. At low temperatures, this has, as discussed above, been mostly attributed to geometric confinement or lack of dislocation sources. At higher temperatures, experimental results indicate that diffusional creep may play an increasingly important role [8, 10, 12, 14].

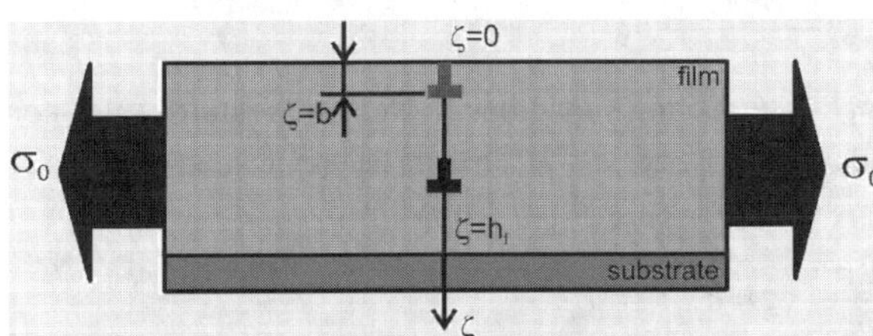

Figure 2: Geometry and coordinate system for the continuum model of constrained diffusional creep. Diffusional mass transport along the grain boundary is modeled by climb of edge dislocations.

The model of constrained diffusional creep [13, 15] has recently been further developed in light of experimental results to directly model the average stress in the film during the thermal cycling experiments by introducing a convolution procedure [14]. Weiss *et al.* [14] invoked the constrained diffusional creep model to explain the occurrence of a stress drop observed during the first heating cycle of thin copper films. They also used constrained diffusional creep to model the stress-temperature curves measured during thermal cycling of several films and found reasonable agreement for a 500 nm copper film, albeit by assuming a very large grain size.

In this section we discuss a modification of this continuum model including a threshold stress for diffusion initiation. The assumption of the threshold stress is motivated by theoretical considerations about diffusion initiation as discussed in earlier work on this topic [16, 17], in particular mesoscopic discrete dislocation simulations [18], as well as experimental results [10,

12]. We will show that the modified model including the threshold stress provides better agreement of the experimental results with the theory particularly for higher temperatures [19].

In the following, we report a modified continuum model to capture the threshold stress, and compare thermal cycling curves predicted by the new continuum mechanics model featuring a threshold stress with the experimental results for two different film thicknesses.

Continuum model with threshold stress

Here we review a solution to the problem reported in [19]. Compared to earlier analysis of the problem of constrained diffusional creep [13, 14], the boundary conditions are modified to account for a threshold stress σ_t below which no diffusive stress relaxation occurs. The coordinate system and the geometry of the thin film are shown in Figure 2. We assume that $\sigma_{xx}(\varsigma = 0,t) = \sigma_t$, and

$$\frac{\partial \sigma_{xx}}{\partial \varsigma}(\varsigma = h_f, t) = \frac{\partial^2 \sigma_{xx}}{\partial \varsigma^2}(\varsigma = h_f, t) = 0. \tag{2}$$

The initial conditions remain the same, thus $\sigma_{xx}(\varsigma, t = 0) = \sigma_0$. We consider the possibility that σ_t may be different under tension and compression. That is, no grain boundary diffusion can occur when $\sigma_t^- < \sigma_{xx}(0,t) < \sigma_t^+$ where $\sigma_{xx}(0,t)$ is the stress at the site of initiating climb edge dislocations near the entrance to the grain boundary. Using superposition $\sigma_{xx}(\varsigma,t) = \sigma_t + \tilde{\sigma}(\varsigma,t)$, we obtain the governing equation for $\tilde{\sigma}(\varsigma,t)$ as

$$\frac{\partial \tilde{\sigma}_{xx}}{\partial t} = \frac{E D_{\mathrm{GB}} \delta_{\mathrm{GB}} \Omega}{4\pi(1-\nu^2)\mathrm{k}T} \int_0^{h_f} S(\varsigma,\xi) \frac{\partial^3 \tilde{\sigma}_{xx}(\xi,t)}{\partial \xi^3} \mathrm{d}\xi \tag{3}$$

where $S(\varsigma,\xi)$ is a kernel function [13], E Young's modulus, ν Poisson's ratio, Ω the atomic volume, $D_{\mathrm{GB}}\delta_{\mathrm{GB}}$ is the grain boundary diffusivity, and $\mathrm{k}T$ has its usual meaning (see further details in [13-15, 19]). The boundary conditions are $\tilde{\sigma}_{xx}(\varsigma = 0,t) = \sigma_t$,

$$\frac{\partial \tilde{\sigma}_{xx}}{\partial \varsigma}(\varsigma = h_f, t) = \frac{\partial^2 \tilde{\sigma}_{xx}}{\partial \varsigma^2}(\varsigma = h_f, t) = 0, \tag{4}$$

and the initial condition is $\tilde{\sigma}_{xx}(\varsigma, t = 0) = \sigma_0 - \sigma_t$. These equations are identical to those of [13-15], except that the initial condition is effectively reduced. Therefore, we can simply use the previous solution to obtain the average stress along the grain boundary as

$$\hat{\sigma}_{\mathrm{GB}}(t) = (\sigma_0 - \sigma_t)\exp(-\lambda_0 t/\tau) + \sigma_t, \tag{5}$$

with a geometry dependent constant $\lambda_0 = 8.1 + 30.65 h_f / d$, where d is the grain size [14]. Note that $\tau(T) = 4\pi(1-\nu^2)\mathrm{k}T h_f^3 / (E D_{\mathrm{GB}} \delta_{\mathrm{GB}} \Omega)$ is a characteristic time for stress decay [13-15]. We can now generalize the convolution procedure described in [14]. We assume that

$$\frac{\mathrm{d}\sigma_0}{\mathrm{d}T} = -\Delta\alpha M_f, \tag{6}$$

where $M_f = E/(1-\nu)$ is the biaxial modulus, and $\Delta\alpha$ refers to the thermal mismatch of film and substrate material. Then, the stress in the film during thermal cycling is

$$\hat{\sigma}_{\mathrm{GB}}(t) = \sigma_t + (\sigma_0^{\mathrm{start}} - \sigma_t)\exp\left[-\lambda \int_{T_{\mathrm{start}}}^{T} \frac{\mathrm{d}\varsigma}{\dot{T}\tau(\varsigma)}\right] - M_f \Delta\alpha \dot{T} \int_{T_{\mathrm{start}}}^{T} \exp\left[-\frac{\lambda}{\dot{T}}\left(\int_{\varsigma}^{T} \frac{\mathrm{d}\xi}{\tau(\xi)}\right)\right] \frac{1}{\dot{T}} \mathrm{d}\varsigma \qquad (7)$$

where $\sigma_0^{\mathrm{start}}$ is the stress when the thermal cycling is initiated, and T_{start} is the starting temperature. The average stress in the film σ is related to σ_0 and $\hat{\sigma}_{\mathrm{GB}}$ as

$$\sigma = \sigma_0 - (\sigma_0 - \hat{\sigma}_{\mathrm{GB}})\theta, \qquad (8)$$

where

$$\theta = \frac{4h_f}{d} \tanh\left(\frac{d}{4h_f}\right). \qquad (9)$$

For a given experimentally measured stress σ, the average stress along the grain boundary $\hat{\sigma}_{\mathrm{GB}}$ is given by

$$\hat{\sigma}_{\mathrm{GB}} = \theta^{-1}(\sigma - \sigma_0) + \sigma_0. \qquad (10)$$

This equation shows that the average film stress depends on the ratio of grain size to film thickness. Equation (7) is numerically solved using MATLAB.

Numerical examples and comparison with experiment

Figure 3a shows a numerical and experimental example of the decay of the average stress in the film with and without threshold stress (the characteristic time is roughly fitted to experimental results to allow qualitative comparison). From these experimental results, the parameter σ_t^+ is assumed to be 65 MPa (and $\sigma_t^- = -\sigma_t^+$ for the studies below) [19]. The experimental results show that the film stress does not decay to zero but instead approaches a plateau value, thus supporting the model of a threshold stress for diffusion. Note that even

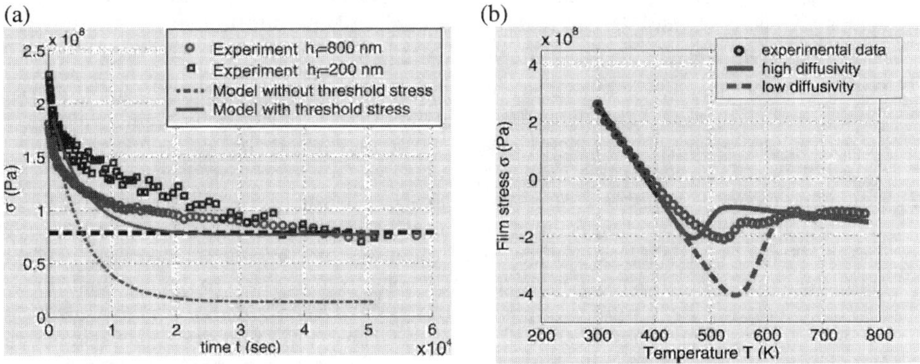

Figure 3: (a) Stress decay in a thin film with and without threshold stress. Comparison with the experimental data clearly supports the concept of a threshold stress. From the experimental data [19], the threshold stress is estimated to be around 65 MPa. (b) The plot shows a fit of diffusivities to experimental data during first cooling cycle.

without the threshold stress, the experimentally measured stress in the film is non-zero (albeit quite small), even though the traction along the grain boundary is completely relaxed (as can be verified by equation (8)).

In Figure 4a we show a comparison of experiment and the continuum model with threshold stress for thermal cycling of thin films of thickness $h_f \approx$ 100 nm. In Figure 4b we show a comparison of experiment and the continuum model with threshold stress for thermal cycling of a 600 nm film. In both plots, the diffusivity is assumed to be $D_{\mathrm{GB}}\delta_{\mathrm{GB}} = 5\times10^{-15}\exp(-Q_b / \mathrm{R}T)$ $\mathrm{m^3/s}$, with $Q_b = 104$ kJ/mole (the experimental data is taken from [19]). In both cases, some qualitative agreement is found, especially at elevated temperatures where diffusional creep clearly dominates deformation [8, 10, 12]. For the 600 nm film, the results at low temperatures during the cooling cycle (upper curve) do not match as good as for high temperatures. This could partly be explained by the fact that, due to the rather large film thickness, not only parallel glide dislocations are present. In addition, threading dislocations are nucleated and relieve stresses as has been confirmed experimentally [8, 10, 12]. Therefore, during cooling of the film, diffusional creep dominates at high temperatures, but threading dislocations dominate at low temperatures. Since the continuum model does not account for threading dislocations, the stresses at low temperatures (upper left corner in the plot) are overestimated. At high temperatures during the heating cycle, the stress relaxes to the value associated with the threshold stress σ_t^-, and the experimental measurements and continuum modeling agree quite well.

Another interesting aspect that we explored is the estimation of diffusivities from the experimental data. It can be observed in Figure 4b that the stress decay in the first heating cycle (blue dashed line) is larger in the experiment than in the simulation. When the classical diffusivities as reported in the literature are used, the maximum compressive stress is overestimated by a factor of two (similar observation was also made previously [14]). This could be due to the fact that the diffusivities used in the continuum mechanics model are smaller than the actual experimental value. Figure 3b shows a fit of the continuum mechanics model to the experimental data or a 600 nm film as obtained from experiment. Fitting the maximum compressive stress to the experimental results by adjusting the diffusivity yields a diffusivity that is 80 times higher than in the literature, therefore $D_{\mathrm{GB}}\delta_{\mathrm{GB}} = 4\times10^{-13}\exp(-Q_b / \mathrm{R}T)$ $\mathrm{m^3/s}$. Possible reasons for this could be the fact that the grain boundary structure strongly influences the diffusivities, a point that is not addressed in most classical theories of grain boundary diffusion. Finally, we compare thermal cycling experiments with the new estimate for diffusivities to the experimental results. In Figure 5 we show a comparison of experiment and the continuum model with threshold stress for steady-state thermal cycling of thin films with film thickness 100 nm and 600 nm. Although the difference to the previous case (shown in Figure 4) for the 100 nm film is not large, for the 600 nm film we observe some improvement in particular at the low-temperature regime. Further discussion on this topic is left to future publications.

The main result of the work report is that the threshold stress proposed theoretically is supported by experimental results [8, 10, 12]. This is illustrated in Figure 3a (creep experiment) and Figure 4 (thermal cycling experiment). The occurrence of the thermal slope at high temperatures around 700-800 K could not be reproduced by the continuum model without the threshold stress [14] (see dotted line in Figure 4b showing the solution without threshold stress).

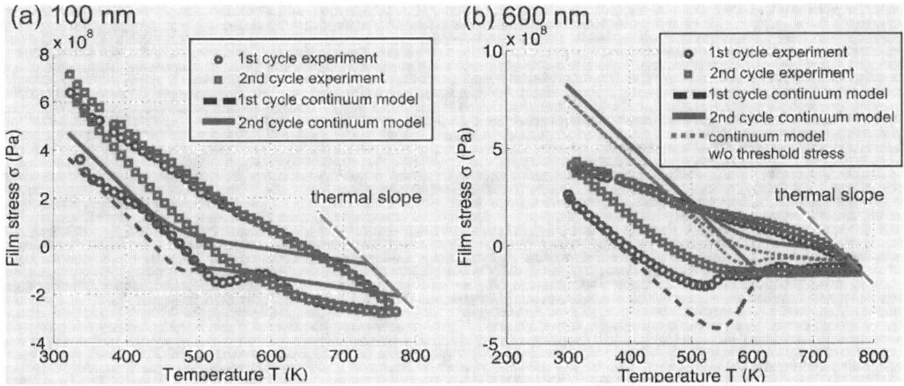

Figure 4: Modeling of thermal cycling experiments with the modified continuum theory, in comparison with experimental data [19]. Plot (a) depicts the results for a 100 nm film, and plot (b) shows the stress versus temperature for a 600 nm film. Grain boundary diffusivities are chosen as in [14]. In subplot (b), the solution without the threshold stress is shown by a dotted line for comparison. The deviation from the experimental results at $T \approx 700..800$ K is obvious.

MODELING OF CONSTRAINED DIFFUSIONAL CREEP IN POLYCRYSTALLINE THIN FILMS

Atomistic modeling of plasticity in thin films is just at its beginning, and few studies of such systems have been reported in the literature. Recently, atomistic simulations of two-dimensional thin film systems were discussed [20]. In this study, the yield strength was investigated and nucleation and motion of threading dislocations was in the focus. However, the model did not contain grain boundaries despite the fact that grain boundaries can serve as fertile sources for dislocations and could play an important role in plasticity of thin films. Some previous studies focused on constrained diffusional creep in quasi-two dimensional plane strain models, where nucleation of parallel glide dislocations was studied [16, 21]. In contrast to these simplistic models, we propose a three-dimensional model of thin films with a more realistic microstructure and focus on modeling of constrained diffusional creep and parallel glide dislocation nucleation.

The model studied in this section is a polycrystalline thin film consisting of hexagonal shaped grains, as shown in Figure 6. The choice of this geometry is motivated by the grain microstructure found in experiments [8, 10, 12]. An advantage of this model over the geometry used in previous models [16] is that fully-periodic boundary conditions in the x- and y-direction can be assumed, avoiding effects by rigid boundaries during application of loading [16].

We will focus on dislocation nucleation and motion from grain boundaries and the grain boundary-substrate interface. One of the important objectives will be to study the effect of grain boundary traction relaxation by diffusional creep on the dislocation mechanism that operates in the film. As known from previous studies [17], the structure of the grain boundaries has significant influence on the nucleation of dislocations. Here we investigate the role of the grain boundary structure on diffusional creep and associated dislocation mechanisms.

The outline of this section is as follows. After presenting details about the atomistic modeling procedure, we continue with a discussion of the results of large-scale atomistic studies. We will show that grain boundary relaxation by diffusional creep gives rise to dominance of parallel glide

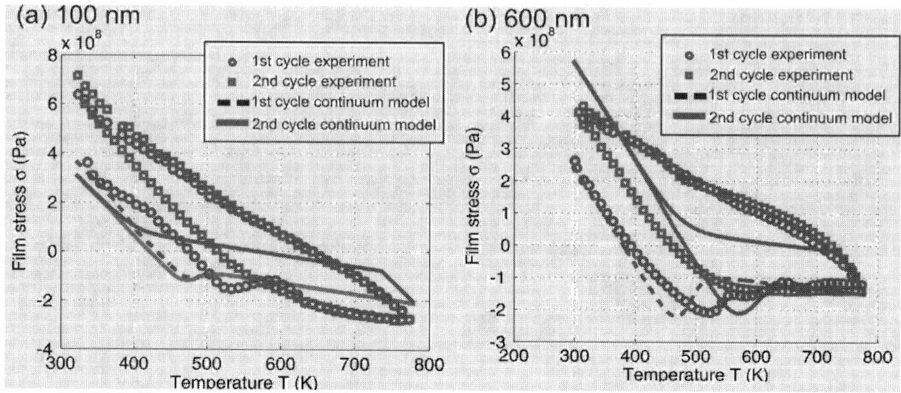

Figure 5: Modeling of thermal cycling experiments with the modified continuum theory, in comparison with experimental data [19]. Plot (a) shows the results for a 100 nm film, and plot (b) plots the stress versus temperature for a 600 nm film. Grain boundary diffusivities are modified according to the first cycle (see also Figure 3b and associated discussion).

dislocations, in accordance with experiment. In contrast, in grains where grain boundary tractions are not relaxed, threading dislocations prevail.

Atomistic modeling

We extend the two-dimensional studies reported earlier [16, 21] to the three-dimensional case. We apply a biaxial loading rate on the order of 1 percent total strain per nanosecond. The temperature is, as in the two-dimensional studies [16], chosen around 90 % of the melting temperature. At this temperature, the strain rate is low enough such that diffusional creep can be observed. Our simulation tool is classical molecular dynamics, which has proven to be capable of describing small-scale materials phenomena associated with materials failure in the past, including diffusional creep [16, 17, 19, 23, 26, 27]. Simulations are carried out using the ITAP-IMD molecular dynamics code [24] suitable for large-scale simulations on parallel computing platforms. We use a multi-body EAM potential to model the thin copper film [25].

We model a film of thickness $h_f \approx 11$ nm with a grain diameter of about 22 nm in the x-direction (experimentalists find a similar grain size to film thickness ratio between two and three). The simulation sample is constructed such that there are high-energy as well as low-energy grain boundaries. This is motivated by our objective to investigate the effect of grain boundary structure on the deformation mechanisms. As shown in Figure 6, grain #1 is completely surrounded by high-energy grain boundaries, and the other grains feature low-energy grain boundaries (grain #1 is in its reference configuration, grain #2 is rotated by 35.4 degrees, grain #3 by 44.7 degrees and grain #4 by 53.4 degrees). The system contains about 2 million atoms.

Constrained grain boundary diffusion

At elevated temperatures, high-energy grain boundaries transform into liquid-like intergranular layers [26, 27], whereas low-energy grain boundaries establish as arrays of misfit dislocations [17]. Formation of grain boundary diffusion wedges is accompanied by surface grooving at the grain boundary interface. Therefore, the surface height profile provides an indication of diffusive activities in the grain. Figure 7 plots the surface profile of a polycrystalline sample during different stages of the simulation (similar to AFM results). We observe that, in agreement with the predictions in the literature [26, 27], high-energy grain boundaries provide very fast diffusion paths, in contrast to low-energy grain boundaries. This strongly underlines the notion that the grain boundary structure needs to be taken account when diffusivities are determined. Figure 8 plots the surface profile in two different 3D views, clearly showing the development of surface grooves due to constrained diffusional creep. The observation of surface grooves is in perfect agreement with experimental reports [14].

Compared with all other diffusion paths, grain triple junctions provide the fastest paths for diffusion. This is verified since at grain triple junctions, the surface grooves are deepest (see Figure 7b-d and Figure 8 showing the surface profile at later stages). As in the two-dimensional case [16], we find that the grain boundaries become curved along the z-direction.

Nucleation of parallel glide dislocations

According to the hypothesis by continuum theory [13, 15], parallel glide dislocations should exclusively be nucleated along grain boundaries whose tractions are relaxed by diffusional creep, and threading dislocations should dominate elsewhere. Since high-energy grain boundaries are predominant paths for diffusion [16, 26, 27], in grains neighboring such high-energy grain boundaries, parallel glide dislocations should dominate plasticity. This prediction is verified by our atomistic simulations. In our sample, we find no other dislocation activity than parallel glide dislocations in grain #1. We observe nucleation of first parallel glide dislocations (see Figure 7c) at a biaxial strain of about 1.6 %. In other grains where little grain boundary relaxation is possible by diffusional creep, threading dislocations are easily nucleated (due to generation of a surface step, they appear as lines on the surface in Figure 7). Predominant nucleation site are, in agreement with previous results of studies of plasticity in thin films at low temperature [17], misfit dislocations at the grain boundary.

The parallel glide dislocations are shown in Figures 7c and d as a black line. Their shape is determined using the energy method as exemplified in the blow-up of Figure 7c. Additional analysis was performed based on geometrical methods identical to those applied in [16]. After the first parallel glide dislocation is nucleated, additional dislocations appear as the loading is increased and the dislocations form a complex network (Figure 7d). Note that even at higher loads, there are exclusively parallel glide dislocations in grain #1 (see Figure 7d, and Figure 8). In other grains, we observe parallel glide dislocations at later stages in addition to the threading dislocations (these are not shown in the figures).

We summarize the main findings. Using a polycrystalline model, we have studied constrained diffusional creep. The most important result is that constrained diffusional creep leads to nucleation of parallel glide dislocations. Another important result is that the grain boundary structure has significant influence on diffusivities, thus explaining the fact that parallel glide dislocations are only nucleated at specific grain boundaries. This finding is not contradicting experimental results, since parallel glide dislocations only appear at specific grain boundaries [8, 10, 12].

(a)

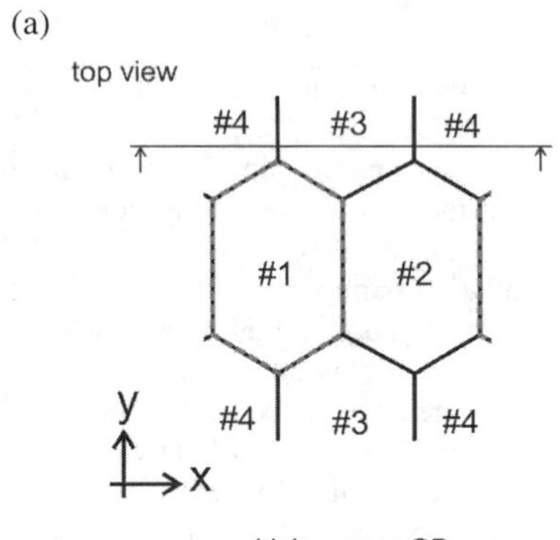

(b)

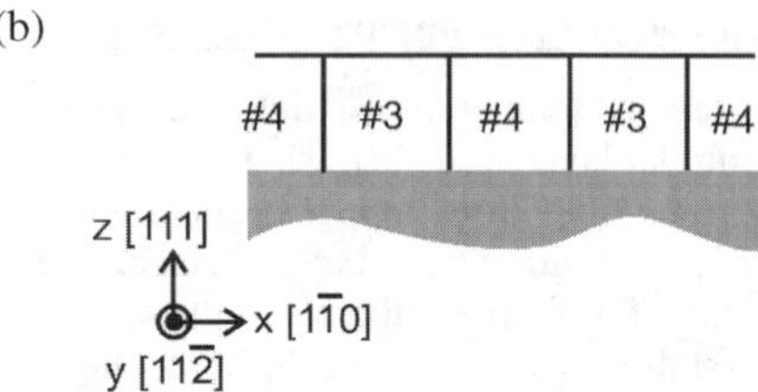

Figure 6: Geometry of the atomistic model of polycrystalline thin films, (a) view from the top, and (b) cross sectional view on the polycrystalline thin film. The high-energy GB is indicated in subplot (a) by the dashed line. We assume periodic boundary conditions in the *x*- and *y*-direction.

DISCUSSION AND CONCLUSIONS

We have summarized recent progresses in joint atomistic-continuum modeling of constrained diffusional creep in ultra thin films. The results helped to understand the deformation mechanisms in thin metal films deposited on substrates and similar small-scale materials systems.

Modeling of thermal cycling experiments with a threshold stress for diffusion initiation

We have extended the continuum model of thermal cycling experiments [15] to incorporate a threshold stress for diffusion. This approach was motivated by experimental creep relaxation experiments (Figure 3a), where a remaining athermal stress is found in the long-time limit. Further, the prediction of the stress during thermal cycling provides qualitative agreement of theory and experimental results for film thicknesses ranging from 100 nm to 600 nm (Figure 4). In summary, the new concept of a threshold stress for diffusion is supported by experiment and seems to be a useful approach. The modeled thermal cycling curves show that the introduction of a threshold stress based on experimental data represents some improvement of the previous model. Unlike in [14] where the grain size was changed in order to improve agreement, the parameters used here are identical to those in the experiments. The most important difference is that without the threshold stress, the numerically estimated curves are very "thin" in contrast to the experimental results where the curves are wider (see Figure 4b for a comparison of the solution with and without threshold stress). Most importantly, the thermal slope, which is apparent at high temperatures during the cooling cycle, can only be reproduced with a model featuring a threshold stress. Additional discussion was devoted to estimates of the grain boundary diffusivity from experimental data. It was found that diffusivities may be higher than reported in the literature [14] (see Figure 5). It was proposed that this may be attributed to the dependence of diffusivities on the grain boundary structure. This observation is supported by the MD results of diffusion in polycrystalline thin films shown in Figures 7 and 8, also suggesting a strong dependence of diffusivities on grain boundary structure.

Effect of grain boundary structure on constrained grain boundary diffusion

It was shown earlier that the GB structure has significant influence on the deformation mechanisms by dislocation motion [17]. Our results extend this understanding, since the governing character of the grain boundary structure in thin films is found also when deformation is mediated by diffusional creep. This is in qualitative agreement with the discussion in [26, 27], where it was proposed that the grain boundary structure has a significant influence on the diffusivities. In light of the model of constrained diffusional creep [13, 15], the grain boundary structure determines how fast the tractions along the grain boundaries are relaxed and a singular stress field develops. In Figures 7 and 8 we have shown that the depth of grain boundary grooves is deepest at high-energy grain boundaries corresponding to the fastest diffusion paths. We find that at those grain boundaries, parallel glide dislocations are nucleated suggesting rapid development of singular stress field.

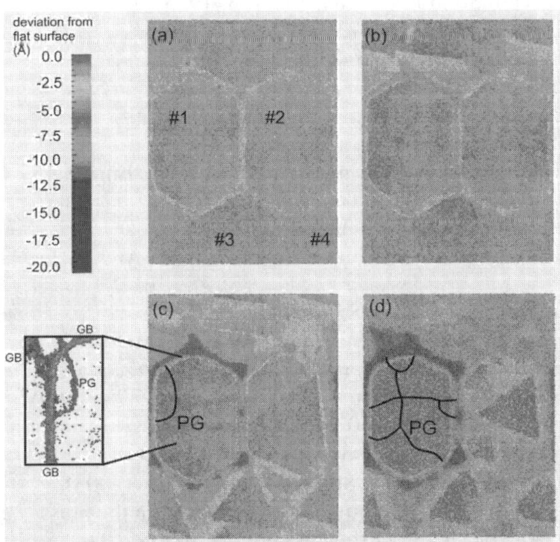

Figure 7: Plot of the surface profile of the polycrystalline thin film. The color refers to the surface height (red=high, blue=low). This color scheme allows determining diffusive deformation along specific grain boundaries surrounding grain #1. The results reveal existence of surface grooves at high-energy grain boundaries that are up to 2 nm deep (that is 18 % of film thickness). The results therefore indicate that parallel glide dislocations are only nucleated at grain boundaries whose tractions are relaxed by diffusional creep (see exclusively PG dislocations in grain #1). This result strongly supports theoretical predictions [13, 15].

Deformation mechanisms of small-scale materials

The dominance of grain boundary processes during deformation of ultra thin films is in qualitative agreement with recent investigations of other small-scale materials, such as nanostructured materials [26, 27]. The results reported in this paper support the notion that interfaces in small-scale materials could play a dominating role. Further research could be directed to develop a deeper understanding of this complex interplay of diffusional, dislocation and fracture mechanisms.

ACKNOWLEDGMENTS

We highly acknowledge helpful discussions with John Balk, Gerhard Dehm and Eduard Arzt. We thank the Institute for Theoretical and Applied Physics at the University of Stuttgart for providing the molecular-dynamics code [24].

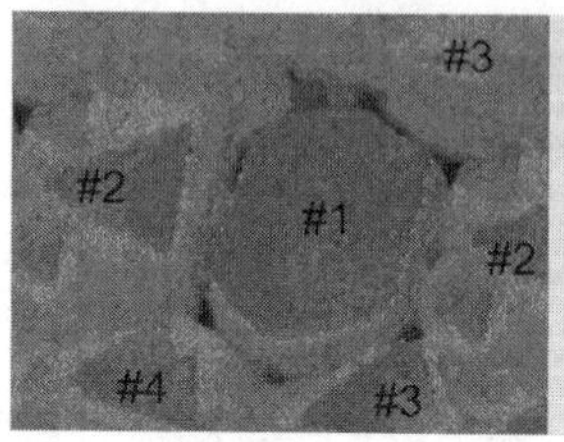

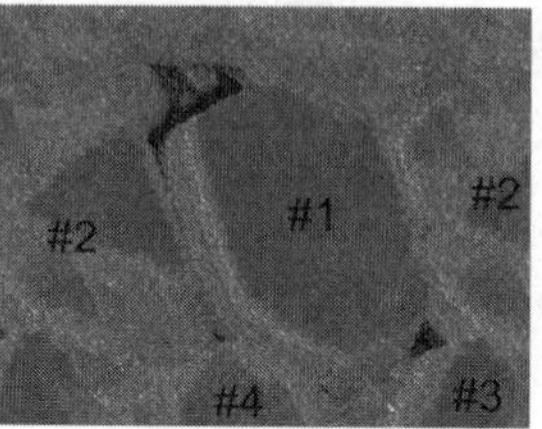

Figure 8: The figure shows three-dimensional views onto the surface. The surface grooves at the high-energy grain boundaries neighboring grain #1 can clearly be observed. The plot also confirms that no threading dislocation activity is found in grain #1 (same color map as in Figure 7).

REFERENCES

1. R. Vinci, E. Zielinski, J. Bravman, Thin Solid Films **262**, 142-153 (1995)
2. M.J. Kobrinsky, C.V. Thompson, Appl. Phys. Letts. **73**, 2429-2431 (1998).
3. R.-M. Keller, S. P. Baker, E. Arzt, Acta Mat. **47**, 415-426 (1999).
4. L.B. Freund, J. Appl. Mech. 54, 553-557 (1987)
5. W. Nix, Metall. Trans. A. **20**, 2217-2245 (1989)
6. W. Nix, Scripta Mater. **39**, 545-554 (1998)
7. B. von Blanckenhagen, P. Gumbsch, E. Arzt, Modelling Simul. Mater. Sci. Eng. **9**, 157-169 (2001).
8. T. J. Balk, G. Dehm, E. Arzt, Z. Metallk. **93** (5), 383-391 (2002)
9. M. Legros, K.J. Hemker, A. Gouldstone, S. Suresh, R.-M. Keller-Flaig, E. Arzt, Acta Mat. **50**, 3435-3452 (2002).
10. T.J. Balk, G. Dehm, E. Arzt, Acta Mat. **51**, 4471-4485 (2003).
11. B. von Blanckenhagen, P. Gumbsch, E. Arzt, Phil. Mag. Lett. **83**, 1-8 (2003).
12. T.J. Balk, G. Dehm, E. Arzt, Mat. Res. Soc. Symp. Proc. **673**, 2.7.1-2.7.6 (2001).
13. H. Gao, L. Zhang, W. Nix, C. Thompson, E. Arzt, Acta Mater. **47**, 2865-2878 (1999)
14. D. Weiss, H. Gao, E. Arzt, Acta Mater. **49**, 2395-2403 (2001)
15. L. Zhang and H. Gao, Zeitschrift fuer Metallkunde **93** (5), 417-427 (2002)
16. M.J. Buehler, A. Hartmaier, H. Gao, J. Mech. Phys. Solids **51**, 2105-2125 (2003).
17. M.J. Buehler, A. Hartmaier, H. Gao, Modelling Simul. Mater. Sci. Eng., in the press (2004).
18. A. Hartmaier, M.J. Buehler, H. Gao, Defect and Diffusion Forum **224-225**, 107-128 (2003).
19. M.J. Buehler, T.J. Balk, E. Arzt, H. Gao, in: *Handbook of Computational and Theoretical Nanotechnology* (to appear).
20. Y.L. Shen, J. Mater. Res. **18**, 2281-2284 (2003)
21. M.J. Buehler, A. Hartmaier, H. Gao, Mat. Res. Soc. Symp. Proc. **779**, 103-108 (April 2003).
22. M.J. Buehler, H. Gao, in: *Handbook of Computational and Theoretical Nanotechnology* (to appear).
23. M.J. Buehler, F.F. Abraham, H. Gao, Nature **426**, 141-146 (2003).
24. J. Stadler, R. Mikulla, H.-R. Trebin, Int. J. Mod. Phys. C **8**, 1131-1140 (1997).
25. Y. Mishin, M.J. Mehl, D.A. Papaconstantopoulos, A.F. Voter, J.D. Kress, Phys. Rev. B **63**, 224106 (2001).
26. V. Yamakov, D. Wolf, W. Phillpot, H. Gleiter, Acta Mater. **50**, 61-73 (2002)
27. D. Wolf, Y. Yamakov, S.R. Phillpot, A.K. Mukherjee, Z. Metallk. **94**, 1052-1061 (2003).
28. L. Nicola, E. van der Giessen, A. Needleman, J. Appl. Phys. **93**, 5920-5928 (2003).

Mat. Res. Soc. Symp. Proc. Vol. 821 © 2004 Materials Research Society P6.4

A Molecular Dynamics Study of Short Nanoparticle Chains under Mechanical Strain: Linear and Kinked Configurations

Adamos Dalis and Sheldon K. Friedlander
Department of Chemical Engineering, University of California in Los Angeles
Los Angeles, CA 90095-1592

ABSTRACT

Previous studies in our laboratory have shown that individual nanoparticle chain aggregates (NCA) exhibit remarkable mechanical behavior when under strain inside the transmission electron microscope. NCA made of various materials (e.g. carbon, metal oxides, metals, etc.) were strained by as much as 100% when tension was applied to them. After breaking, the NCA rapidly contracted to form more compact structures. In this study, molecular dynamics (MD) computer simulations are employed to investigate, at the atomic scale, the behavior of short nanoparticle chains under strain and to obtain quantitative information of the forces involved in chain straining and fracturing. The interaction potential used is that of copper obtained with the embedded atom method (EAM). Although the methodology is generally applicable, copper was selected as a test material because reliable interatomic potentials are available. Seven single-crystal nanoparticles, each 2.452 nm in diameter, are placed in contact in two chain configurations, linear and kinked. The structures are initially relaxed adiabatically with MD steps for 225 ps, at a starting temperature of 300 K. The bonding energy between any two nanoparticles in contact ranges from about 20 eV to 30 eV at 0 K. The two relaxed chain configurations are strained along their longest dimension, to the breaking point, at strain rates spanning from 0.3 m/s to 10 m/s. We identify mechanisms of stress accommodation that lead to plastic deformation and eventually fracture for both chain configurations, linear and kinked, and we construct the corresponding stress-strain curves. The two chain configurations exhibit different mechanical behavior. Applications of our experimental and simulation studies on NCA are to the behavior of nanocomposite materials, including carbon black reinforced rubber, sampling of aggregates by high speed impactors and the formation of flexible coatings of nanoparticles.

INTRODUCTION

Previous studies in our laboratory have shown that individual nanoparticle chain aggregates (NCA) exhibit remarkable elastic behavior when under strain inside the transmission electron microscope (TEM) [1-3]. NCA made of various materials (e.g. carbon, metal oxides, metals, etc.) were strained by as much as 100% when tension was applied to them at strain rates from 0.5 nm/s to 300 nm/s. Much of this strain was probably associated with rotation and grain boundary sliding of the particles that led to chain straightening. After breaking, the straightened NCA rapidly contracted to form smaller compact structures. The purpose of the current work is to obtain atomic scale understanding of straining a nanoparticle chain to its breaking point. Within the time and length scales attainable by molecular dynamics (MD), this is a direct simulation attempt to describe our experiments on NCA. Here we extend our simulation studies [4] to chains of seven Cu nanoparticles. Copper is selected as a test material because of the availability of reliable interatomic potentials such as the ones based on the embedded atom method (EAM) [5]. Our experimental and simulation studies may have applications to the break-up of aggregates during aerosol sampling by instruments such as impactors [6], the design of nanocomposite materials including carbon black reinforced rubber [7] and the understanding of the behavior of aggregates as chemical-mechanical planarization agents for wafers [8].

COMPUTATIONAL METHODS

In an extension of our previous MD work [4], we probed the atomic-scale behavior of two short chains of seven Cu nanoparticles subject to strain to the breaking point. The description of the interatomic Cu interactions was based on the embedded-atom method (EAM) [5],[9]. More fitting parameters than usual, including *ab initio* energies, were used for constructing the EAM potential used in this study. It was shown [9] that this particular EAM potential gives a correct description of the energies not only of equilibrium, but also of several nonequilibrium Cu structures (including surfaces and interfaces, and a diatomic molecule (dimer) of Cu at very short atomic separations). The Newton's equations of motion were integrated with a 5th order predictor-corrector algorithm; thus the system was involved in time. The integration time step was $\Delta t = 1.5$ fs.

Seven identical Cu single-crystal particles, each containing 675 atoms and having a diameter of 2.452 nm, were placed in contact at 300 K in two configurations shown in figure 1. In both cases the nanoparticles were initially brought into contact along {100} and {111} planes so that all atoms were arranged in the same face-centered cubic (fcc) lattice. The first arrangement of the seven nanoparticles in figure 1 is in this work referred to as *linear* chain configuration, while the second one is referred to as *kinked*. The Cartesian x, y, and z axes initially corresponded to the [100], [010], and [001] crystallographic directions of the fcc lattice. In both cases the structures were initially relaxed for 225 ps. In this first part of the simulation (sintering), all atoms were unconstrained to move as directed by their equations of motion in an adiabatic simulation. Following the sintering period, strain was applied to the nanoparticle chains in the x-direction as explained below. Each one of the circles in figure 1 is a Cu atom. Black circles correspond to atoms that interact with the rest of the system with the same interatomic potential, but are not moved according to the MD algorithm. These atoms belong to the three edge (100) layers on each side of the two chain configurations and they are referred to as *static layers* [10]. Tensile strain is applied to the systems by changing the position of the right and left static layers by +0.05 Å and –0.05 Å respectively (both in the x-direction). After each such length increment the systems are relaxed, that is, dynamically evolved for a period ranging from about 30 ps to 1 ps depending on the applied strain rate (the corresponding strain rates are from 0.33 m/s to 10.67 m/s). This process of strain application and subsequent relaxation is repeated until the chains eventually fracture. The (100) atomic layers in contact with the static layers on each side of the chains are the *thermostat* layers through which heat is dissipated. Atoms belonging to thermostat layers are marked white in figure 1. These atoms move according to the MD algorithm, but during the strain application part of the simulation they undergo velocity rescaling after each time step so that their temperature is 300 K. The rest of the atoms (shown as gray circles in figure 1) are free to move according to their equations of motion without imposed constraints.

(a) *linear* chain configuration

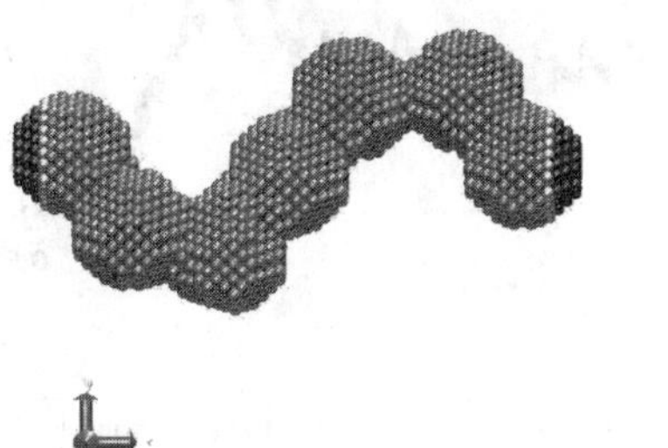

(b) *kinked* chain configuration

Figure 1. Projection on the xy plane of the two nanoparticle systems being strained. (a) is the linear configuration and (b) is the kinked configuration. Black atoms correspond to the static (100) atomic layers, white are the thermostat (100) layers and gray are non-constrainted atoms.

Most of the analysis in this work (e.g. stress and atomic displacement calculations) was performed on atomic arrangements that correspond to potential energy local minima. These local minima were calculated by using the same MD code; starting from the MD snapshot (positions and velocities) whose energy local minimum is sought, the temperature was gradually reduced to 0.01 K by velocity rescaling on all atoms after each MD time step.

DISCUSSION

Strain was applied to two different configurations of seven identical Cu crystal nanoparticles. Each of the seven primary particles was composed of 675 Cu atoms forming a truncated octahedron (TO). There is one atom in the center of the TO and all atoms are arranged in a Cu ffc lattice with their centers lying within 1.226 nm from the center atom. The TO exposes only {111} and {100} facets and it is a stable structure; during a 300 ps MD simulation at a temperature of 900 K surface atoms moved slightly closer to the center atom, but no atoms otherwise changed their lattice positions.

Sintering

Seven of the nanoparticles described above were brought into contact and MD steps were performed for 225 ps at a starting temperature of 300 K; thus the structures were relaxed [11], [12]. Sintering was a fast process and ended in both cases in less than 100 ps. The distances between the center atoms of the nanoparticles decreased, while the neck regions thickened. The average interparticle bond energies at 0 K for the linear and kinked chain configurations after sintering were 21.18 eV and 29.25 eV respectively.

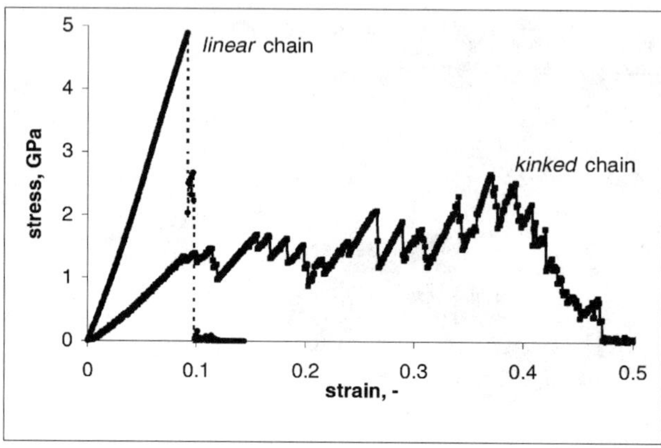

Figure 2. Stress- strain curves (σ_{xx}-ε_{xx} components) of the linear and kinked chain configurations of seven nanoparticles subject to strain rate of 0.33 m/s. All points correspond to energy local minimum atomic arrangements.

Strain Application

The two sintered chain configurations shown in figure 1 were subject to strain in the x-direction by extending the distance between the two sets of static layers as described above. Several different strain rates were tried ranging from 0.33 m/s to 10.67 m/s. The different strain rates did not affect the qualitative deformation behavior of the strained system. In figure 2 the stress-strain curves (σ_{xx}-ε_{xx} components) for the linear and kinked configurations corresponding to the lowest strain rate (0.33 m/s) are shown. The two curves show different mechanical behavior. The curve that corresponds to the linear configuration exhibits an almost linear elastic regime up to a strain of 0.091 (ε_{xx} component). The maximum stress that corresponds to the maximum elastic strain of 0.091 was around 4.88 GPa (σ_{xx} component). At higher strains the linear chain yields plastically and eventually breaks at a 0.098 strain (ε_{xx} component). The breaking occurs at the neck region between two nanoparticles. The curve that corresponds to the kinked chain configuration of the seven nanoparticles shows a more complex mechanical behavior. In figure 3 the kinked chain at different strains is presented. After an initial almost purely elastic straining of about 0.090 (ε_{xx} component) the kinked portions of the chain start to straighten. The mechanism of the chain straightening is a combination of particle-particle rotation and simultaneous interface sliding. During the chain straightening process the stress along the straining axis (x-axis) fluctuates around about 1.5 GPa (σ_{xx} component). The straightening compares well with the experimental observations [2]. Once the nanoparticle rotation is completed, the stress increases further to reach a maximum of 2.658 GPa at a strain of 0.369 (σ_{xx} and ε_{xx} components). At higher strains, repeated slip events on {111} planes cause the neck region between two nanoparticles to decrease in size to the point where it reduces to a single atom contact. Eventually, the single atom contact yields and the chains break at a strain of 0.472 (ε_{xx} component).

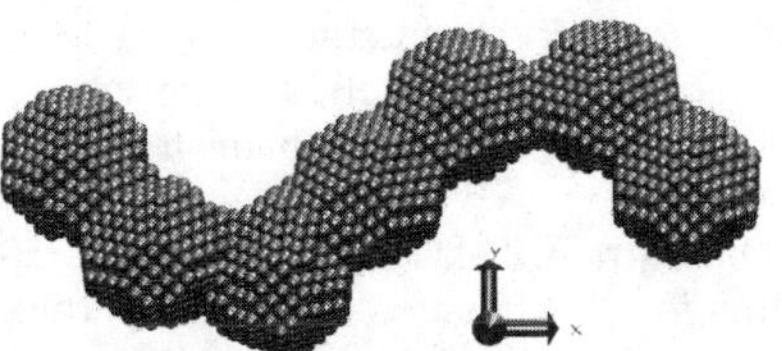

(a) Strain 0.079; elastic straining only.

(b) Strain 0.325; chain straightening through nanoparticle rotation.

(c) Strain 0.412; interparticle neck weakening.

(d) Strain 0.472; single atom contact.

(e) Strain 0.476; chain breaks.

Figure 3. The kinked chain configuration at different strains (all strains correspond to the ε_{xx} component). The strain rate was 0.33 m/s.

The simulation strain rates are orders of magnitude higher than the experimental ones [1-3]. Nevertheless, the relaxation period after each length increment was long enough for the systems not to exhibit discernable variations in their properties beyond natural fluctuations. Furthermore, performing the numerical calculations at strain rates that differed by almost three orders of magnitude did not affect the qualitative behavior for the strained systems.

CONCLUSIONS

Two different configurations (referred to as linear and kinked chain configurations) of seven crystalline Cu nanoparticles each 2.452 nm in diameter were strained to the breaking point at high strain rates (0.33 m/s to 10.67 m/s) using MD. Prior to the strain application the two configurations were sintered for 225 ps in an adiabatic MD simulation starting at 300 K. The average interparticle bond energies after the sintering were 21.18 eV (0 K) and 29.25 eV (0 K) for the linear and kinked configuration respectively. During strain application, the two chains revealed different mechanical behavior. The linear chain strained elastically up to $\varepsilon_{xx} = 0.091$. The maximum stress (σ_{xx} component) was 4.885 GPa, and fracture occurred at $\varepsilon_{xx} = 0.098$ (0.33 m/s strain rate). The kinked chain strained elastically up to $\varepsilon_{xx} = 0.090$. At higher strains the chain started to straighten through a combination of particle-particle rotation and interface sliding events on {111} planes. At a strain of about 0.350 the chain was straightened completely and then reached the maximum stress of 2.658 GPa at a strain of 0.369 (σ_{xx} and ε_{xx} components). At even higher strains the neck region between two particles weakened and eventually reduced to a single atom contact. The kinked configuration finally fractured at a strain of 0.473. The reported qualitative deformation results (type and sequence of plastic deformation and fracture mechanisms) were not affected by the different strain rates. Furthermore, for small strains the different strain rates gave the same quantitative results. The large differences in properties of linear and kinked chains may play a significant role in the effects of chain aggregates (e.g. carbon black) on nanocomposite materials (such as rubber), and on the break-up of aggregates when sampled from high speed flows.

ACKNOWLEDGMENTS

The computational time involving simulations and result analyses was provided by the UCLA Chemical Engineering Department Dalton cluster and by the UCLA Academic Technology Services (ATS) Beowulf cluster. This research was supported in part by National Science Foundation grant CTS 9911133.

REFERENCES

[1] S. K. Friedlander, H. D. Jang, K. H. Ryu, *Appl. Phys. Lett.* **72**, 173 (1998).
[2] K. Ogawa, T. Vogt, M. Ullmann, S. Johnson and S. K. Friedlander, *J. Appl. Phys.* **87**, 63 (2000).
[3] Y. J. Suh and S. K. Friedlander, *J. Appl. Phys.* **93**, 3515 (2003).
[4] A. Dalis and S.K. Friedlander *Mat. Res. Soc. Symp. Proc.* **778**, U9.8.1 (2003).
[5] S. M. Foiles, M. I. Baskes and M. S. Daw, *Phys. Rev. B* **33**, 7983 (1986).
[6] M. Seipenbusch, S. Froeschke, A.P. Weber and G. Kasper, *P. I. Mech. Eng. E-J. Pro.* **216**, 219 (2002).
[7] M. Gerspacher, C. P. O'Farrell, *Kaut. Gummi Kunstst.* **54**, 153 (2001).
[8] R.K. Singh, S.M. Lee, K.S. Choi, G.B. Basim, Z. Chen, B.M. Moudgil, *Mater. Res. Bull.* **27**, 752 (2002).
[9] The EAM parameterization used was the EAM1 of: Y. Mishin, M.J. Mehl, D.A. Papaconstantopoulos, A.F. Voter and J.D. Kress, *Phys. Rev. B* **63**, 224106 (2001).
[10] U. Landman, W. D. Luedtke, N. A. Burnham, R. J. Colton, *Science* **248**, 454 (1990).
[11] H. Zhu and R. S. Averback, *Phil. Mag. Lett.* **73**, 27 (1996).
[12] P. Zeng, S. Zalac, P. C. Clapp, J. A. Rifkin, *Mater. Sci. Eng.* **A252**, 301 (1998).

Part V: Deformation, Texture Development, and Ductility

A. Experiments

Mat. Res. Soc. Symp. Proc. Vol. 821 © 2004 Materials Research Society P9.8

Grain Size Distribution Effect on Mechanical Behavior of Nanocrystalline Materials

A.V. Sergueeva, N.A. Mara, and A.K. Mukherjee
Department of Chemical Engineering and Materials Science, University of California, Davis, CA 95616, USA

ABSTRACT

Grain size distribution effect on the mechanical behavior of NiTi and Vitroperm alloys were investigated. Yielding at significantly lower stresses than found in equiaxed counterparts, along with well defined strain hardening was observed in these nanocrystalline materials with large grains embedded in the matrix during tensile deformation at temperatures of $0.4T_m$. At higher temperature the effect of grain size distribution on yield stress was not revealed while plasticity was increased in 50% in NiTi alloy with bimodal grain size structure.

INTRODUCTION

Reducing the grain size in a polycrystalline material is known to have beneficial effects on its mechanical properties [1]. A material's strength is expected to increase in a manner described by, for example, the well-known Hall-Petch relationship. A concurrent rise in toughness is often expected for many engineering alloys after grain refinement. Ductility, the ability of a material to plastically deform without failure under tensile stresses, was predicted to be improved as well [1]. An extrapolation of grain size dependence of important mechanical properties all the way to the nanocrystalline regime has the exciting prospect of achieving unique mechanical properties never before possible. The promised gain in strength (typically 5-10 times) has indeed been realized in nanocrystalline materials (NCM) [2]. Their ductility at room temperature, on the other hand, is often disappointingly low compared with their coarse-grained counterparts [3]. More importantly, the useful uniform tensile deformation, i.e., the plastic strain before localized deformation sets in at or near the peak in the stress-strain curve, is close to zero for almost all NCM and ultra-fine grained (UFG) metals at ambient temperatures [4].

Obtaining a high strength from UFG or NCM seems to be straightforward, but that alone may not be sufficient for many applications if the material suffers from ductility problems. Even in UFG materials that show better ductility than NCM, strain hardening and strain rate hardening remain insignificant for normal strain rates [5]. The lack of a work hardening mechanism is entirely compatible with the lack of accumulation of dislocations, as suggested by both simulation and experiment [6].

It was noticed that for most of the NCM that showed some strain hardening and respectable ductility, the microstructure usually contained a distribution of grain sizes even though the volume fraction of the larger grains may be low [4,7-12]. In this paper the effect of the presence of large grains in a nanocrystalline matrix on the mechanical behavior of NiTi and Vitroperm alloys was investigated.

EXPERIMENTAL DETAILS

Nitinol containing 50at.%Ni and 50at.%Ti and strips of melt-spun Vitroperm ($Fe_{73.5}Cu_1Nb_3Si_{15.5}B_7$) ribbon were used in this investigation. The initial rod of nitinol with diameter of 20 mm was quenched from 800°C in water. Initial grain size of the material after quenching was about 50 μm. Discs were cut from the quenched rod and subjected to high pressure torsion (HPT) under a pressure of 5 GPa to a true logarithmic strain of 7. Details of the HPT processing were reported earlier [13] The as-processed material was in the form of discs with diameter of 12 mm and thickness of about 0.3 mm. The torsion strain of 7 corresponds to the part of the disk situated at the distance of 2 mm from the center. From this part of the disc tensile specimens with a 1-mm gage length × 1 mm width were electro discharge machined for tensile testing. Vitroperm strips approximately 2 mm wide, ~20 microns thick, and with a gage length of 4 mm were also tested in tension. The tensile tests were performed on a custom-built computer controlled constant strain rate tensile test machine with a displacement resolution of 5 μm and a load resolution of 0.1 N. Scanning electron microscopy (SEM) was used to examine the fracture surfaces of failed specimens. Transmission electron microscopy (TEM) was performed on a Philips CM-200 FEG microscope operating at 200 kV. The TEM samples were prepared through electrochemical polishing.

RESULTS

Mechanical behavior of nanocrystalline nitinol

HPT of nitinol leads to amorphization of the alloy [14]. Two types of annealing were applied to crystallize the material. In order to obtain a homogeneous structure the material was subjected to two-step annealing: at 200°C for 1h and at 500°C for 0.5h. In this case the first annealing step at low temperature was used to increase the number of possible nucleation sites for the following crystallization step at higher temperature [14]. The obtained structure with mean grain size of 70nm is shown in Fig.1a. Annealing at 500°C for 0.5 h right after HPT results in formation of a microstructure with bimodal grain size distribution (Fig.1b) where large grains (d=100-200 nm) are embedded in a nanocrystalline matrix with d=40-70 nm. Histograms of grain size distribution for both annealed materials are presented in Fig.2.

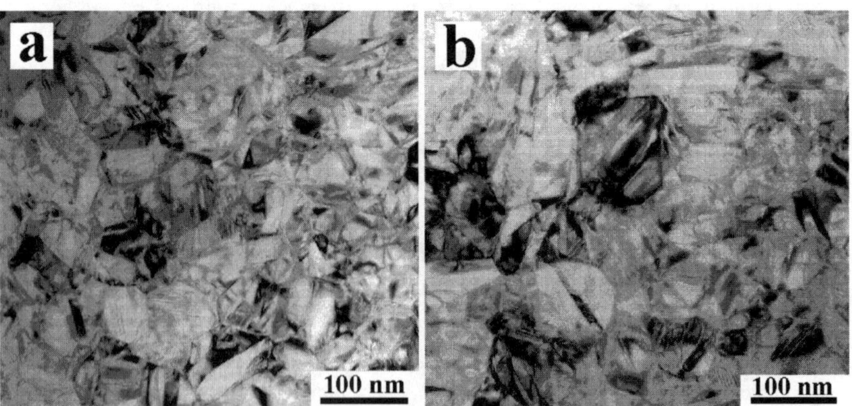

Figure 1. Microstructure of NiTi alloy after HPT and annealing **(a)** at 200°C for 1 h and at 500°C for 0.5 h and **(b)** at 500°C for 0.5

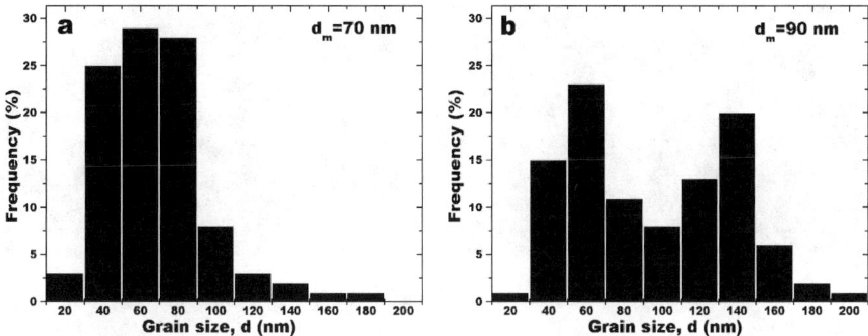

Figure 2. Histograms of grain size distribution in NiTi alloy after HPT and annealing **(a)** at 200°C for 1 h and at 500°C for 0.5 h and **(b)** at 500°C for 0.5 h.

Annealed material was tested in tension at 400° and 500°C and at a strain rates of $10^{-3}s^{-1}$. At lower temperature (400°C) more pronounced strain hardening was observed in the sample with bimodal grain size distribution (Fig.3). The ultimate tensile strength of this sample was 13% less than that of the sample with homogeneous nanostructure. The material also revealed extensive necking. SEM images of the deformed samples with homogeneous and bimodal grain size distributions are shown in Fig.4. Uniform plastic strain is approximately the same in both materials (<5%).

An increase in test temperature leads to a more pronounced difference in plasticity of the material with different grain size distribution while the difference in strength characteristics is insignificant (Fig.5). In both cases NiTi alloy displayed yielding at 220 MPa and ultimate strength of more than 900 MPa. At the same time the ductility was increased more than twice in the sample with initial bimodal grain size distribution (170% vs. 367%). Extensive grain growth was detected during these tests. Mean grain size was about 300 nm in both samples after deformation.

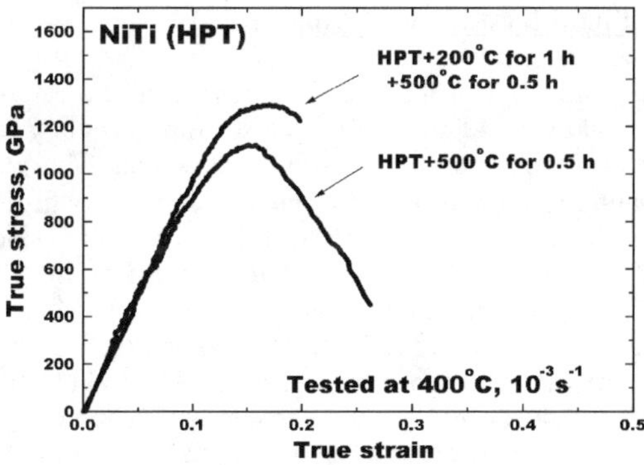

Figure 3. True stress – true strain tensile curves for NiTi alloy after HPT and annealing. Different annealing results in different grain size distribution in the material (see text for details).

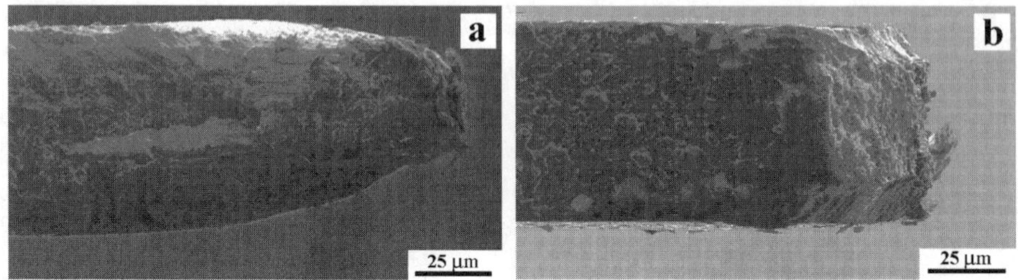

Figure 4. Gage view of post-deformed samples with **(a)** homogeneous and **(b)** bimodal grain size distribution.

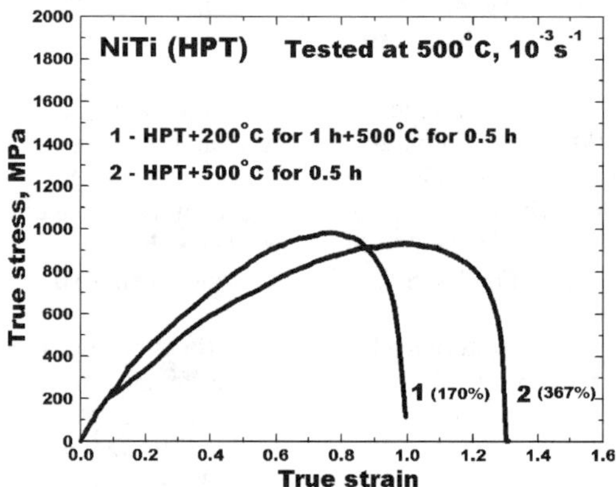

Figure 5. True stress – true strain tensile curves at 500°C for NiTi alloy after HPT and annealing.

Mechanical behavior of nanocrystalline Vitroperm

Similar trends in grain size distribution effect on mechanical behavior were observed in Vitroperm alloy tested at 600°C ($0.4T_m$). As-received strips of this alloy were amorphous and its annealing at different temperatures resulted in different structures. Thus, homogeneous one phase (α-Fe) structure with mean grain size of 15 nm was formed in this alloy after annealing at 600°C for 1 h (Fig.6a). An increase in annealing temperature to 650°C leads to formation of large nonequaxied grains (100-200 nm in length) embedded in nanocrystalline matrix with mean grain size of 15 nm (Fig.6b). The appearance of a second phase was detected in SAED pattern. True stress – true strain curves at 600°C and at $10^{-4}s^{-1}$ are shown in Fig.7.

Yielding of the alloy in the presence of large grains started at about 400 MPa with an extensive strain hardening while a nanocrystalline material with homogeneous structure has demonstrated elastic perfectly plastic behavior. No strain hardening was revealed after yielding at 1270 MPa. The alloy exhibited low plasticity in both states (2% in the case of homogeneous structure and 4% at wide grain size distribution). Wide grain size distribution led to necking (not shown) similar to that observed in NiTi.

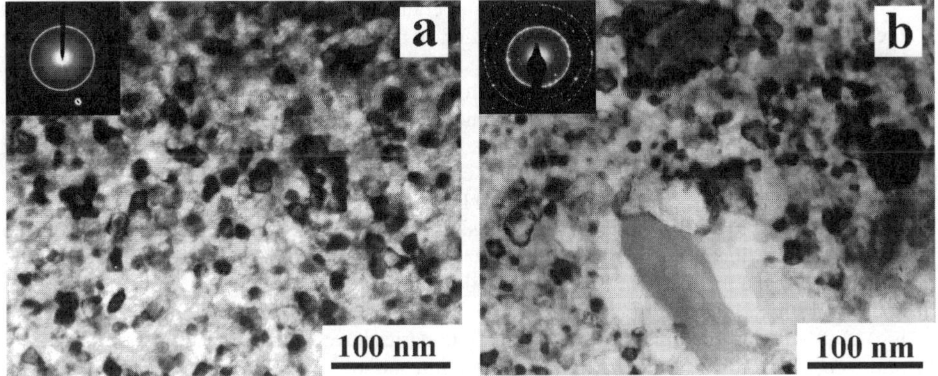

Figure 6. Microstructure with corresponding SAED patterns of Vitroperm alloy after crystallization (**a**) at 600°C and (**b**) at 650°C.

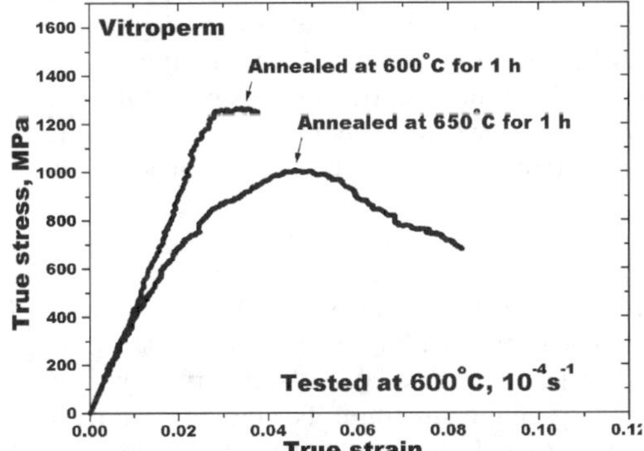

Figure 7. True stress – true strain tensile curves for Vitroperm alloy after crystallization at different temperatures.

DISCUSSION

The results of the current study confirmed a strong effect of the grain size distribution on mechanical behavior of nanocrystalline materials. At the same time, our work represents a different approach to understanding of deformation processes at extremely diminished length scale. First of all, both our materials were obtained by crystallization from amorphous state that assumes a formation of equilibrium structures with high angle grain boundaries of random missorientations. Secondly, the size of the large grains embedded into nanocrystalline matrix is twice of the mean grain size of the matrix and might be considered to be at the upper level of the definition of a nanocrystalline range (<500 nm).

It was shown that at relatively low temperatures (<0.4T_m, where T_m is a melting point) the presence of such grains leads to the yielding of the materials at significantly lower stresses with well defined strain hardening. A homogeneous matrix of nanocrystalline grains (d=15 nm) had displayed almost zero strain hardening in Vitroperm alloy (Fig.7). Nitinol had demonstrated some hardening (Fig.3) but this alloy has a more wide grain size distribution with mean size of 70 nm (Fig.2a). Formation of a bimodal structure in this alloy results in lower yielding stress and significant hardening during deformation as compared to the alloy with homogeneous structure (Fig.3). One of the possible reasons for such behavior might be related to an increase in dislocation activity in the grains larger than some critical grain size. Such a critical size can be different depending on the material. For Vitroperm alloy d=15 nm seems to be below such critical value and the material exhibited elastic perfectly plastic behavior. Some dislocation activity is suggested to occur in nanocrystalline NiTi alloy with homogeneous structure (d=70 nm). Actually, dislocations formed at grain boundaries have been observed in the materials with grain size down to 30-50 nm by molecular dynamic simulations [15] and experimentally [16,17]. The presence of larger grains seems to accelerate dislocation activity. Extensive necking (which is common during creep deformation) of the materials with wide grain size distribution also confirms this assumption.

At higher temperatures (T>0.4T_m) diffusion processes, especially grain boundary diffusion, start to play an important role, leading to extensive grain growth during deformation. The initial grain size distribution seems not to be such important for the material yield stress while a significant increase in ductility was observed. Additional experiments are needed for more detailed analysis of the grain size distribution effect on mechanical properties of the nanocrystalline materials at ambient and elevated temperatures.

CONCLUSIONS

In the case of NiTi and Vitroperm alloys tested at 0.4T_m, the presence of large grains embedded in a nanocrystalline matrix leads to yielding of the material at significantly lower stresses with more pronounced strain hardening compared to the deformation behavior of the same material with a homogeneous nanocrystalline structure. For the NiTi alloy, at a higher test temperature little difference is seen in the yield strength between the material with two type of grain size distribution but the ductility is 50% higher in the alloy with the bimodal distribution

ACKNOWLEGEMENTS

The authors would like to acknowledge the National Science Foundation, Division of Materials Research for its financial support under grant NSF-DMR-0240144.

REFERENCES

1. R.W. Hertzberg, *Deformation and fracture mechanics of engineering materials*, 3rd ed. New York: John Wiley and Sons; 1989. p.392.
2. J.R. Weertman, in *Nanostructured materials; processing, properties and applications,* edited by C.C. Koch, Norwich: William Andrews Publishing; 2002. p.397-.
3. C.C. Koch, *J. Metast. Nanocryst. Mater.* **18**, 9 (2003).
4. Y.M. Wang, M.W. Chen, F. Zhou, E. Ma, *Nature* **419**, 912 (2002).

5. Y.M. Wang, E. Ma, *Mater. Sci. Eng.* A, in press.
6. H. Van Swygenhoven, Z. Budrovich, P. M. Derlet, and A. Hasnaoui, in *Processing and Properties of Structural Nanomaterials*, MS&T 2003, Chicago, IL; USA; 9-12 Nov. 2003. pp.3-10.
7. X. Zhang, H. Wang, R.O. Scattergood, J. Narayan, C.C. Koch, *Acta Mater.* **50**, 3995 (2002).
8. X. Zhang, H. Wang, R.O. Scattergood, J. Narayan, C.C. Koch, *Acta Mater.* **50**, 3527 (2002).
9. X. Zhang, H.Wang, R.O. Scattergood, J. Narayan, C.C. Koch, A.V. Sergueeva, A.K.Mukherjee, *Acta Mater.* **50**, 4823 (2002).
10. V.L. Tellkamp, A. Melmed, E. Lavernia, *Metall. Mater. Trans.* A**32**, 2335 (2001).
11. D. Witkin, Z. Lee, R. Rodriguez, S. Nutt, E. Lavernia, *Scripta Materialia* **49**, 297 (2003).
12. M. Legros, B.R. Elliot, M.N. Rittner, J.R. Weertman, K.I. Hemker, *Phil. Mag.* A**80**, 1017 (2000).
13. R.Z. Valiev, R.K. Islamgaliev, I.V. Alexandrov, *Prog. Mater. Sci.* **45**, 103 (2000).
14. A.V. Sergueeva, C. Song, R.Z. Valiev, A.K. Mukherjee, *Mater. Sci. Eng.* A**339**, 159 (2003).
15. V. Yamakov, D. Wolf, S.R. Philpot, A.K. Mukherjee, and H. Gleiter, *Nature Materials* **1**, 45 (2002).
16. A. K. Mukherjee, in *Creep Deformation: Fundamentals and Applications,* edited by R. S. Mishra, TMS-AIME, Pittsburgh, PA (2002), p.3.
17. C.J. Youngdahl, J.R. Weertman, R.C. Hugo, and H.H. Kung, *Scripta Mater.* **44**, 1475 (2001).

Mat. Res. Soc. Symp. Proc. Vol. 821 © 2004 Materials Research Society P9.10

On the Influence of the Misorientation of Grains, Grain Size and Boundary Volume on the Strength and Ductility of Ultrafine Grained Cu

Florian H. Dalla Torre, Rimma Lapovok , Peter F. Thomson, James D. Sandlin, Chris H.J. Davies and Elena V. Pereloma.
School of Physics and Materials Engineering, Monash University, VIC, 3800, Australia

ABSTRACT

The influence of the microstructure and the misorientation relationship of grains on mechanical properties is investigated in specimens of ultrafine-grained copper processed by equal channel angular extrusion (ECAE) route B_c for 1, 4 and 12 passes. XRD texture analyses have shown that the major texture component is developed during the first pass of ECAE, and remains approximately constant with greater number of passes. EBSD measurements indicate that the majority of grain boundaries are still of low angle (>15°), while after four and twelve passes more than 50 % of all boundaries are high angle ones. TEM analyses have shown that the microstructure evolves from microbands and elongated cells towards a more equiaxed homogenous microstructure. On the microscale observed by TEM the degree of misorientation among subgrains/cells increases and the width of boundaries decreases while the cell/subgrain size remains approximately constant as the number of passes increases. The mechanical properties show a saturation level with a maximum in the yield stress and UTS after 4 passes. The strength of the material decreases between the fourth and the twelfth passes and the uniform elongation increases. It is suggested that the increase in ductility (and decrease in strength) is associated with the decrease in width of boundaries leading to an increase in the mean free path of dislocations.

1. INTRODUCTION

Generally, it is well known that nanostructured (NC) and ultrafine grained (UFG) materials exhibit a high strength but only limited ductility. Recently different approaches have been considered to overcome the limitation in ductility [1-4]. One of the approaches involves a change in the deformation mechanism of a microstructure, which is mainly characterized by high angle grain boundaries and a large volume fraction of 'non-equilibrium' grain boundaries [1,3]. Such a microstructure can be achieved after a very large deformation via equal channel angular extrusion (ECAE). The deformation mechanism of UFG metals is often considered to be governed not only by lattice dislocation activity, but also by other mechanisms such as grain rotation and grain boundary sliding, especially for an equiaxed randomly textured microstructure [3]. The latter type of deformation mechanisms are often adapted for materials exhibiting even smaller NC grain sizes (below 100 nm), where dislocation activity is strongly reduced and where the deformation is more concentrated in the grain boundaries. In materials processed by ECAE it is, however, not clearly understood, if and how these grain boundary mechanisms act and whether they are beneficial to the mechanical properties. In this respect it is important to record the evolution of the microstructure and the associated change in the mechanical behaviour.

The purpose of the present work is to describe the microstructural evolution from 0 to 12 passes of ECAE using XRD texture, EBSD (electron backscatter diffraction) and TEM (transmission electron microscopy) analyses on one hand and to investigate the gain in ductility with an increasing number of ECAE passes on the other hand.

2. EXPERIMENTAL DETAILS

Annealed Cu with a purity of >99.5 % was processed via ECAE, route B_C using a die angle of 90°, a back-pressure of 25 MPa and a forward speed of 2 mm/sec. Cu bars of 20 x 20 mm cross section were subjected to 1, 4 and 12 passes of ECAE. All specimens for XRD texture, TEM and EBSD analyses were cut from transverse sections perpendicular to the extrusion direction and 45° to the shear plane. Electropolished specimens were used for TEM observations with a Philips CM20 operated at 200 kV. 150 to 200 cells/subgrains were counted on each specimen to provide statistical significance. EBSD measurements were performed with a LEO 1530 FEG SEM using the EBSD detector and software package of HKL. An area of 10 x 18 μm on mechanically polished specimens was scanned with a step size of 40 nm. The raw data were post processed using VMAP V8 [5]. The same specimens were used for XRD texture analysis using a GBC texture goniometer equipped with a Cu anode and a polycapillary beam enhancer resulting in a collimated beam of 10 x 10 mm. Tensile tests were performed at room temperature with a screw driven Instron tensile tester operating at a strain rate of $1.6 \times 10^{-3}\ s^{-1}$. The dog bone shaped tensile specimens had a gauge length of 25 mm and a diameter of 4 mm.

3. RESULTS

3.1 Texture analysis

Texture analyses were performed using the {111}, {200} and {220} reflections. In Fig. 1 only the pole figures for the {111} planes of the specimen subjected to 1, 4 and 12 passes are shown, because they describe well the macroscopic texture evolution in the material.

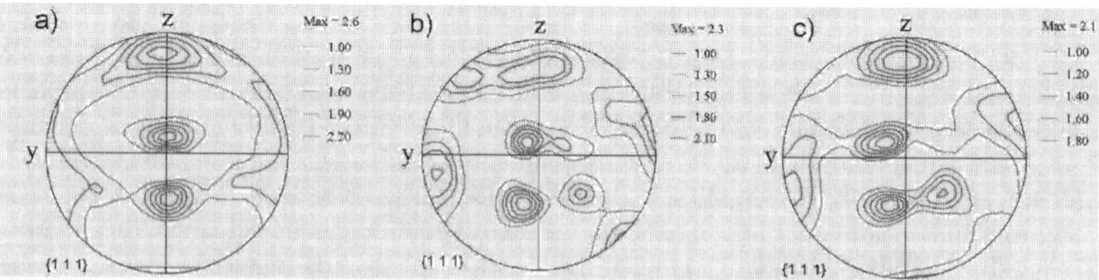

Figure 1. {111} pole figures of one, four and twelve passes of ECAE.

The y - axis indicated in Fig.1 points towards the inner and outer planes of the die channel making the 90° angle, while the x – axis, which is sticking out of the plane, is parallel to the extrusion direction. As shown in Fig. 1 no mirror symmetry is found about the y - axis of the three specimens. This has previously been observed in ECAE processed Cu [6,7]. Sample symmetry about the z - axis is only preserved after the first pass of ECAE, but not after multiple passes (see Figs. 1b and 1c). During the first pass the dominant texture with a three point intensity along the z-axis develops and is preserved up to 12 passes. The mirror asymmetry about the y - axis evolves between 1 and 4 passes, which again remains stable up to 12 passes. This indicates that the first passes are crucial for the development of texture. Once a macroscopic texture is developed it is preserved even after multiple passes of ECAE. Hence on a macroscopic scale UFG copper processed by ECAE using multiple passes does show a strong texture component.

It has been shown by Agnew *et al.* [6] and Huang *et al.* [7] that the deformation introduced by the ECAE process can be, in analogy to torsional deformation, well described by simple shear. To describe the texture components in our samples, we adopted the ideal orientations, which are developed by torsion. Apart from the asymmetry about the y - axis, the ideal orientations obtained by torsion rotated by 45° represent satisfactorily the texture deformation up to 12 passes, which indicates a major shear along the <110> direction.

3.2 EBSD analysis

The texture analysis obtained by EBSD measurements has shown that the scan-size chosen does not represent the bulk or macro texture, but a micro-texture, which changes depending on the chosen area. This can be explained by the scan size, which is approximately of the same size as the original grain size before ECAE. After 1 pass it can be assumed that the original grain size has decreased only by a small amount, incorporating however many cells/subgrains with a much smaller size. This difference between the grain size and the cell/subgrain size is reflected in Fig. 2, where in Fig. 2a only grains with high angle grain boundaries (HAGB > 15°) are depicted compared with Fig. 2b where only low angle boundaries (LAGB) with misorientation angle smaller than 15° are taken into account. The grey scale values are chosen arbitrarily. The black areas indicate areas, which are not indexed. To clarify the view of grains and subgrains the micrographs have been cleaned by increasing the original indexing quota of 66% to 95%.

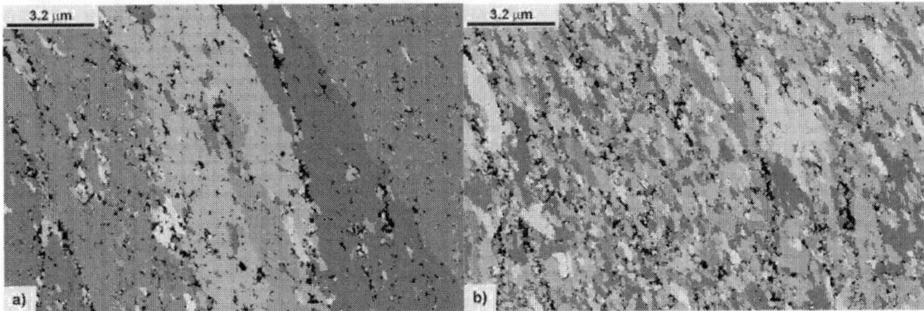

Figure 2. EBSD measurement on 1 pass specimen showing grains separated by the HAGB (>15°) (a), and the subgrains/cells divided by LAGB (<15°). Grains and subgrains are elongated along the shear direction.

Statistical calculations of the misorientation relationship indicate that after 1 pass still more than 80 % of all cells are confined by small angle grain boundaries (see Fig. 2 and

Table I). After 4 passes, more than 50% of the boundaries show a high angle misorientation. The degree of misorientation remains essentially stable after 12 passes (see Table I). However, in order to confidently confirm this observation more measurements have to be undertaken. EBSD grain size measurements vary greatly after 1 pass between 200 and 2400 nm, with grains elongated along the shear direction (see Fig. 2 and Table I). The subgrain size after 1 pass lies between 60 and 340 nm. After 4 passes the grain size has an average value of ~200 nm and interestingly is larger after 12 passes with an average at 260 nm. The subgrain size at 1 and 4 passes is very similar, but increases for 12 passes.

3.3 TEM analysis

Figure 3 presents TEM bright field (BF) micrographs and SAD patterns of the as-processed Cu, viewed from the <110> zone axis. After the first pass, as shown in Fig. 3a the microstructure mainly consists of parallel microbands with strongly elongated subgrains with a mean spacing of ~400 nm (see Table I). The microbands are mainly low-angled, oriented along the traces of the {111} planes within an angular distribution of ±15°, which follows the major operating slip system along {111}/<110> in agreement with the texture measurements. Inside the microbands, dislocation cell boundaries are present. The microstructure after 4 passes is shown in Fig. 3b. After 4 passes the fraction of the strongly elongated lamellar boundary structure with respect to the equiaxed subgrains is notably lower and the width is less (Table I). Approximately half of the area investigated is characterized by an equiaxed subgrain structure. Previous microbands are often truncated by different shear planes, which lead to a more homogeneous microstructure. The increase in the misorientation relationship is also shown in the SAD pattern, where diffraction spots are distributed more regularly. The microstructure after 12 passes is depicted in Fig. 3c. It shows a homogeneous microstructure, where previous slip systems are not visible any more. The subgrains are mainly equiaxed. Only a few elongated grains were detected, which however do not show any preferential orientation. The SAD pattern shows a more random misorientation and more distinct spots compared with the pattern taken from the sample subjected to 4 pass ECAE. On the scale of the TEM this indicates that the microstructure after 12 passes shows a larger fraction of high angle grain boundaries and a state of higher recovery than after 4 passes. The greater recovery is also indicated by the larger fraction of sharp grain boundaries (see Table I), which show a closer approach to equilibrium. The macroscopic XRD texture analyses in comparison are too coarse to indicate a difference in misorientation between 4 and 12 passes.

Table I shows the size measurements made on the different passes by TEM. The microband spacing decreases gradually up to 4 passes. It is interesting to note that the cell size parallel to the microbands decreases gradually to a value of about 180 nm, while perpendicular to it a minimum size has been reached already after 1 pass, which remains approximately constant up to 12 passes. In comparison with subgrain size measurements obtained by EBSD analysis, a similar subgrain size is measured after one pass. This is due to the fact that most of the cells are separated by LAGB. However, with an increasing number of passes and consequently increasing misorientation of subgrains, the TEM size measurements encompass a mixture of grains with LAGB and HAGB and thus result in values roughly between those measured by EBSD.

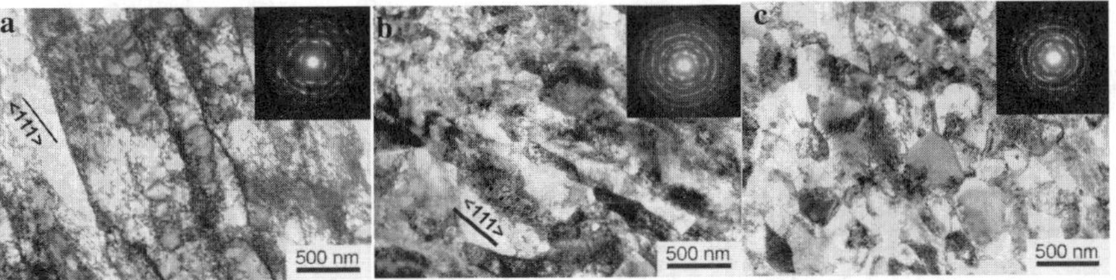

Figure 3. TEM bright field image with SAD patterns (aperture ~6 μm) of 1 (a), 4 (b) and 12 (c) passes and the traces of the {111} plane.

Table I. TEM and XRD measurements of various microstructural features.

Passes	1	4	12
EBSD			
Grain size [nm]	600±2400	190±120	260±270
Subgrain size [nm]	200±140	120±60	180±120
Volume fraction HAGB [%]	19±2	60	50
TEM			
Microband spacing [nm]	400±250	220±130	no microbands
Subgrain/cell size parallel microbands [nm]	280±190	180±100	180±100
Subgrain/cell size perp. microbands [nm]	160±70	160±70	150±70
Subgrain/cell wall thickness [nm]	52±48	48±31	23±8
Number of sharp grain boundaries per 10 μm^2	0	13	63

The evolution of the dislocation cell walls with an increasing number of passes shows a decrease in the boundary/wall width. After 1 pass the boundaries are often not yet well defined so that single dislocations can be distinguished. The boundary/wall thicknesses characterized by a high density of dislocations extends over several tens of nanometers. The width of the cell walls decreases between 1 to 12 passes gradually from ~50 nm to ~23 nm. At the same time the fraction of atomically sharp boundaries increases with the number of passes. These observations are very important; it means that the actual boundary volume decreases from 4 to 12 passes, which can be taken as an increase in the mean free path of dislocations. The reduction of the cell/subgrain walls is strongly coupled to the increase in the misorientation from cells over subgrains into grains.

3.4 Tensile properties

Figure 4 shows the tensile curves for 0, 1, 4 and 12 passes, while the tensile property data are summarized in Table II. After 1 pass of ECAE, a strong increase in strength with a significant decrease in ductility was observed. After 4 passes the yield strength (σ_y) and the ultimate tensile stress (UTS) show maximum values and decrease for the 12 pass specimen. The total elongation remains approximately constant within 8-9 % over all passes. However, the uniform elongation increases with the number of passes from 0.75% up to 2.5 % at 12 passes. The increase in the uniform plastic elongation with increasing number of passes can be explained by a lower decrease in the work hardening rate with increasing stress. The lower yield strength and the increase in uniform elongation with increasing number of passes is attributed to the decrease in the subgrain/grain boundary volume, which leads to regaining the ability to work harden by dislocation accumulation.

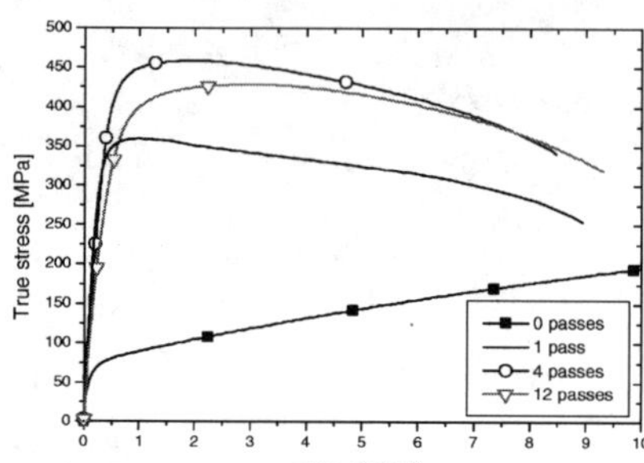

Figure 4. Tensile data for zero, 1, 4 and 12 passes tested at room temperature with a strain rate of 1×10^{-3} s^{-1}.

Table II. Tensile properties of specimens subjected to 1, 4 and 12 paths of ECAE.

Passes	0	1	4	12
Yield strength (0.2.%) [MPa]	62	342±5	415±1	348±0
UTS [MPa]	344	358±2	455±4	423±8
Total elongation [%]	48.5	8.9±0.0	8.2±0.4	8.5±1.1
Uniform elongation [%]	28.7	0.75±0.09	1.6±0.3	2.5±0.5

CONCLUSIONS

XRD texture analyses have shown development of a strong macro texture after 1 pass, which remains roughly constant with subsequent passes. The fraction of HAGB increases between 1 and 4 passes from below 20 % to more than 50 % and remains so to 12 passes. TEM observations indicate a constant cell/subgrain size, but a decrease in the cell wall width and an increase in more equilibrium boundaries. The decrease in the strength and increase in uniform elongation observed after the twelfth pass compared with that after the fourth pass is caused by an increase in the mean free path of dislocations due to the decrease in the width of boundaries, which is associated to the increase in the misorientation relationship.

REFERENCES

1. R.Z. Valiev, R.K. Islamgaliev, I.V. Alexandrov. *Prog. in Mat. Sci.* **45**, 103 (2000).
2. Y.Wang, M. Chen, F. Zhou, E.Ma, *letters to Nature*, **419**, 912 (2002).
3. R.Z. Valiev, I.V. Alexandrov, Y.T. Zhu, T.C. Lowe, *J. Mater. Res.* **17**, 5 (2002).
4. Y.M. Wang, E. Ma. *Acta Mater.* **52**, 1699 (2004).
5. F.J. Humphreys, *J. Mater. Sci.* **36**, 3833 (2001).
6. S.R. Agnew, J.R. Weertman, *Mater. Sci. Eng.* **A242**, 174 (1998).
7. W.H. Huang, L. Chang, P.W. Kao, C.P. Chang *Mater. Sci. Eng.* **A307**, 113 (2001).

Mat. Res. Soc. Symp. Proc. Vol. 821 © 2004 Materials Research Society P9.11

Experiment and FEM Analysis of Tensile Behavior of Bimodal Nanocrystalline Al-Mg Alloys

Zonghoon Lee, Johnathan Lee, Enrique J. Lavernia[1] and Steven R. Nutt
Department of Materials Science, University of Southern California, Los Angeles, CA 90089-0241, U.S.A.
[1]Department of Chemical Engineering and Materials Science, University of California at Davis, Davis, CA 95616, U.S.A.

ABSTRACT

The tensile behavior of bimodal nanocrystalline Al-7.5Mg alloys was investigated using experiments and two-dimensional axisymmetric elastic-plastic finite element method (FEM). Cryomilled nanocrystalline powders blended with 15% and 30% unmilled coarse-grained powders were consolidated by hot isostatic pressing followed by extrusion to produce bulk bimodal nanocrystalline Al-7.5Mg alloys, which were comprised of nanocrystalline grains separated by coarse-grain regions. The calculated stress-strain curves have acceptable agreement with experimental curves of the bimodal structures. The bimodal Al-7.5Mg alloys show reasonable ductility while retaining enhanced strength compared to conventional alloys and nanocrystalline metals.

INTRODUCTION

Nanocrystalline metals exhibit remarkable improvements in strength and the possibility of weight savings in structural applications. However, these materials generally suffer from insufficient ductility and reduced toughness compared to conventional alloys with coarser grains. Ductility enhancement of nanocrystalline metals has been tried in nanostructured metals through the incorporation of coarser grains in a fine-grained matrix by annealing [1-3]. Previous reports suggest that the presence of coarser grains within the nanocrystalline matrix may enhance the ductility of nanocrystalline materials [4-6]. However, the selective grain growth and achievement of target size and phase of embedded coarse grains may be difficult or impossible to control by thermal treatment in practice. In contrast, a bimodal grain structure can be achieved by consolidation of blended powders. In this process, cryomilled nanocrystalline (NC) and coarse-grained (CG) powders can be combined in specific proportions, enabling the design and manufacture of materials that achieve the desired balance of enhanced strength with acceptable ductility and toughness. Moreover, mechanical alloying is possible to produce bulk samples in sufficient size while several other techniques have fabricated small samples to explore the intrinsic properties of the materials [7-9].

In the present work, ductile phase toughening in bimodal structured Al-7.5Mg was achieved by deliberate blending of CG powders with cryomilled powders in select proportions. The stress-strain behavior was investigated using uniaxial tensile tests and finite element method (FEM). In the FEM work, plain strain two-dimension solid axisymmetric nonlinear (elastic-plastic) anyalysis was used to reveal the stress-strain behaviors and deformation mechanisms of the bimodal nanostructures.

EXPERIMENTAL DETAILS

Prealloyed Al-7.5%Mg (in weight percent) alloy powders were used to produce materials for this work. Nanocrystalline powder was produced using low-energy mechanical attrition with stainless steel balls at a cryogenic temperature (cryomilling). The ball-to-charge ratio was 36:1, with stearic acid added at 0.25% of the powder weight to moderate the cold welding process. The cryomilling was operated at 180 rpm for eight hours, and maintained at a temperature of –190 ºC using flowing liquid nitrogen in the vessel, submerging the powder and balls. Cryomilled NC Al-7.5Mg powders (average grain size of ~20 nm) were combined with 0, 15 and 30 weight percent unmilled CG Al-7.5Mg powders (particle size of 4 - 40 µm) and blended. These samples were loaded in aluminum cans for vacuum degassing at 400 ºC. The cans were sealed and consolidated in HIPing at 325 ºC and 25 ksi to produce bimodal samples. The consolidated billets were extruded using an extrusion ratio of 6.5:1 and proprietary rate and temperature parameters.

The resulting extrusions were machined to produce cylindrical tensile specimens (tensile axis parallel to the extrusion direction) with a gauge length of 13.5 mm and a gauge diameter of 3.5 mm. Tensile tests were performed on a universal testing machine at room temperature, using a nominal strain rate of 10^{-3} s^{-1}. The polished and chemically etched surfaces were observed by optical microscopy. Plain strain two-dimension solid axisymmetric nonlinear (elastic-plastic) finite element method was used to calculate stress-strain curves in tension.

RESULTS AND DISCUSSION

Bulk nanocrystalline samples containing 0, 15, and 30% unmilled (CG) powder fractions were consolidated by HIPing and extrusion. Polished sections were chemically etched to reveal NC and CG regions. Global views of the as-extruded bimodal microstructures are shown in figure 1. The alloy containing 0% unmilled powder exhibited uniform NC grains. In contrast, Al-7.5Mg samples containing 15 and 30% unmilled powders exhibited CG regions (light) and NC regions (dark). CG regions formed elongated bands that extended along the extrusion direction, while the NC regions formed a continuous matrix. In the sample produced from 15% unmilled powder, the CG bands were shorter and narrower than those in the 30% CG sample. Also, the spacing between the CG bands was greater in the 15% CG sample than in the 30% CG sample. The CG bands exhibited grains and sub-grains ~1 µm in diameter, compared with 100 - 300 nm grains in NC regions. No intermetallic phases were detected in either region. The interfaces between NC region and CG bands are remarkably discrete and abrupt. The composition within the CG bands and the NC regions was consistent with the alloy composition of Al-7.5Mg.

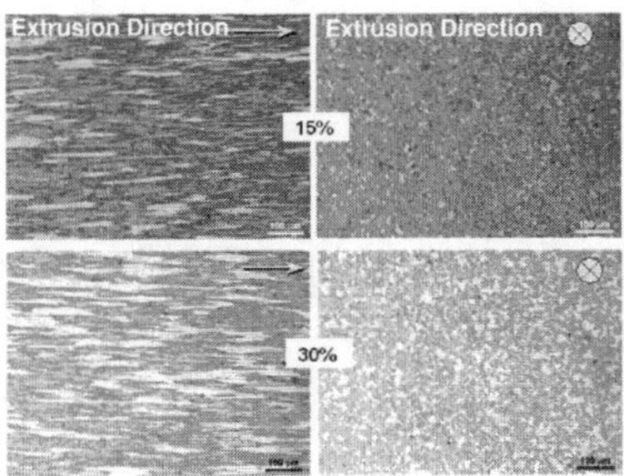

Figure 1. Microstructure of as-extruded bimodal Al-7.5Mg alloys (15% and 30% CG content). The bright and dark parts indicate CG and NC regions, respectively.

In the FEM study, plain strain two-dimension solid axisymmetric nonlinear (elastic-plastic) finite element method and the law of mixture were used to investigate the initial stress-strain behaviors of bimodal nanostructures. The unit cell, which was based on measurement of the microstructure, contains an ellipsoidal CG band at the center and NC region as shown in figure 2. The aspect ratio of CG was assumed identical in all bimodal structures with 15%, 30% and 50% CG content. By virtue of axisymmetry, one fourth of the unit cell is considered. The four node two-dimensional solid elements were used in the finite element calculations. Total numbers of the nodes and the quad elements were around 22000 and 11900, respectively. This nonlinear analysis based on classical metal plasticity theory followed Von-Mises yield criterion.

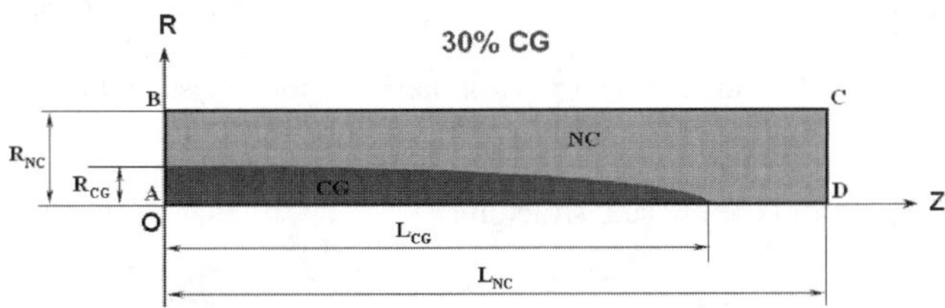

Figure 2. Schematic of unit cell of bimodal structure (30% CG) for FEM analysis.

The comparison between calculated and experimental tensile stress-strain curves of 15% CG bimodal Al-7.5Mg alloys is shown in figure 3 along with experimental curves of all cryomilled (i.e. 0% CG) and all coarse-grained (i.e. 100% CG) samples. The sample produced solely from cryomilled powder exhibited a yield strength more than four times that of conventional Al 5083 alloy. There were very little ductility, and negligible work-hardening compared to ductile conventional Al 5083 alloy. The calculated stress-strain curve of 15% CG has acceptable agreement with experimental curves of the bimodal structured sample with the same content.

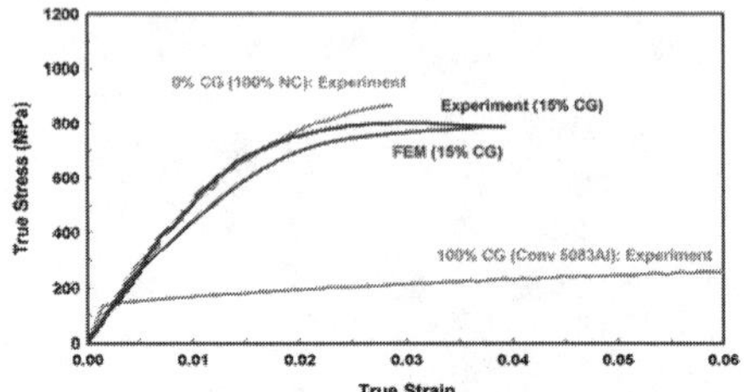

Figure 3. The comparison between calculated and experimental tensile stress-strain curves of bimodal Al-7.5Mg alloys comprised of 15% CG. 0%CG is all cryomilled Al-7.5Mg alloy and 100% CG is conventional 5083 aluminum alloys.

The calculated tensile stress-strain curves of bimodal Al-7.5Mg alloy comprised of 15%, 30% and 50% CG are shown in figure 4. Experimental curves of 0% CG (i.e. all cryomilled Al-7.5Mg alloy) and 100% CG (i.e. conventional 5083 aluminum alloy) are accompanied with the calculated curves for comparison. The 30% CG alloy exhibited elastic-nearly perfectly plastic stress-strain behavior, which is consistent with the reported tensile behavior for NC Al-Ti-Cu [4-6] and Al alloy [10]. The ductility and toughness of bimodal Al-7.5Mg alloy with 30% CG increased enormously compared with the 0% sample, while the yield stress decreased small. The peak flow stresses of the Al-7.5Mg with 0%, 15%, 30% and 50% CG were about 850 MPa, 790 MPa, 700 MPa and 560MPa at room temperature, respectively. These values are notable for an aluminum alloy, given that the flow stress for conventional 5083 Al. Especially noteworthy is the elongation-to-failure of Al-7.5Mg with 30% CG fraction, which was about 7% at room temperature. These results indicate a correlation between toughness and CG content. However, unlike alloys with a single grain size, the strength and ductility of the present material cannot be attributed solely to grain size refinement. It also is related to a change in deformation mechanism stemming from the unusual structure.

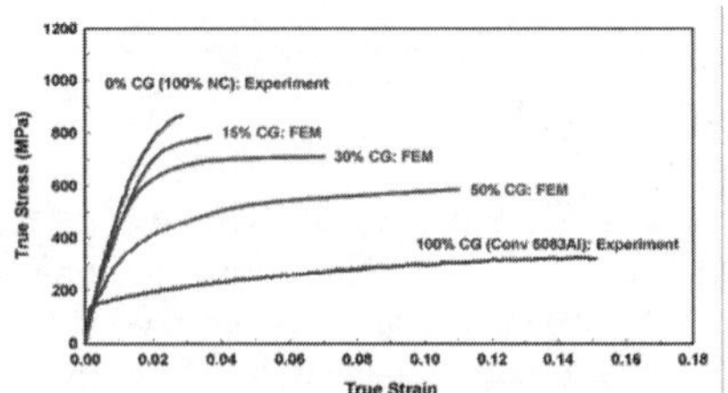

Figure 4. The calculated tensile stress-strain curves of bimodal NC Al-7.5Mg alloys comprised of 15%, 30% and 50% CG in red. Experimental stress-strain curves of 0%CG (all cryomilled Al-7.5Mg alloy) and 100%CG (conventional 5083 Al alloy) are also shown in blue.

The Von Mises stress contours of 15% CG alloy in tension are shown in figure 5 (a) and (b). At the early stage (figure 5(a)) of uniaxial tensile deformation of the bimodal structure, the Von Mises stress of ductile CG was higher than that of less ductile NC. After applying more displacement (figure 5(b)), this tendency was reversed, thus the less ductile NC started to carry higher stress than ductile CG. It may imply that NC region carries more tensile stress than CG over a certain level of tensile deformation while the ductile CG region elongates more than less ductile NC region.

The tensile deformation of bimodal materials exhibits considerably inhomogeneous features due to difference in strength and ductility between the NC and CG regions. The presence of CG region in NC region can result in inhomogeneous distributions of plasticity and stress upon a tensile loading. The dependency of yield strength and ductility of bimodal Al-7.5Mg alloys did not follow a simple rule-of-mixture based on the proportion of NC and CG. This implies that CG additions affect deformation mechanisms of bimodal Al-7.5Mg alloys in a more subtle way.

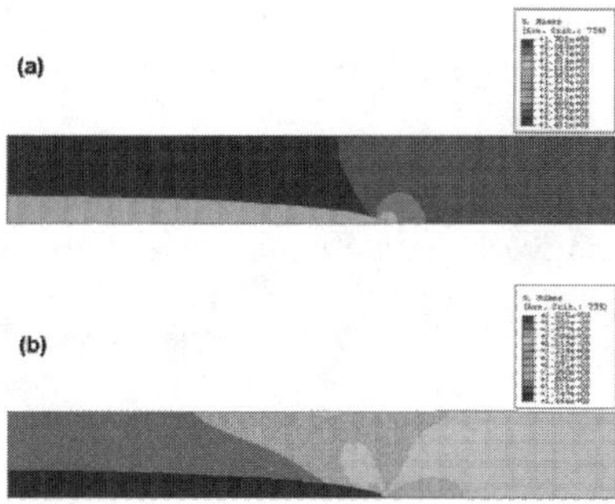

Figure 5. The contours of the Von Mises stress of 15% CG bimodal structure in tension. The (a) and (b) indicate the contours of Von Mises stress in a time step and the next time step of tensile deformation, respectively.

CONCLUSIONS

Bimodal Al-7.5Mg alloys comprised of NC grains and CG bands were produced successfully by consolidation of cryomilled and unmilled powders. The calculated tensile stress-strain curves have acceptable agreement with experimental curves of the bimodal structures. Bimodal Al-7.5Mg alloys show balanced mechanical properties between enhanced strength and reasonable ductility compared to conventional and nanocrystalline alloys. The experimental and FEM investigations suggest unusual deformation mechanisms and interactions between ductile CG bands and NC regions. The present work has shown that the yield strength can be substantially increased while retaining reasonable ductility. The bimodal nanostructure and its ductility toughening effect inspire us to design microstructures and to select processing parameters leading to optimal performance characteristics on various nanocrystalline metals.

ACKNOWLEDGEMENTS

This work was supported by a MURI grant from the Office of Naval Research, N00014-01-1-0882.

REFERENCES

1. Y. Ming, M. Chen, F. Zhou and E. Ma, Nature, 2002, 419, pp. 912-915
2. M. Legros, B.R. Elliot, M.N. Rittner, J.R. Weertman and K.J. Hemker, Phil. Mag. A, 2000, 80(4), pp. 1017-1026.
3. V.L. Tellkamp, A. Melmed, and E.J. Lavernia, Met. Mat. Trans. A, 2001, 32, pp. 2335-2343.
4. R.W. Hayes, R. Rodriguez and E.J. Lavernia, Acta Mater., 2001, 49, pp. 4055-68.
5. Z. Lee, R. Rodriguez, E.J. Lavernia and S.R. Nutt, *Ultrafine Grained Materials II*, edited by Y.T. Zhu, T.G. Langdon, R.S. Mishra, S.L. Semiatin, M.J. Saran, and T.C. Lowe, TMS, Seattle, WA, 2002, pp. 653-59.
6. Z. Lee, R. Rodriguez, R. W. Hayes, E. J. Lavernia and S. R. Nutt, Metall. Mater. Trans. A, 2003, 34A, pp. 1473-1481.
7. R.W. Siegel, MRS Bull., 1990, pp.60-67.
8. C.C. Koch, NanoStructured Materials, 1997, 9, pp.13-22.
9. C. Suryanarayana and C.C. Koch in *Non-Equilibrium Processing of Materials*, C. Suryanarayana, ed., Pergamon, New York, 1999, pp.313-344.
10. X.K. Sun, H.T. Cong, M. Sun and M.C. Yang: Metall. Mater. Trans. A, 2000, 31A, pp. 1017-24.

Mat. Res. Soc. Symp. Proc. Vol. 821 © 2004 Materials Research Society P9.1

Deformation Mechanisms and Ductility of Nanostructured Al Alloys

Bing Q. Han [a], Farghalli A. Mohamed [b] and Enrique J. Lavernia [a]
Department of Chemical Engineering and Materials Science
[a]University of California, Davis, CA 95616
[b]University of California, Irvine, CA 92697

ABSTRACT

Low tensile ductility is one of the critical challenges facing the science and technology of nanostructured materials. As an example, despite the fact that high strength is frequently observed in bulk nanostructured Al alloys, ductility and work hardening are often observed to decrease with decreasing grain size. In the present study, the tensile ductility of bulk nanostructured aluminum alloys processed via severe plastic deformation and consolidation of mechanically milled powders is analyzed. Adding coarse grains to the nanostructured matrix is proposed as an approach to improve ductility.

INTRODUCTION

The successful synthesis of large-scale nanostructured materials with a grain size in the range of 10 – 200 nm represents a major achievement in the broad field of nanotechnology. In this field the synthesis of the nanostructured aluminum alloys is attractive because of the high ratio of strength to density in aluminum. Currently, the two most widely used techniques for synthesizing bulk nanostructured aluminum alloys are severe plastic deformation and consolidation of mechanically milled powders [1-4].

Data reported for the mechanical behavior of bulk nanostructured aluminum alloys have shown two trends. First, for a grain size ranging from 20 to 300 nm, the grain size softening phenomenon (e.g., reverse Hall-Petch relationship [5,6]), which is sometimes reported, is absent, and the flow strength follows the regular Hall-Petch relationship: $\sigma = \sigma_o + k \cdot d^{-1/2}$. Second, in some nanostructured aluminum alloys [2,7,8], extensive strain hardening is not observed during tensile deformation. This trend is manifested in the presence of a longer steady-stage extending from the ultimate stress to the failure point (for instance, cryomilled Al alloys [3,9]).

The purpose of the present study is to provide insight into the deformation behavior of nanostructured aluminum alloys processed via the above two approaches.

TENSILE BEHAVIOR OF NANOSTRUCTURED ALUMINUM ALLOYS

Figure 1 shows the tensile elongation as a function of grain size for several aluminum alloys processed by two approaches, i.e., consolidation of mechanically milled powders and severe plastic deformation [2-4,8-17]. Despite considerable scatter in the data, the ductility decrease with decreasing grain size is evident. Also, the trend of data appears to indicate that the processing approach has an influence on the value of elongation. For the same grain size, the values of elongation in nanostructured aluminum alloys by severe plastic deformation are higher than those by consolidation of mechanically milled powders, probably because of the absence of processing defects in the material produced by the former technique. As reported elsewhere [18], the low tensile ductility of many nanostructured materials is often attributed to processing defects and flaws (porosity or poor interfacial bonding). Finally, Figure 1 illustrates an interesting finding: grain refinement via severe plastic deformation and consolidation of mechanically milled powders is limited to approximately 200 nm and 20 nm, respectively. For instance, the minimum grain size in nanostructured materials processed via consolidation of mechanically milled powders is about 20 – 30 nm (reported in a mechanically alloyed pure Al [12]), compared to about 100 - 300 nm via severe plastic deformation [1]. This difference most likely is the result of (a) the higher level of severe plastic deformation introduced during milling procedure, and (b) the limitation of grain growth by the Zener pinning of dispersions in the former approach. Also, this difference is consistent with the general observation that the final grain size of nanostructured materials is largely determined by the inherent thermal stability of the microstructure in combination with the parametric space used during processing [19].

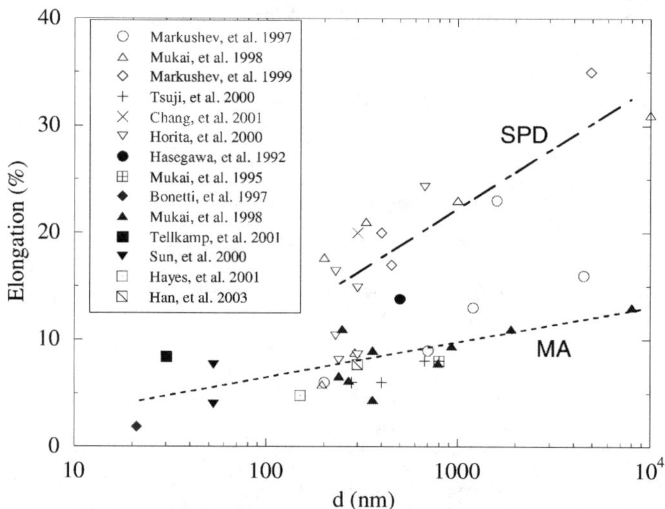

Fig. 1 Tensile elongation as a function of grain size in aluminum alloys processed by consolidation of mechanically milled powders and severe plastic deformation [2-4,8-17].

By comparing tensile behavior of nanostructured aluminum alloys with coarse-grained aluminum alloys, guiding information regarding the underlying deformation mechanisms in nanostructured aluminum alloys can be obtained. Figure 2 (a) shows the plots of true tensile stress as a function of true strain for five 5083 Al alloy specimens, i.e., a cryomilled nanostructured 5083 Al alloy consolidated via cold isostatic pressing (CIP) and extrusion, and a cryomilled nanostructured 5083 Al alloy consolidated via heat isostatic pressing (HIP) and extrusion (Han and Lavernia, unpublished data, 2003), a cryomilled nanostructured 5083 Al alloy [3], an ultrafine-grained 5083 Al processed via equal-channel angular pressing (ECAP) [16], and an as-received coarse-grained 5083 Al [16]. The cryomilled 5083 Al alloy consolidated via CIP and extrusion has a grain size of approximately 150 nm. The cryomilled 5083 Al alloy consolidated via HIP and extrusion has a broad range of grain size distribution with an average grain size of about 500 nm. The cryomilled 5083 Al alloy [3] is reported to possess bimodal grains with some of grains of about 30 nm, and some large grains of about several hundred nanometers in microstructure after extrusion at elevated temperatures. The ECAP 5083 Al alloy has a homogeneous uniform grain size of about 300 nm [16]. The coarse-grained 5083 Al alloy has a grain size of 200 μm [16]. Inspection of the figure shows two observations. The first observation is related to yield strength. The yield strength of cryomilled 5083 alloys is higher than that of the ECAP 5083 Al alloy. The highest value of yield strength of 713 MPa is observed in the cryomilled 5083 Al alloy consolidated via CIP and extrusion. Such a high strength is mostly attributed to the grain size effect compared to other 5083 Al alloys. The second observation is pertinent to ductility, which decreases with decreasing grain size. An extremely low ductility of 0.3% is observed in the cryomilled 5083 Al alloy consolidated via CIP and extrusion.

Figure 2 (b) shows the plots of true tensile stress as a function of true strain for three Al-Mg alloys, i.e., a cryomilled nanostructured Al-7.5%Mg-0.3%Sc alloy [20], a cryomilled ultrafine-grained Al-7.5%Mg alloy [9], and a coarse-grained Al 6.61%Mg alloy [21]. The cryomilled Al 7.5%Mg-0.3%Sc alloy has a grain size of approximately 200 nm. The cryomilled Al-7.5%Mg alloy has a grain size of approximately 300 nm. The strength of the cryomilled Al-7.5%Mg-0.3%Sc alloy is higher than that of the cryomilled Al-7.5%Mg alloy, for the two possible reasons, one is the grain size effect and the other is the dispersion strengthening. Another observation is related to the low strain hardening ability in cryomilled Al alloys. The strain hardening exponent, n ($n = d\ln\sigma/d\ln\varepsilon$), is approximately 0.04 in the cryomilled Al-7.5Mg alloy, whereas, the strain hardening exponent is relatively high (about 0.36) in the coarse-grained Al-Mg alloy.

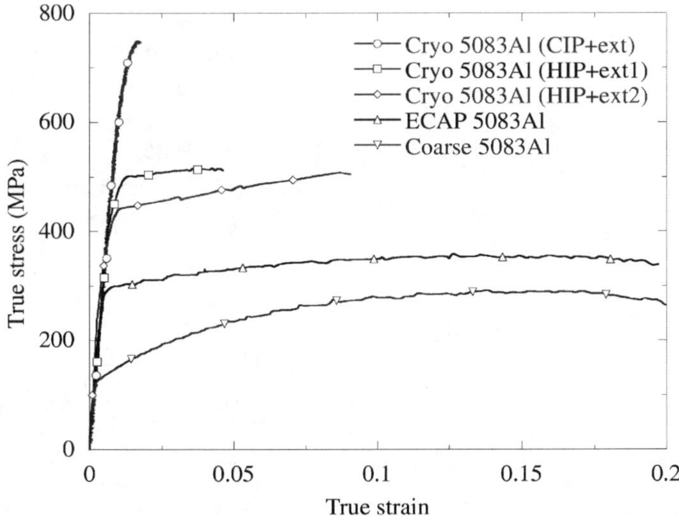

(a)

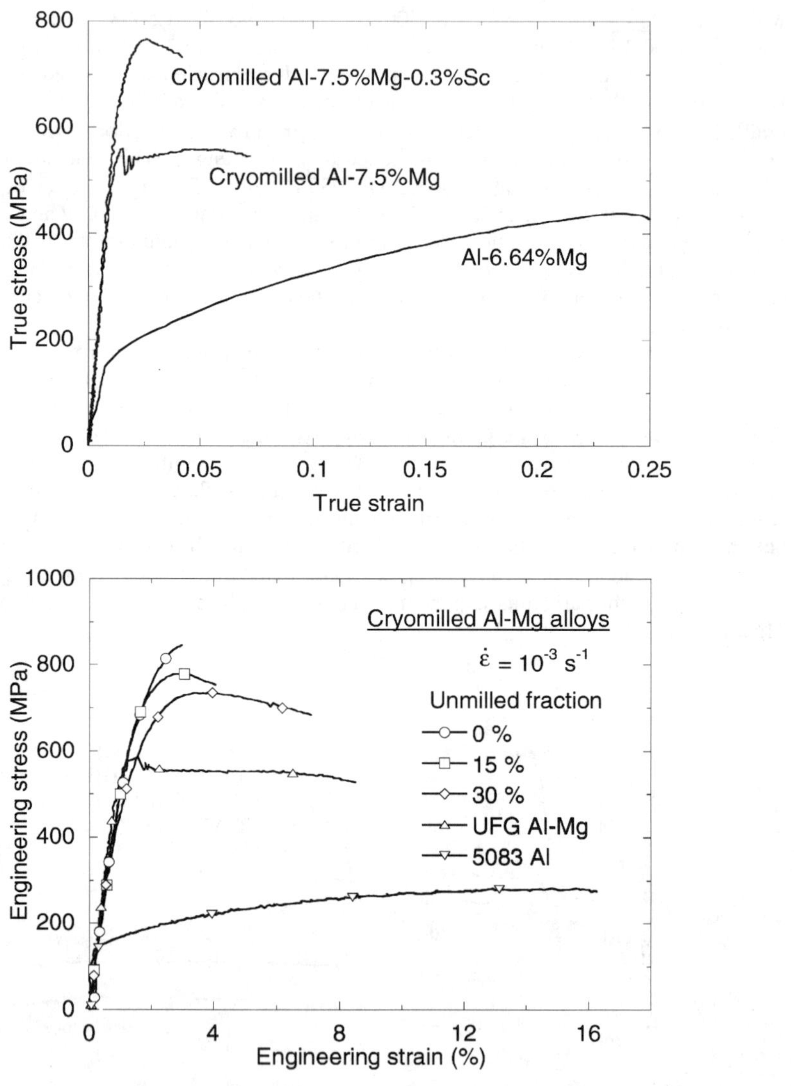

Fig. 2 (a) True tensile stress as a function of true strain of cryomilled 5083Al alloys (cryomilled 5083 Al (CIP+ext) and cryomilled 5083 Al (HIP+ext1) (Han and Lavernia, unpublished data 2003), cryomilled 5083 Al (HIP+ext2) [3] and ECAP 5083 Al and coarse 5083 Al alloys [16]). (b) True tensile stress as a function of true strain of a cryomilled Al-Mg alloy and a cryomilled Al-Mg-Sc alloy [9,20]. (c) Engineering tensile stress as a function of engineering strain of bimodal cryomilled Al-Mg alloys.

In addition to revealing the aforementioned two observations, the data of Figures 2 (a) (b) show that the cryomilled ultrafine-grained Al-7.5%Mg alloy [9], the cryomilled bimodal 5083 Al alloy [3], and the ECAP 5083 Al alloy [16] exhibit relatively high ductilities (several to nearly 20 percent of elongation). This observation contrasts with those reported for nanostructured materials in which low tensile ductility, typically no more than 2% elongation for grain sizes less than 25 nm [22] was reported. It is noteworthy that the elongation (20 %) of the ECAP 5083 Al alloy is essentially equal to that of the as-received alloy [16].

Realizing the benefits derived from the presence of submicron grains on ductility of nanostructured aluminum alloys suggests the development of bimodal cryomilled Al-Mg alloys, which are processed with the same processing routes and parameters except containing the different volume fractions (15 and 30 wt pct) of unmilled coarse-grained powders, with an intention to improve ductility. Figure 2 (c) shows the plots of engineering stress as a function of engineering strain of several cryomilled Al-7.5%Mg alloys with a bimodal microstructure. According to previous studies on two-phase cryomilled Al alloys, there is insignificant change in the microstructure of nanostructured regimes after extrusion [23]. However, there is a directional distribution of coarse-grained regions after extrusion, which is elongated along the extrusion direction [23]. The coarse grain size is about 1 μm in the coarse-grain bands. Obviously, the bimodal cryomilled Al alloys show an improved ductility without much decrease in strength.

DISCUSSION

If the necking occurs when $d\sigma/d\varepsilon = \sigma$ at the maximum stress, the true strain at the onset of necking is given by $\varepsilon_u = n$ [24]. According to an empirical relation, $n = 5/(10 + d^{-0.5})$ where d is in mm and grain size (d) (less than 10 μm) in carbon steels [25], $\varepsilon_u = 5/(10 + d^{-0.5})$. Although it is not certain whether the above relationship can be applicable to materials other than carbon steels or whether can be extrapolated to the nanometer range, it at least shows that the tensile strain decreases with decreasing grain size. The processing approach has an obvious influence on ductility. There are two reasons for a lower ductility in nanostructured materials processed by consolidation of mechanically milled powders than that of nanostructured materials processed by severe plastic deformation. One is due to the possible presence of processing flaws (porosity) or poor interfacial bonding [26]. The other is due to a lower density of mobile dislocations (ρ_m). The value of uniform strain (ε_u) can be expressed as: $\varepsilon_u = \rho_m \cdot b \cdot L$ where b the Burgers vector and L is the average distance of dislocation movement. In nanostructured materials processed via consolidation of mechanically milled powders, most of the dislocations are most likely to be immobilized due to their strong interactions of dislocations with nanoscale dispersions. Adding coarse grains into the nanostructured matrix will bring both enhanced dislocation activity as well as longer distance of dislocation movement in the materials, resulting in a higher ductility.

SUMMARY

In the present study, data on tensile behavior of nanostructured aluminum alloys processed via consolidation of mechanically milled powders and severe plastic deformation are analyzed. Ductility decreases with decreasing grain size. The aluminum alloys processed via severe plastic deformation exhibit a higher ductility than those processed via consolidation of mechanically milled powders. The lower ductility in the latter materials can be the result of one or a combination of the following possibilities: the low density of mobile dislocations, and the presence of processing flaws (porosity) or poor interfacial bonding. Bimodal nanostructured aluminum alloys show an improved ductility without much decrease in strength.

ACKNOWLEDGEMENTS

The work is supported by the Office of Naval Research under Grant No. N00014-03-1-0149 with Dr. Lawrence Kabacoff as program officer.

REFERENCES

1. R.Z. Valiev, R.K. Islamgaliev and I.V. Alexandrov, *Prog. Mat. Sci.* **45,** 103 (2000).
2. X.K. Sun, H.T. Cong, M. Sun and M.C. Yang, *Metall. Mater. Trans. A* **31A,** 1017 (2000).
3. V.L. Tellkamp, A. Melmed and E.J. Lavernia, *Metall. Mater. Trans. A* **32A,** 2335 (2001).
4. R.W. Hayes, R. Rodriguez and E.J. Lavernia, *Acta mater.* **49,** 4055 (2001).
5. H. Conrad and J. Narayan, *Scrip. mater.* **42,** 1025 (2000).
6. W.W. Milligan, Mechanical Behavior of Bulk Nanocrystalline and Ultrafine-grain Metals. *In:* Comprehensive structural integrity, vol. 8. Elsevier Pergamon, 2003, p. 529.
7. R.Z. Valiev, A.V. Korznikov and R.R. Mulyukov, *Mater. Sci. Eng.* **A168,** 141 (1993).
8. M.V. Markushev, C.C. Bampton, M.Y. Murashkin and D.A. Hardwick, *Mater. Sci. Eng.* **A234-236,** 927 (1997).
9. B.Q. Han, Z. Lee, S.R. Nutt, E.J. Lavernia and F.A. Mohamed, *Metall. Mater. Trans. A* **34A,** 603 (2003).
10. T. Hasegawa, T. Miura, T. Takahashi and T. Yakou, *ISIJ Int.* **32,** 902 (1992).
11. T. Mukai, K. Ishikawa and K. Higashi, *Mater. Sci. Eng.* **A204,** 12 (1995).
12. E. Bonetti, L. Pasquini and E. Sampaolesi, *NanoStruct. Mater.* **9,** 611 (1997).
13. T. Mukai, M. Kawazoe and K. Higashi, *NanoStruct. Mater.* **10,** 755 (1998).
14. M.V. Markushev, M.Y. Murashkin, P.B. Prangnell, A. Gholinia and O.A. Maiorova, *NanoStruct. Mater.* **12,** 839 (1999).
15. Z. Horita, T. Fujinami, M. Nemoto and T.G. Langdon, *Metall.Mater.Trans.* **31A,** 691 (2000).
16. S.Y. Chang, J.G. Lee, K.T. Park and D.H. Shin, *Mat. Trans.* **42,** 1074 (2001).
17. N. Tsuji, Y. Saito, Y. Ito, H. Utsunomiya and T. Sakai, *In:* Ultrafine grained materials, TMS (The Minerals, Metals & Materials Society), 2000, p. 207.
18. J.R. Weertman, *Mater. Sci. Eng.* **A166,** 161 (1993).
19. C.C. Koch, *NanoStruct. mater.* **9,** 13 (1997).
20. B.Q. Han, E.J. Lavernia and F.A. Mohamed, *Metall. Mater. Trans. A* **submitted** (2003).
21. E.M. Taleff, D.R. Lesuer and J. Wadsworth, *Metall. Mater. Trans. A* **27A,** 343 (1996).
22. C.C. Koch, D.G. Morris, K. Lu and A. Inoue, *MRS Bulletin* **Feb.,** 54 (1999).
23. Z. Lee, D.B. Witkin, E.J. Lavernia and S.R. Nutt, MRS Proc., vol. 740, 2003 pp. I1.7.1-6.
24. T.H. Courtney, *Mechanical Behavior of Materials*, Second Edition (McGraw-Hill Higher Education, 2000).
25. W.B. Morrison and R.L. Miller, *In:* Ultrafine-grain metals (Syracuse University Press, Syracuse, NY, 1970), p. 182.
26. M.N. Rittner, J.R. Weertman, J.A. Eastman, K.B. Yoder and D.S. Stone, *Mater. Sci. Eng.* **A237,** 185 (1997).

Mat. Res. Soc. Symp. Proc. Vol. 821 © 2004 Materials Research Society P5.8

Elevated Temperature Mechanical Properties of Devitrified Metallic Glass

Nathan A. Mara, Alla V. Sergueeva, and A.K. Mukherjee
Materials Science Division, University of California, Davis
One Shields Avenue
Davis, CA 95616, USA

ABSTRACT

Elevated temperature tensile tests of different microstructures arising from different heat treatments of the Fe-based metallic glass Vitroperm ($Fe_{73.5}Cu_1Nb_3Si_{15.5}B_7$) are presented. An anneal at 600^0C for 1h yields a single phase α-Fe microstructure with equiaxed, randomly oriented 15 nm grains, which is an ideal candidate for study of material properties at diminishing length scale. This microstructure has good stability during tensile testing at 600^0C, showing a strain rate exponent correlating to grain boundary sliding (m=0.5), but little ductility, and strengths to 1250 MPa. The brittle behavior could be attributed to the lack of dislocation activity at such length scales. At temperatures up to 725^0C, grain growth occurs, leading to elongations as large as 65% at flow stresses of 250 MPa. Precipitation of a second Nb-rich phase accompanies the grain growth. This investigation is supported by NSF, Division of Materials Research, grant NSF-DMR-0240144.

INTRODUCTION

High-temperature mechanical properties of nanocrystalline materials can often be quite challenging to ascertain due to the inherent instablility of the microstructure at length scales below 100nm. At the test temperatures necessary for diffusion-driven mechanisms to play a significant role in deformation (in excess of $0.5T_m$), grain growth that occurs during testing can cause the microstructure to coarsen to the point that the sample is no longer truly nanocrystalline. Based on findings in the literature, there is a large scatter in the available experimental data, with wide ranges given for characteristic strain rate exponent values and for the activation energy of deformation [1-5]. These quantities give important information as to the dominant deformation mechanism at hand, and the scatter in their values reflects an important point—that a nanocrystalline structure does not point clearly to a single dominant deformation mechanism.

Although there are many methods of producing nanocrystalline materials, devitrification of amorphous materials can offer some attractive advantages, so long as heat treatments are carefully carried out. With the correct chemistry, this method allows for many microstructures of uniform overall chemistry to be evolved by simply altering the annealing parameters, effectively avoiding some of the shortfalls associated with other production methods. In recent years, alloys have been developed that after crystallization yield a fully crystalline matrix with grain sizes as low as 20 nm [6]. Fe-based alloys with composition close to that of FINEMET show some promise in this area [7]. Commercially available alloys such as Vitroperm yield extremely fine one-phase microstructures, or bimodal grain size distributions consisting of several phases depending on the annealing parameters chosen.

In this paper, a commercially available metallic glass (Vitroperm, $Fe_{73.5}Cu_1Nb_3Si_{15.5}B_7$) was crystallized into several different microstructures. Of particular note are a single phase structure of α-Fe with a grain size of approximately 15 nm, and a multiphase structure consisting of a bimodal distribution of grain sizes. These crystallized structures were then subjected to constant strain rate tensile testing in an argon atmosphere at different temperatures and strain rates. Analysis of the tensile data and the material microstructure before and after testing using analytical transmission electron microscopy (TEM), high-resolution TEM (HRTEM), and X-ray diffraction (XRD) sheds some light on the definition of the rate-controlling deformation mechanism at test temperatures.

EXPERIMENTAL DETAILS

Strips of melt-spun Vitroperm ribbon approximately 2 mm wide, ~20 microns thick, and with a gage length of 4 mm were used in this investigation. Tensile testing was carried out at a variety of strain rates using a custom-designed mini tensile tester with displacement resolution of 5 microns and load resolution of ~1 gram that is capable of testing materials with a gage length and width as small as 1 mm, and thicknesses as small as to 5 microns. All testing was carried out in a high-purity (99.999%) argon atmosphere to minimize effects of oxidation. A 600^0C or 650^0C anneal for 1 hour was performed *in-situ* in the tensile tester prior to tensile testing.

Microstructural analysis was conducted using a Philips CM-200 at 200 kV for analytical (EDS, EELS), high-resolution, diffraction contrast microscopy except where noted. In these cases, a Topcon 002B TEM at 190 kV was employed. X-ray analysis was performed using a Scintag XDS 2000 X-ray diffractometer with a 3 axis goniometer and Cu K_α radiation. Differential Scannning Calorimeter (DSC) scans were made on a Perkin-Elmer DSC 7 from ambient to 725^0C at a heating rate of 40^0C/minute. This heating rate was chosen because it mimics the heating rate encountered in the tensile testing furnace.

RESULTS

As-received microstructure

Figure 1 presents XRD and DSC traces of Vitroperm in the as-received state. As is evident from Figure 1a, the as-received structure of Vitroperm is not completely glassy, with some notable peaks resulting from α-Fe nanocrystallites. The DSC scan in Figure 1b exhibits a glass transition temperature at 470^0C, and an onset of crystallization at 635^0C.

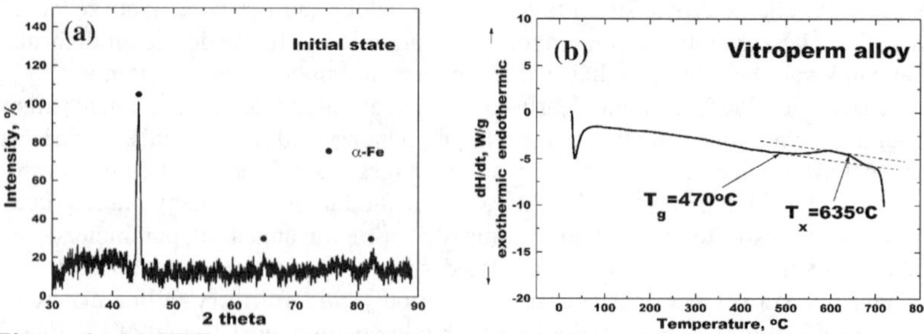

Figure 1. **(a)** XRD pattern from the as-received Vitroperm alloy. **(b)** DSC trace of the as-received Vitroperm alloy at a heating rate of 40^0C/min.

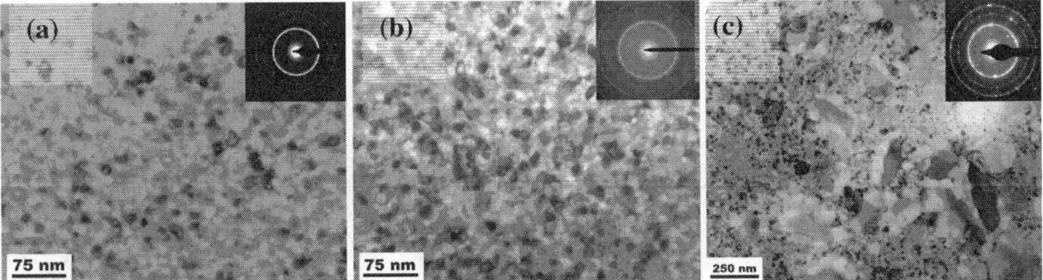

Figure 2. Bright field micrographs and selected area electron diffraction (SAED) patterns resulting from various 1 hour anneals at different temperatures. **(a)** 550^{0}C for 1h. **(b)** 600^{0}C for 1h. **(c)** 650^{0}C for 1h.

From the information deduced from figure 1, a series of short 1-hour anneals were carried out to create a variety of microstructures. Figure 2 shows 550^{0}C, 600^{0}C, and 650^{0}C microstructures, respectively. The 550^{0}C anneal gave an extremely fine single-phase α-Fe microstructure with mean grain size of 10-15 nm (figure 2a). Annealing for 1 hour at 600^{0}C also gives a single-phase α-Fe structure, with slightly larger grains in the 15-20 nm range (figure 2b). Further increasing the annealing temperature to 650^{0}C yields the single-phase microstructure with bimodal grain size distribution exhibited in Figure 2c. The one-phase structures were verified by XRD analysis (not shown).

Tensile tests of annealed microstructures

The microstructures seen in figures 2b and 2c form the basis for the mechanical testing carried out in the rest of this work. Figure 3a contains the tensile curves resulting from testing the post-annealed microstructure seen in Figure 2b (600^{0}C anneal) at 600^{0}C for varying strain rates. Of interest is that the strain rate sensitivity of this material under these testing conditions is 0.5, which will be discussed later. The post-tensile tested microstructure at a strain rate of $1x10^{-4}$ s^{-1} shown in figure 3b reveals little grain growth, and the retention of a one-phase microstructure.

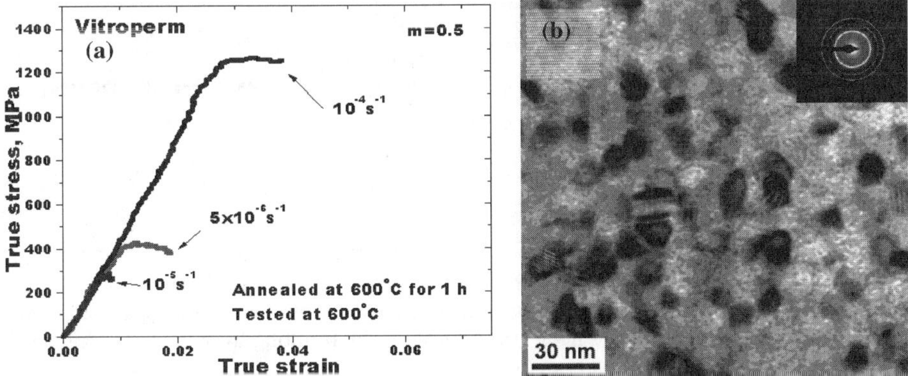

Figure 3. **(a)** Tensile curves for samples tested at 600^{0}C after a 1 hour anneal at 600^{0}C. **(b)** Bright field micrograph of post-tested microstructure at a strain rate of $1x10^{-4}$ s^{-1}.

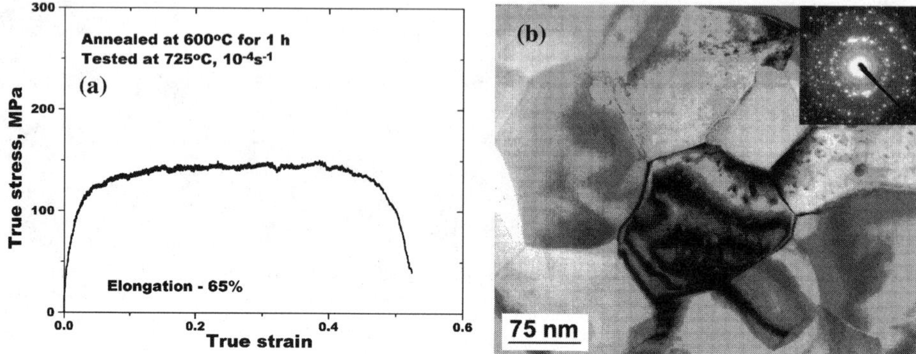

Figure 4. **(a)** Tensile curve for sample tested at 725^{0}C after a 1 hour anneal at 600^{0}C. **(b)** Corresponding bright field micrograph of post-tested microstructure (micrograph taken on Topcon 002B at 190kV).

At higher test temperatures, the mechanical behavior is decidedly different (figure 4a). Rather than the high strengths seen in the 600^{0}C tests, testing at strain rate 10^{-4} s^{-1} results in an UTS of only about 150 MPa with appreciable elongation. As would be suspected, the microstructure is decidedly different, with a Nb-rich phase nucleating at the triple points as evidenced in the bright-field and SAED micrographs in figure 4b, and supported by EDS and EELS analysis (not shown).

Through the use of several isotemperature strain-rate jump tests (not shown), a logarithmic true stress vs. true strain plot can be constructed as seen in figure 5a. Figure 5b depicts an Arrhenius plot derived from figure 5a, where strain rate vs. temperature values were taken at an isostress condition at 170 MPa across three test temperatures (675-725^{0}C) and two orders of magnitude of strain rate. There is a definite change in mechanism from the 600^{0}C tests and those at temperatures at and above 675^{0}C, with strain rate sensitivity values changing from 0.5 to 0.2, respectively.

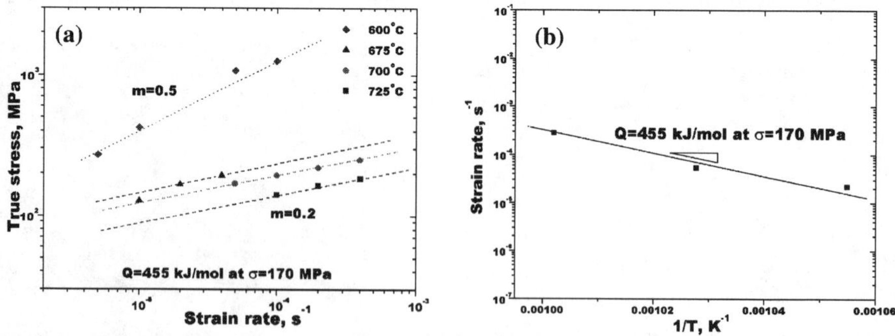

Figure 5. **(a)** Summary of tensile tests conducted at various temperatures and strain rates after the initial 600^{0}C anneal for 1 hour. **(b)** Arrhenius plot resulting from the tests conducted at 675^{0}C and above.

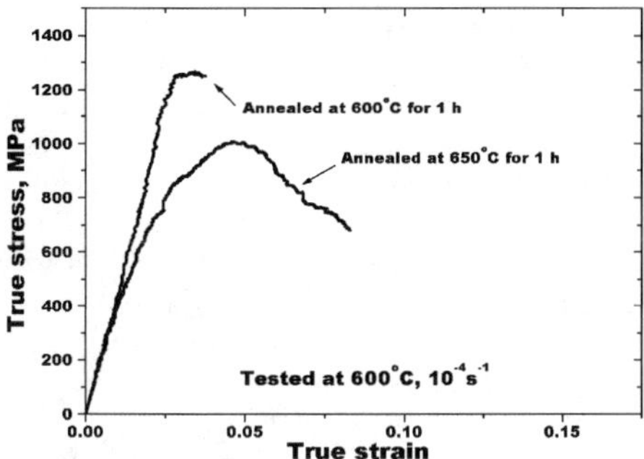

Figure 6. Tensile curves for samples tested at 600^0C after a 1 hour anneal at 650^0C and after a 1 hour anneal at 600^0C.

To this point in the study, the sample that underwent the 600^0C anneal for 1 hour has served as the basis for mechanical testing. However, the bimodal grain size distribution depicted in figure 2c after a 1 hour anneal at 650^0C can also show some interesting tensile results as seen in figure 6. Both tensile curves in this figure are conducted at 600^0C at a strain rate of 10^{-4} s^{-1}, but the second sample is annealed at 600^0C for 1 hour as a reference. There is a clear difference between the two curves in that the bimodal distribution microstructure (annealed at 650^0C for 1h) exhibits strain hardening, enhanced ductility, and reduced strength during tensile testing as compared to the equiaxed, very fine grained sample annealed at 600^0C.

DISCUSSION

For the 600^0C tensile tests, the microstructure consists of a one-phase, 15-20 nm grain size material throughout the duration of the test. Usually, for mechanisms corresponding to m=0.5, grain boundary sliding occurs, and at stress concentrations such as triple points, dislocations are emitted and slip and glide their way to another grain boundary to accommodate the deformation. This model assumes that the grain size is sufficient to permit dislocations to form and propagate through the grain interior or along grain boundaries. At grain sizes as low as 15 nm, dislocation generation is so difficult that there is no way to accommodate the strain after sliding has occurred unless by diffusion-based mechanisms. These are probably not working fast enough in this case to alleviate the stress concentrations at triple points, so a void nucleates and quickly grows, leading to early fracture. The strain rate sensitivity of m=0.5 corresponds to the actual sliding of the grains as the rate-limiting step for deformation, since other mechanisms, especially those operated by dislocations, seem to be diminished. That is, the conditions of the test (temperature, strain rate, etc.) are such that the deformation mechanism falls within the bounds of superplastic deformation, but occurs in the absence of any sufficiently rapid method of accommodation.

For the 675^0C to 725^0C tests, there is evolution of another phase at the triple points of the original grains. If clean, defect-free grains are formed at stress concentrations such as triple points through the diffusion of Nb-containing phases, the growth of these phases could play an important role in the accommodation of the strain seen in the tensile tests. If it is assumed that grain boundary sliding occurs, which is reasonable since grains remain equiaxed after deformation, stress concentrations such as triple points and particles at grain boundaries can help to emit dislocations. Once emitted, the dislocations must move their way through the matrix to aid in accommodation of the resulting strain. In this case, it is possible for either grain boundary sliding, dislocation glide, or dislocation climb to be the rate-limiting step. With a strain rate sensitivity value of m=0.2, this rate limiting step should be the same as seen in power-law creep—the climb of dislocations within the grain interior.

The Arrhenius plot in Figure 5(b) shows an activation energy value of 455 kJ/mol that is higher than any literature values for lattice self-diffusion or the other present species in α-Fe. This value reflects the complexity of diffusion-based processes in FINEMET alloys that have been found in other studies [8]. Due to the large number of alloying elements contained in the α-Fe initial structure, and the formation of secondary phases during tensile testing, it is difficult to attribute this activation energy to any particular diffusion mechanism at this time. Analysis of the secondary phases is to be conducted in future work, and this may shed some light on the significance of this value.

Other authors have found that in FINEMET metallic glasses, Nb tends to reside at the grain boundaries in the as-crystallized structure and will pin the boundaries, leading to a structure that is quite stable at high temperatures. [9] This coincides with the above results, in that the grain size is very stable up to a certain point, and once Nb is segregated to its own phase away from the α-Fe grain boundaries, grain growth of all phases quickly proceeds. Furthermore, if the 15 nm grains are too small to have dislocation activity to accommodate the deformation at triple points, it is logical that once the grains have grown sufficiently to allow dislocation motion that the strength should drop, and plasticity should increase. This is in fact exactly the trend that is exhibited in the data.

The tests of the bimodal grain size distribution seen in figure 2c reflect a trend that could lead to additional tailoring of existing nanomaterials simply by introducing additional plasticity at the expense of some drop in strength. Since nanomaterials are generally quite strong due their small grain size, for many applications, a certain decrease in strength may be worthwhile if it is made up for in material toughness. An important question that remains to be answered is: "At what critical grain size do dislocations begin to take part in deformation?" Experiments are planned to tailor the grain size to try to find an answer to this question.

CONCLUSION

Two distinct microstructures were tensile tested at elevated temperatures in this work. Firstly, a single-phase α-Fe matrix with grain size of ~15 nm was produced through the devitrification of Vitroperm metallic glass. Secondly, a bimodal grain size distribution consisting of multiple phases was delivered through a heat treatment at higher temperatures. For the uniform fine grain size work, the grains are stable during tensile testing at 600^0C, and show great strength, but little ductility. However, an analysis of strain rate sensitivity shows a strain rate exponent of 0.5, which usually corresponds to grain boundary sliding as the rate controlling step. Testing at higher temperatures up to 725^0C reveals formation of at least one other phase, grain growth to approximately 150 nm, 65% elongation at much lower flow stresses, and a transition to a possible dislocation-climb based deformation mechanism, although no grain elongation was observed. TEM investigation shows that Nb has segregated to new grains mainly located at triple points of the original structure. Nucleation and growth and the associated grain boundary migration of these new phases is expected to play an important role in deformation accommodation processes. For the bimodal distribution work, enhanced ductility and lowered strength was witnessed in comparison to its fine-grained counterpart. This behavior is expected to be explained through the introduction of greater dislocation activity in the larger 250 nm grains, where they allow for free dislocation plasticity, while the smaller 30 nm grains give the material high strength.

ACKNOWLEDGEMENTS

The authors would like to acknowledge the National Science Foundation, Division of Materials Research for its financial support under grant NSF-DMR-0240144.

REFERENCES

[1] Mishra RS, Mukherjee AK. Superplasticity and Superplastic Forming 1998. Proceedings of a Conference held as part of the TMS Annual Meeting. TMS. 1998, pp.109-16.
[2] Sergueeva AV, Stolyarov VV, Valiev RZ, Mukherjee AK., Scripta Materialia, vol.45, no.7, (10 Oct. 2001), 747-52.
[3] H. Hahn, R.S. Averback, Journal of the American Ceramics Society, 74 (1991) 2918.
[4] K. Taketani, A. Uoya, K. Ohtera, T. Uehara, K. Higashi, A. Inoue, T. Masumoto, Journal of Materials Science, 29 (1994) 6513.
[5] K. Higashi, Materials Science and Engineering A, 166, (1993) 109-118.
[6] Xue-Dong Liu, Met. Trans., Vol. 39, No. 8 (1998), pp. 783-794.
[7] Y. Yoshizawa, S. Oguma, K. Yamauchi, Journal of Applied Physics. 64 (1988) 6044.
[8] E. Illekova, Thermochemica Acta, 387, (2002) 47-56.
[9] Kulik T, Journal of Non-crystalline Solids, 287 (2001) 145-161.

Mat. Res. Soc. Symp. Proc. Vol. 821 © 2004 Materials Research Society P8.28

Role of Deformable Fine Spinel Particles in High-Strain-Rate Superplastic Flow of Tetragonal ZrO_2

Koji Morita, Keijiro Hiraga, Byung-Nam Kim and Yoshio Sakka
National Institute for Materials Science, 1-2-1 Sengen, Tsukuba, Ibaraki 305-0047, Japan

ABSTRACT

The role of $MgAl_2O_4$ spinel particle dispersion in high-strain-rate superplasticity (HSRS) of tetragonal ZrO_2 was examined by characterizing microstructural changes during deformation. The dispersed spinel particles elongate with strain along tensile direction and the elongation tends to be pronounced with increasing strain rate. In the elongated spinel particles, intragranular dislocations lying along the elongated direction were observed, suggesting that the elongation relates to the dislocation motion. The flow behavior characterized by a stress exponent of ≈2.0 suggests that grain boundary sliding (GBS) is the predominant flow mechanism. The dislocation-induced plasticity in the spinel particles may assist the relaxation of stress concentrations exerted by GBS, leading to HSRS in tetragonal ZrO_2.

INTRODUCTION

Recently, we attained high-strain-rate superplasticity (HSRS) in fine particle dispersed ceramic composites; Al_2O_3-spinel-ZrO_2 [1,2] and ZrO_2-spinel [3-5]. For tetragonal ZrO_2 dispersed with 30vol% spinel, the fine spinel particle dispersion can improve strain rate by 10^2-10^3 times in attaining similar tensile elongation in monolithic tetragonal ZrO_2 [6-8]. The tensile elongation, e_f, exceeds ≈250% at 1823 K and at an initial strain rate of $\dot{\varepsilon}_0 \approx 0.7\ s^{-1}$ [3-5].

A common microstructural feature in the high-strain-rate superplastic deformed ceramic composites is that the dispersed particles appear to be plastically deformable to some extent during deformation. Particularly in the ZrO_2-spinel composite, the dispersed spinel particles elongate along tensile direction after deformation. This aspect indicates that the dispersed particles play an important role in the attainment of HSRS in ceramics.

The present study was therefore performed to examine the role of deformable particle dispersion in HSRS of ZrO_2-spinel composite. In this study, we lay an emphasis on characterizing quantitatively microstructural changes during high-strain-rate superplastic flow.

EXPERIMENTAL PROCEDURES

The specimens used in this study was prepared by a method described elsewhere [3]. Briefly, 3mol%-Y_2O_3-stabilized tetragonal ZrO_2 (TZ-3Y, Tosoh Co., Ltd.,) mixed with 30vol% spinel powders (SP-12, Iwatani Co., Ltd.,) were cold-isostatically pressed at about 400 MPa and sintered at 1673 K for 2 h in air. Constant displacement-rate tensile tests were conducted using an Instron-type tensile machine at 1773 K and at $\dot{\varepsilon}_0 \approx 2.0\times10^{-3}$-$0.3\ s^{-1}$ under vacuum.

For scanning electron microscopy (SEM) observation, the surface of the specimens were mechanically polished and thermally etched at 1573 K for 10 min. The average grain size, d, was determined as 1.56 times of the average intercept lengths of grains [9]. The grain aspect ratio (GAR) was determined from the intercept lengths measured in the directions parallel and perpendicular to the tensile axis. For transmission electron microscopy (TEM) observation, thin sheets with a thickness of about 500 μm were cut with a low-speed diamond cutter, mechanically polished to about 100 μm in thickness and further thinned with an Ar ion-milling machine.

EXPERIMENTAL RESULTS

Figure 1 shows a SEM image of the as-sintered ZrO_2-spinel composite. The spinel particles with $d \approx 0.4$ μm disperse homogeneously among the ZrO_2 grains with $d \approx 0.3$ μm. It should be noted that, before deformation, the ZrO_2-spinel composite has equiaxed grains.

Figure 2 shows the superplastic flow behavior of the ZrO_2-spinel composite at 1773K [4]. As shown in Fig. 2(a), the composite exhibits high-strain-rate superplastic flow of $e_f \geq 200$ % even at $\dot{\varepsilon}_0 \geq 0.02\ s^{-1}$. The spinel dispersion heightens the available strain rate by about 10^2-10^3 times for attaining a similar tensile elongation in monolithic ZrO_2 [6-8].

The stress-strain rate relationship divided into three categories as shown in Fig. 2(b). Although the flow behavior tends to deviate from a straight relationship at high and low stresses, the general trend at the intermediate stress can be characterized by a stress exponent, n, of ≈2.0, which is a typical value for superplastic flow [10].

In order to examine microstructural changes during the high-strain-rate superplastic flow, the representative two strain rates were selected in this study as indicated by arrows in Fig. 2(b). The deformed ZrO_2-spinel composite reveals an interesting microstructural feature as shown in Fig. 3. In most superplastic ceramics, grains retain initial equiaxed shape even after large deformation [10]. Although the ZrO_2 grains retain almost the initial equiaxed shape even after deformation up to $e_f \approx 430\%$, the dispersed spinel particles clearly show anisotropic grain morphology. For the plane perpendicular to the tensile direction, the spinel particles retain equiaxed shape similar to those of the as-sintered specimen [Fig. 1], whereas for the planes parallel to the tensile direction, the particles tend to elongate along the tensile direction. This indicates that the particle elongation occurs only along the tensile direction. Since no elongation in the spinel particles was observed only by static annealing, the elongation must occur through high-strain-rate superplastic flow.

The change in GAR during superplastic flow is plotted against local strain, ε_l, in Fig. 4 [4,11]

The GAR-value is apparently different between the ZrO_2 and spinel grains. For the ZrO_2 grains, although GAR increase gradually with ε_1, the value stays less than ≈1.4. For spinel grains, on the other hand, it rapidly increases with ε_1 up to ≈1.6. In addition, for the ZrO_2 grains, the increment in GAR is independent of the $\dot{\varepsilon}_0$-level, whereas for the spinel particles, the increment becomes pronounce with increasing the $\dot{\varepsilon}_0$-level.

The deformed substructure was examined by TEM. In order to preserve the substructures formed during high-strain-rate superplastic flow, the specimens were deformed up to a desirable true strain of $\varepsilon \approx 0.4$ and cooled rapidly under loading. TEM observation reveals many intragranular dislocations within the spinel grains. An important feature is that the dislocations in the spinel particles tend to lie along the elongated direction as typically shown in Fig. 5. This suggests that the dislocation motion may contribute to the grain elongation. Since those dislocation substructures were not observed before deformation, they must be developed during deformation.

DISCUSSION

Superplastic deformation has generally been characterized by the n-value defined in the following empirical creep equation

$$\dot{\varepsilon} = A\sigma^n, \qquad (1)$$

where $\dot{\varepsilon}$ is the steady-state strain rate, σ is the true stress and A is a constant. For the most ceramics without intergranular amorphous phases, the superplastic flow characterized by $n \approx 2.0$ has been ascribed to grain boundary sliding (GBS) [10,12]. For the present ZrO_2-spinel composite, it is therefore reasonable to explain that the high-strain-rate superplastic flow also occurs primarily through GBS.

GBS is constrained by the surrounding grains at multiple grain junctions. To occur the deformation through GBS, stress concentrations around the grain junctions exerted by GBS should be relaxed. At high-temperatures, diffusion processes along grain boundaries or through grains would, of course, work and thereby those may contribute to the relaxation of the stress concentrations. The existence of intragranular dislocations within the elongated spinel particles, however, suggests that, in the present material, the stress concentrations cannot be fully relaxed only by diffusional processes.

The flow stress is 10-15 times smaller than the yield stresses estimated from the data on spinel single crystals [13]. This suggests that significant stresses higher than the flow stresses should be generated in the ZrO_2-spinel composite during superplastic flow. The activated dislocation motion strongly suggests that the significant stress concentrations must reach a critical stress for dislocation nucleation even at the present loading. The dislocations observed within the elongated spinel grains must contribute to the relaxation of the significant stress concentrations exerted by the predominant GBS mechanism.

The elongation of spinel grains along tensile direction after deformation suggests that grain strain of spinel particles, ε_g, contributes to the total strain, ε_{total}. The ε_g-value is given by the following equation

$$\varepsilon_g = 2/3 \ln(d_{//}/d_{\perp}) , \qquad (2)$$

where $d_{//}$ and $d_{\perp}$ are the grain sizes parallel and perpendicular to the tensile direction, respectively.

For ZrO_2 grains, the contribution of ε_g to ε_{total}, ξ_g^Z ($=\varepsilon_g/\varepsilon_{total}$), stays less than 20%, whereas for spinel grains, ξ_g^S exceeds ≈30%. The ξ_g^S-value tends to increase with an increase in $\dot{\varepsilon}_0$ [4,11]. The difference in the ξ_g-values between the ZrO_2 and spinel grains can be ascribed to a difference in the contribution to the accommodation process. The high ξ_g^S-value suggests that the enhanced accommodation process would significantly take place in the spinel grains rather than in the ZrO_2 grains. Microstructural observation, providing activated dislocations within the elongated spinel grains, suggests that the relaxation process caused primary by dislocation motion would be enhanced in the spinel grains in the present composite.

CONCLUSIONS

The role of fine spinel particle dispersion in high-strain-rate superplasticity was examined in tetragonal ZrO_2. The 30vol% spinel particle dispersion into ZrO_2 heightened the available strain rate by 10^2-10^3 times for attaining similar elongation in monolithic ZrO_2. The enhanced elongation can be attributed to the accelerated stress relaxation of stress concentrations exerted by the predominant GBS mechanism. The dislocation motion and high ξ_g^S-value suggests that the deformable spinel grains would significantly contribute to the enhanced accommodation process of the stress concentrations, leading to the attainment of HSRS in tetragonal ZrO_2.

ACKONOWLEDGEMENTS

The authors are grateful to the Mitsubishi Foundation for supporting the present work.

REFERENCES

1. B.-N. Kim, K. Hiraga, K. Morita and Y. Sakka, Nature **413**, 288 (2001).
2. B.-N. Kim, K. Hiraga, K. Morita, Y. Sakka and T. Yamada, Scripta Mater. **47**, 775 (2002).
3. K. Morita, K. Hiraga and Y. Sakka, J. Am. Ceram. Soc. **85**, 1900 (2002).
4. K. Morita, B. N. Kim, K. Hiraga and Y. Sakka, Mat. Sci. Forum **447-448**, 329 (2004).
5. K. Morita, K. Hiraga, B. N. Kim and Y. Sakka, Mater. Trans., in press (2004).
6. T. G. Nieh, C. M. McNally and J. Wadsworth, Scripta Mater. **24**, 763 (1989).

7. D. J. Schissler, A. H. Chokshi, T. G. Nieh and J. Wadswort, Acta Mater. **39**, 3227 (1991).
8. T. Kondo, Y. Takigawa, Y. Ikuhara and T. Sakuma, Mat. Trans. JIM **39**, 1108 (1998).
9. J. C. Wurst and J. A. Nelson, J. Am. Ceram. Soc. **55**, 109 (1972).
10. T. G. Nieh, J. Wadsworth and O. D. Sherby, "Superplasticity in Metals and Ceramics," (Cambridge University Press, 1997) pp. 91-119.
11. K. Morita, K. Hiraga, B. N. Kim and Y. Sakka, J. Am. Ceram. Soc., to be submitted.
12. M. Jiménez-Melendo, A. Domínguez-Rodríguez and A. Bravo-León, J. Am. Ceram. Soc. **81**, 2761 (1998).
13. T. E. Mitchell, J. Am. Ceram. Soc. **82**, 3305 (1999).

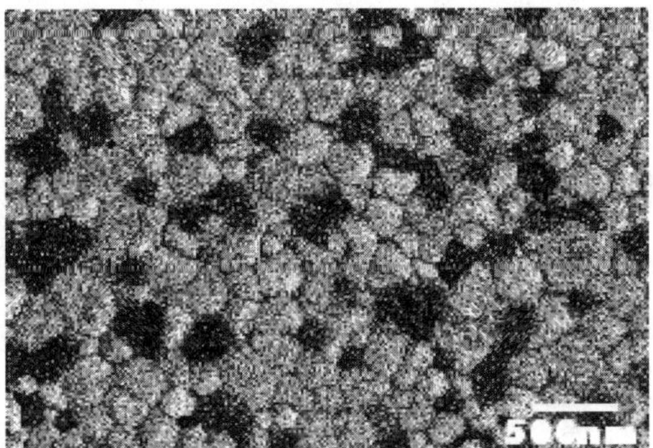

Figure 1. SEM image of the as-sintered ZrO_2-spinel composite.

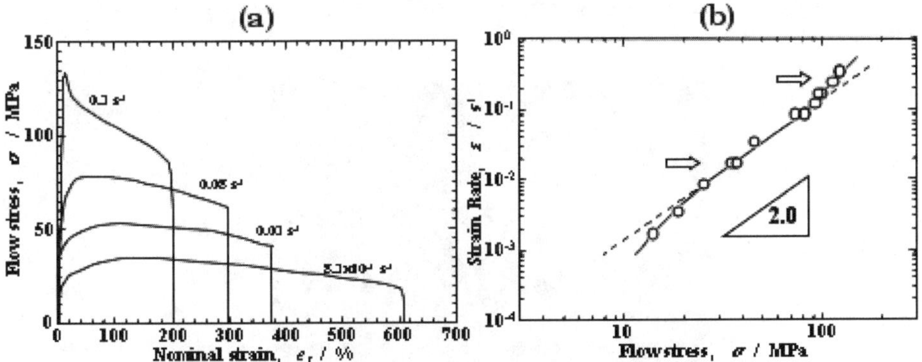

Figure 2. High-strain-rate superplastic flow behavior of the ZrO_2-spinel composite at 1773 K; (a) the stress-strain curves and (b) the flow stress (σ)-strain rate ($\dot{\varepsilon}_0$) relationship [4].

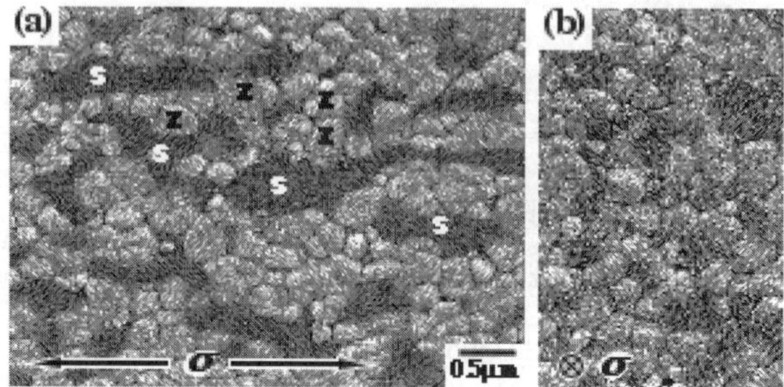

Figure 3. SEM images of a typical microstructure in the directions (a) parallel and (b) perpendicular to the tensile axis after deformation of ≈430% at 1773 K and at $\dot{\varepsilon}_0$ = 0.02 s^{-1} [11].

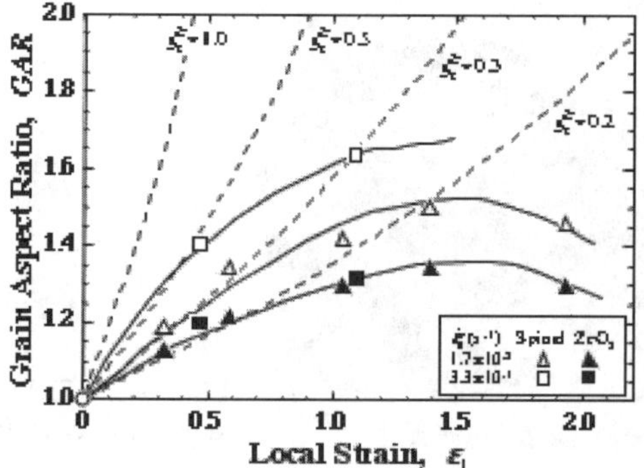

Figure 4. Grain aspect ratio (GAR) plotted as a function of local strain, ε_l, at 1773 K [4,11]. The broken-lines represent GAR expected from the contributions of grain strain to total strain, ξ_g.

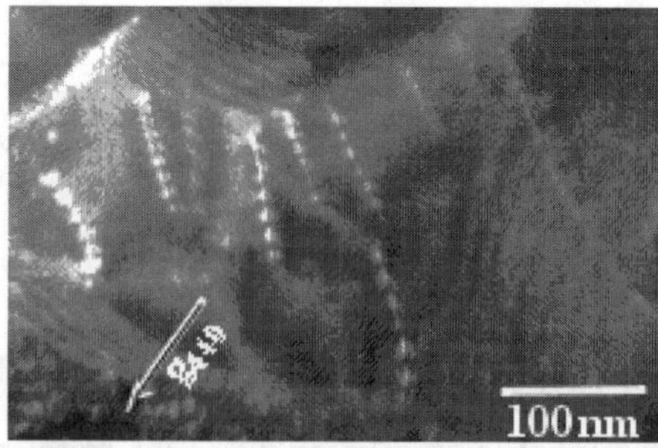

Figure 5. Dark-field TEM image of spinel grains deformed up to ε = 0.2 at 1773 K and at $\dot{\varepsilon}_0$ = 0.02 s^{-1} [4].

B. Modeling

Mat. Res. Soc. Symp. Proc. Vol. 821 © 2004 Materials Research Society P4.2/N4.2

Nanocrystalline fcc metals: bridging experiments with simulations

H. Van Swygenhoven, P. M. Derlet; A. G. Frøseth, S. Van Petegem; Z. Budrovic; A. Hasnaoui

Paul Scherrer Insitute, ASQ/NUM – Computer Modelling and Experiments, PSI-Villigen, Switzerland.

ABSTRACT

Atomistic simulations have provided unprecedented insight into the structural and mechanical properties of nanocrystalline materials, highlighting the role of the non-equilibrium grain boundary structure in both inter- and intra-grains deformation processes. One of the most important results is the capability of the nanosized grain boundary to act as a source and sink for dislocations. However the extrapolation of this knowledge to the experimental regime requires a clear understanding of the temporal and spatial scales of the modelling technique and a detailed structural characterisation of the simulated samples. In this contribution some of the synergies that can be developed between atomistic simulations and experiments for this research field are briefly discussed by means of some typical examples.

INTRODUCTION

The understanding of the mechanical properties of nanocrystalline (nc) materials (with grain sizes less than 100 nm) poses a fundamental challenge to materials science research [1, 2]. With decreasing grain sizes, it is generally believed that there exists a transition from dislocation mediated plasticity within the grains to a plasticity that is primarily accommodated by the GB structure. Many controversial and conflicting results have been reported and despite the extensive efforts over the past decade no complete and consistent picture exists concerning the relation between the geometrical grain boundary (GB) network, including details of GB structure, and overall nc mechanical behaviour.

Tensile deformation studies of nc materials demonstrate a significant increase in strength and decrease in ductility when compared to their coarse grained counterparts, in a way that strongly depends on the synthesis method and the different obtained microstructures [1, 3, 4]. Other typical features characterizing the deformation mechanism of nc-metals is an increased strain rate sensitivity (up to 10 times higher than the value of the coarse grained material but still low compared to what is observed during superplastic deformation) [5], a relatively low activation volume measured by strain rate jump tests, and a fast decrease in the work hardening leading to limited uniform deformation and the onset of instabilities, resulting in shear bands at higher strain rates [3,4].

THE SYNERGY BETWEEN SIMULATION AND EXPERIMENT

Atomistic simulation of nc systems provide a unique and complementary approach to the ongoing experimental investigation of the mechanical and structural properties of nc systems, demonstrating that nc materials cannot be characterized by grain size alone, and that many other structural parameters play an important role in their mechanical behaviour [6, 7]. Such modelling can therefore be very helpful in providing better insight into the structural and mechanical properties of nc metals, especially since the method provides three-dimensional atomic scale resolution allowing direct visualisation of GB and triple junction (TJ) structure, their stored excess energy, the free volume in GBs and TJs, and the internal stress distribution. However a good synergy between experiments and modelling requires a full understanding and careful interpretation of the results obtained by the applied techniques. For simulations, this means firstly a good understanding of the basic caveats of the technique and secondly, as in experiments, a systematic and complete characterization of the resulting simulated GB network.

Atomistic methods carry with them many caveats that must be appreciated to interpret results in the context of experiment. For example to determine the numerical solution to the atomic equations of motion for each atom, a discrete time step of a few femto seconds is typically needed for good convergence. Thus a simulation that runs for one million iterations corresponds to a physical time of only 1ns. Even when using advanced parallel computing, large systems containing millions of atoms can only be realistically simulated of a few nanoseconds. For condensed matter atomistic systems, short and long range mass transport occurs primarily via discrete stochastic atomic activity such as migration via a diffusive hop, nucleation of a dislocation, and atomic shuffling within the GB region. Such activated processes usually occur with a frequency that is much larger than the nano-second range. So to measure significant diffusion activity simulations would have to be performed over much longer time periods. Attempts have been made to overcome this problem by performing simulations at temperatures close to the melting point [8] where diffusion is greatly enhanced, but this in turn introduces concerns since experimentally it is known that nc-metals are thermodynamically unstable at temperatures slightly above room temperature. Therefore, it has to be emphasized, that MD techniques can not catch the rate limiting deformation process seen in experiment and can only be used to gather information on possible deformation mechanisms. Moreover, since no significant structural migration can be observed in the timeframe of a nanosecond, the measured mechanical and structural properties of computer generated nc configurations will depend strongly on the artificial synthesis procedure.

Despite these concerns simulation can give insights into the possible deformation processes and structure that might be found in real nc materials.

OVERVIEW OF COMPUTER GENERATED NC STRUCTURE AND GB STRUCTURE

There exist a number of techniques to construct computer generated nc systems [6, 7]. In the work that follows all nc structures contain a fully three dimensional GB structure according to the Voronoi geometry. For construction details see refs. [9, 10]. As in experiment a complete microstructural characterisation is needed to fully understand the observed deformation properties, especially when comparing the simulated deformation properties between different simulation groups [8, 11, 12].

To fully characterise the resulting GB structures many different criteria can and should be used. For example a medium range order analysis allowing the determination of a local crystallinity class [13], positional disorder to establish those atoms which cannot be assigned to a particular lattice site of the neighbouring fcc grains [10], local atomic potential energy and also local stress at the scale of the GB [6]. Such detailed visualisation can simply not be obtained via experimental methods and this is one of the strong virtues of simulation, which can be exploited further by the simulation of experimental characterisation techniques such as the various spectroscopies involving photons, neutrons, phonons and positrons. Moreover with the combined knowledge gained from the static structural classification of the GB and GB network, the simulation of experimental probes can result in a much more fundamental understanding of how and to what extent these spectroscopies probe the nc GB structure. In what follows, we briefly outline applications of this approach for the interaction of xrays, phonons and positrons with computer generated nc systems.

Xrays

For an atomic configuration consisting of N atoms, a (Debye) powder X-ray spectrum can be calculated via [14],

$$I(k) = |f|^2 \left(N + \sum_{i \neq j}^{N} \frac{\sin kr_{ij}}{kr_{ij}} \right) , \qquad (1)$$

where r_{ij} represents the distance between atoms i and j, f is the atomic form factor (assumed to be the same for each atom and in the present work independent of the scattering vector), and k is the scattering vector magnitude. Equation 1 is the spherically averaged structure factor of the atomic configuration and involves a double summation over all atoms.

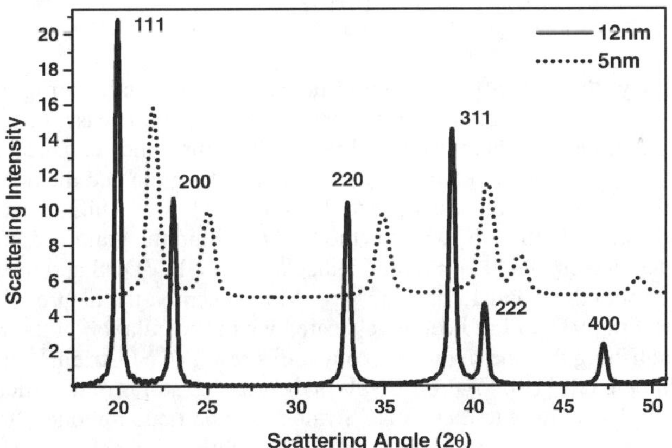

Figure 1: Numerical X-ray spectra of 12 and 5 nm average grain size computer generated nc samples. The 5nm grain size spectrum has been artificially shifted in both angle and intensity for better viewing.

Figure 1 displays the two-theta spectrum for two computer generated nc samples, one containing 15 grains with a computed average grain size of 12 nm and another containing 125 grains with a computed average grain size of 5 nm [15]. The wavelength for which this was calculated corresponds to 17.5 KeV synchrotron photons, which has been used to measure in situ deformation peak profiles of nc systems [16]. An immediate observation is that there is increased broadening for the smaller grain size sample. The integral widths of the hhh and h00 family of peaks were determined by fitting each peak to a PearsonVII function, from which a Williamson-Hall analysis assuming a Cauchy-Gaussian convolution was performed (shown in table 1). We see that the resulting grain sizes are in good agreement with the 12 nm and 5 nm values determined geometrically from the Voronoi construction. The resulting inhomogeneous strains are in a range typically seen for real nc systems.

Sample	D(nm)	$\overline{\delta\varepsilon^2}$ (%)
5 nm	5.3±0.2	0.34±0.25
12 nm	10.1±0.2	0.10±0.06

Table 1: The results of a Williamson-Hall analysis of the simulated Xray spectra

Phonons

The phonon properties of nc systems have been extensively investigated experimentally via inelastic thermal neutron scattering, which allows for the determination of the phonon (or vibrational) density of states (VDOS). For nc systems it has been established that there is both low and high frequency anomalous behaviour when compared to the perfect crystalline VDOS. The VDOS of a computer generated nc system can be determined in a number of ways. For example, the VDOS is the Fourier transform of the atomic velocity-velocity auto-correlation function that can be determined via the ensemble average of

$$\left\langle \tilde{v}(0)\cdot\tilde{v}(t)\right\rangle = \frac{1}{N}\sum_i \tilde{v}_i(0)\cdot\tilde{v}_i(t) \ . \qquad (2)$$

Figure 2a displays the VDOS for the same 5 nm and 12 nm atomic configurations introduced in the previous section, as well as that for the perfect fcc crystalline case. Inspection reveals that with decreasing grain size, there is both a high and low frequency enhancement. By considering the partial VDOSs arising from fcc atoms constituting the grain and the non-fcc atoms constituting the GB, it has been established that both the low and high frequency enhancement arises from the atoms in the GB, whereas the VDOS within the grain is essentially the same as that for a perfect fcc crystal [17]. Figure 2b displays the GB VDOS and the grain VDOS (which is the same for both the 5 and 12 nm samples). More recently, the nature of the high frequency enhancement of the VDOS has been investigated via the calculation of the phonon Green's function, establishing that the high-frequency tail is result of a distribution of localised modes forming within the GB [18]. Figure 3 displays the non-fcc atoms of a particular GB, where the atoms are shaded according to there VDOS value for a particular phonon frequency [19]. Inspection of the free volume indicated by positron annihilation calculations (see section below) reveals that there exist localised modes in the vicinity of the free volume and more generally in regions of reduced coordination. This can be understood from the unstaturated nature of the metallic bond where a reduced coordination stiffens the force matrix of the remaining bonds, an effect which is known to lead to localised modes.

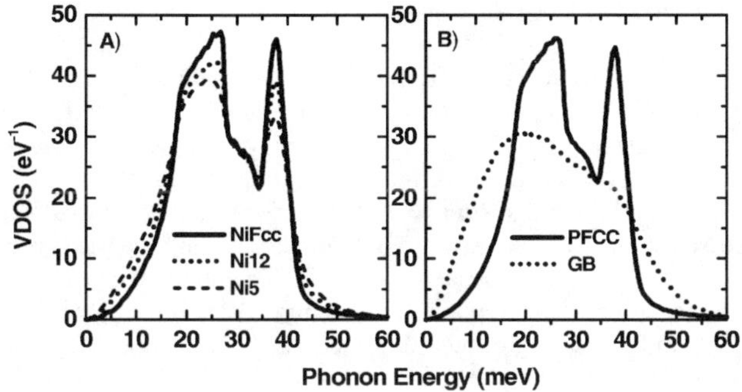

Figure 2: A) Phonon density of states for a 5nm and 12nm nc system, and also that for a perfect fcc structure. B) The partial density of states for the GB and the interior of the grains.

Positrons

The free volume content of an nc system can significantly effect the structural and mechanical properties of the material. Even for samples that have been produced with methods that allow the synthesis of fully dense materials (as determined by the Archimedes principle), free volume defects of sizes of the order of 1-2 nm or less are measured via positron lifetime spectroscopy – suggesting that such small free volume defects are an intrinsic property of the nanosized grain boundaries [20]. Although computer generated samples do not *a priori* contain voids of 5-20 vacancies, they do contain free volume of the order of a vacancy. For computer generated nc samples, this can be identified via the calculation of the positron wave function in the GB region and the determination of the corresponding trapped positron lifetime [21]. Figure 3 displays a grain boundary region for which the positron wave function has been calculated. The shaded region indicates a free volume region where the positron wave function peaks in value – corresponding to a positron lifetime of 150 psecs, which is slightly less than the 160 psecs normally seen for a Ni fcc lattice vacancy.

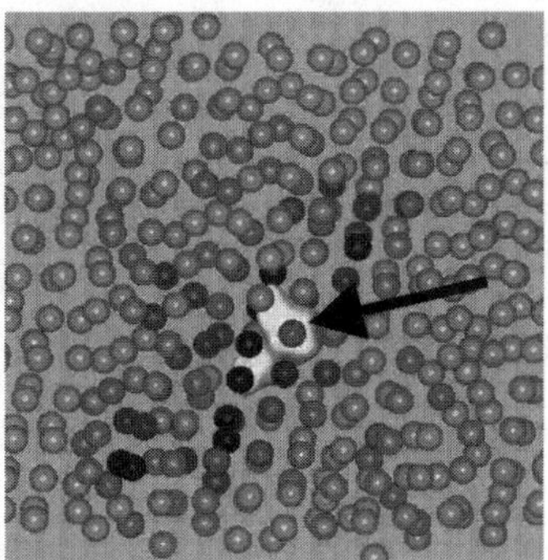

Figure 3: Non-fcc atoms belonging to a GB. Atoms are shaded according to their local vibrational oscillator strength where black represents atoms with oscillator strengths above 10^4 eV^{-1} and white less than 1 eV^{-1}. The light region (pointed to by an arrow) represents a peak in the positron wave function indicating free volume.

OVERVIEW OF DEFORMATION MECHANISMS

In all samples with mean grain sizes up to 20 nm, grain boundary sliding (GBS) and dislocation activity have been observed as plastic deformation mechanisms, the latter being of increasing importance with increasing grain size. Careful analysis of the GB structure during sliding under constant tensile load shows that GBS includes a significant amount of discrete atomic activity through uncorrelated shuffling and stress assisted free volume migration [22]. More recently, collective processes such as cooperative grain boundary sliding via the formation of shear planes spanning several grains have been observed [23, 24]. The underlying mechanisms that were observed for the formation of shear planes are (1) pure GBS induced migration of parallel and perpendicular GBs to form a single shear interface, (2) coalescence of neighbouring grains that form a low angle GB facilitated by the propagation of Shockley partials and (3) continuity of the shear plane by intragranular slip.

In fully three dimensional Ni GB networks, which have been modelled now up to 30 nm grain sizes, only partial dislocations that extend from GB to GB have been observed [22, 25], in which the partial is found to nucleate at regions of misfit in the GB [26, 27]. Extended partial dislocations have also been observed in nc Cu up to grain sizes of 50 nm [12], whereas in nc-Al, primarily full dislocations are emitted from the GBs at these grain sizes [11, 28]. The latter has been explained via the assumption that it is the splitting distance (determined solely by the stable stacking fault energy) between the leading and trailing partials and its relation to grain size that determines whether or not full or partial dislocations are emitted [29]. However this theory can not explain why extended partials are observed for both Ni and Cu, one having a higher and the other, a lower stable stacking fault energy. The type of slip activity as a function of fcc material can be understood from the generalised planar defect curves in which, both the stable (γ_{sf}) and unstable stacking (γ_{usf}) fault energies play the crucial role in determining whether or not full or partial dislocations are seen within the simulation time [30]. Following a Peierls model for dislocation nucleation, it is the unstable stacking fault energy which forms the energy barrier for the nucleation of the leading partial and it is the difference between the stable and unstable stacking fault energies that forms the energy barrier for the nucleation of the trailing partial. Figure 4 displays the generalised planar fault energy curves for the used Al, Ni and Cu empirical potentials (see ref. [30]), and we see that once the first partial is created, the energy barrier for the second to be nucleated is small for Al but much larger for Ni and Cu. This explains why simulations that span only very short time intervals reveal full dislocation activity for Al and extended partial activity for both Ni and Cu [30]. Quantitatively the nucleation rates will also depend on the applied and internal stresses, GB structure, and the stress field of the already nucleated leading partial.

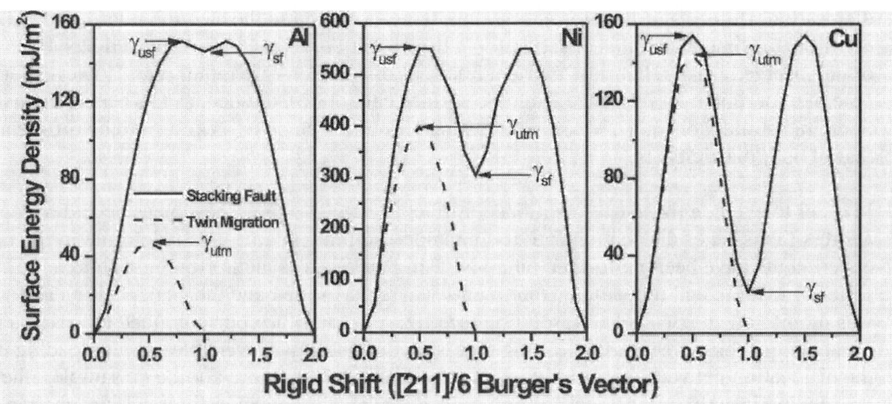

Figure 4: Generalised planar default energy curves for twin boundary migration, and partial and full dislocation slip activity for Al, Ni and Cu.

Such an approach has also been used to understand the enhanced deformation properties of nc-Al with grown-in twins [28], where it is shown via the generalised planar defect energy curves that the energy barrier to twin boundary migration (γ_{utm}) is significantly lower than that of partial dislocation nucleation for the empirical potentials used to model Al. The generalised twin migration curves are also shown in figure 4. Figure 5 displays a cross-section of a pre-twinned nc-Al configuration under loading. The circled region indicates twin boundary migration in progress, where a partial dislocation nucleates at the intersection of a twin boundary and GB, and propagates on a (111) plane adjacent to the pre-existing twin, resulting in the latter migrating to this adjacent plane. Similar simulations for nc-Cu and nc-Ni, revealed not such a strong preference for twin boundary migration over partial dislocation nucleation. Inspection of the generalised planar defect curves of Cu and Ni confirmed that in both cases the energy barrier for partial dislocation nucleation and twin boundary migration is not so different (see figure 4) and therefore there exists no strong preference for twin boundary migration [31].

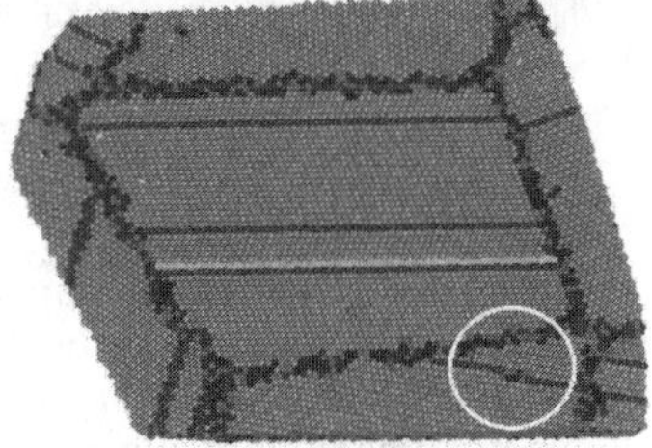

Figure 5: Section of the pre-twinned nc-Al sample. Twin planes are indicated by the darkened (111) planes. The white plane represents the original (111) plane of the adjacent migrated twin, and the circled region displays a twin migration as it happens.

Assuming that the observation of extended partial dislocations extending over the entire grain cross section is an artifact from the short time span typical for an atomistic simulation, the most important suggestion from the simulations is that the GB acts as a source and sink for lattice dislocations during plastic deformation. In-situ deformation in the electron microscope [32, 33] confirms to some extent this observation in the vicinity of a crack front, since dislocations are often observed in-situ, but post-mortem traces of dislocations are rather rare. However for this type of experiment an influence from the surface cannot be excluded, as has been demonstrated in atomic simulations [34]. Moreover plasticity around a crack tip is known to be different from that of bulk plasticity.

In order to exclude the above mentioned artifacts a new in-situ approach has been developed to address the issue of dislocation activity in nanocrystalline metals. In situ deformation X-ray experiments have been performed on nc-Ni using the Swiss synchrotron light source [16]. A series of loading and unloading experiments where performed and the peak position and width was monitored at a 30 second resolution (see figure 6). For the course grained material, a peak broadening is observed during loading that is predominantly irreversible upon unloading due to the dislocation network that is built up during plastic deformation. For the nc samples, the peak broadening is entirely recoverable upon unloading. This is shown in figure 6b. Inspection of the peak shift as a function of strain, reveals a linear dependency during elastic deformation that saturates soon after yield (see figure 6c), where as the peak width constantly increases indicating that a significant portion of the broadening is due to plasticity with the rest being due to elastic anisotropic broadening. For more details of this experiment, we refer to [16].

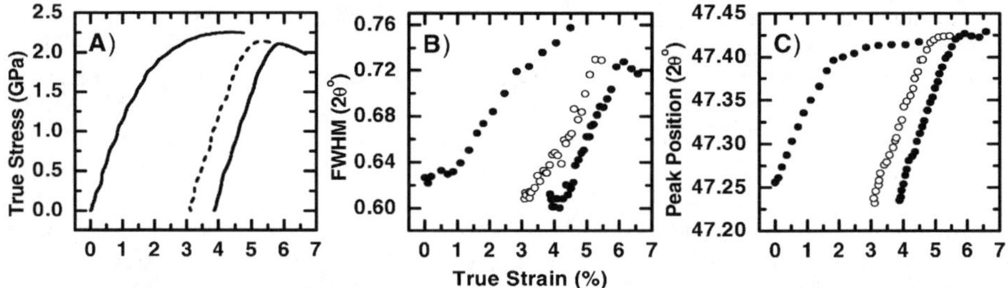

Figure 6: A) Stress-Strain Curve of three loading/unloading cycles, with corresponding 400 B) peak width and C) peak position as a function of strain.

CONCLUSIONS

Powerful synergies can be developed between atomistic simulations and experiments that address both the structural and mechanical properties of nc systems. In order to fully benefit from this synergy, the results have to be evaluated with care, taking into account the possible artifacts due to the high-stress/short time characteristics of simulations.

The central discovery of atomistic simulation of fcc nc materials has been that in the high-strain rate regime, there is a transition with increasing grain size, from an inter-grain plastic deformation mechanism to a mixture of inter- and intra-grain mechanisms where the latter takes the form of dislocation nucleation/annihilation at GBs and propagation through the grains. Simulations suggest the possibility of partial dislocations, full dislocations or twinning; the activity that will be observed within the simulation time can be determined by the generalized planar defect energy curves, but also by the GB structures and GB network since they will determine the criteria that have to be reached in the GB in order to nucleate a particular dislocation activity. This subject has not been fully addressed until now and needs, in addition to the issue of GB structure, also other aspects such as free volume and impurities to be considered.

Regardless of the quality of the atomistic simulations, their interpretation and understanding should be considered primarily as a guidance to explain experimental results or to develop new experimental approaches, as has been successfully demonstrated by the development of the in-situ deformation X-ray diffraction peak profile analysis technique.

REFERENCES

1. J. R. Weertman, *Mechanical behaviour of nanocrystalline metals.* In: Nanostructured materials: processing, properties, and potential applications. Norwich (NY): William Andrew Publishing; 2002, Chapter 10 (2002).
2. K. S. Kumar, H. Van Swygenhoven and S. Suresh., *Acta Mater.* **51**, 5743 (2003).
3. F. Dalla Torre, H. Van Swygenhoven and M. Victoria, *Acta Mater.* **50**, 3957 (2002).
4. F. Dalla Torre, Ph.D. Thesis entitled "Microstructure and Mechanical Properties of Nanocrystalline Ni produced by Three Different Synthesis Techniques", EPFL, Switzerland, 2003.
5. Y. M. Wang and E. Ma, *Acta Mater.* **52**, 1699 (2004).
6. H. Van Swygenhoven, P. M. Derlet, Z. Budrovic and A. Hasnaoui, *Z. Metallkd.* **10**, 1106 (2003).
7. P. M. Derlet, A. Hasnaoui and H. Van Swygenhoven, *Scripta Mater.* **49**, 629 (2003).
8. V. Yamakov, D. Wolf, S. R. Phillpot and H. Gleiter, *Acta. Mater.* **40**, 61 (2002).
9. H. Van Swygenhoven, D. Farkas and A. Caro, *Phys. Rev. B* **62**, 831 (2000).
10. P. M. Derlet and H. Van Swygenhoven, *Phys. Rev. B* **67**, 014202 (2003).
11. V. Yamakov, D. Wolf, S. R. Phillpot, A. K. Mukherjee and H. Gleiter, *Phil. Mag. Lett.* **83**, 385 (2003).
12. J. Schiøtz and K. W. Jacobsen, *Science* **301**, 1357 (2003).
13. D. J. Honeycutt and H. C. Andersen, *J. Phys. Chem.* **91**, 4950 (1987).
14. B. E.Warren, X-ray Diffraction, (Addison-Wesley, Massachusetts, 1969), Chapter 1.
15. P. M. Derlet, Z. Budrovic and H. Van Swygenhoven, in preparation.
16. Z. Budrovic, H. Van Swygenhoven, P. M. Derlet, S. Van Petegem and B. Schmidt, *Science* **304**, 273 (2004).
17. P. M. Derlet, R. Meyer, L. J. Lewis, U. Stuhr, and H. Van Swygenhoven, *Phys. Rev. Lett.* **87**, 205501 (2001).
18. P. M. Derlet and H. Van Swygenhoven, *Phys. Rev. Lett.* **92**, 035505 (2004).
19. P. M. Derlet, S. Van Petegem and H. Van Swygenhoven, in preparation.
20. S. Van Petegem, F. Dalla Torre, D. Segers and H. Van Swygenhoven, *Scripta Mater.* **48**, 17 (2003).
21. Van Petegem S, Positron annihilation study of nanocrystallinematerials, Doctoral Thesis, University of Ghent, 2003.
22. H. Van Swygenhoven and P. M. Derlet, *Phys. Rev. B* **64**, 224105 (2001).
23. A. Hasnaoui, H. Van Swygenhoven and P. M. Derlet, *Phys. Rev. B* **66**, 184112 (2002).
24. A. Hasnaoui, H. Van Swygenhoven and P. M. Derlet, *Science* **300**, 1550 (2003).
25. M. J. Caturla, H. Van Swygenhoven and P. M. Derlet, unpublished
26. H. Van Swygenhoven, P. M. Derlet and A. Hasnaoui, *Phys. Rev. B* **66**, 024101 (2002).

27. P. M. Derlet, H. Van Swygenhoven and A. Hasnaoui, *Phil. Mag.* **83**, 3569 (2003).
28. A. G. Frøseth, H. Van Swygenhoven and P. M. Derlet, *Acta Mater.* **52**, 2259 (2004).
29. V. Yamakov, D. Wolf, S. R. Phillpot, A. K. Mukherjee and H. Gleiter, *Nature Mater.* **3**, 43 (2004).
30. H. Van Swygenhoven, P. M. Derlet and A. Frøseth, *Nature Mater.* in press, (2004).
31. A. G. Frøseth, P. M. Derlet and H. Van Swygenhoven, submitted (2004).
32. R. C. Hugo, H. Kung, J. R. Weertman, R. Mitra, J. A. Knapp and D. M. Follstaedt, *Acta Mater.,* **51**, 1937 (2003).
33. K. S. Kumar, S. Suresh, M. F. Chisholm, J. A. Horton, P. Wang, *Acta Mater.* **51**, 387 (2003).
34. P. M. Derlet and H. Van Swygenhoven, *Phil. Mag. A.* **82**, 1 (2002).

Mat. Res. Soc. Symp. Proc. Vol. 821 © 2004 Materials Research Society P8.2

Analysis of a one-billion atom simulation of work-hardening in ductile materials

Markus J. Buehler[1], Alexander Hartmaier[1], Mark Duchaineau[2], Farid F. Abraham[3] and Huajian Gao[1]
[1]Max Planck Institute for Metals Research, 70569 Stuttgart, Germany
[2]Lawrence Livermore National Laboratory, Livermore, CA 94550-9234, USA.
[3]IBM Almaden Research Center, San Jose, CA 95120-6099, USA

ABSTRACT

We analyze a large-scale molecular dynamics simulation of work hardening in a ductile model material comprising of 500 million atoms interacting with a Lennard-Jones pair potential within a classical molecular dynamics scheme. With tensile loading, we observe emission of thousands of dislocations from two sharp cracks. The dislocations interact in a complex way, revealing three fundamental mechanisms of work-hardening. These are (1) dislocation cutting processes, jog formation and generation of point defects; (2) activation of secondary slip systems by cross-slip; and (3) formation of sessile Lomer-Cottrell locks. The dislocations self-organize into a complex sessile defect topology. Our analysis illustrates mechanisms formerly only known from textbooks and observed indirectly in experiment. It is the first time that such a rich set of fundamental phenomena has been seen in a single computer simulation.

INTRODUCTION

The plastic or non-reversible deformation of materials occurs immediately after a regime of recoverable elastic deformation and is governed by the nucleation and motion of defects in the crystal lattice [1-3]. In experiment, scientists often rely on indirect techniques to investigate the crystallographic structure of defects. In theory, predictions are primarily based on continuum theory with phenomenological assumptions. While the continuum description has been very successful in the past, some key features of plasticity can only be understood when the atomistic viewpoint is adapted. This may be achieved using molecular dynamics (MD) simulation [4-9]. However, most MD studies consider a small number of dislocations and are limited to specific dislocation mechanisms [9, 10]. Here we present the analysis of ultra-large scale simulations of work hardening in system sizes of 500,000,000 and 1,000,000,000 atoms in a rare-gas atom FCC crystal modelled by a LJ potential fitted to copper within a classical MD framework.

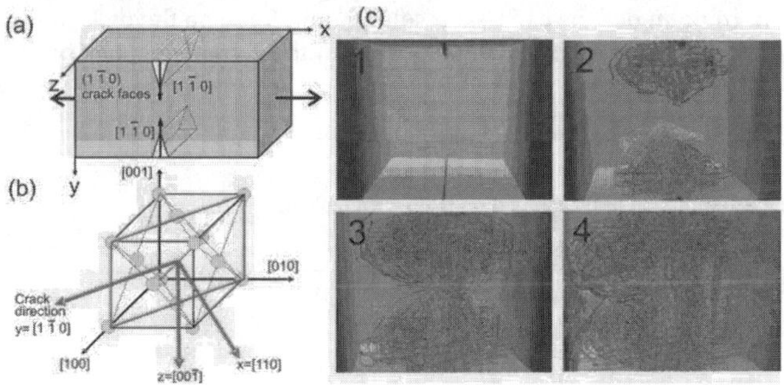

Figure 1: Simulation geometry (a), lattice orientation (b) and time-sequence of the work-hardening simulation (c). The snapshots present an overview over the total simulation and show build-up of a complex, entangled dislocation network.

MODELING AND ANALYSIS TECHNIQUES

The simulation geometry, lattice orientation and coordinate system are presented in Fig. 1a and b. We apply tensile mode I loading in the x-direction as indicated in the figure. Dislocations are visualized based on the energy method, that is, only atoms with energy a few percent higher than the bulk are displayed. An alternative method we use is the centrosymmetry method that allows distinguishing stacking faults in addition. Both techniques do not provide quantitative information about the Burgers vector and the line direction. This information is extracted using geometrical methods (by doing a Burgers circuit in the crystal).

SIMULATION RESULTS

Upon application of loading to the system, the cracks serve as fertile sources for dislocations and we observe nucleation of dislocations from the crack tips. Within a few picoseconds, thousands of dislocations appear and flow into the interior of the solid. The dislocations from the cracks glide on two primary glide planes (111) and $(11\bar{1})$, in agreement with the fact that dislocations are nucleated in the direction of largest shear stress and thus largest Schmid factor in the singular stress field of the cracks [1, 2]. We only observe the nucleation of partial dislocations because of the very small stacking fault energy in our ductile model material [1, 4, 6]. The Burgers vectors of the emitted partials are predominantly $[\bar{2}11]$ and $[1\bar{2}1]$ on the (111) glide plane, as well as $[\bar{2}1\bar{1}]$ and $[1\bar{2}\bar{1}]$ on the $(11\bar{1})$ glide plane[1]. Figure 1c shows a time-sequence of the defect structure evolution over the whole simulation time based on the energy method.

Dislocation cutting processes

When two screw dislocations intersect, it is commonly known that each acquires a jog with a direction and length equal to the Burgers vector of the other dislocation [1, 2]. Upon intersection, the dislocations cannot glide conservatively since each jog has a sessile edge segment. However, if the applied stress is large, the dislocation will glide, and the moving jogs will leave a trail of vacancies, or a trail of interstitials depending on the line orientation and the Burgers vector of the reacting dislocations [1, 2]. Generation of trails of interstitial atoms is energetically very expensive and not observed frequently [1, 2]. Another possible mechanism of dislocation cutting is when a partial dislocation moves through the stacking fault generated by another partial dislocation. Intersection of a partial dislocation with the stacking fault of another dislocation is observed in early stages when dislocations within the same cloud intersect the stacking fault generated by other dislocations. The dislocation line forms a jog that features a sessile edge component. Creation of a trail of point defects causes a resulting drag force on the dislocation that result in a bowing out of the segment. Because of the values of the Burgers vectors of the

---

[1] We encourage the reader to view an animation of the simulation that is available at:

http://www.almaden.ibm.com/st/computational_science/msmp/fractures/ductile/frost_disl3med.avi.

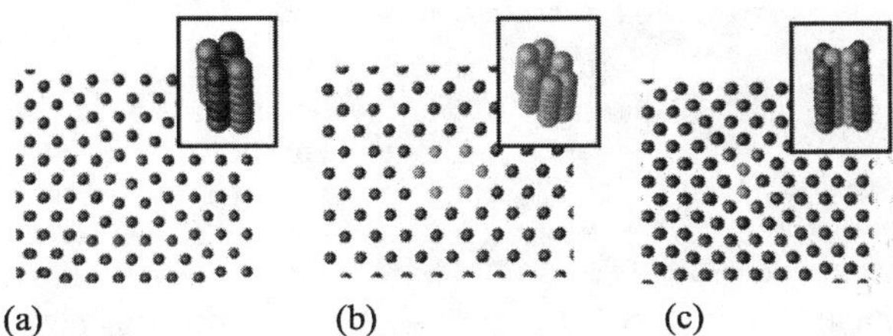

Figure 2: Panels (a)-(c) show atomistic simulation results of different types of point defects: (a) Trail of partial point defects, (b) vacancy tube, and (c) trail of interstitials.

intersecting dislocations (as determined by geometrical analysis), and the line orientation, the screw component of the intersecting dislocations is small, only $\sqrt{6}/12a_0$ (half the length of the Burgers vector of a partial dislocation) [1, 2]. Therefore, not a complete point defect is generated but rather a trail of local lattice distortion, a phenomenon we will refer to as a *trail of partial point defects*. The defect features a dipole structure as depicted in Fig. 2a. From the energy per unit length of the trail of partial point defects and comparing to complete vacancy tubes, the dragging force is estimated to be 20 % of that of generation of complete vacancy tubes. The defect thus causes a significant dragging force on the dislocations, in particular when the dislocation density is as large as in the present simulation. The forces on the jog segment are a combination of tensile and compressive forces in the two <112> directions, explaining the dipole structure in Fig. 2a. The bowing effect on the dislocations is shown in Fig. 3a based on an energy analysis. To our knowledge, the trail of partial point defects has not been described previously. Since generation of trails of partial point defects appears in reactions of partials only, such deformation mode could become important in nanostructured materials (as there is evidence that partial dislocations dominate plastic deformation in such materials [7]).

Thus far we have discussed the reactions that take place when dislocations originating from the same crack react. Numerous dislocation reactions occur when dislocations of the two different clouds start to interact, as shown in Fig. 3b. Due to the Burgers vectors and the dislocation line orientation, when the first dislocations nucleated at different cracks meet straight ahead of the crack, the reactions are very similar to those observed in the previous stages when dislocations of the same clouds cut each other's stacking fault and thus, trails of partial point defects are generated [1, 2]. In Fig. 3b, the significant effect of jog dragging on the motion of the dislocations is clearly observed. The elastic interaction of dislocations repelling each other causes a decrease in the dislocation velocity.

Due to the crystal orientation and the Burgers vector of the dislocations, only trails of partial

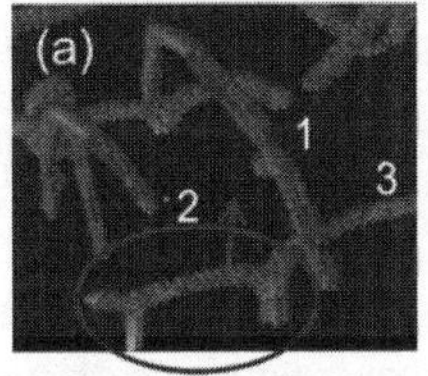

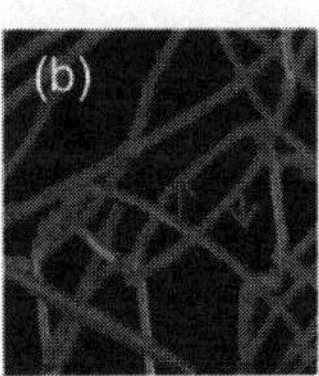

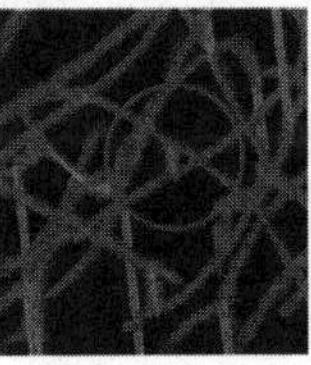
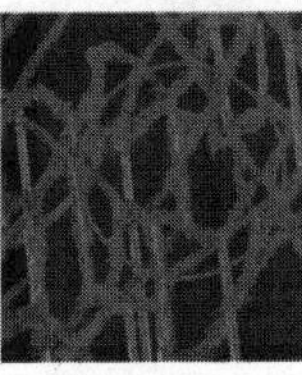

Figure 3: (a) Dislocations #1 and #2 leave behind a stacking fault plane, which is subsequently cut by dislocation #3. Two trails of partial point defects are generated resulting in bowing of dislocation #3. Panel (b) shows the reaction of the two dislocation clouds originating from opposing crack tip causing generation of numerous point defects.

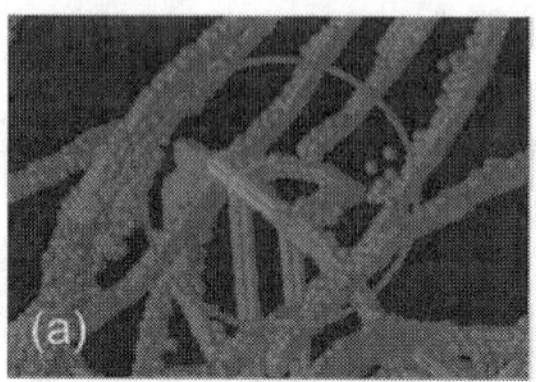

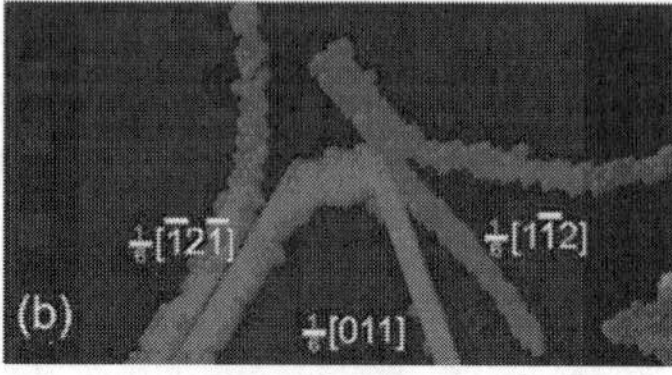

Figure 4: Plot (a) shows details of activation of secondary slip systems. Plot (b) shows a detailed view on the formation of sessile Lomer-Cottrell locks with its typical shape of a straight sessile arm connected to two partial dislocations.

point defects as well as interstitials can be generated from the cutting processes on the primary glide systems. Complete vacancy tubes are not generated until later stages of the simulation when the dislocation density becomes very large and secondary slip systems are activated. A large number of such defects is created and appears as straight, thick lines in the plot of the potential energy of atoms. The geometry of complete vacancy tubes is shown in Fig. 2b. The vacancy tubes observed in the simulation are only several nanometres long. We also observe formation of some trails of interstitial atoms. The number of such processes is rather small since the energy to create such defects is extremely large [1, 2, 12]. A trail of interstitials is depicted in Fig. 2c.

The observation of generation of vacancies is in agreement with the understanding that heavy plastic deformation causes generation of vacancies. Mott [12] was the first to predict the vacancy formation by motion of jogged screw dislocations. The production of vacancies and their influence on plasticity has been studied in experiment and theory [8, 12] has also been included in discrete dislocation modelling recently [11]. The excess of vacancy concentration is also found in experiments of materials under heavy plastic deformation [1, 8, 11, 12].

Activation of secondary slip planes

We observe that secondary slip systems are activated once the dislocation density is above a critical value. Figures 4a illustrates activation of secondary $(\bar{1}1\bar{1})$ and $(1\bar{1}1)$ slip systems. We find that dislocations on secondary slip systems are generated by cross slip mechanisms. This is unexpected because cross slip is only possible along constricted screw segments [1, 2]. Due to the low stacking fault energy of the model material used here, only partial dislocations exist in our simulation. Thus, each occurring cross-slip event leaves behind a straight, stair-rod dislocation to conserve the Burgers vector. The sessile segment can clearly be observed in Fig. 4a. The main result is that by this mechanism it is possible to observe cross-slip in a situation when only partial dislocations, together with high stresses are present. We observe formation of complete vacancy tubes when secondary slip systems are activated.

Formation of sessile locks

Another important mechanism of dislocation interactions is formation of Lomer-Cottrell locks. Formation of sessile locks is not observed until dislocations on secondary glide planes are activated. Such locks provide a serious burden for further dislocation glide through the material, since other dislocations approaching the sessile dislocations cannot glide through this defect agglomerate [1, 2]. In Fig. 4b we show a snapshot when such defects are generated. The typical structure of the Lomer-Cottrell lock is characterized by a straight sessile arm connected with two partials on different glide planes [1, 2]. This can be verified in Fig. 4.

The work-hardening regime

The simulation reveals a final sessile structure of a large number of dislocations in the late stages of the plastic deformation. When a situation is reached where the plastic deformation of a solid has generated such a high dislocation density that dislocation motion is hindered by their mutual interactions, one generally speaks of the work-hardening regime of deformation. In the MD simulation the deformation is large enough that this work-hardening regime is reached. In this stage, we see a structure composed of point defects, sessile dislocations and partial dislocations. Several views onto this final stage are shown in Fig. 5a. The blow-up in Fig. 5a shows a more detailed analysis of the network. The characteristic structure of the network is due to the fact that all sessile defects (trails of point defects as well as $\frac{1}{6}[110]$ -sessile dislocations as part of the Lomer-Cottrell locks assume tetrahedral angles and lie on the edges of Thompson's tetrahedron. This immediately explains the particular structure of the sessile network that is characterized by straight defect segments under tetrahedral angles (sessile dislocations, and trails of point defects) as well as wiggly partial dislocations. In contrast to the first stages of plastic deformation, where dislocation glide occurred easily, dislocation motion is essentially stopped due to the work hardening and plastic relaxation.

DISCUSSION AND SUMMARY

In this work we were able to show the fundamental, classical mechanisms of hardening [1-3] in computer simulations: (1) dislocation cutting processes, (2) cross-slip, and (3) formation of sessile dislocation locks. A large ensemble of defects self-organizes into a complex defect network with a regular structure leading to a final defect network composed out of trails of partial point defects and complete vacancy tubes as well as some trails of interstitials, sessile dislocations, and partial dislocations. The characteristic structure of the final network is given by the geometrical condition that the sessile defects appear as straight lines lying at the intersection of stacking fault planes, thus along the sides of Thompson's tetrahedron. An interesting observation is the observation of cross-slip of partial dislocations: This mechanism was proposed theoretically almost 50 years ago by Fleischer [13] and is here observed for the first time in MD simulation.

Figure 5b plots the development of the density of different defects during the simulation. The different defects are distinguished based on the different energies associated with them. At the first stage, partial dislocations are emitted from cracks with a high rate. As a consequence of cutting processes trails of point defects are produced. Finally, the activation of secondary glide systems results in additional dislocation cutting processes including the generation of complete vacancy tubes. Another important hardening mechanism is formation of sessile locks. The rapid production of partial dislocation ceases, with the creation of cutting debris (and the plastic relaxation), and the density of partial dislocation finally goes into an equilibrium.

The analysis validates and underlines the power of continuum mechanics theories [1, 2] in predicting the properties and interactions of topological defects in lattices, as the observed effects can all be explained by the classical theories. We emphasize that such detailed atomistic views

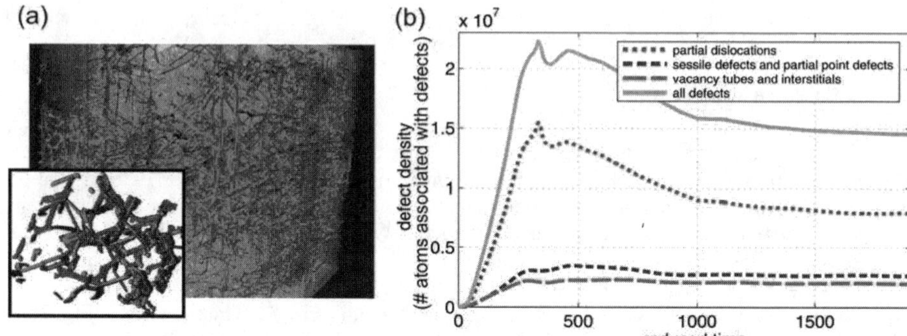

Figure 5: The final network from a distant view (a), including a blow-up to show the details of the network. Subplot (b): Development of the density of different defects during the simulation (based on an energy method).

on fundamental aspects of plasticity, as shown in our study, can be obtained neither from experimental techniques, nor can they currently be calculated with continuum mechanics methods. It is apparent that MD simulations are a unique way to obtain such information. This is the first time that such a rich set of fundamental phenomena of work-hardening is being seen within a single computer simulation. Due to page limitation, we are not able to report all the details of the analysis in the paper and leave more details to a full length paper [6].

ACKNOWLEDGEMENTS

FFA wishes to acknowledge the generous support provided by the Humboldt Research Award for Senior US Scientists and a computer grant at the San Diego Supercomputer Center. We also acknowledge very helpful discussion with Dr. V. Bulatov from LLNL.

REFERENCES

1. J.P. Hirth and J. Lothe. *Theory of Dislocations* (Wiley-Interscience, New York, 1982)
2. D. Hull and D.J. Bacon. *Introduction to Dislocations* (Butterworth Heinemann, 2002)
3. T.H. Courtney. *Mechanical Behavior of Materials* (McGraw-Hill, New York, 1990)
4. F.F. Abraham, R. et al. Proc. Natl. Acad. Sci. USA **99** (2002) 5783.
5. M.J. Buehler, F.F. Abraham and H. Gao. Nature **441** (2003) 141.
6. M.J. Buehler, A. Hartmaier, M. Duchaineau, F.F. Abraham and H. Gao. Under submission.
7. D. Wolf, S.R. Phillpot, and A.K. Mukherjee. Z. Metallk. **94** (2003) 1052.
8. J. Schiøtz, T. Leffers, and B.N. Singh. Phil. Mag. Lett. **81** (2001) 301.
9. S.J. Zhou, D.L. Preston, P.S. Lomdahl, and D.M. Beazley. Science **279** (1998) 1525.
10. V. Bulatov, F.F. Abraham, L. Kubin, B. Devincre, and S. Yip. Nature **391** (1998) 669.
11. M. Rhee, H.M. Zbib, J.P. Hirth, H. Huang, and T.D. de la Rubia. Modeling Simul. Mater. Sci. Eng. **6** (1998) 467.
12. N. F. Mott. Trans. Met. Soc. AIME **218** (1960) 962
13. R.L. Fleischer. Acta Met. **7** (1959) 134-135.

Mat. Res. Soc. Symp. Proc. Vol. 821 © 2004 Materials Research Society P6.1

Mechanism-based strain gradient crystal plasticity

Chung-Souk Han[⋆◇], Huajian Gao[⋆], Yonggang Huang[†], and William D. Nix[◇]

[⋆] Max Planck Institute for Metals Research, Heisenbergstr. 3, D-70569 Stuttgart, Germany
[◇] Dept. of Mater. Sci. and Engng., Stanford University, Stanford, CA 94305, USA
[†] Dept. of Mech. and Indust. Engng., University of Illinois, Urbana, IL 61801, USA

ABSTRACT

To model size dependent plastic deformation at micron and submicron length scales the theory of mechanism-based strain gradient plasticity (MSG) was developed. The MSG approach incorporates the concept of geometrically necessary dislocations into continuum plastic constitutive laws via Taylor hardening relation. This concept is extended here to develop a mechanism-based strain gradient theory for crystal plasticity (MSG-CP) based on the notions of dislocation density tensor and resolved density force corresponding to the Peach-Koehler force in dislocation theory. An effective density of geometrically necessary dislocations is defined on the basis of resolved density force for specific slip systems and is incorporated into the plastic constitutive laws via Taylor relation.

INTRODUCTION

Size effects in plasticity at micron and submicron scales has been modeled by strain gradients by various authors, e.g. [3]. Analyzing micro-indentation experiments Nix & Gao [11] found that the measured indentation hardness strongly suggests a linear dependence between the square of plastic flow stress and strain gradient. This has led to the development of mechanism-based strain gradient (MSG) plasticity [4]. The core idea of MSG has been to incorporate the concept of geometrically necessary dislocations into the plastic formulation by the Taylor relation $\tau = \alpha\mu b\sqrt{\rho}$, where μ denotes the shear modulus, b the Burgers vector length, α is an empirical coefficient usually taken to be 0.2-0.5, and ρ the dislocation density.

Most of the theories on strain gradient plasticity have been developed for isotropic and polycrystalline materials. Crystal plasticity models, however, offer a higher resolution of plastic deformation at sizes comparable to single grains, which are often on the order of microns where significant size effects have been observed. The anisotropy of size effects has for example been observed in micro-indentation experiments of silver crystals in the (100) and (110) orientations [8]. In order to obtain a better resolution a MSG-type crystal plasticity model is presented.

Notation. To set the stage, the notation has to be provided. The decomposition of the deformation gradient is adopted as, $\mathbf{F} = \mathbf{F}_e\mathbf{F}_p$, where the lattice rotation is contained in $\mathbf{F}_e$ and the plastic slip is included in $\mathbf{F}_p$. Further expressions in the intermediate configuration $\tilde{\mathcal{B}}$ are given in Table 1 (see [6] for details). The scalar product used therein is defined by $\mathbf{M} \cdot \mathbf{N} = \sum_{ij} M_{ij}N_{ij}$.

DISLOCATION DENSITY TENSOR

Conventional crystal plasticity assumes that the lattice remains essentially undistorted in the deformation process. Lattice distortions can be described with the dislocation density tensor which can be associated with the concept of geometrically necessary dislocations and size dependent deformation [1, 4]. In the following Grad, Div, and Curl denote gradient, divergence, and curl differential operators with respect to a material point in the reference configuration $\mathcal{B}_o$, and grad, div, and curl denote operators with respect to a material point in the current configuration $\mathcal{B}$. The compatibility conditions Curl $\mathbf{F} = \mathbf{0}$ and curl $\mathbf{F}^{-1} = \mathbf{0}$, yield [5]

$$\tilde{\mathbf{A}}_e = det(\mathbf{F}^e)\ \mathbf{F}^{e-1}(\text{curl } \mathbf{F}^{e-1})^T \qquad \text{and} \qquad \tilde{\mathbf{A}}_p = \frac{1}{det(\mathbf{F}^p)}\ \mathbf{F}^p(\text{Curl } \mathbf{F}^p)^T \tag{1}$$

where $\tilde{\mathbf{A}} = \tilde{\mathbf{A}}_e = \tilde{\mathbf{A}}_p$ is defined as the dislocation density tensor in $\tilde{\mathcal{B}}$.

velocity gradient:	$\mathbf{l} = \mathbf{l}_e + \mathbf{F}_e\tilde{\mathbf{L}}_p\mathbf{F}_e^{-1}$ with $\mathbf{l}_e = \dot{\mathbf{F}}_e\mathbf{F}_e^{-1}$	(2)
plastic velocity gradient:	$\tilde{\mathbf{L}}_p = \dot{\mathbf{F}}_p\mathbf{F}_p^{-1} = \sum \dot{\gamma}^\alpha\, \tilde{\mathbf{s}}^\alpha \otimes \tilde{\mathbf{m}}^\alpha$	(3)
Lagrangian strain tensor:	$\tilde{\mathbf{E}} = \tilde{\mathbf{E}}_e + \tilde{\mathbf{E}}_p$	(4)
	$\tilde{\mathbf{E}}_e = \frac{1}{2}(\mathbf{F}_e^T\mathbf{F}_e - \mathbf{1})$ and $\tilde{\mathbf{E}}_p = \frac{1}{2}(\mathbf{1} - \mathbf{F}_p^{-T}\mathbf{F}_p^{-1})$	
2nd Piola-Kirchhoff stress:	$\tilde{\mathbf{S}} = \mathbf{F}_e^{-1}\boldsymbol{\tau}\mathbf{F}_e^{-T}$	(5)
Mandel stress:	$\tilde{\mathbf{P}} = (\mathbf{1} + 2\tilde{\mathbf{E}}_e)\tilde{\mathbf{S}} = \tilde{\mathbf{C}}_e\tilde{\mathbf{S}}$	(6)
elastic relation:	$\tilde{\mathbf{S}} = \tilde{\mathbb{K}}_e\, \tilde{\mathbf{E}}_e$	(7)
shearing rate:	$\dot{\gamma}^\alpha = \dot{\gamma}_o \left\lvert \frac{\tau^\alpha}{g^\alpha} \right\rvert^n sign(\tau^\alpha)$	(8)
linear silp resistance:	$g^\alpha = g_o + c_h \sum \lvert\gamma^\alpha\rvert$	(9)
resolved shear stress:	$\tau^\alpha = \tilde{\mathbf{P}} \cdot (\tilde{\mathbf{s}}^\alpha \otimes \tilde{\mathbf{m}}^\alpha)$	(10)

Table 1: Crystal plasticity relations in the intermediate configuration $\tilde{\mathcal{B}}$.

Relation of $\tilde{\mathbf{A}}$ to dislocation theory. The dislocation density tensor $\tilde{\mathbf{A}}$ has been defined above in the continuum theory of crystal plasticity. It is helpful to investigate the correlating expressions in the discrete dislocation theory. For a dislocation segment, $\tilde{\mathbf{A}}$ is defined by

$$\tilde{\mathbf{A}}\ dv = \tilde{\mathbf{l}} \otimes \tilde{\mathbf{b}}\ dl\ , \tag{11}$$

where $\tilde{\mathbf{l}}$ denotes the unit tangent vector and dl the infinitesimal element of the dislocation line, $\tilde{\mathbf{b}}$ the Burgers vector, and dv the associated volume element. In the case of multiple dislocations linear combination yields $\tilde{\mathbf{A}} = \sum_k \eta_k^A\ \tilde{\mathbf{l}}_k \otimes \tilde{\mathbf{b}}_k$, where $\tilde{\mathbf{b}}_k$ and $\tilde{\mathbf{l}}_k$ are unit vectors of dislocation k, i.e., $|\tilde{\mathbf{b}}_k| = |\tilde{\mathbf{l}}_k| = 1$. The linear combination is, however, in strict sense only valid for small strains. A coordinate expansion [1] may motivate $\tilde{\mathbf{A}} = \sum_{ij} \eta_{ij}^A\ \tilde{\mathbf{e}}_i \otimes \tilde{\mathbf{e}}_j$, with an orthonormal basis set $\tilde{\mathbf{e}}_i$. In terms of slip systems $\tilde{\mathbf{A}}$ can be expressed as $\tilde{\mathbf{A}} = \sum_\alpha(\eta_\odot^\alpha\ \tilde{\mathbf{s}}^\alpha \otimes \tilde{\mathbf{s}}^\alpha + \eta_\vdash^\alpha\ \tilde{\mathbf{t}}^\alpha \otimes \tilde{\mathbf{s}}^\alpha)$, where $\tilde{\mathbf{t}}^\alpha = \tilde{\mathbf{s}}^\alpha \times \tilde{\mathbf{m}}^\alpha$ and $\eta_\odot^\alpha$, $\eta_\vdash^\alpha$ are interpreted in [1, 5] as the densities of screw and edge dislocations on slip system α. Note that $\eta_\odot^\alpha$ and $\eta_\vdash^\alpha$ are signed scalars that can be positive or negative. The dimensions of $\eta^A = |\tilde{\mathbf{A}}|$, η_{ij}^A, η_k^A in continuum crystal plasticity are $1/m$ because the length of the Burgers vector is incorporated. The usual dislocation density can be recalculated with $\rho = \eta^A/b$.

PEACH-KOEHLER AND CORRESPONDING CONTINUUM FORCES

An essential concept in dislocation theory is the Peach-Koehler force. The Peach-Koehler force on a dislocation line element in the intermediate configuration is defined here by [6]

$$\tilde{\boldsymbol{\xi}}_L = -\tilde{\mathbf{l}} \times \tilde{\mathbf{P}}^T \tilde{\mathbf{b}} \,, \tag{12}$$

where $\tilde{\mathbf{P}}$ represents the applied stress field. Note that the notation departs from [7], in that $\tilde{\boldsymbol{\xi}}_L$ is used to denote force/length rather than line direction and that $\tilde{\boldsymbol{\xi}}_L$ has units of force/length, since $\tilde{\mathbf{l}}$ is a unit vector and $\tilde{\mathbf{b}}$ has dimensions of length. The Peach-Koehler force describes the energetic force acting on a dislocation line segment. For multiple dislocations the Peach-Koehler force on a dislocation k can be described by $\tilde{\boldsymbol{\xi}}_L^k = -\tilde{\mathbf{l}}^k \times (\tilde{\mathbf{P}}^T + \sum_{j\neq k} \tilde{\mathbf{P}}_j^T)\, \tilde{\mathbf{b}}^k$ which can be obtained via superposition of external stress fields $\tilde{\mathbf{P}}$ and stress fields associated with the j dislocation $\tilde{\mathbf{P}}_j$ [13]. Note again that this is in strict sense only valid for small strains. The component ξ_L^k of $\tilde{\boldsymbol{\xi}}_L^k$ in the direction $\tilde{\mathbf{l}}^k \times \tilde{\mathbf{m}}^k$, i.e. normal to the dislocation line direction and parallel to the α slip plane is $\xi_L^k = \tilde{\mathbf{m}}^k \cdot (\tilde{\mathbf{P}}^T + \sum_{j\neq k} \tilde{\mathbf{P}}_j^T)\, \tilde{\mathbf{b}}^k$, which determines the force and motion of the glide dislocation in its slip plane. The net Peach-Koehler force of a volume element can be written as $\tilde{\boldsymbol{\xi}}_L^{net} = \sum_k \tilde{\boldsymbol{\xi}}_L^k$.

Corresponding forces in continuum plasticity. In view of (11) and (12), the corresponding force induced by geometrically necessary dislocations in a crystalline continuum – referred to in the following as the density force – can be defined by

$$\tilde{\boldsymbol{\zeta}} = -\,\tilde{\mathbf{A}} \times \tilde{\mathbf{P}}^T \,, \tag{13}$$

where $[\mathbf{A} \times \mathbf{B}]_k = A_{mi}\, B_{ni}\, e_{mnk}$ defines a tensor cross product with the permutation tensor e_{ijk}. $\tilde{\boldsymbol{\zeta}}$ can be understood as the force acting on geometrically necessary dislocations - described by $\tilde{\mathbf{A}}$ - because of the applied stress $\tilde{\mathbf{P}}$. Note that $\tilde{\boldsymbol{\zeta}}$ in (13) is identical to $\tilde{\boldsymbol{\xi}}_L$ in (12) for a single dislocation and can be directly derived on the basis of (11), see [10]. In the general case of multiple dislocations the dislocation-dislocation interactions are however not reflected in (13).

Resolved density force on a slip system. $\tilde{\zeta}$ is in general not parallel to the slip plane and the relevant components of $\tilde{\zeta}$ have to be projected. For a given slip system, the shear rate $\dot{\gamma}^\alpha$ is related to the plastic velocity gradient by (3). The associated density force of its thermodynamic conjugate, τ^α, resolved onto the slip system α is suggested in [6] as

$$\tilde{\boldsymbol{\zeta}}^\alpha = -\,\tau^\alpha\, \tilde{\mathbf{A}} \times (\tilde{\mathbf{s}}^\alpha \otimes \tilde{\mathbf{m}}^\alpha)^T. \tag{14}$$

The norm yields $|\tilde{\boldsymbol{\zeta}}^\alpha| = |\,\tau^\alpha|\; \eta_G^\alpha$, with

$$\eta_G^\alpha = |\tilde{\mathbf{A}} \times (\tilde{\mathbf{s}}^\alpha \otimes \tilde{\mathbf{m}}^\alpha)^T| \,, \tag{15}$$

where η_G^α can be seen as a projection of $\tilde{\mathbf{A}}$. In slip system terms suggested in [1,5], we can write (14) as $\tilde{\boldsymbol{\zeta}}^\alpha = \tau^\alpha\, \tilde{\mathbf{m}}^\alpha \times \sum_\beta \tilde{s}^{\alpha\beta}(\eta_\odot^\beta\; \tilde{\mathbf{s}}^\beta + \eta_\vdash^\beta\; \tilde{\mathbf{t}}^\beta)$, where $\tilde{s}^{\alpha\beta} = \tilde{\mathbf{s}}^\alpha \cdot \tilde{\mathbf{s}}^\beta$, for single slip this yields $\tilde{\boldsymbol{\zeta}}^\alpha = \tau^\alpha\, (\tilde{\eta}_\vdash^\alpha\; \tilde{\mathbf{s}}^\alpha - \tilde{\eta}_\odot^\alpha\; \tilde{\mathbf{t}}^\alpha)$. It may be worth noting that in general specific types of dislocations can not be explicitly quantified on the basis of $\tilde{\mathbf{A}}$ and that the components $\tilde{\eta}_\vdash^\alpha$,

$\tilde{\eta}^{\alpha}_{\odot}$ can only be seen as net dislocation densities induced by lattice distortions of a volume element, see [1] for more details.

In dislocation theory the dislocation velocity is a direct function of the slip component of the Peach–Koehler force ξ^k_L and is therefore directly linked to the plastic strain rate via Orowan relation $\dot{\varepsilon}_p = b\rho_m v_m$, where ρ_m denotes the mobile dislocation density and v_m the average velocity of dislocations contained in ρ_m. Similarly, the projected density force $\tilde{\boldsymbol{\zeta}}^{\alpha}$ in continuum plasticity can be interpreted as the force acting on the geometrically necessary dislocations because of the shear stress τ^{α}. The dislocations of the density η^{α}_G are however in essence immobile because of their geometrically necessary characteristics. The dislocations contained in η^{α}_G will therefore hinder the dislocation motion in the α slip plane similar as dislocations in a dislocation pile up.

TAYLOR BASED HARDENING MODEL

The relation $\tau = \alpha\mu b\sqrt{\rho}$ links the dislocation density ρ to the shear flow stress. It should be noted that this relation has been experimentally verified for different deformation stages materials, e.g. single crystals in single and double slip, and polycrystals, and has been found to agree well for dislocation densities in the range of 10^9 to $10^{15}\,\frac{1}{m^2}$ [2]. Dividing ρ into the density of statistically stored ρ_S and geometrically necessary dislocations ρ_G the Taylor relation has been suggested in the MSG approach [4] as

$$\tau = \alpha\mu b\sqrt{\rho_S + \rho_G}\ . \tag{16}$$

Following MSG, ρ_S is related to the strain hardening which is described in crystal plasticity by the slip resistance g^{α}. With the notion that the lattice is unaffected by the deformation in conventional crystal plasticity we can write $g^{\alpha} = \alpha\mu b\sqrt{\rho^{\alpha}_S}$ which yields the density ρ^{α}_S effective on slip system α as $\rho^{\alpha}_S = (\frac{g^{\alpha}}{\alpha\mu b})^2$.

The basic assumptions of the suggested adaptation of the MSG approach to crystal plasticity are: (A) the hardening mechanisms of dislocation–dislocation interactions can be approximated by (16) and (B) the anisotropy of size effects can be described by the projected density of geometrically necessary dislocations η^{α}_G. With these assumptions the total slip resistance effective on slip system α can be described by

$$g^{\alpha}_T = g_o\sqrt{(g^{\alpha}/g_o)^2 + \ell\eta^{\alpha}_G} \tag{17}$$

where g_o denotes a reference slip resistance and $\ell = \alpha^2\mu^2 b/(g_o)^2$ the intrinsic length scale [6]. Assuming b to be around one tenth of a nanometer and $\mu/g_o \approx 100$, the intrinsic length is estimated on the order of microns. The shear rate of a rate-dependent material is then modified to

$$\dot{\gamma}^{\alpha} = \dot{\gamma}^{\alpha}_o\left(\frac{\tau^{\alpha}}{g^{\alpha}_T}\right)^n \mathrm{sign}(\tau^{\alpha})\ . \tag{18}$$

It should be noted that also other projections of $\tilde{\mathbf{A}}$ could be applied. The projection η^{α}_G incorporates the property that gradients normal to the slip plane are not contained (this is actually also true for $\tilde{\mathbf{A}}$, however only for the single slip case). Forest dislocations intersecting a slip plane α can for example expressed by $\eta^{\alpha}_{\times} = |\tilde{\mathbf{A}}^T\tilde{\mathbf{m}}^{\alpha}|$ whose hardening mechanisms are not accounted for in the suggested material model which may limit the suggested model a priori for the deformation stage of easy glide.

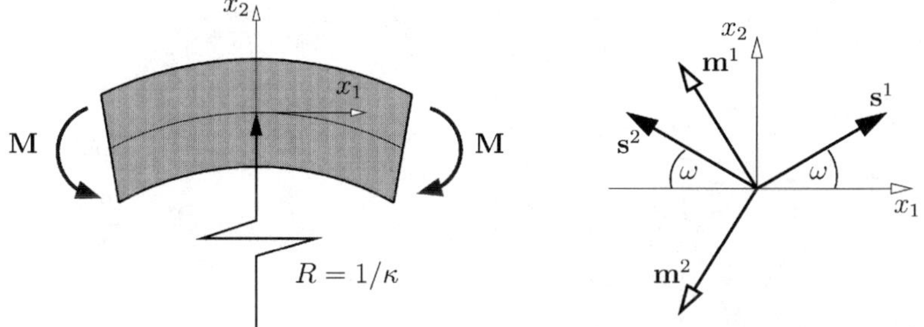

Figure 1: Micro-bending of a thin crystalline beam with two active slip systems.

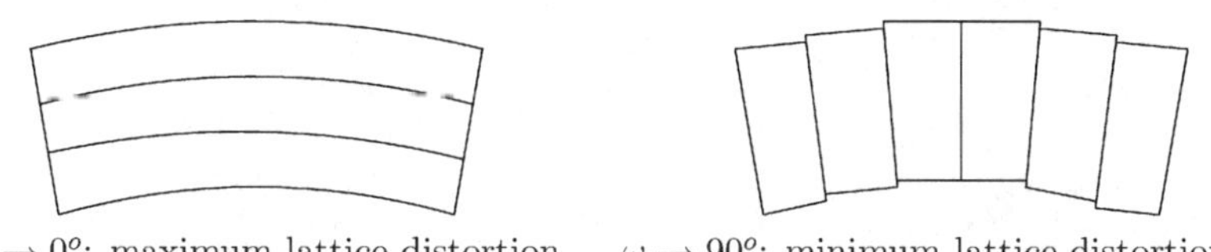

Figure 2: Lattice distortion and orientation of slip systems.

While simple hardening models like forest dislocations [12] or super-dislocations [7, 8] containing dislocations in a pile up may be helpful for theoretical considerations, in real materials, specific dislocation distributions seldom exist and real distributions are mixed and complex and may vary with the metal which limits the applicability of such simple models as suggested in [7, 8, 12]. It is therefore not surprising that even if strain gradients are excluded various different hardening descriptions have been suggested in the literature which can be of high complexity and may nether the less be limited to specific materials and deformations stages. A relative simple description of strain gradient crystal plasticity motivated by the notion of geometrically necessary dislocations as intended here can therefore only be as good as the applied strain hardening description and may be subjected to even stronger limitations than many models for strain hardening in the slip resistance.

EXAMPLE: Micro-bending of a beam in plane strain

For the case of small deformation and negligible elastic strains, the micro-bending of a thin beam under plain strain, Fig. 1 (left) is considered in the following. For a detailed discussion of this problem it is referred to [6]. The strain components are obtained by assuming the plain strain condition $\varepsilon_{i3} = 0, i = 1, 2, 3$ and deformation by pure bending $\varepsilon_{11} = \kappa x_2$. The remaining strain components are obtained by the incompressibility of plastic deformation yielding $\varepsilon_{22} = -\kappa x_2$ and the classical Kirchhoff theory for thin beams assuming $\varepsilon_{12} = 0$. Two active slip systems are assumed, Fig. 1 (right). Applying relation $\boldsymbol{\varepsilon} = \sum_\alpha \gamma^\alpha (\mathbf{s}^\alpha \otimes \mathbf{m}^\alpha)_{sym}$ the non-zero components of the dislocation density tensor are obtained after some algebra as $A_{31} = -\kappa$ and are therefore independent of the orientation of slip systems. The resolved strain gradient measure yields $q^\alpha = \kappa \cos \omega$, where ω describes the orientation of the slip

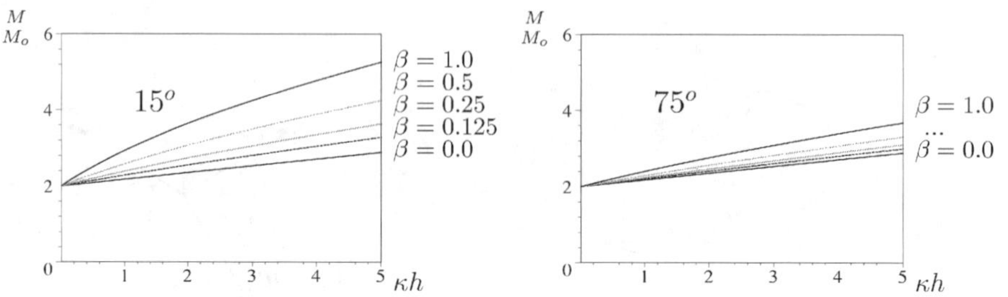

Figure 3: Bending moments versus κh for $\omega = 15^o$ and 75°, and several values of $\beta = \ell/h$.

systems as indicated in Fig. 1.

To illustrate how the anisotropy of gradient hardening affects the deformation of the beam, the bending moment $M = \int_{-h/2}^{h/2} \tau_{11}\, x_2\, dx_2$ is considered. The exponent n in (18) with a given $\dot{\gamma}_o$ is assumed to be large enough for the relation $\tau^\alpha = g_T^\alpha$ to be valid. Thus rate effects are assumed to be negligible. The normalized moments M/M_o, where $M_o = \tau_o h^2/4$, are plotted in Fig. 3 for $\omega = 15^o, 75^o$ and several values of $\beta = \ell/h$. Here $\beta = 0$ corresponds to macroscopic values of the beam thickness h and is equivalent to solutions without strain gradients. With decreasing size h, the normalized moment M/M_o required to deform the beam increases. The effect of the orientation of the slip systems can be assessed by comparing normalized moments for $\omega = 15^o$ and 75°. While the moment to initiate deformation is the same the hardening induced by the resolved dislocation density shows significant differences with increasing deformation. This is consistent with the notion of lattice distortions induced by geometrically necessary dislocations, as indicated in Fig. 2.

ACKNOWLEDGMENTS

The authors gratefully acknowledge support from an NSF-NIRT Project: "Mechanism Based Modeling and Simulation in Nanomechanics," through grant No. NSF CMS-0103257, under the direction of Dr. Ken Chong. Helpful discussions with Alexander Hartmaier are also gratefully acknowledged.

REFERENCES

1. Arsenlis, A., Parks, D.M., *Acta Mater.*, **47**, 1597–1611 (1999).
2. Courtney, T.H., *Mechanical Behavior of Materials* (1990).
3. Fleck, N.A., Muller, G.M., Ashby, M.F., Hutchinson, J.W., *Acta Mater.*, **42**, 475 (1994).
4. Gao, H., Huang, Y., Nix, W.D., Hutchinson, J.W., *J. Mech. Phys. Sol.*, **47**, 1239 (1999).
5. Gurtin, M.E., *J. Mech. Phys. Sol.*, **50**, 5 (2002).
6. Han, C.-S., Gao, H., Huang, Y., Nix, W.D., *J. Mech. Phys. Sol.* (accepted), (2004).
7. Hirth, J. P., Lothe, J., *Theory of Dislocations* (1982).
8. Ma, Q., Clarke, D.R., *J. Mater. Res.*, **10**, 853 (1995).
9. Mott, N.F., *Phil. Mag.*, **43**, 1151 (1952).
10. Mura, T., *Micromechanics of Defects in Solids* (1987).
11. Nix, W.D., Gao, H., *J. Mech. Phys. Sol.*, **46**, 411 (1998).
12. Taylor, G.I., *Proc. Roy. Soc. Lon. A*, **145**, 362 (1934).
13. Van der Giessen, E., Needleman, A., *Model. Sim. Mater. Sci. Eng.*, **3**, 689 (1995).

Mat. Res. Soc. Symp. Proc. Vol. 821 © 2004 Materials Research Society P8.14

Texture and Microstructure Evolution in Silver Thin films with FACET

Asit Rairkar[1], Jie Zhang[2] and Jim Adams[1]
[1]SEM Program , Arizona State University, Tempe, AZ 85287-1704
[2]Research and Development, Cypress Semiconductor, San Jose, CA, 95134

ABSTRACT

Metallic polycrystalline thin films are used in a plethora of applications, including especially metallic interconnects for the microelectronics industry. Despite the numerous approaches at modeling and simulating processing of these films, there have not been any simulation tools, which can accurately predict texture and microstructure evolution in these films. We have developed a novel 2D model called FACET for the simulation of polycrystalline thin film growth at realistic spatial and time scales. The basic idea is to use the results of smaller-scale atomic simulations (density functional theory, molecular dynamics, and lattice Monte Carlo) to provide input and guidance on the evolution of grain structure and texture on a micron scale. The feature scale model is based on describing grains in terms of two-dimensional faceted surfaces and grain boundaries. The model includes the major phenomena involved in film growth, including deposition, nucleation, surface diffusion (on the substrate and on the growing film), inter-facet diffusion, and grain growth and coarsening. In addition, the texture of each grain is treated individually, so that the texture evolution of the system can be simulated. Predictions of the FACET code are compared with previous experimental studies of texture and microstructure in Silver films.

INTRODUCTION

The need for prediction of thin film profiles and microstructure based on their deposition condition is well understood. As the semiconductor industry moves towards smaller devices and faster speeds, reliability constraints on thin films as interconnects are becoming more and more stringent. Copper has replaced Aluminum as an interconnect material due to its higher thermal conductivity and resistance to electromigration resistance[3]. Silver[4] is also being investigated as a better alternative, though some reliability issues remain. It is well known that electromigration lifetimes of fcc films are dependent on columnar shape[5] and texture and that <111> textured films are 4 times more reliable than <100> or random textured films[6]. Understanding the mechanisms of thin film growth and being able to predict thin film structure in terms of experimental parameters such as temperature, deposition rate, substrate materials, etc, are very useful for optimizing processing parameters so as to increase production yield and to reduce cost. Typically models for thin film growth can be classified into atomic and feature scale models. Atomic scale models include Lattice Monte Carlo, Molecular Dynamics and its derivatives Hyperdynamics and temperature accelerated Dynamics (TAD) Feature scale models like EVOLVE, SUPREM-DSMC and level-set method. A good review of these techniques can be found in [7] and {Adams, 2000 #24}. Each method has its own advantages and drawbacks. There is no method for realistically modeling microstructure evolution (texture and grain distribution) of micron-scale polycrystalline films deposited at realistic deposition rates. Also, several of the models have been developed with little or no experimental validation; they are often just models of generic materials behavior.

Considering the importance of texture and microstructure in thin film processing, computer models capable of predicting the final film microstructure would be very useful in helping process engineers to efficiently optimize processing conditions, provided that the models are reliable, fast and easy to use. FACET offers a unique ability to provide realistic predictions of microstructural evolution for micro-scale systems for realistic deposition rates.

Initial Development of FACET

In our previous work we developed an initial version of FACET[8, 9], a model for simulating polycrystalline thin film growth. Figure 1 shows a snapshot of a silver film grown for 2400 secs. at a deposition rate of 60 nm/min using FACET. The methodology of FACET has been discussed in detail in our previous work[8] but will be briefly explained here. It is based on the fundamental physical processes, which occur during thin film growth, including deposition, nucleation, surface diffusion, surface energy minimization and grain growth. Within certain approximations, this model is capable of predicting, in 2D, thin film structures, which include grain size, shape, orientation and surface roughness as a function of deposition and material parameters. The model is extremely fast and computationally efficient, and is easy to use by non-experts due to its integrated Graphical User Interface (GUI). It can run on a standard windows PC, so that it can be made widely available. FACET can be downloaded freely from our website http://ceaspub.eas.asu.edu/cms/research/facet_growth.htm. We plan to also make future versions freely available, so that it can be widely used for both research and educational purposes.

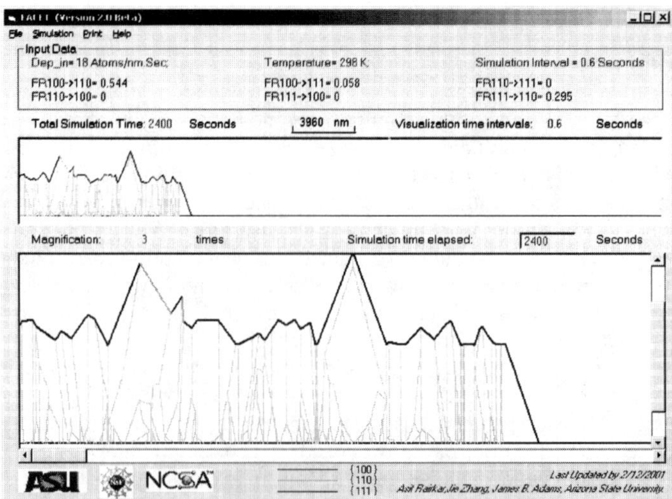

Figure 1: Snapshot of Silver films deposition for 2400 secs. at ~60 nm/min with FACET

The key assumptions of the model are:-

1. 2D simulation with facets and grain boundaries described as line segments. Each facet is described by one line segment while multiple line segments describe grain boundaries in order to depict the proper grain shape and film structure. (The assumption of 2D produces some inaccuracies, but greatly simplifies the methodology, and is a sensible first step prior to development of a 3D model).
2. Current version assumes an FCC crystal structure (valid for aluminum, copper, silver, gold, and many other metals). Future versions will allow for other crystal structures.
3. Each nucleus has its own orientation, which affects facet growth rates. The zone axes are assumed to have only 0, 1, or -1 as their indices, allowing for 26 distinct zone axes in 3D. Additionally, each grain is allowed to rotate by any angle around the zone axis.
4. The input parameters include: nuclei density, initial nuclei size, texture, surface energy, diffusion rates, temperature, and deposition flux conditions.
5. Growth of each facet is based on 3 factors; deposition directly onto the facet, flux due to diffusion from the surface and adjacent facets, and flux due to surface energy minimization of facets.

Description of Thin film Structure

Every facet is described by one line function and 2 ending nodes. Uncovered substrate surfaces are treated as a special type of facet. Grain boundaries are described by multiple line segments using a an array of singly linked lists of (x,y) co-ordinates. The nodes are described with a doubly linked list, which makes for a highly efficient data structure. All such data types are dynamically initialized to conserve system memory. Virtual picture box and pixel handling techniques are used to visualize the growth process.

Nucleation and grain growth

Nucleation is a very complex phenomenon, and heterogeneous nucleation as it depends upon various factors such as temperature, deposition rate, substrate material, substrate orientation, deposition material, surface roughness, surface defects (grain boundaries, dislocations, etc.), impurities, contamination, surfactants, etc., is extremely difficult to predict with an analytical model. Thus, in the absence of a generally accepted nucleation model, we instead treat nucleation rate and initial nuclei size as an input parameter. FACET assumes that the initial nuclei are non-overlapping, and an exclusion region around a nucleus can be included. The orientation of the nuclei can be determined randomly, or an initial degree of texturing can be assumed, depending on the materials system (deposition material and substrate material). Similarly, the shape and facet orientation is determined in a crystallographically correct manner based on the initial assumption regarding the crystallographic orientation of the nuclei.

Grain coarsening is modeled with the help of merging nodes and facets and formation of grain boundaries as necessary. The basic growth and coarsening mechanisms in the code are 1) Diffusion on the substrate- where Deposited atoms which land on the substrate are assumed to diffuse to one of the adjacent nuclei. (there is also a small chance that they can form a new nuclei, but that is assumed to be incorporated into the overall nucleation model, which assumes that different nuclei have nucleated at different times, resulting in a distribution of nuclei sizes and shapes) and 2) Diffusion between facets- Atoms which land on a growing film may become incorporated into that facet, or diffuse to adjacent facets. We have carried out the first calculations of the activation energies for diffusion between facets, and we have incorporated that information into a 3-D KLMC simulation of diffusion between facets. We have also developed a simple 1-D KLMC method, which closely reproduces the full 3-D results and we can use that model for diffusion between any two facets. So, using that model, we can determine the fraction of the deposited atoms that become incorporated into the facet, and the fraction that diffuse to adjacent facets. In addition to this and in an attempt to model surface energy mechanisms, which contribute to texturing and grain growth, we have incorporated another algorithm, which directs flux from high-energy facets to adjacent low energy facets. The diffusive flux here is calculated on the assumption of equilibrium angles between the 2 surfaces (of known surface energy) and the grain boundary (assumed to have an average energy substantially lower than the surface energy). Figure 2 shows a schematic explaining this process.

Based on these 2 mechanisms, the growth rate of each facet according to its facet type is calculated and so is the position of nodes based on it. Thus some facets may get annihilated in the interaction and the grain boundaries and grain structure evolves.

Comparison with Experimental Data

Comparison of the predictions of FACET with experimental results is needed to validate the model. Unfortunately there is a scarcity of published experimental work on as-deposited thin films, which can be appropriately used for the validation of FACET poses a problem. However, We have used FACET to simulate conditions in published experimental work of Greiser et al[1, 2] for deposition of Silver thin films. Figure 3 shows the result of 5 simulations, which we did with our FACET software to match the Experimental conditions. Most of the input data required by FACET is well known, such as the lattice constant, substrate temperature, and deposition rate. To estimate the surface diffusion rates, we rescaled the activation energies for Cu[8] by the relative melting points to approximate surface diffusion for Ag, which is reasonable. The only other major unknowns are the initial number of nuclei and the initial fraction of nuclei with a <111> texture. The initial number of nuclei is estimated using the nuclei density designed to match the initial grain size. The initial texture fraction is set to 50 % <111> which is reasonable for fcc films. However, as the films grow, the effect of those initial assumptions becomes less important due to the coarsening process.

In general, we find that the FACET simulations of silver deposition compare well with experimental data. Specifically, the grain size and grain size distribution agree with experimental data. The grain distribution in FACET evolved with a lognormal and monomodal grain size distribution, which has been reported and is shown in figure with the median grain size within agreeable limits. The texture fraction in FACET increases from 0.57 to 0.63, but the experimental texture fraction was not determined. However it is generally well accepted that <111> textures generally develop during deposition of fcc films.

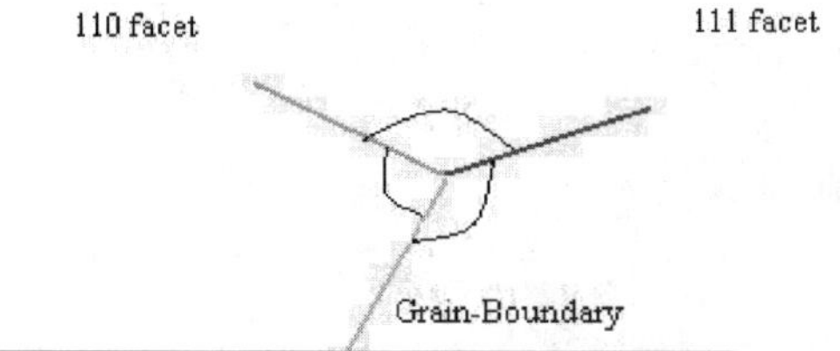

Figure 2: {111} being the lower surface energy facet will receive flux from {110} at the grain boundary

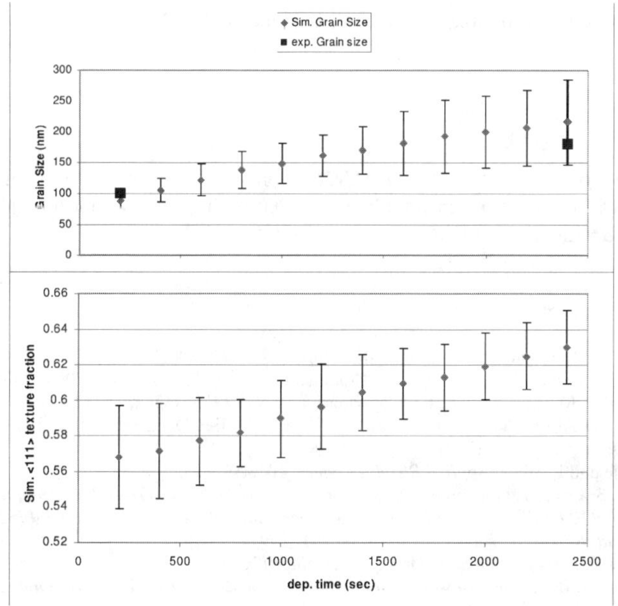

Figure 3: Comparison of silver film data for FACET simulations with data from [1, 2](top); Chart comparing experimental and simulated grain sizes for the as-deposited films, only 2 experimental data points are available;(bottom) Chart of the simulated <111> texture fraction

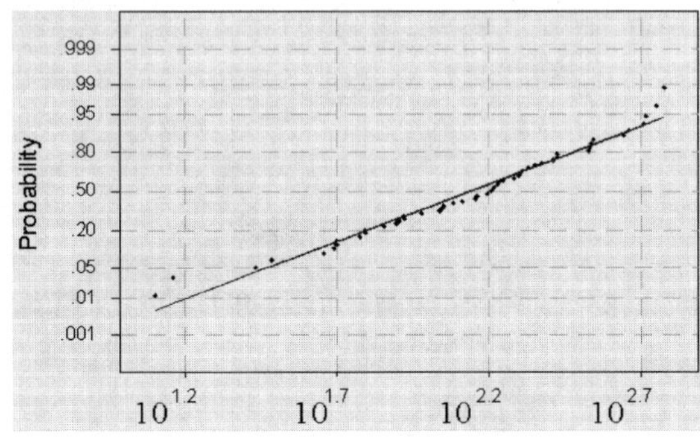

Figure 4: Normal probability plot of the log of grain size in silver films, showing that the grain size distribution is lognormal and monomodal.

CONCLUSION

A 2D model of thin film growth, FACET, is presented. It's predictions of grain size & grain size distribution are in reasonable agreement with limited experimental data, Further testing & validation of FACET is in progress.

REFERENCES

1 J. Greiser, P. Mulner and E. Arzt, *Acta Materialia,* **49,** 1041 (2001)

2 J. Greiser, P. Mulner, E. Arzt and C. V. Thompson, *Scripta Materialia,* **41,** 709 (1999)

3 R. P. Vinci, E. M. Zielinski and J. C. Bravman, *Thin Solid Films,* **262,** 142 (1995)

4 Y. Zeng, T. L. Alford, Y. L. zou, A. Amali, U. B. Manfred, F. Deng and S. S. Lau, *Journal of Applied Physics,* **83,** 779 (1998)

5 F. D'Heurle and I. Ames, *Applied Physics Letters,* **16,** 80 (1970)

6 C. Ryu, A. S. Loke, T. Nogami and S. S. Wong, *Proceedings of the 1997 35th Annual IEEE International Reliability Physics Symposium/Proceedings of the 1997 35th Annual IEEE International Reliability Physics Symposium* 201 (1997)

7 Y. Li, PhD Thesis, Arizona State University, Tempe (2003)

8 J. Zhang and J. B. Adams, *Modelling and Simulation in Materials Science and Engineering,* **10,** 381 (2002)

9 J. Zhang, *Computational Materials Science,* accepted for publication (2004)

Mat. Res. Soc. Symp. Proc. Vol. 821 © 2004 Materials Research Society **P8.15**

Deformation and Failure of Nanostructured Materials with Bimodal Grains

R.Q. Ye, B.Q. Han and E.J. Lavernia*
Department of Chemical Engineering & Materials Science
University of California, Davis, CA 95616

Abstract

The low ductility of nanostructured materials is attributed to the deficit of dislocation activity in the nanometer range. Recent scientific interest in nanostructured materials stems from reports of alterative combinations of mechanical properties, although a low ductility is typically reported. One promising approach based on the concept of multiple length scales is illustrated by a "bimodal" microstructure, i.e. containing a mixture of nanostructured and coarse grains. The present work reports a numerical study of the tensile deformation and fracture of a nanostructured Al alloy with a bimodal microstructure. In the theoretical framework used in the present study, the elastic-plastic behavior and deformation processes are approximated by Ramberg-Osgood formula and finite element method, respectively. The numerical results are found to be in a good agreement with the experimental behavior.

1. Introduction

The processing and properties of nanostructured materials have been intensively investigated in the last two decades [1]. Available data suggest that bulk nanostructured materials usually display very high strength and hardness, but very poor ductility[1-3]. An early study by Tellkamp et al. [4] on the behavior of nanostructured Al alloys processed via a cryomilling (i.e. milling in liquid nitrogen) and consolidation approach suggested that the presence of coarse-grained regions may contribute to the improved ductility [5-7].

Most recently, cryomilled aluminum alloys with a mixture of nanostructured grains and submicron grains (e.g., so-called "bimodal" microstructure) have been successfully fabricated [4,8]. In these studies, cryomilled aluminum alloys, processed with a predetermined fraction of coarse grains have revealed an improved ductility without much sacrifice in strength. Despite encouraging results reported for bimodal materials, our fundamental understanding of the elastic and plastic deformation mechanisms that governs the behavior of bimodal nanostructured materials remains inadequate.

In order to understand the behavior of bimodal materials, the micromechanical behavior should be studied if one considers a broad range of length scales [9-13]. For example, the stress field influences failure processes (yielding, crack initiation and propagation) and determines the final fracture modes. Accordingly, numerical simulation of the stress distribution during deformation process will provide information on understanding the deformation mechanisms in bimodal materials.

In the present work, the tensile deformation behavior and failure process in a cryomilled nanostructured aluminum alloys with a bimodal microstructure are simulated via a unit-cell model, under uniaxial tensile loading. The unit-cell model is based on fundamental continuum mechanics at the meso-scale and provides an ideal approach that

* Corresponding author, Email: lavernia@ucdavis.edu

can be used to quantify the stress distribution and deformation behavior in bimodal materials. The elastic-plastic behavior and the dynamic fracture process of bimodal materials are approximated via a Ramberg-Osgood formula and finite element method, respectively. The possible explanations of deformation mechanisms are then discussed.

2. Fundamental model and numerical solutions

2.1 Fundamental formula

As observed from the experimental results, the ultra-fine grains are well bonded to coarse grains. The interface between ultra-fine grains and coarse grains is then assumed to be perfect so that there is a continuity of strain across the interface. Therefore, a continuum mechanics approach can be applied to model the deformation characteristics of a bimodal microstructure and predict mechanical behavior. The uniaxial stress-strain curve of typical work-hardening materials without a sharply defined yield stress, is described by Ramberg-Osgood formula [14,15]

$$\sigma = \begin{cases} E\varepsilon, & \varepsilon \leq \varepsilon_{0.2} \\ \sigma_{0.2}\left(\dfrac{\varepsilon}{\varepsilon_{0.2}}\right)^n, & \varepsilon > \varepsilon_{0.2} \end{cases} \tag{1}$$

Where σ and ε are the uniaxial stress and strain, E is Young's modulus, $\sigma_{0.2}$ is the yield stress related to $\varepsilon_{0.2}$ and n is the strain hardening exponent.

2.2 Unit-cell model and deformation process

In the present study, a two-dimensional self-consistent cell model is applied to model the uni-axial tensile behavior during the elastic-plastic deformation. The unit-cell which is extracted from the regular array of uniform particulate composite materials can exactly model the elastic-plastic response [16-25]. An inclusion/matrix cell embedded in a rectangular equivalent composite material can be used to characterize the bimodal materials, where the spherical inclusion (coarse grains) is surrounded by a circular matrix (cryomilled nanostructured grains). Because of the symmetry of the cell, a quarter of a schematic plane strain cell model is illustrated in Fig.1. The volume fraction of inclusions in the matrix, f is expressed in $f=(d/D)^3$, where d is the diameter of inclusions and D is the diameter of the matrix.

The ABAQUS code was employed to carry out the calculations. The self-consistent unit-cell model has a high efficiency to simulate the elastic-plastic behavior of bimodal materials via only several iterative processes. During the calculations, an initially assumed bi-linear stress-strain curve is applied to the composite as the first iteration step. An updated stress-strain relationship is then obtained by calculating the overall mechanical response via the above equation from the assumed stress-strain curve. The stress-strain curve of entire bimodal materials is averaged over the equivalent composite. The procedure is repeated until the calculated result is identical to that of the previous iteration step.

During the failure analysis, the stress-strain curve follows the Ramberg-Osgood formula during the elastic-plastic deformation. Perfect bonding is assumed to exist until first failure takes place. With increasing stress, the cell is broken when its maximum normal stress reaches the tensile strength, which has been specified during the calculation setup. The stress/strain distribution is then calculated according to the revised structure.

This deformation procedure proceeds until the fraction of failure cells reaches a previously specified critical value.

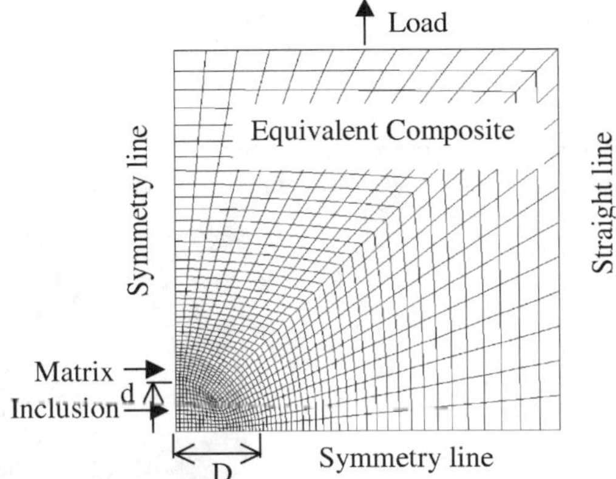

Fig.1 Typical symmetry and boundary conditions of meshed geometry.

3. Results and discussions

The mechanical response of Al-Mg bimodal materials is calculated using the self-consistent unit-cell model. The stress-strain curves of Al-Mg bimodal materials with different volume fractions of coarse grains are shown in Fig.2. Although there is no significant effect of volume fraction of coarse grains on the elastic response of Al-Mg bimodal alloys, the elastic limit increases with decreasing volume fractions of coarse grains. For instance, the yield stress decreases from 664 MPa to 482 MPa when the volume fraction of coarse grains increases from 5% to 50%. The comparison of numerical calculations with experimental data is shown in Fig.3. It is found that the numerical and experimental results match very well.

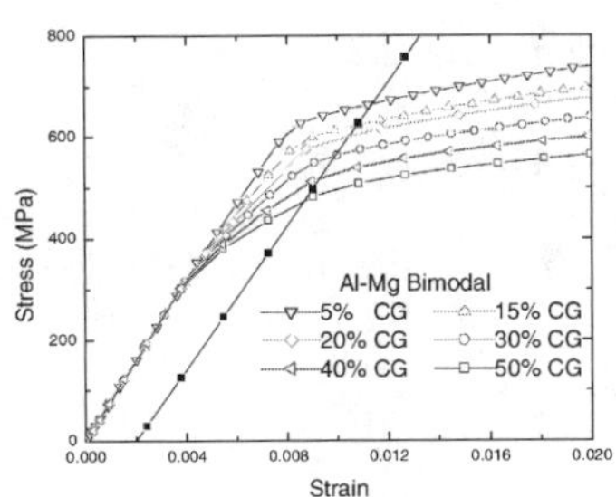

Fig.2 The simulated stress-strain curve of entire Al-Mg bimodal alloys.

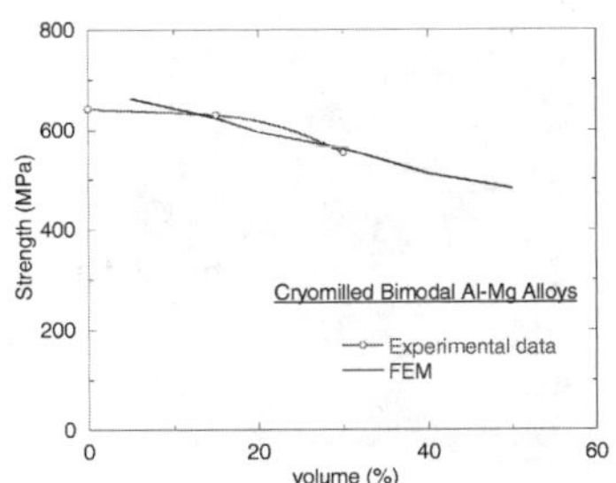

Fig.3 Comparison of simulation results with experimental yield strength.

Since there is a difference in strength between cryomilled nanostructured grains and coarse grains, the stress distribution is heterogeneous in the bimodal materials, which significantly affects failure and fracture processes. When the applied stress reaches the local yield strength, damage in the form of microcracks occurs. After the nucleation of microcracks, the stress distribution is rearranged, resulting in the changes of the fracture preferring path and fracture modes. In the present simulation, the maximum principal stress criterion is applied to determine the stress status of Al-Mg bimodal materials during failure and fracture processes. The stress states during the different failure processes are extracted from the FEM simulation.

1. Cryomilled bimodal Al-Mg with 15% coarse grains

It is interesting to note that the meshes still remain perfect when stress reaches yield strength. In addition, the stress/strain distribution is uniform right after yielding in the materials, as shown in Fig.4. During the early plastic straining stage, no domain in the materials is distorted. There is no strain concentration in the simulated materials. The strain field is homogeneous over the entire bimodal materials.

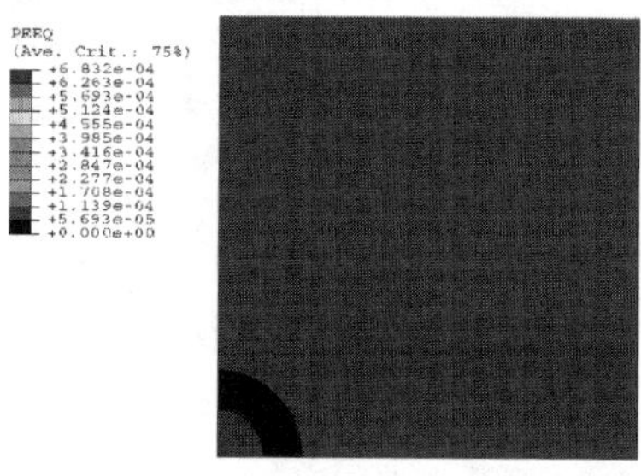

Fig.4 Plastic strain for the 15%CG Al-Mg alloy right after yielding.

As the development of deformation, the plastic strain tends to concentrate and the stress state changes. The damage firstly takes place at the weakest domain. The Von Mises stress distribution of Al-Mg alloys with the 15% volume fraction coarse grains (15%CG) with the extension of plastic deformation is shown in Fig.5. The stress field in the equivalent composite has some heterogeneity. It is seen from Fig.5 that the largest stress occurs in the matrix (cryomilled nanostructured grains), which is adjacent to inclusions (coarse grains). This suggests that the coarse grains may play a role to accommodate its local stress state. The stress field after fracture initiation for bimodal materials with the 15% volume fraction of coarse grains is presented in Fig.6. It is found that the stress heterogeneity is even worse after fracture initiation.

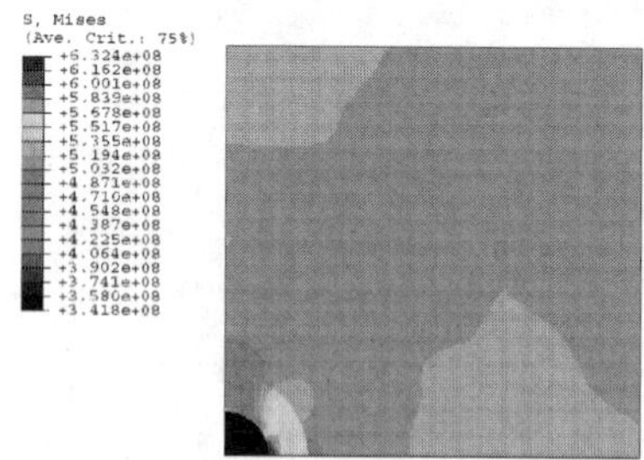

Fig.5 Von Mises stress distribution for the 15%CG Al-Mg alloy before failure initiation.

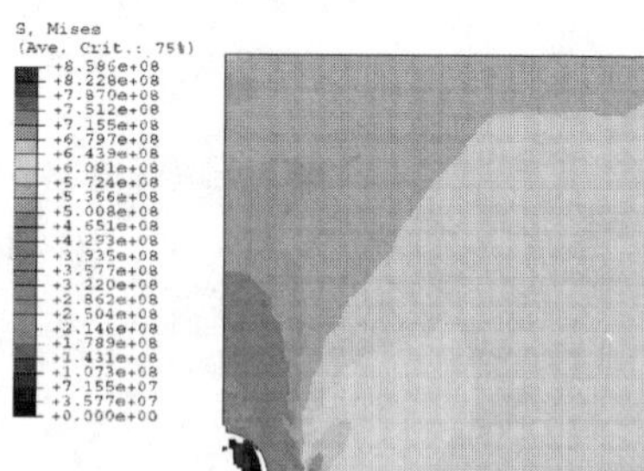

Fig.6 Von Mises stress distribution for the 15%CG Al-Mg alloy after fracture initiation.

2. Cryomilled bimodal Al-Mg with 30% coarse grains

The stress heterogeneity in the Al-Mg bimodal alloys with the 30% volume fraction of coarse grains (30%CG) when the deformation beyond yielding is shown in Fig.7. Similarly, the failure also initiates in the nano-phase grains, which are close to coarse grains. However, the nucleation and propagation of cracks are different from those in the 15%CG Al-Mg alloy.

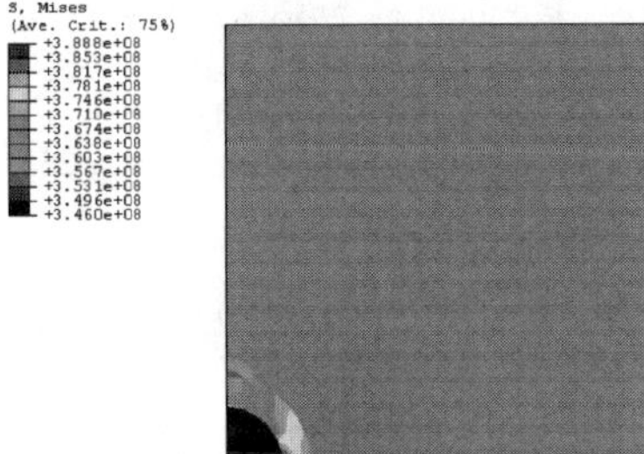

Fig.7 Von Mises stress distribution for the 30%CG Al-Mg alloy before failure initiation.

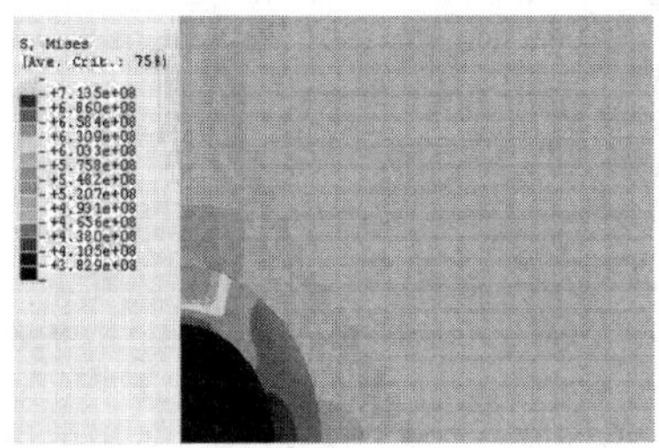

Fig. 8 Von Mises stress distribution for the 30%CG Al-Mg alloy after fracture initiation.

The stress distribution after the initiation of failure in the 30%CG Al-Mg alloy is shown in Fig.8. It is found that although the stress field in the composites is heterogeneous, it is more uniform than that in the 15%CG Al-Mg alloy. With the increase of volume fraction of coarse grains, the significant stress concentration has been released by the coarse-grained regions. Therefore, the stress level of local scale and entire materials is much smaller than that in the 15%CG Al-Mg alloy. Furthermore, the strain fields increase with increasing volume fraction of the coarse-grained regions. As a result, the material with higher volume fractions of the coarse-grained regions has a lower strength but a higher ductility.

4. Conclusions

In the present work, the tensile deformation behavior and dynamic fracture process of Al-Mg alloys with a bimodal structure are simulated via self-consistent unit-cell model. The tensile elastic-plastic behavior of bimodal materials is approximated by Ramberg-Osgood formula combined with Finite element method. It is found that the yield stress decreases with increasing volume fraction of coarse grains. The stress/strain fields become more uniform with increasing volume fraction of coarse grains. The stress concentration occurs in the cryomilled nano-phase grains adjacent to coarse grains. Failure process initiates in those stress concentration regions. Microcracking occurs more easily in bimodal materials with a lower volume fraction of coarse grains, resulting in high strength and poor ductility. The coarse grains left at the wake zone of the cracks play the bridging role in the fracture process for bimodal alloys with more coarse grains.

Acknowledgement

This work was supported by the Office of Naval Research under Grant No. N00014-03-1-0149. The useful discussions with Dr. Y. Liang of University of California, Santa Barbara are gratefully acknowledged.

References

1. C.C. Koch, Nanostructured Materials: Processing, Properties and Potential Applications. Noyes Publications (William Andrew Publishing), Norwich, NY, 2002.
2. C.C. Koch, D.G. Morris, K. Lu and A. Inoue, *MRS Bulletin* **February,** 54 (1999).
3. B.Q. Han, F.A. Mohamed and E.J. Lavernia, *J. Mater. Sci.* **38,** 3319 (2003).
4. V.L. Tellkamp, A. Melmed and E.J. Lavernia, *Metall. Mater. Trans.* **32A,** 2335 (2001).
5. S.L. Semiatin, K.V. Jata, M.D. Uchic, et al., *Scri. mater.* **44,** 395 (2001).
6. R.W. Hayes, R. Rodriguez and E.J. Lavernia, *Acta mater.* **49,** 4055 (2001).
7. B.Q. Han, Z. Lee, S.R. Nutt, et al., *Metall. Mater. Trans. A* **34A,** 603 (2003).
8. D. Witkin, Z. Lee, R. Rodriguez, et al., *Scri. mater.* **49,** 297 (2003).
9. E. Van der Giessen and A. Needleman, *Annu. Rev. Mater. Res.*, **32**, 141 (2002).
10. S. Schmauder, *Annu. Rev. Mater. Res.*, **32**, 437 (2002).
11. F. Radjai, D.E. Wolf, M. Jean et al., *Phys. Rev. Lett.*, **80**, 61 (1998).
12. W.W. Gerberich, J.M. Jungk, M. Li, et al., *Int. J. Fract.*, **119/120**, 387 (2003).
13. Y.J. Wei and L. Anand, *J. Mech. Phys. Solids*, submitted (2003).
14. W. Ramberg and W.R. Osgood, *NACA TN 902*, 1943.
15. J. Lubliner, Plasticity Theory, Macmillan Publishing Co., New York (1990).
16. G. Bao, J.W. Hutchinson and R.M. McMeeking, *Acta Metall Mater.*, **39**, 1871 (1991).
17. Y.-L. Shen, M. Finot, A. Needleman et al., *Acta Metall Mater.*, **42**, 77 (1994).
18. A.A. Mammoli and M.B. Bush, *Acta Metall Mater.*, **43**, 3743 (1995).
19. J.W. Leggoe, X.Z. Hu and M.B. Bush, *Eng. Fract. Mech.*, **53**, 873 (1996).
20. M.K. Kiser, F.W. Zok and D.S. Wilkinson, *Acta Metall Mater.*, **44**, 3465 (1996).
21. D.S. Wilkinson, E. Maire and J.D. Embury, *Mater.Sci.Eng.*, **A233**, 144 (1997).
22. L. Farrissey, S. Schmauder, M. Dong, et al., *Comp. Mater. Sci.*, **15**, 1 (1999).
23. E. Soppa, S. Schmauder, G. Fischer, et al., *Comp. Mater. Sci.*, **16**, 323 (1999).
24. R. Estevez, E. Maire, P. Franciosi et al., *Eur. J. Mech. A/Solids*, **18**, 785 (1999).
25. D. S. Wilkinson, W. Pompe and M. Oeschner, *Prog. Mater. Sci.*, **46**, 379 (2001).

Mat. Res. Soc. Symp. Proc. Vol. 821 © 2004 Materials Research Society P3.27

A Molecular Dynamics Simulation Framework for an Al+Fe_2O_3 Reactive Metal Powder Mixture

Vikas Tomar and Min Zhou
The George W Woodruff School of Mechanical Engineering
Georgia Institute of Technology
Atlanta, Georgia 30332, U.S.A.

ABSTRACT

This work focuses on the development of a classical molecular dynamics (MD) framework that can be used to analyze the quasi-static strength, elastic-constants, and dynamic strength of nanostructured Fe_2O_3+Al reactive metal powder (NRMP) mixtures. An interatomic potential is developed for the Fe_2O_3+Al NRMP system, accounting for the behavior of Al, Fe, Fe-Al intermetallics, Fe_2O_3, and Al_2O_3. This potential can be regarded as a generalization and combination of several existing potentials for the individual components in the system. The potential incorporates electronegativity equalization to account for pressure-induced phase transformation. In addition, the parameter set of the potential is capable of predicting the elastic properties of components over a range of 0 - 900 K.

INTRODUCTION

Molecular level modeling and characterization using classical molecular dynamics (MD) can provide vital input for the desired nano-construction of advanced materials. The most important input required for a classical MD analysis is a phenomenological description of the interatomic interactions. This description is calibrated by *ab initio* and/or experimental values of material properties at a specific temperature and pressure. Interatomic potentials fitted to properties at a specific temperature and pressure have been shown to perform satisfactorily over a range of temperatures and pressures [1, 2]. The current work focuses on the development of a classical MD framework that can be used to characterize the behavior of a nanostructured Fe_2O_3+Al reactive metal powder (NRMP) thermite mixture. The potential should also be able to describe the pressure-induced structural changes, and it should predict the system properties over a range of temperature. The primary interest is to use this potential to analyze nanoscopic morphological characteristics that affect the elastic constants, quasi-static strength, and dynamic strength as functions of applied loading with and without the possibility of phase transformation.

The MD model must account for the behavior of the products of the Fe_2O_3+Al thermite reaction as well as that of the reactants. The interatomic potential proposed here meets this requirement by providing a unified description for the reactants (Fe_2O_3 and Al), the products (Al_2O_3 and Fe) and the Al-Fe intermetallics. The potential incorporates electronegativity equalization to update the charges of atoms according to changes in their environment during pressure-induced deformation. The parameters for the potential obtained by fitting to the properties at room temperature (300 K) are shown to predict the system properties over a temperature range of 0 K to 900 K. The potential is expected to provide a reasonable account of the pressure-induced mechanochemical deformation of the NRMP mixture.

INTERATOMIC POTENTIAL

In order to capture deformation behavior of the Fe_2O_3+Al NRMP mixture, the potential should model the interactions related to all products and intermediate states of chemical reactions between Fe_2O_3+Al. Development of such an interatomic potential is difficult because (1) the structural details of the NRMP system undergoing reactions under pressure loading are not known experimentally and (2) there is a great number of combinations of possible intermediate structures between the initial Fe_2O_3+Al mixture and the final Al_2O_3+Fe state (many of which are even not well characterized). We therefore choose to develop a potential that captures the behavior of the reactants (Al and Fe_2O_3), the most common Fe-Al intermetallic (B2 Fe-Al), and the products (Fe and Al_2O_3) of the reaction. The reaction during pressure-induced mechanochemical deformation is modeled as a structural transition through changes of bonding from one type to another. Subsequent changes in the interaction energy imply a change in the bonds of an atom with its neighbors; therefore can be used to identify the occurrence of reactions. Electronegativity equalization is incorporated in the potential as a means to calculate the changes in charges of atoms according to their environments. This approach has been used in the past to describe pressure-induced reactions between Al and Oxygen [3] and pressure-induced phase changes [4, 5] in ferroelectric $Ba_xPb_{1-x}TiO_3$ and Silica. This treatment is motivated by similarities between the bonds in our system and the bonds in the systems in the above references.

The potential includes a term (E_{glue}) similar to the modified embedded atom method (MEAM) potential which is a generalized cluster functional [6] and another term (E_{es}) for electro-static interactions with environment-dependent charges. Specifically, the total energy of the system is a function of interatomic distance **r** and charge $\boldsymbol{q}$ in the form of

$$E(\mathbf{r},\boldsymbol{q}) = E_{glue}(\mathbf{r}) + E_{es}(\mathbf{r},\boldsymbol{q}) \cdot \tag{1}$$

Here,

$$E_{glue}(\mathbf{r}) = \sum_i F_i(\rho_i) + \frac{1}{2}\sum_{i(\neq j)} \phi_{ij} \tag{2}$$

is the glue potential with generalized cluster functional $F(\rho)$ which depends on embedding electron density [7] ρ; and

$$E_{es}(\mathbf{r},q) = \sum_i q_i \chi_i + \frac{1}{2}\sum_{i(\neq j)} q_i q_j V_{ij} \tag{3}$$

is the electrostatic energy with χ being the instantaneous electronegativity given as

$$\chi_i = \chi_i^0 + \sum_{j\neq i} Z_j\left([j \mid f_i] - [f_i \mid f_j]\right) = \chi_i^0 + \sum_{j\neq i} Z_j\left(\int d^3r \frac{f_j(r)}{|r-r_i|} - \int d^3r_1 \int d^3r_2 \frac{f_i(r_1) f_j(r_2)}{r_{12}}\right). \tag{4}$$

In the above equation, χ_i^0 is the ground state electronegativity. Also in Eq. (3), V is the electrostatic pair interaction potential in the form of

$$V_{ij} = J_i^0\delta_{ij} + [f_i | f_j] = \begin{cases} J_i^0 & i = j, \\ \int d^3r_1 \int d^3r_2 \dfrac{f_i(r_1) f_j(r_2)}{r_{12}} & i \neq j \cdot \end{cases} \quad (5)$$

J_i^0 is referred to as the atomic hardness [8] or self-Coulomb repulsion [9]. Function $f_i(r)$ specifies the distribution of diffused charges. The form chosen for f_i is,

$$f_i(r) = \frac{\varsigma_i^3}{\pi}\exp(-2\zeta_i r) \cdot \quad (6)$$

The approach for electronegativity equalization is based on the work in [9] and [10]. The values of q_i are chosen such that electrostatic energy E_{es} is minimized under the constraint that the total charge of the system be constant [10]. The form of the cluster functional is chosen as

$$F(\rho) = E_0 A(\rho \ln(\rho))^B \cdot \quad (7)$$

Here, A and B are fitting parameters. The formulation in [7] uses an arbitrary reference structure and is a nearest-neighbor model. The cluster functional in Eq. (7) has an extended range of interaction and does not use an equilibrium reference structure like the ones used in [7] and [11]. The forms for the atomic density distributions include angular dependence upto three-body interactions and are directly taken from [7]. The functional form of pair interaction potential ϕ_{ij} in Eq. (2) is based on that in [12] for Fe-Al alloys, i.e.,

$$\phi_{ij}(r) = \psi(r) - \psi(D_p) + \frac{D_p}{20}\left[1 - \left(\frac{r}{D_p}\right)^{20}\right]\psi'(D_p), \quad \text{for } r < D_p \cdot \quad (8)$$

The interaction implied here vanishes at the cutoff distance D_p. The form for $\psi(r)$ above is expressed as

$$\psi(r) = \psi_0\left[\exp(-2\gamma(r-t)) - 2\exp(-\gamma(r-t))\right] \cdot \quad (9)$$

Excluding the glue potential and short range electrostatic cutoff radii, the total number of parameters needed to describe the behavior of the model system is 72. These include pair potential parameters for six (Al-Al, Fe-Fe, Al-Fe, Al-O, Fe-O, and O-O) interactions and mono-atomic glue and electrostatic parameters for Al, Fe, and O.

Table 1: Experimental and Predicted Properties of Components

		Exp	Predicted	Exp	Predicted	Exp	Predicted	Exp	Predicted
		0 K		**300 K**			**600 K**	**900 K**	
Al (fcc)	a (Å)	4.05	4.05	4.078188	4.078188	4.106376	4.106376	4.134564	4.134564
	C_{11}(GPa)	114.3	112.44	106.43	111.65	94.57	57.63	80.81	54.78
	C_{12} (GPa)	61.9	63.16	60.35	61.1	57.58	48.93	53.92	46.14
	C_{14}(GPa)	31.6	29.04	28.21	30.02	23.72	4.32	18.55	4.11
Fe (bcc)	a (Å)	2.87	2.87	2.8804181	2.8804181	2.8908362	2.8908362	2.90125413	2.90125413
	C_{11}(GPa)	252	233.74	225.9	237.62	209.4	197.2	196.9	199.52
	C_{12}(GPa)	138	151	139.8	149.66	128.3	117.1	110.1	114.15
	C_{14}(GPa)	122	116.61	115.9	117.97	109.46	81.33	102.98	80.724
α-Al_2O_3	a (Å)	4.7504	4.7504	4.7602	4.7602	4.77194	4.77194	4.78499	4.78499
	c (Å)	12.964	12.964	12.9912	12.9912	13.0192	13.0192	13.0522	13.0522
	C_{11}(GPa)	500.07	503.1	497	499.97	486.8	263.2	475.4	496.3
	C_{12} (GPa)	161.67	180.1	164	183.7	165.01	354.7	167.1	197.4
	C_{13}(GPa)	111.4	90.5	111	90.62	109.32	80.3	105.7	89
	C_{33}(GPa)	502.4	490.2	498	490.97	487.53	391.9	473.6	451.3
	C_{14} (GPa)	-23.3	-5.8	-24	-7.02	-24.22	-259.8	-24.7	-20.35
	C_{44}(GPa)	151.0173	147.3	147	147.93	139.99	-457.96	132.1	93.72
α-Fe_2O_3	a (Å)	5.0252	5.0252	5.035	5.035	5.048	5.048	5.0675	5.0675
	c (Å)	13.716	13.716	13.747	13.747	13.7844	13.7844	13.8294	13.8294
	C_{11}(GPa)	249.9	273.54	242	237.5	234.1	201.7	226.2	192.41
	C_{12}(GPa)	56.8	84.02	55	86.7	53.2	90.97	51.4	95.7
	C_{13}(GPa)	16.52	27.08	16	18.4	15.5	11.2	14.96	6.4
	C_{33}(GPa)	235.4	184.05	228	178.9	220.6	174.8	213.13	197.34
	C_{14} (GPa)	-12.6	56.93	-13	51.8	-13.4	48.32	-13.85	51.01
	C_{44}(GPa)	**87.8**	106.6	**85**	103.9	**82.2**	101.35	**79.5**	97.7

Fitting of the potential involves least square minimization of the difference between the fitted values and actual values of the properties [13]. A lattice program [14] 'GULP 1.3' is used for the fitting calculations. As a practical issue, it is very difficult to obtain a set of parameters for a classical interatomic potential that can yield satisfactory predictions of material properties at all temperatures and pressures. Fitting based on the use of Helmholtz free-energy instead of the static lattice energy offers an attractive approach to take into account the temperature dependence of properties up to approximately one half of the melting temperature of solids [15]. However, such an approach is inapplicable to the NRMP mixture system under consideration since the system includes metals Al and Fe since it neglects the electronic contribution to Helmholtz free energy owing to its classical nature. Therefore, a number of potential parameter sets were obtained by fitting to the component properties over a range of temperature of 0-900 K.

Each parameter set was then further tested to predict the properties of components over a range of temperature of 0-900 K. It was found that the parameter set obtained by fitting to the properties at room temperature (300 K), best predicts the properties at other temperatures (0-900 K). The prediction of this parameter set is shown in Table 1, along with the experimental properties at different temperatures. The values of the properties at different temperatures was obtained from the data provided by [16]. The potential model with the parameter set obtained is therefore valid over a range of temperature and is suitable to describe pressure-induced mechanochemical deformation of the Fe_2O_3+Al mixture system. Tables 2 and 3 show the value of potential parameters obtained after fitting at 300 K.

Table 2. Values of Monatomic Potential Parameters

	A	B	E_0 (eV)	r_0 (Å)	β_0 ($Å^{-1}$)	t_0	β_1 ($Å^{-1}$)	t_1	β_2 ($Å^{-1}$)	t_2	β_3 ($Å^{-1}$)	t_3	χ (eV)	J (eV)	ζ ($Å^{-1}$)	Z (eVÅ)
Al	0.97	1	4.769	0.19	1.604	5	37.55	7.5	6.824	8.6	37.59	9.6	0	9.63	0.94	0.75
Fe	0.63	1	7.659	0.28	1.506	1	4.658	1.9	4.925	1.1	4.5	4.1	1.58	2.6	0.6	6.79
O	-1.8	1	2.6	1.1	2.5	0.3	2.7	2.1	0.032	2.1	0.696	3.1	5.48	14.0	2.14	0.0

Table 3. Values of Pair Potential Parameters

	Ψ (eV)	t (Å)	γ ($Å^{-1}$)	Cutoff (Å)
Al-Al	12.8	0.230	0.0273	5
Al-O	14.1	5.3	-0.17	5
Al-Fe	-1.48	5.178	-2.70	5
Fe-Fe	0.331	2.40	1.82	5
O-O	1207.5	-26.98	0.206	5
Fe-O	-1.84	4.11	-0.36	5

The potential is currently being implemented in a MD code to analyze the mechanical behavior of the NRMP mixture at a range of loading and boundary conditions. For this purpose, a range of Fe_2O_3+Al NRMP composites have been developed using widely used technique of voronoi tessellation. The distribution of grains sizes in the developed nanophase composites is close to the widely observed log-normal grain size distribution in nanophase materials, see [17] and [18]. Figure 1 shows one such nanophase composite with Al volume fraction kept at 0.4 and Fe_2O_3 volume fraction kept at 0.6 according to the standards observed for Fe_2O_3+Al thermite reactions, see [19].

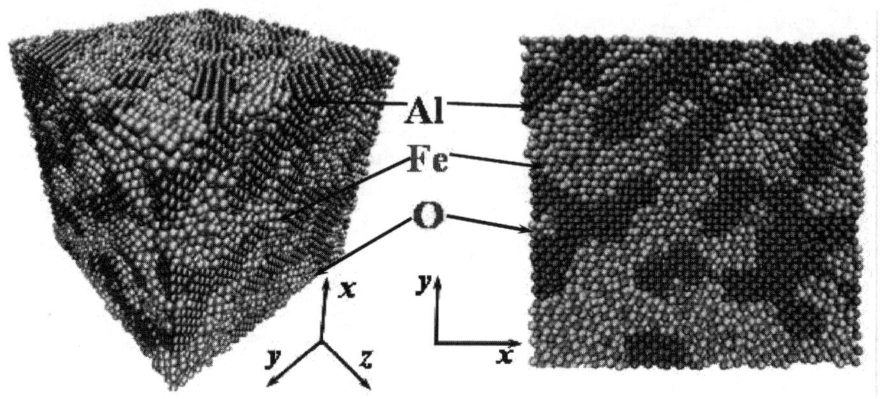

Figure 1 A Fe_2O_3+Al nanophase composite with volume fractions of Al and Fe_2O_3 at 0.4 and 0.6 respectively.

CONCLUSION

A framework for nanoscale analysis using classical MD is developed as a tool for the design of nanostructured reactive metal powder mixtures. An MD potential to describe Fe, Al, Fe-Al intermetallic, α-Fe_2O_3, and α-Al_2O_3 simultaneously over a range of temperature is obtained. The potential includes electronegativity equalization to account for pressure-induced phase changes. It is expected that MD calculations with the developed potential will be helpful in obtaining useful insights into the mechanical behavior of the NRMP mixture considered.

ACKNOLEDGMENTS

Funding through an AFOSR MURI grant to Georgia Tech and a NaST fellowship from Georgia Tech is greatly appreciated. The authors thank Dr. Julian D. Gale for providing the GULP 1.3 code used in the fitting calculations for interatomic potential. Computations are carried out at ERDC and NAVO MSRCs.

REFERENCES

1.S. M. Foiles and J. B. Adams, Phys. Rev. B **40** 5909 (1989).
2.Y. Mishin, M. J. Mehl, D. A. Papaconstantopoulos, A. F. Voter and J. D. Kress, Phys. Rev. B **63** 224106 (2001).
3.S. Ogata, T. J. Campbell, K. Tsuruta, A. Nakano, R. K. Kalia, P. Vashishta and C.-K. Loong, in *Oxidation dynamics of nanophase aluminum clusters: A molecular dynamics study*, edited by E. Ma and P. Bellon, (Materials Research Society Pittsburgh, 1998), p. 625-630.
4.A. C. T. Van Duin, A. Strachan, S. Stewman, Q. Zhang, X. Xu and W. A. Goddard III, J. Phys. Chem. A **107** (19), 3803 (2003).

5.W. A. Goddard III, Q. Zhang, M. Uludogan, A. Strachan and T. Cagin, in *The reaxFF polarizable reactive force fields for molecular dynamics simulation of ferroelectrics*, edited by R. Cohen and T. Egami, (American Institute of Physics, Washington DC, 2002), p. 45-50.
6.A. E. Carlsson Solid State Physics **43** 1 (1990).
7.M. I. Baskes Phys. Rev. B **46** (5), 2727 (1992).
8.R. G. Parr and R. G. Pearson, J. Am. Chem. Soc. **105,** 7512 (1983).
9.A. K. Rappe and W. A. Goddard III, J. Phys. Chem. **95,** 3358 (1991).
10.F. H. Strietz and J. W. Mintmire, Phys. Rev. B **50** (16), 11996 (1994).
11.M. I. Baskes and R. A. Johnson, Mod. Sim. Mat. Sci. Engg. **2,** 147 (1994).
12.R. Besson and J. Morillo, Phys. Rev. B **55** (1), 193 (1997).
13.V. Tomar and M. Zhou, in *Molecular dynamics simulation of shock induced detonation*, edited by Y. M. G. M.D. Furnish, W. Forbes (AIP Press, Portland, 2003), (In Press).
14.J. D. Gale and A. L. Rohl, Mol. Sim. **29** (5), 291 (2003).
15.J. D. Gale J. Phys. Chem. B **102** 5423 (1998).
16.H. B. Huntington in *The elastic constants of crystals*, edited by F. Seitz and D. Turnbull (Academic Press, New York, 1958), p. 213.
17.J. Schiøtz, T. Vegge, F. D. Di Tolla and K. W. Jacobsen, Phys. Rev. B **60** 11971 (1999).
18.A. J. Haslam, S. R. Phillpot, D. Wolf, D. Moldovan and H. Gleiter, Mat. Sci. Engg. **A318** 293 (2001).
19.M. B. Boslough J. Chem. Phys. **92** (3), 1839 (1990).

Part VI: Fracture, Fatigue, and Degradation

A. Experiments

Mat. Res. Soc. Symp. Proc. Vol. 821 © 2004 Materials Research Society P2.5

UTILIZING ON-CHIP TESTING AND ELECTRON MICROSCOPY TO STUDY FATIGUE AND WEAR IN POLYSILICON STRUCTURAL FILMS

D.H. Alsem[1,2,3], E.A. Stach[3], C.L. Muhlstein[4], M.T. Dugger[5], and R.O. Ritchie[1,2]

1 Department of Materials Science and Engineering, University of California, Berkeley
2 Materials Sciences Division, Lawrence Berkeley National Laboratory
3 National Center for Electron Microscopy, Lawrence Berkeley National Laboratory
4 Department of Materials Science and Engineering, Pennsylvania State University
5 Materials and Process Sciences Center, Sandia National Laboratory, Albuquerque

Abstract

Wear and fatigue are important factors in determining the reliability of microelectromechanical systems (MEMS). While the reliability of MEMS has received extensive attention, the physical mechanisms responsible for these failure modes have yet to be conclusively determined. In our work, we use a combination of on-chip testing methodologies and electron microscopy observations to investigate these mechanisms. Our previous studies have shown that fatigue in polysilicon structural thin films is a result of a 'reaction-layer' process, whereby high stresses induce a room-temperature mechanical thickening of the native oxide at the root of a notched cantilever beam, which subsequently undergoes moisture-assisted cracking. Devices from a more recent fabrication run are fatigued in ambient air to show that the post-release oxide layer thicknesses that were observed in our earlier experiments were not an artifact of that particular batch of polysilicon. New *in vacuo* data show that these silicon films do not display fatigue behavior when the post release oxide is prevented from growing, because of the absence of oxygen. Additionally, we are using polysilicon MEMS side-wall friction test specimens to study active mechanisms in sliding wear at the microscale. In particular, we have developed *in vacuo* and *in situ* experiments in the scanning electron microscope, with the objective of eventually determining the mechanisms causing both wear development and debris generation.

Introduction

Recent advances in micromachining have increased the demand for more reliable microscale structures. Although silicon is an excellent construction material at the microscale and is most widely used in these structural thin film applications, it is very brittle. Consequently, reliability is often the limiting factor as far as commercial applications are concerned. Since the surface to volume ratio of these structural films is very large, classical models for failure modes cannot always be applied. At these size scales, surface effects may become dominant in determining the mechanical properties of the material. The main reliability issues for microelectromechanical systems (MEMS) are stiction, fatigue and wear. Stiction occurs when freestanding parts stick to the substrate or to each other when releasing the device (release stiction), or while operating, due to particular surface contact conditions (in-use stiction) [1]. Fatigue is important in cases where parts get strained a large number of times below their fracture stress, whereas wear occurs at contacting moving surfaces. While the reliability of MEMS has received extensive attention, the physical mechanisms responsible for these failure modes have yet to be conclusively determined [2-7]. The goal of this research is to find the mechanisms responsible for wear and fatigue in polysilicon structural thin films. In this work, on-chip testing methods are combined with electron microscopy. Specifically, fatigue specimens were cycled and observed using high-voltage transmission electron microscopy (HVTEM). For the wear studies, an *in situ* scanning electron microscope (SEM) method was developed to permit the collection of wear debris while the films are wearing.

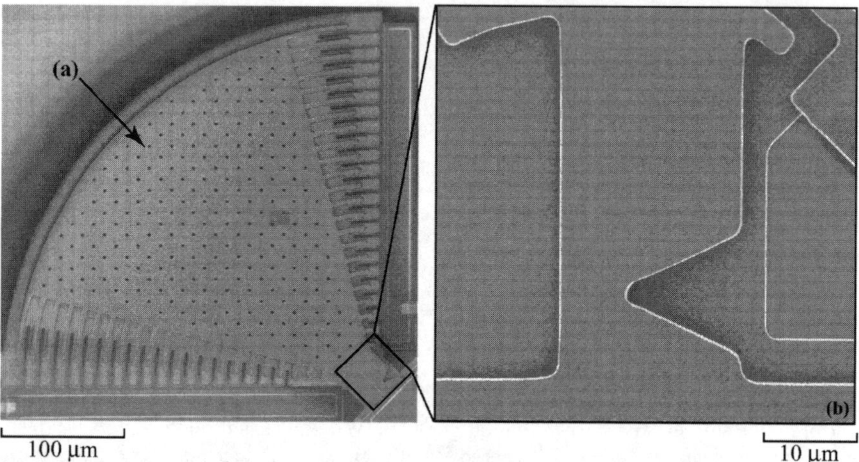

Figure 1: Scanning electron micrographs of the polysilicon MEMS fatigue life characterization resonator fabricated at the MEMSCAP MUMPs foundry (formerly CRONOS) [8]. a) Triangular proof mass with interdigited comb drive actuator and capacitive displacement sensor; b) notched cantilever beam connecting the resonator mass and anchor.

Fatigue

Our previous work [9-12] has shown that micron-scale silicon structural films, unlike bulk silicon, display high cycle fatigue behavior at room temperature. Fatigue resonator devices, actuated by an interdigited comb drive (Figure 1a) at ~40Khz and a load ratio (R) of -1 at ~25°C with 30-50% relative humidity (RH), have been used to measure the fatigue behavior of a notched cantilever beam (Figure 1b). The devices are actuated by one of the comb drives, while using the other one to monitor displacements using capacitive sensing electronics. With this system, displacements can be kept constant during the entire test. With known dimensions, finite element calculations are used to calculate the stresses at the notch root. To summarize our previous results, a "metal-like" stress versus number of cycles to failure (*S/N*) curve was obtained for 2 μm thick polysilison. Additionally, HVTEM of fatigued and fractured fatigue test devices revealed a thickening of the post-release oxide layer at the notch root for fatigued samples, in contrast to (manually) fractured specimens which showed no such thickening. This led us to propose a 'reaction layer' fatigue mechanism consisting of three steps: (i) the initial post-release oxide layer grows while mechanically stressed during cycling, (ii) moisture-assisted cracking of the oxide layer results in stable crack growth, and (iii) provided the oxide layer is thick enough, at some point the critical crack size for the entire structure is reached and the structure fails catastrophically.

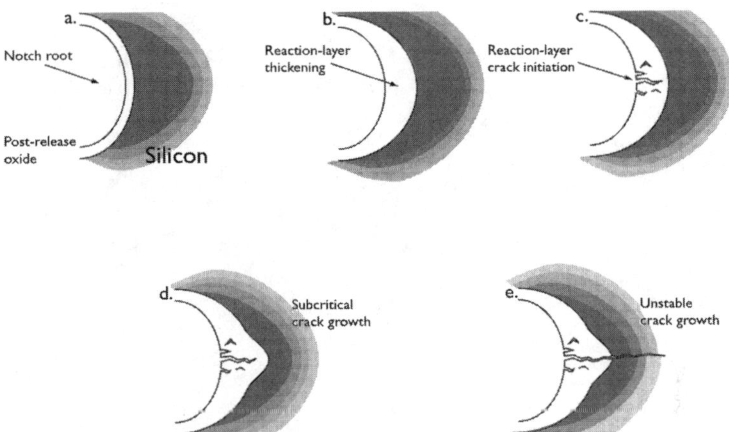

Figure 2: Schematic of the reaction-layer fatigue mechanism at the notch of the polycrystalline silicon cantilever beam (a). (b): Localized oxide thickening at the notch root. (c): Environmentally assisted crack initiation in the native oxide at the notch root. (d): Additional thickening and cracking of reaction layer. (e): Unstable crack growth in the silicon film.

In this paper, we report on fatigue data acquired from devices from a different run at the Multi User MEMS Process (MUMPs) foundry, as well as new *in vacuo* observations. This allows us to check the reproducibility of our previous findings and to characterize the changes in fatigue behavior and development of reaction layers in a relatively oxygen-free environment. Figure 3 shows a *S/N* curve from previously obtained fatigue data [9,10] in ambient air from devices from MUMPs run 18 (best fit line). These data are compared to the current data from MUMPs fabrication run 50, acquired in both ambient air (~25°C, 30-40% RH) and vacuum (~2.0 x 10^{-7} mbar). No samples failed *in vacuo*; correspondingly, the data points plotted are run-outs (i.e., the number of cycles when the tests were stopped). From these results, two important deductions can be made: the fatigue tests performed with specimens from a different fabrication run display identical fatigue behavior to previous tests. Next to that, the attempts to fatigue polysilicon in vacuum showed that the devices did not fail at stresses up to the fracture strength, even after cycling up to a high number of cycles. HVTEM of fatigued and fractured devices in ambient air and in vacuum reveals a thickening of the oxide layer for the specimen fatigued in ambient air of up to 100-150 nm. However, this local thickening at the notch root is not visible for the fractured specimens in ambient air and in the specimens that were loaded monotonically. For the latter specimens the post release oxide layer (~30 nm) is not thick enough to accommodate the critical crack size for the structure (~50 nm) [11]. In order to see the oxide in the notch root, the samples that did not fatigue in vacuum were either fractured manually or by applying large displacements electrostatically. Thereafter, they were lifted from the chip onto a TEM clamshell grid. Both the newly acquired fatigue data as well as the TEM micrographs of the oxide layers point towards the 'reaction layer' fatigue mechanism being the governing mechanism for fatigue failure in polycrystalline silicon thin films.

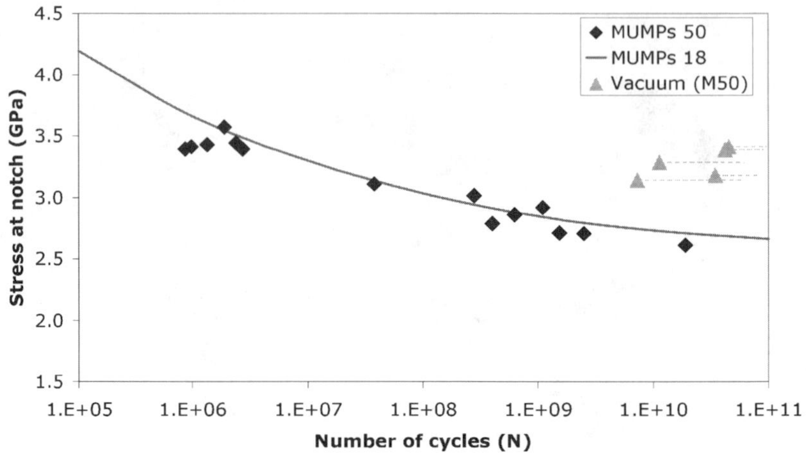

Figure 3: Polysilicon *S/N* curve. Fitted line for previous data from MUMPs run #18 in ambient air [9,10] compared with current fatigue data from MUMPs run 50 in ambient air and vacuum (~2.0 x 10^{-7} mbar). As no fatigue failures are seen *in vacuo*, data points are run-outs.

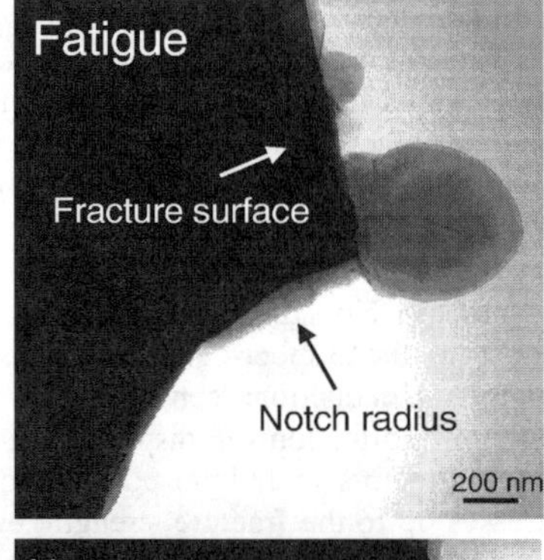

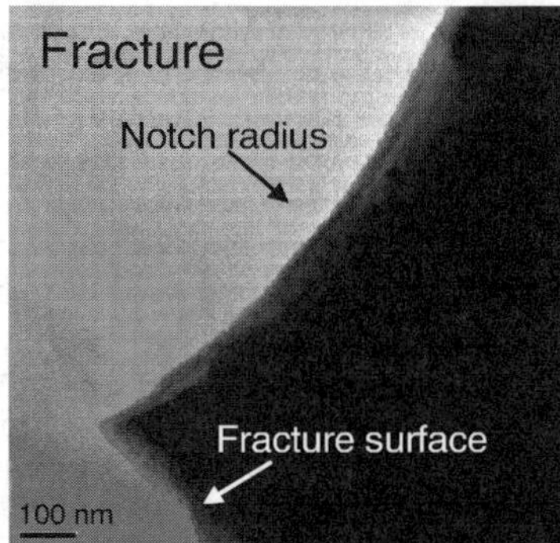

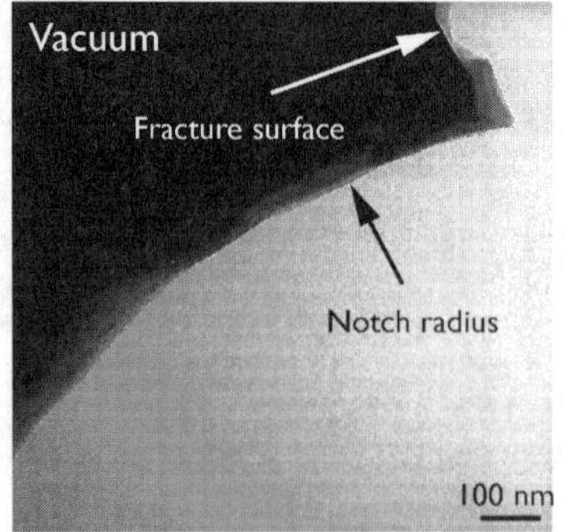

Figure 4: HVTEM images from failed resonator devices showing surface oxide layers in the notch. *UPPER LEFT*: fatigued in ambient air (notice the sliver of Si that got attached to the fracture surface after being fatigued (twinning is visible), similar slivers were found on comb drive teeth; principle stress at the notch root: 2.86 GPa; number of cycles at failure: 6.28 x 10^{8}).
UPPER RIGHT: Fractured specimen in ambient air: no (local) oxide thickening.
LEFT: Device after fatigue attempt *in vacuo* and manual fracture: no oxide layer thickening (principal stress at the notch root: 3.29 GPa, number of cycles when stopped: 1.14 x 10^{10}).

Wear

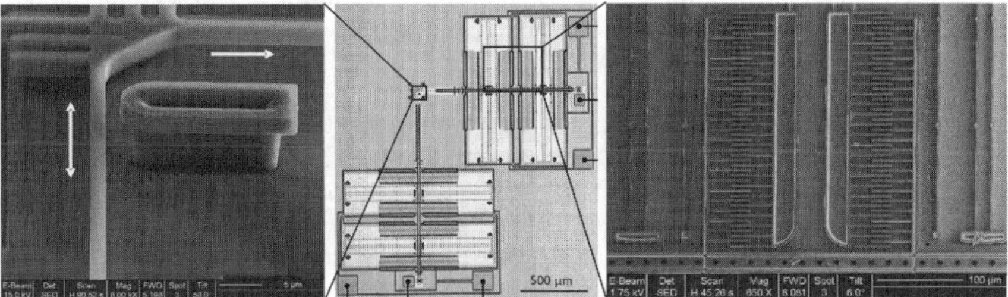

Figure 5: Polysilicon side wall friction test device fabricated at the Sandia SUMMIT process. The device produces two-axis motion provided by electrostatic actuation of interdigited comb drives used to pull a beam against a post to wear.

To study active mechanisms in sliding wear we have used polysilicon MEMS side-wall friction test specimen fabricated at the Sandia National Labs SUMMIT process (Figure 5) [13]. *In vacuo* and *in situ* experiments in the scanning electron microscope (SEM) were conceived, with the objective of determining the mechanisms causing both wear development and debris. Wire feed-troughs and special stages built for a dual beam focused ion beam (DB-FIB) system and a field emission scanning electron microscope (FESEM) allowed the side-wall friction devices to be run *in situ*. Imaging actuated devices is a trade-off between the amount of detail that can be resolved (resolution) and the stability of the images. The alternating charges on the comb drives, induced by voltages from ±80 V, influence the electrons coming off the sample. A high kV beam provides the most stable image, but it has less surface detail visible (Figure 6b). A low

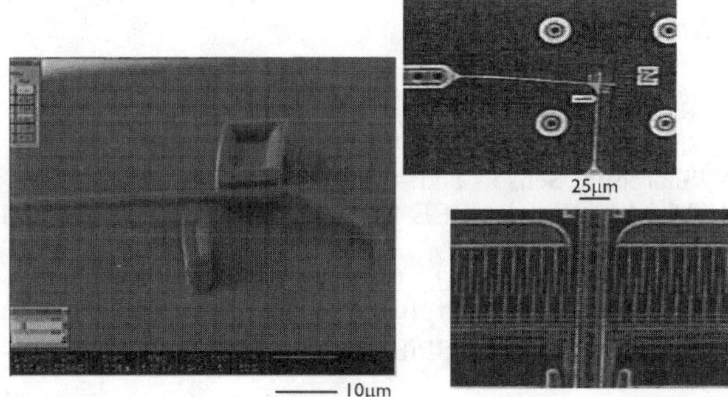

Figure 6: Screenshots from *in situ* SEM wear experiments (a: *LEFT*) synchronized driving frequency and image acquisition at lower beam acceleration (5 kV) (b: *RIGHT*) wear (*TOP*) and comb drive (*BOTTOM*) at higher beam current (20 kV).

kV beam results in a heavily vibrating image but with more detail. To optimize the information that we could obtain, the devices were run at the frequency of the electron detector (around 3Hz; depending on the magnification). This results in a stable synchronized image that can reveal the development of wear debris with time (An example of the type of video acquired can be seen in Figure 6a). Current research including FIB lift-off sample preparation techniques to make TEM samples of worn beams are used to complement these results and promises a fruitful way of learning more about the mechanism(s) that cause wear on the micro-scale.

Conclusions

Fatigue and wear in silicon structural films were investigated. To illustrate the reproducibility of earlier acquired fatigue data and to verify the relevance of the used material [9,10], devices from MUMPs run 50 were actuated in ambient air. When compared with earlier data on MUMPs run 18, identical fatigue behavior was observed and HVTEM images revealed the same post-release oxide layer thickening after fatiguing as in earlier tests. Most importantly, for fatigue studies at stresses very close to the fracture strength and at a large number of cycles, no fatigue failures at all were observed in vacuum (~2.0 x 10^{-7} mbar). In addition, the local oxide layer thickening at the notch root was absent in the vacuum tests. These new results serve as additional evidence to support the validity of the 'reaction layer' fatigue model.

An *in situ* SEM method of collecting wear data was developed. Synchronizing the device's actuation with the SEM detector refresh rate results in a stable image that shows a high amount of detail. Preliminary results were obtained and in combination with FIB TEM lift off sample preparation techniques this will serve as a powerful tool to obtain more detailed knowledge about the mechanisms that cause wear in polysilicon at these length scales.

Acknowledgements

The authors would like to thank Bob Ashurst for sharing his experience of operating the wear devices. This work was supported by the Laboratory Directed Research and Development Program of Lawrence Berkeley National Laboratory under the Department of Energy Contract No. DE-AC03-76SF00098.

References

1. R. Maboudian, W.R. Ashurst *et al.* Tribology Letters **12**: p95–100, (2002).
2. S.M. Allameh, P. Shrotriya, et al, J. MEMS **12**(3), 313-324, 2003.
3. J. Bagdahn, W.N. Sharpe Jr., Sensors and Actuators A **103**, 9-15, 2003.
4. D. M. Tanner, W. M. Miller, *et al.* . IEEE 42th Annual International Reliability Physics Symp. Proc., p26-35, 2003.
5. D. M. Tanner, J. A. Walraven, *et al.*, IEEE 37th Annual International Reliability Physics Symposium Proceedings, 189-197, 1999.
6. D. C. Senft and M. T. Dugger, Proc. SPIE Micromachined Devices and Components III, **3224**, 31-38, 1997.
7. S. T. Patton and J. S. Zabinski, Tribology International **15**: p373-379, 2002.
8. More information on the MUMPS process at:
http://www.memscap.com/memsrus/crmumps.html
9. C.L. Muhlstein, E.A. Stach, R.O. Ritchie, Appl. Phys. Lett. **80**(9), 1532-1534, 2002.
10. C.L. Muhlstein, E.A. Stach, R.O. Ritchie, Acta Mat. **50**, 3579–3595, 2002.
11. C.L. Muhlstein, S.B. Brown, R.O. Ritchie, Sensors and Actuators A **94**: P177-188, 2001.
12. C.L. Muhlstein, S.B. Brown, R.O. Ritchie, J. MEMS **10**(4): 593-600, 2001.
13. More information on the SUMMIT process at: http://mems.sandia.gov

Mat. Res. Soc. Symp. Proc. Vol. 821 © 2004 Materials Research Society P7.5

Improved Fracture Toughness in Advanced Nanocrystalline Ceramic Composites

Joshua D. Kuntz, Guo-Dong Zhan, and Amiya K. Mukherjee
Department of Chemical Engineering and Materials Science
University of California - Davis
Davis, CA 95616

ABSTRACT

Nanocrystalline materials have demonstrated very interesting changes in physical, chemical and mechanical properties at severely diminished length scales. This article focuses on the topic of nanocrystalline ceramic composites specifically designed for applications requiring improved fracture toughness. The difficulty in producing fully consolidated ceramic composites that retain a nanocrystalline structure is the main hurdle for thorough investigations in this area. This obstacle has been overcome in the current investigation through the use of a fast, comparably lower temperature, sintering technique e.g., Spark Plasma Sintering. Alumina based nanocomposites incorporating carbon nanotubes and additionally incorporating nanocrystalline niobium have yielded fracture toughness values that have exceeded that for pure nanocrystalline alumina by more that 300%. This introduces the question of whether this improvement is merely additive or evidence of a synergistic toughening mechanism involving ductile phase and fiber toughening.

INTRODUCTION

Advanced ceramic materials are attractive because of their low density, chemical inertness, high strength, high hardness and high temperature capability. Nanocrystalline ceramics are commonly defined as having a grain size of 100nm or less. While this definition is rather arbitrary, nanocrystalline ceramics are known to possess unique physical and mechanical properties, including 1) enhanced superplasticity, 2) superior strength, and 3) optical transparency in usually opaque ceramics. Pilot-scale facilities for nanopowder synthesis and the commercialization of sizable quantities of certain types of nanosize powders have been achieved. The fabrication of nanopowders into fully dense components that retain a nanocrystalline grain size has lagged behind powder synthesis and characterization. In part, the gap between powder synthesis and fabrication is related to an incomplete theoretical understanding of the mechanisms and consequences of densification and sintering when grain interfacial regions dominate. Equally relevant is the incomplete understanding of the particular experimental conditions that yield high-density compacts without microstructural coarsening. The available experimental work in this area clearly demonstrates that the conditions for densification and sinterability of nanocrystalline ceramics is system-specific and is not readily deduced from theory alone at the present time.

Nanocrystalline ceramics do not appear to possess higher fracture toughness than conventional microcrystalline ceramics. Therefore, it is very likely that, in spite of the significant advantage in other properties, toughness is likely to be the bottleneck that controls the application of nanocrystalline ceramics in both structural and functional applications. In the last two decades significant advancements have been made in the understanding of toughening mechanisms in microcrystalline ceramics. Unfortunately many of the microcrystalline material systems with improved fracture toughness have shown significant degradation of strength and hardness. This has lead to interest in nanocrystalline ceramic composites where the possibility exists to offset this property degradation through the increases in strength and hardness typically seen in nanocrystalline materials.

FIBER TOUGHENING IN CERAMICS

Certain toughening mechanisms in microcrystalline ceramic systems have been reasonably well developed and investigated. It has yet to be determined whether these same mechanisms will be germane to the corresponding nanocrystalline systems. The most common toughening mechanism that is associated with the incorporation of fibers into a ceramic matrix is fiber bridging. This toughening mechanism involves the bridging of the crack wake by the second phase fibers (Figure 1). The toughening effect is achieved when the fibers either shed load from the crack tip while remaining intact (Figure 1(c)), the interface debonds between the fiber and the matrix followed by pull out (Figure 1(b)), and/or the individual fibers fracture followed by energy adsorption through pull out of the broken fiber (Figure 1(a)). These effects lead to increased extrinsic toughness. Another possible toughening mechanism as seen in fiber reinforced ceramic composites is crack deflection. When the fiber is of a particular orientation (Figure 1(d)), the crack propagation direction can be deflected away from the axis of highest stress to a less efficient orientation directed by the longitudinal orientation of the fiber. This leads to increased fracture energy through increased fracture surface area and lower driving forces due to the reduced resolved normal stresses at the crack tip. This phenomenon will enhance the intrinsic toughness. Both crack deflection and fiber bridging should be pertinent to nanoceramic composites, since neither method is inherently dependent on the matrix grain size.

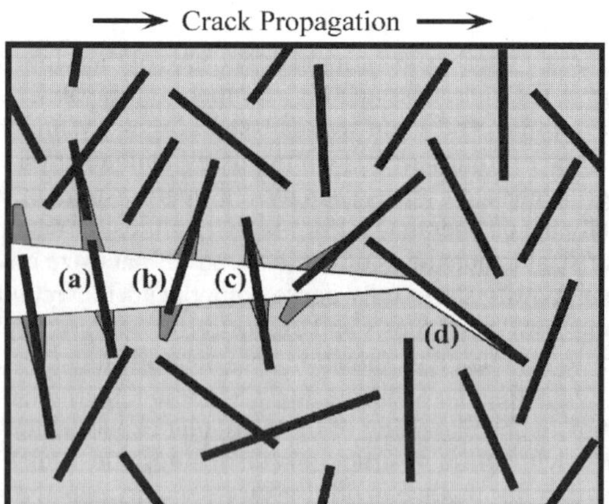

Figure 1. Schematic of the fiber toughening mechanism in a ceramic matrix composite: (a) fiber rupture followed by pull-out, (b) de-bonding and fiber pull-out, (c) fiber bridging in the crack wake, and (d) crack deflection along fiber-matrix boundary.

Carbon nanotubes (CNTs) are attracting increasing interest as constituents of nanostructured materials for a wide range of applications. [1-3] Single-walled carbon nanotubes (SWCNTs) have a particularly desirable combination of mechanical, thermal, and electrical properties. They have an elastic stiffness comparable to that of diamond ~1.5 TPa, and they are several times as strong (yield strength 52 GPa).[1] Additionally, CNTs conduct electricity along their length with very little resistance. Depending on bonding orientation within the nanotube, a SWCNT may theoretically be either metallically conducting or semiconducting. Experimental results show that a SWCNT rope has a longitudinal conductivity of 106 S/m at 300 K.[2] Their size, shape, and properties make them prime candidates for use in the development of potentially revolutionary composite materials, especially in combined multifunctional/structural applications.

CONSOLIDATION OF NANOCERAMICS

The system of nano-Al_2O_3-based composites is one of the most difficult to produce, since Al_2O_3 possesses one of the highest homologous temperatures for full-density sintering. The high sintering temperatures lead to aggressive grain growth as the compacts reach full density. The realization of nano-Al_2O_3-based nanocomposites, e.g., successful suppression of Al_2O_3 grain growth, depends on the application of lower sintering temperature and/or shorter sintering time. Conventional processing routes such as hot-pressing or pressureless sintering will not serve this purpose. Unconventional processing routes, including spark plasma sintering (SPS), have demonstrated significantly reduced sintering temperatures and sintering times, and have achieved remarkable success in producing nanocomposites

SPS is a pressure assisted sintering method based on the theory of a high temperature plasma momentarily generated in the gaps between powder materials by electrical discharge during DC pulsing. It has been suggested[4] that the ON-OFF DC pulse energizing method could generate: (1) spark plasma, (2) spark impact pressure, (3) Joule heating, and (4) an electrical field diffusion effect. In this process, powders are loaded into a graphite die and were heated by passing an electric current through the assembly. The low heat capacity of the graphite die allows rapid heating. Therefore, SPS can rapidly consolidate powders to near theoretical density through the combined actions of a rapid heating rate, pressure application, and proposed powder surface cleaning. In our research, spark plasma sintering is carried out under vacuum in a Dr. Sinter 1050 apparatus (Sumitomo Coal Mining Co., Japan). Typical SPS processing parameters are: (1) an applied pressure between 50 and 100 MPa, (2) pulse cycles with a period of 2.5 ms and follow the pattern of 12 cycles on and 2 cycles off, and (3) maximum pulse parameters of 10,000 A and 10 V. After applying the given pressure, samples are heated to 600 $^{\circ}C$ in 2 minutes and then ramped to their sintering temperature at rates of 150 $^{\circ}C/min$ to 500 $^{\circ}C/min$.

RESULTS & DISCUSSION

We have investigated the applicability of fiber toughening to nanocomposites in two different material systems. The first material system under investigation for was alumina combined with single walled carbon nanotubes, SWCNT [5]. These composites were consolidated using SPS at 1150°C for only 3 minutes. In this system the best results to date have a toughness value of 9.7 $MPam^{1/2}$ with only 10 vol. % SWCNT added. The remarkable toughening seen in the current work stands out as the only ceramic composite to realize the promise of significant toughening through the addition of carbon nanotubes, Figure 2. This can be attributed to the ability to produce fully consolidated composites at temperatures low enough to prevent damage to the carbon nanotubes while producing an intertwining and interconnected network at alumina grain boundaries, as evident in TEM investigation, Figure 3. The 13 orders of magnitude increase in electrical conductivity in this composite is further evidence of the continuous SWCNT structure.[6]

The addition of a ductile phase to examine the effect of ductile phase toughening in a nanocomposite has been investigated through the addition of niobium to an alumina matrix. Nb was added through high-energy ball milling (HEBM) with the nano-Al_2O_3 powder and sintered by SPS at only 1100°C for 3 minutes. The 10 vol.% Nb nanocomposite has a fracture toughness value of 7 $MPam^{1/2}$ without a decrease in hardness[7]. This increase in toughness may be attributed to the novel microstructure in the nanocomposite. TEM (XDS/HRTEM) results show the Nb phase distribution of ~20 nm particles along with a continuous 3-4 nm layer at boundaries between Al_2O_3 grains, Figure 4. This microstructure should lead to toughening by increasing ductility at the crack tip along with of the traditional ligament bridging in the crack wake which is typical of micron-scaled metallic-phase toughened ceramics.

We investigated the possibility of simultaneous fiber and ductile phase toughening through the production of a 5 vol.% SWCNT and 5 vol.% Nb composite using the combined consolidation processes from above. As seen in Figure 2, this composite demonstrated the best fracture toughness in the present study. Future work is intended to investigate whether the theories of synergistic toughening proposed by Becher, Tiegs and Stump[8, 9] can be expanded to include the present work.

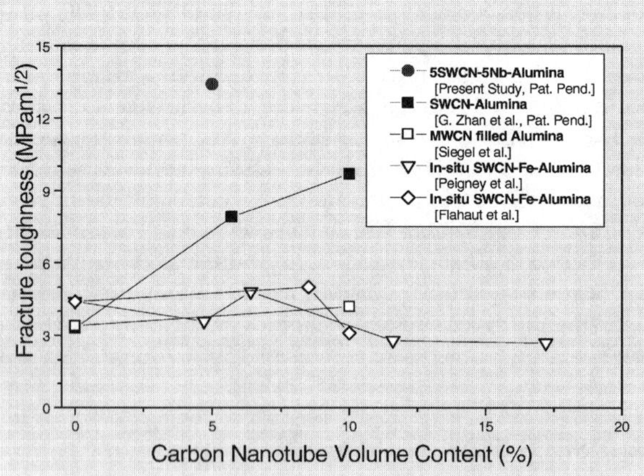

Figure 2. Fracture toughness versus carbon nanotube volume content in alumina-based composites[5, 10-13].

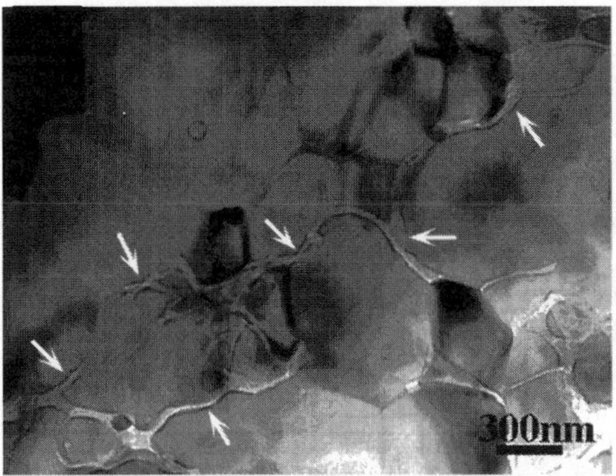

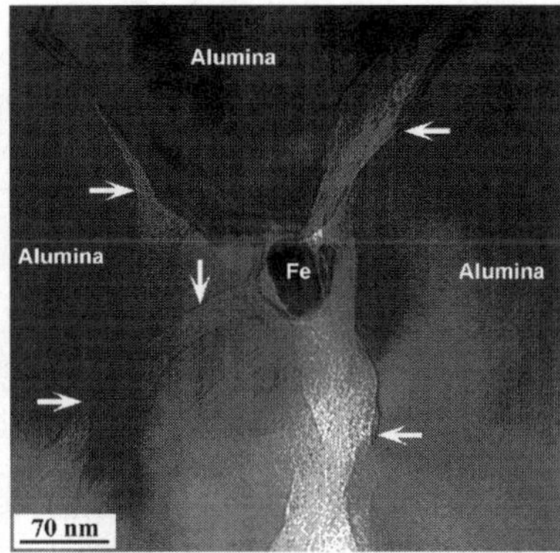

Figure 3. Bright field TEM images of the 5.7 vol.% SWCNT – alumina composite.

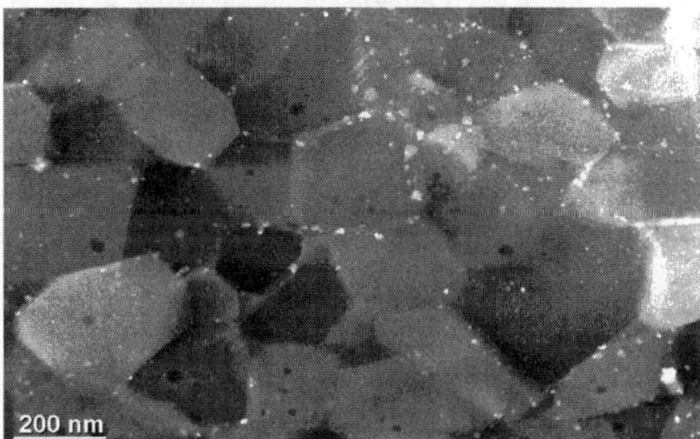

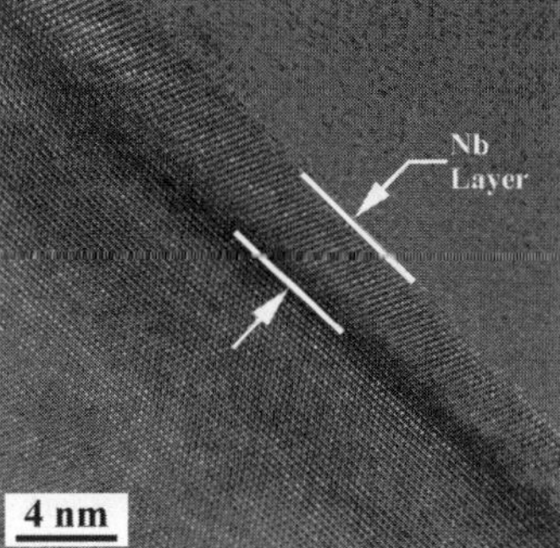

Figure 4. Dark field STEM and HRTEM images of 10 vol.% Nb-alumina composite.

CONCLUSIONS

In order for the SWCNT to be incorporated as effective reinforcing and functional agents there are several requirements. The SWCNT must be effectively bonded to the matrix. The nanotubes must be retained without damage after consolidation. Development of an intertwine network structure is also necessary. SPS is an attractive technique for achieving these objectives. The addition of Nb to the CNT-alumina composites leads to possible synergistic toughening effects between fiber and ductile phase toughening.

ACKNOWLEDGEMENTS

This material is based upon work supported by the U. S. Army Research Laboratory and the U. S. Army Research Office under grant number G-DAAD19-00-1-0185.

REFERENCES

1. Yu, M.-F., et al., *Tensile loading of ropes of single wall carbon nanotubes and their mechanical properties.* Physical Review Letters, 2000. **84**(24): p. 5552-5.
2. Thess, A., et al., *Crystalline ropes of metallic carbon nanotubes.* Science, 1996. **273**(5274): p. 483-7.
3. Baughman, R.H., A.A. Zakhidov, and W.A. de Heer, *Carbon Nanotubes--the Route Toward Applications.* Science, 2002. **297**(5582): p. 787-792.
4. Omori, M., *Sintering, consolidation, reaction and crystal growth by the spark plasma system (SPS).* Materials Science & Engineering A-Structural Materials Properties Microstructure & Processing, 2000. **A287**(2): p. 183-8.
5. Zhan, G.D., et al., *Single-wall carbon nanotubes as attractive toughening agents in alumina-based nanocomposites.* Nature Materials, 2003. **2**(1): p. 38-42.
6. Zhan, G.D., et al., *Electrical properties of nanoceramics reinforced with ropes of single-walled carbon nanotubes.* Applied Physics Letters, 2003. **83**(6): p. 1128-1230.
7. Kuntz, J.D., et al. *Properties and microstructure of alumina-niobium nanocomposites made by novel processing methods.* in *Ultrafine Grained Materials II. Proceedings. TMS Annual Meeting. TMS-Miner. Metals & Mater. Soc. 2002, pp.225-33. Warrendale, PA, USA.* 2002.
8. Stump, D.M., *Toughening and strengthening of ceramics reinforced by dilatant transformations and ductile particles.* International Journal of Solids & Structures, 1991. **28**(6): p. 669-89.
9. Becher, P.F. and T.N. Tiegs, *Toughening behavior involving multiple mechanisms: whisker reinforcement and zirconia toughening.* Journal of the American Ceramic Society, 1987. **70**(9): p. 651-4.
10. Siegel, R.W., et al. *Mechanical behavior of polymer and ceramic matrix nanocomposites.* in *Elsevier for Board of Directors of Acta Metall. Scripta Materialia, vol.44, no.8-9, 18 May 2001, pp.2061-4. USA.* 2001.
11. Peigney, A., et al., *Carbon nanotubes grown in situ by a novel catalytic method.* Journal of Materials Research, 1997. **12**(3): p. 613-15.
12. Peigney, A., et al., *Carbon nanotubes in novel ceramic matrix nanocomposites.* Ceramics International, 2000. **26**(6): p. 677-83.
13. Flahaut, E., et al., *Carbon nanotube-metal-oxide nanocomposites: microstructure, electrical conductivity and mechanical properties.* Acta Materialia, 2000. **48**(14): p. 3803-12.

Mat. Res. Soc. Symp. Proc. Vol. 821 © 2004 Materials Research Society P2.4

Evaluation of Interface Strength between a Copper Submicron Dot and Silicondioxide Substrate

Hiroyuki Hirakata, Takayuki Kitamura and Yoshitake Yamamoto
Department of Engineering Physics and Mechanics, Kyoto University, Kyoto 606-8501, Japan

ABSTRACT

An experimental evaluation method of interface strength between ductile submicron-dots and a hard substrate is developed. The validity is examined with copper (Cu) cylindrical dots of submicron scale on a silicondioxide (SiO_2) substrate. A hard-layer of tungsten (W) is employed to restrain the deformation and concentrate the stress near the free edge of Cu/SiO_2. A diamond tip is dragged horizontally along the SiO_2 surface and the load is applied to the side edge of the W layer at a constant displacement rate using a modified atomic force microscope. Both the lateral and the vertical loads and displacements are continuously monitored during the test. After the tip hits the W layer, the lateral load, F_l, increases almost proportionally with the lateral displacement, δ_l. The Cu dot with the W layer then is clearly separated from the SiO_2 along the interface. The restraint by the W layer works well so that there are little damages in both the delaminated W/Cu dot and the substrate. The delamination lateral load, F_{lC}, is successfully evaluated.

INTRODUCTION

Since micro-electronics and micro-mechanical devices consist of various components, they have many intrinsic bi-material interfaces. The stress concentration on an interface due to the deformation mismatch sometimes causes delamination, which brings about fatal malfunction of the device. On the other hand, the size required for the components becomes submicron-scale in order to shrink the device. It is necessary, therefore, to evaluate the interface strength among the submicron-components and the substrate.

Two steps are required to evaluate interface strength. The first is development of an experimental method for separating a submicron-component from a substrate, and the second is an investigation of the mechanics dominating the interface fracture.

A significant number of approaches [1 - 12] have been investigated to evaluate the interface strength among thin films and a substrate. However, the stress-concentrated region on an interface, which dominates interface delamination, is affected by the length scale [10, 13]. The delamination behavior might be influenced by not only the thickness but also the size of film. Hence, it is important to elucidate the effect of length scale on the delamination. Then, we developed a method [14] for applying a load to dots, of which three-dimensional scale is small, on a substrate by a modified atomic force microscopy (AFM), and successfully separated a hard tungsten (W) micron-dot from a silicon (Si) substrate. However, it can not be applied on ductile dots such as copper (Cu) because the contacted zone collapses before the delamination. It is necessary, therefore, to develop a new method that can separate ductile dots from a substrate without the collapse.

In this study, a delamination method for ductile submicron-dots on a substrate is proposed, and the validity is examined by a copper (Cu) submicron-dot on silicondioxide (SiO_2).

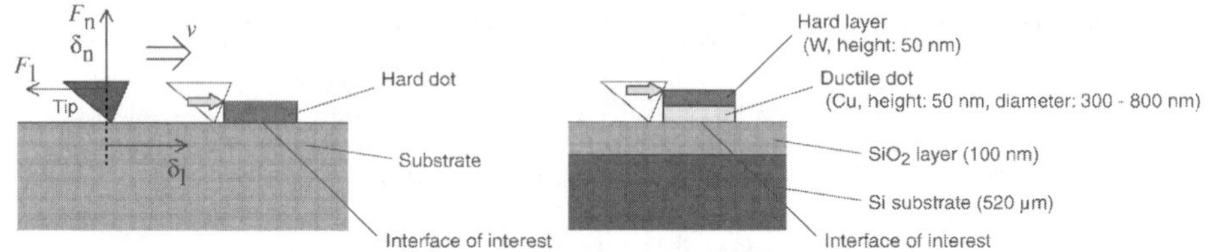

(a) Explanation of delamination test (b) Proposed method for ductile dot

Figure 1. Delamination test for dot on substrate using modified AFM

EXPERIMENTAL PROCEDURE

Proposed method

The delamination test [14] is schematically illustrated in Fig. 1 (a). A dot is prepared on a substrate. A diamond tip is dragged horizontally along the surface of the substrate and a lateral load is directly applied to the side edge of the dot using a modified AFM. Both the lateral and the vertical load and displacement are continuously monitored during the test. This method successfully separates hard dots from a substrate. In this study, a new method shown in Fig. 1 (b) is proposed for ductile dots. A hard-layer is deposited on the dot and a load is applied to the hard-layer. The hard-layer restrains the deformation and collapse of the dot. Moreover, in this method, the influence of a complicated stress field due to the contact of tip is avoided because the contact zone is away from the interface edge at which delamination occurs.

Specimen preparation

A schematic illustration of specimen tested is shown in Fig. 1 (b). A cylindrical Cu dot of submicron-scale with a height of 50 nm is deposited on a Si substrate with a 100 nm thickness of SiO_2. A layer made of W with a thickness of 50 nm is employed as a hard-layer. The effect of adhesion area on the delamination behavior is examined by using specimens (specimens 1 – 5) with different diameters from about 300 to 800 nm. The reasons why the W/Cu/SiO_2 system is selected in this study are: (1) Cu is a ductile material, (2) W and SiO_2 have a high yield stress, and (3) these are important materials in micro-electronic devices. In order to examine the effectiveness of the hard-layer, a specimen without the hard-layer (specimen S, thickness of Cu dot: 150 nm) is also prepared.

The specimens are prepared using a conventional electron-beam (EB) lithography technique. A Si wafer (thickness: 520 μm, surface: (100) plane) is thermally oxidized to grow a 100 nm thickness of SiO_2. After cleaning the wafer by ultrasonic vibration with organic solvents (acetone and isopropyl alcohol), an EB resist with a thickness of about 350 nm is deposited by spin-coating. The dot patterns are drawn using an EB lithography system. The exposed part of the EB resist is removed by developer. After reducing pressure to 4.4×10^{-4} Pa at room temperature, a Cu film of thickness 50 nm is sputtered on the specimen in argon (Ar) gas of 0.8 Pa. Then, a W film of thickness 50 nm is sputtered in the same manner. The EB resist with the W/Cu films is lifted off by remover solvent.

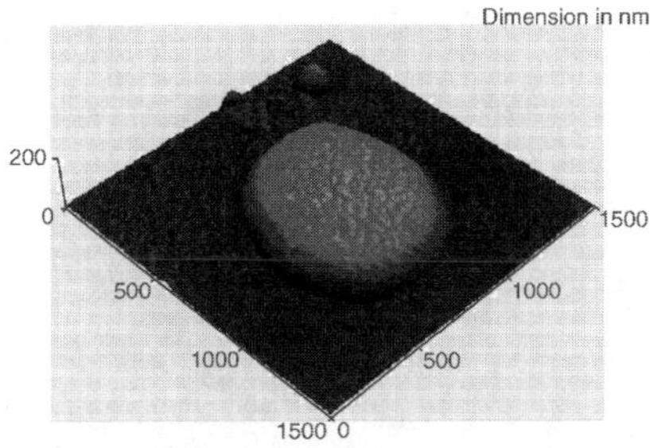

(a) Specimen 4 (Cu dot with W layer)

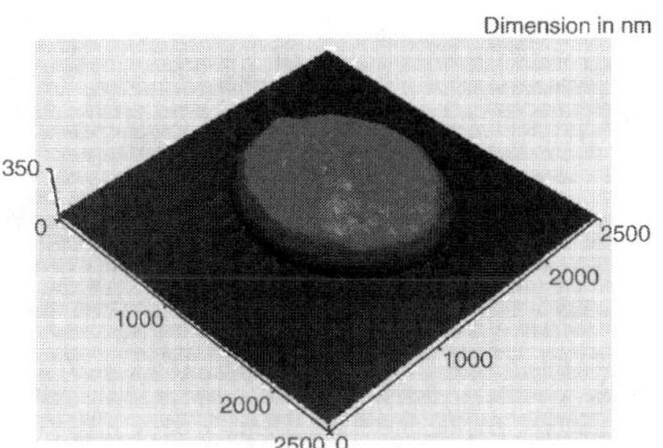

(b) Specimen S (Cu dot without W layer)

Figure 2. AFM images of specimens

Table I. Adhesion area, normal load and delamination load

Specimen No.	Adhesion area A_d, μm^2	Normal load F_n, μN	Delamination load F_{IC}, μN
1	0.089	50	25.8
2	0.208	50	29.7
3	0.269	45	26.4
4	0.457	70	28.9
5	0.603	45	35.1
S	1.44	70	-

The AFM images of the specimens 4 and S are shown in Fig. 2. Although unevenness is observed on the dot side surface, it is fairly smooth in comparison with the dot diameter. Here, the precise angle between the side surface and the interface cannot be measured by means of AFM due to the pyramidal shape of the probe. A field emission scanning electron microscopy observation confirms the sharp corner at the free-edge of interface.

Testing apparatus

A special loading apparatus, which can control precisely the lateral displacement of the tip in the resolution of 4 nm, is attached to an AFM. The apparatus consists of three force/displacement transducers, each of which has two fixed outer plates and a spring-suspended center plate. A loading tip made of diamond is attached to the center plate of the middle transducer. The normal load, F_n, of the tip is electrostatically generated by applying a voltage between the center and the outer plates in the middle transducer, and the resulting normal displacement, δ_n, is detected by change in capacitance. The side transducers generate the lateral load, F_l, in the same manner, and control the lateral displacement, δ_l, using the capacitance signal.

The loading tip is positioned at the SiO_2 surface near the W/Cu dot at first, and it is dragged horizontally along the SiO_2 surface at a constant rate of displacement, v (= 5 nm/s), as shown in Fig. 1. F_n is kept constant during the test. Then, the load is directly applied to the side edge of the W layer.

The tests with several values of F_n (45 - 70 μN) are conducted at room temperature in an air environment. The adhesion area, A_d, and the F_n are summarized in Table I.

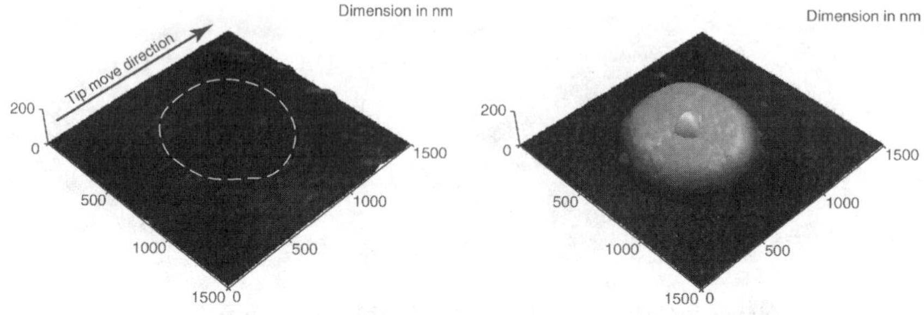

(a) Surface of substrate (b) Delaminated dot

Figure 3. AFM images of delaminated specimen (specimen 4, $A_d = 0.457\ \mu m^2$, $F_n = 70\ \mu N$)

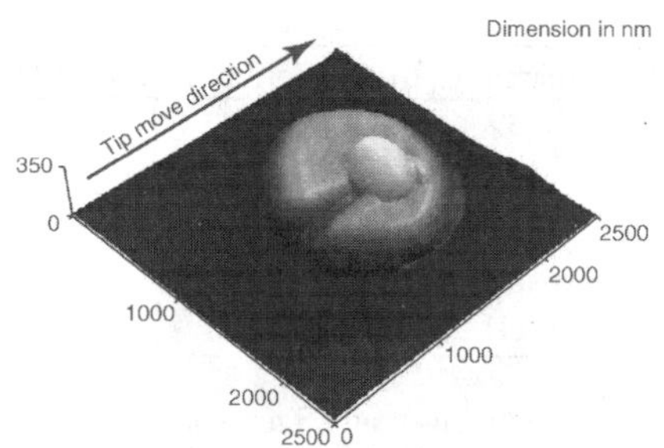

Figure 4. AFM image of collapsed Cu dot without hard-layer on substrate (Specimen S, A_d = 1.44 μm^2, $F_n = 70\ \mu N$)

RESULTS AND DISCUSSION

Figure 3 (a) shows an AFM image of the fracture surface of specimen 4 tested under $F_n = 70$ μN. The region indicated by the dashed line designates the place where the W/Cu dot was located. The fracture clearly takes place along the interface between the Cu dot and the SiO_2 substrate. There is no trace of the scratch on the substrate.

Figure 4 shows an AFM image of the dot without the hard-layer (specimen S) tested under F_n = 70 μN. The contacted area of Cu collapses and delamination is not observed. On the other hand the separated W/Cu dot was carefully examined by AFM. Figure 3 (b) shows the W/Cu dot, which shows no trace of collapse even near the tip-contacted zone. This suggests that the W/Cu/SiO_2 system works well for preventing the fracture of Cu dot.

Figures 5 (a) and (b) show the relationships between the lateral load, F_l, and the lateral displacement, δ_l, and between the normal displacement, δ_n, and δ_l in specimen 2, tested under F_n = 50 μN, respectively. It is clear that F_l remains almost zero up to about δ_l = 100 nm before F_l increases almost in proportion to δ_l up to point "a". This indicates that the tip hits the edge of the W layer at about δ_l = 100 nm as schematically illustrated in Fig. 1 (b). In other words, the load is applied to the dot in the region where F_l increases. Attention must be paid to the fact that δ_n also increases and reaches up to about 10 nm at point "a". The tip contacts only the W layer at point "a" due to the angle of the tip. The abrupt decrease of F_l and δ_n at the point indicates the separation of the W/Cu dot from the SiO_2 substrate, while no abrupt decrease is observed for the dot without the hard-layer (specimen S) as shown in Fig. 6. After then, the tip moves on the SiO_2 substrate again. Then, the pure resistance against the deformation and the subsequent delamination is measured because the friction is negligible.

The peak loads, F_{IC} (the lateral load at point "a"), are listed in Table I. F_{IC} slightly increases with an increase in adhesion area, A_d. Then, the interface strength can be estimated by deformation analysis as the concentrated (intensified) stress at the interface edge on the basis of the obtained values of F_{IC}. However, there are difficulties in performing the precise analysis, because a non-linear analysis is inevitable for the deformation of the Cu dot and the property of minute materials after yield is complex and is different from those of bulk materials [15].

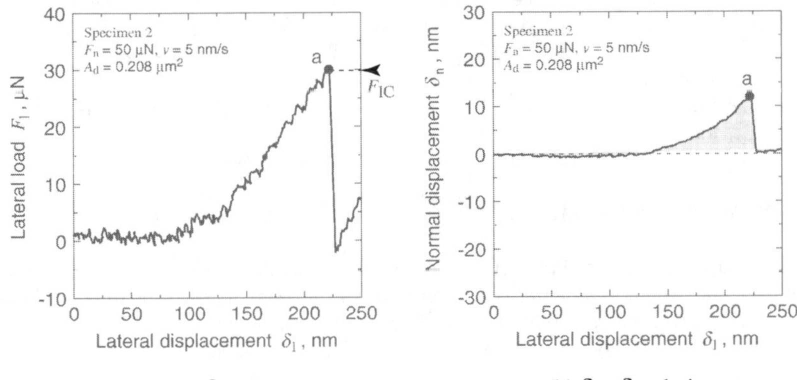

(a) F_l - δ_l relation (b) δ_n - δ_l relation

Figure 5. Relationships between lateral load and lateral displacement and between normal displacement and lateral displacement in specimen 2 (A_d = 0.208 μm^2, F_n = 50 μN)

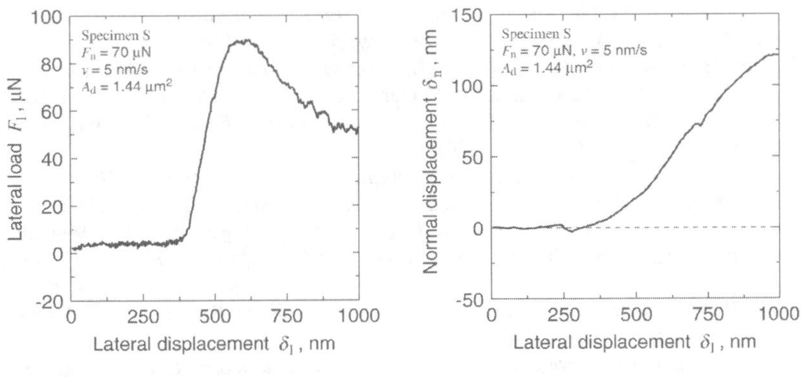

(a) F_l - δ_l relation (b) δ_n - δ_l relation

Figure 6. Relationships between lateral load and lateral displacement and between normal displacement and lateral displacement in specimen S (A_d = 1.44 μm^2, F_n = 70 μN)

Moreover, the effect of the intrinsic internal stress introduced during the fabrication process has to be taken into account [10]. This is next target of this project, and it is out of scope of this paper.

CONCLUSION

We developed a method for separating a ductile submicron-dot from a hard substrate using a modified AFM, which can control and precisely measure the load and tip displacement. The method is applied to Cu submicron-dots on a SiO_2 substrate. The results obtained are summarized as follows:

(1) A W layer is deposited on the Cu dot as a hard-layer to restrain the collapse of the dot and to decrease the influence of a complicated stress field due to the contact of tip.
(2) The diamond tip is dragged horizontally along the SiO_2 surface and a load is applied to the side edge of the W layer. The Cu dot with the W layer is clearly separated from the SiO_2 along the interface.
(3) The restraint by the W layer works well so that there is little damage in both the delaminated W/Cu dot and the substrate.
(4) The delamination lateral load, F_{IC}, can be successfully measured.

ACKNOWLEDGEMENTS

This paper is supported in part by the Center of Excellence for Research and Education on Complex Functional Mechanics (COE program of the Ministry of Education, Culture, Sports, Science and Technology, Japan) and by Grant-in-Aid for Scientific Research of (B)(No.13555026) of Japan Society of the Promotion of Science.

REFERENCES

1. S. Baba, A. Kikuchi and A. Kinbara, *J. Vac. Sci. Technol.* **A4**, 3015 (1986).
2. S. Venkataraman, D. L. Kohlstedt and W. W. Gerberich, *J. Mater. Res.* **11-12**, 3133 (1996).
3. M. D. Kriese, W. W. Gerberich and N. R. Moody, *J. Mater. Res.* **14**, 3007 (1999).
4. D. B. Marshall and A. G. Evans, *J. Appl. Phys.* **56**, 2632 (1984).
5. S. Kamiya, H. Kimura, K. Yamanobe, M. Saka and H. Abe, *Thin Solid Films* **414**, 91 (2002).
6. M. P. de Boer, M. D. Kriese and W. W. Gerberich, *J. Mater. Res.* **12**, 2673 (1997).
7. R. H. Dauskardt, M. Lane, Q. Ma and N. Krishna, *Engng. Fract. Mech.* **61**, 141 (1998).
8. D. W. Butler, *J. Phys.* **E3**, 979 (1970).
9. T. Kitamura, T. Shibutani and T. Ueno, *Engng. Fract. Mech.* **69**, 1289 (2002).
10. T. Kitamura, H. Hirakata and T. Itsuji, *Engng. Fract. Mech.* **70**, 2089 (2003).
11. A. Bagchi, G. E.Lucas, Z. Suo and A. G. Evans, *J. Mater. Res.* **9**, 1734 (1994).
12. A. Kinbara, E. Kusano, T. Kamiya, I. Kondo and O. Takenaka, *Thin Solid Films* **317**, 165 (1998).
13. T. L. Becker Jr., J. M. McNancy, R. M. Cannon and R. O. Ritchie, *Mech. Mater.* **25**, 291 (1997).
14. H. Hirakata, T. Kitamura and Y. Yamamoto, *Int. J. Solids Struct.* **41**, 3243 (2004)
15. M. Hommel and O. Kraft, *Acta Mater.* **49**, 3935 (2001).

Mat. Res. Soc. Symp. Proc. Vol. 821 © 2004 Materials Research Society P2.7

Size Effect on Crack Formation in Cu/Ta and Ta/Cu/Ta Thin Film Systems

P.Gruber[1], J. Böhm[1], A. Wanner[2], L. Sauter[3], R. Spolenak[3] and E.Arzt[1, 3]
[1] Universität Stuttgart, Institut für Metallkunde, Stuttgart, Germany
[2] Universität Karlsruhe, Institut für Werkstoffkunde I, Karlsruhe, Germany
[3] Max-Planck-Institut für Metallforschung, Stuttgart, Germany

ABSTRACT

Layered structures of Cu and Ta thin films on silicon are well established for many technological applications in microelectronics. Electronic circuits used for flexible displays or wearable electronics are becoming increasingly popular. For such applications, the Cu/Ta system must be transferred to flexible substrates, incorporating a design rule for several percent of total strain. We have investigated the deformation behaviour of different Cu/Ta and Ta/Cu/Ta thin film systems on a flexible polyimide substrate subjected to total strains of more than 5%. A novel synchrotron X-ray diffraction technique allowed us to characterize the evolution of mechanical stress in very thin metallic films during isothermal tensile tests. We found that samples with a Cu film thickness below 300 nm showed a sudden stress decrease at a total strain of about 2.5%. This stress drop was attributed to fracture of the entire film system, initiated by cracks in the Ta layers.

INTRODUCTION

The application of thin metallic films in micro- and nanoelectromechanical systems MEMS/NEMS, as well as microelectronics, is widespread. During fabrication and service, large mechanical stresses may develop, which can result from deposition, microstructural changes, thermal mismatch, or external loading. In metallic films, these stresses are partially relaxed by yielding or creep, both of which are strongly influenced by the dimensions of the components. For many years, materials research has therefore focused on the understanding of deformation mechanisms in thin metal films on stiff substrates [1-6]. Current semiconductor technology demands the use of low-resistivity metal layers for integrated circuit conduction lines and contact structures. A well established material system for the CMOS process is a Cu metallization on Si including a Ta diffusion barrier [7]. Recently, flexible electronic circuits have gained widespread interest for numerous applications including flexible displays [8] and wearable electronics [9]. Due to their low Young's modulus, devices on polymeric substrates will have to endure high strains during application. Thus the maximum total strain of such a structure is limited by the extensibility of the metal film.

A recently developed *in situ* synchrotron X-ray diffraction technique [10] allows for the measurement of isothermal stress-strain curves of ultrathin polycrystalline films on a polymeric substrate. Within the scope of this paper, we utilized this novel technique to investigate the deformation behavior and microstructural changes of different Cu/Ta and Ta/Cu/Ta thin film systems during tensile testing to total strains of 5% or higher.

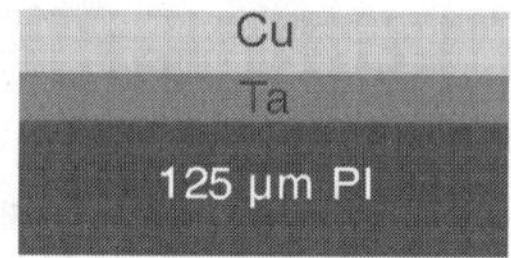

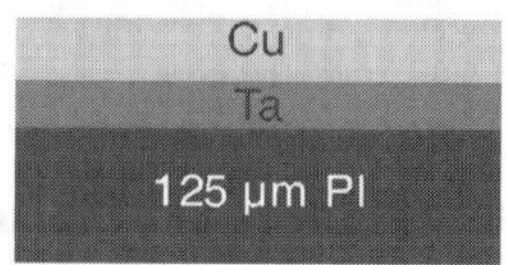

Figure 1. Schematic drawing of the Cu/Ta and Ta/Cu/Ta thin film systems investigated.

EXPERIMENTAL DETAILS

Different Cu/Ta and Ta/Cu/Ta thin film systems were prepared (see Fig 1). For the substrates dogbone-shaped specimens were cut from 125 µm thick polyimide foil (Kapton® HN by DuPont). The gage section was 20 mm long and 6 mm wide. Polycrystalline Ta and Cu films were deposited under UHV conditions (10^{-8} mbar) on one side of each sample by magnetron sputtering. Before sputtering, the substrate surface was cleaned by argon ion bombardment. During deposition, the substrate was electrically grounded. The thickness of all Ta layers was 10 nm, whereas the thickness of the Cu layer was varied between 20 and 1000 nm. All samples were annealed for 0.5 h at 350°C. In addition to serving as a diffusion barrier, the Ta interlayer enhances the adhesion and the texture of the Cu films. These films showed a pronounced <111> fiber texture, which is typical of fcc metal thin films produced in this way.

All *in situ* tensile tests were performed at the Surface Diffraction Beamline operated by the Max Planck Institute for Metals Research (MPI-MF) located at the synchrotron radiation source ANKA (Karlsruhe, Germany). The experimental procedure, data acquisition and data analysis are described in detail for Au specimens in Refs. 10 and 11. For Cu films, one has to consider the different lattice parameter of Cu, which results in different Bragg angles, inclination angles, specimen-to-camera distances and corresponding photon energies. The samples were strained to a maximum total strain of nearly 7.0% and subsequently unloaded. The strain rate was $2 - 4 \cdot 10^{-5}$ s^{-1} and the tensile tests were carried out at room temperature.

SEM micrographs were taken of untested and tested samples to investigate the microstructure of the films. A manual screw driven tensile tester was mounted on the sample stage of a scanning electron microscope (Leo 1530 VP, Oberkochen, Germany) and SEM micrographs were taken at different strains.

RESULTS

In the following, results from tensile tests on different Cu/Ta and Ta/Cu/Ta thin film systems are presented. Fig. 2(a) shows the stress-strain curve of a 300 nm thick Cu film with a 10 nm thick Ta passivation- and interlayer. Initially the Cu layer is subjected to an equibiaxial compressive stress of -88 MPa resulting from preparation and heat treatment. After the linear elastic regime and some strain hardening, the longitudinal stress σ_x reaches a plateau value which stays constant up to a total strain of 2.8%. During unloading, a compressive stress state develops

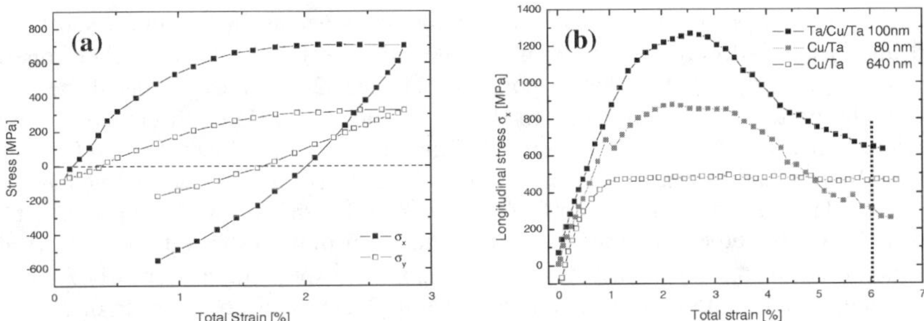

Figure 2. **(a)** Stress-strain curve of a 300 nm thick Cu film with Ta passivation- and interlayer. σ_x and σ_y are the longitudinal and the transverse stresses in the Cu layer.
(b) Longitudinal stress σ_x vs. total strain for different Cu/Ta and Ta/Cu/Ta samples with different Cu film thickness.

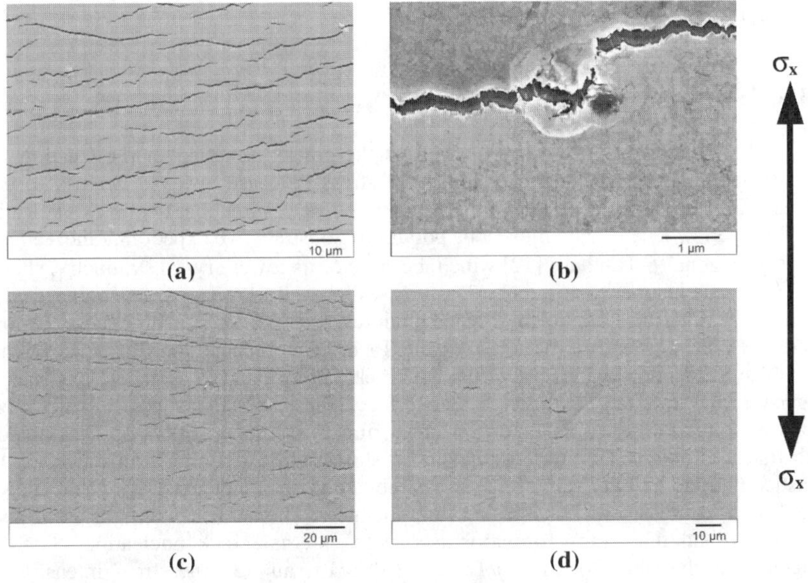

Figure 3. SEM micrographs of the surfaces of different Cu/Ta and Ta/Cu/Ta thin film systems subjected to 6% total strain (see dotted line in Fig. 2(b)). The arrow on the right hand side represents the direction of the external load.
(a), (b) Ta/Cu/Ta sample with 100 nm Cu film thickness.
(c) Cu/Ta sample with 80 nm Cu film thickness.
(d) Cu/Ta sample with 640 nm film thickness.

in the Cu film because it has been plastically deformed, whereas the polyimide substrate behaves elastically. The tensile stress in the transverse direction (σ_y) is due to the higher Poisson´s ratio of Cu (111) in comparison to that of polyimide. In Fig. 2(b) the evolution of the longitudinal stress σ_x during tensile tests to higher total strain is presented. Here, differences in deformation behaviour between samples with different Cu film thickness are evident. For Cu film thicknesses of about 100 nm and lower, a sudden stress drop of 50% is observed at intermediate strain (see Fig. 2(b)). This stress drop occurs for Cu/Ta as well as Ta/Cu/Ta film systems at a total strain of about 2.5%. On the other hand, for the 640 nm thick Cu film no stress drop is observed and the longitudinal stress σ_x stays constant until the maximum total strain of nearly 7%. Results on further samples have shown that the critical Cu film thickness for occurrence of a stress drop is about 300 nm.

In Fig. 3 SEM micrographs of different Cu/Ta and Ta/Cu/Ta thin film systems are shown. All pictures were taken at a total strain of 6% (corresponding to the vertical dashed line in Fig. 2(b)). Both, the Ta/Cu/Ta sample with a Cu film thickness of 100 nm, as well as the Cu/Ta sample with a Cu film thickness of 80 nm, show a crack pattern with open cracks that extend through all metallic layers and lie perpendicular to the loading axis (Fig. 3(a-c); loading axis indicated by the arrow), whereas on the surface of the Cu/Ta sample with a Cu film thickness of 640 nm (Fig. 3(d)), no crack pattern is evident.

DISCUSSION

The measured stress drops as well as the crack formation in the Cu thin films show a significant thickness dependence. Since the stress decrease always occurs at a total strain of about 2.5% and depends on the Cu film thickness, we assume that the crack nucleation occurs in the Ta layer(s). Due to the amorphous polyimide substrate, we expect that the Ta layer(s) consist of the tetragonal β-Ta-phase [12] which, because of its lower crystal symmetry, should behave in a more brittle manner than fcc Cu. The stresses in the Ta layer(s) are also expected to be higher than in the Cu films due to the higher modulus and smaller film thickness. Unfortunately we were not able to measure the stress in the Ta layer(s) because the diffracted intensity was too low. The SEM micrographs show that the crack pattern on a Ta surface looks sharper and more jagged (Fig. 3(a), (b)) than on a Cu surface (Fig. 3(c)). This could indicate some plastic deformation within the cracked area of a Cu film. Preliminary results on Cu films on polyimide substrates without a Ta interlayer have also shown that even an 80 nm thick Cu film deforms plastically up to a total strain of 5% and exhibits no stress decrease [13]. They also show that the yield stress decreases with increasing film thickness. This indicates that the Cu films still deform plastically and that crack formation is due to external crack nucleation. Since the Ta film thickness is the same for all Cu/Ta and Ta/Cu/Ta samples, the stress intensity at the Cu/Ta interfaces is similar and cracks propagate through the interface only if the Cu film cannot yield by plastic flow.

Crack propagation in a film bonded to a substrate is a three-dimensional process. Nevertheless, the energy release rate G of a steady-state channelling crack can be obtained from a two-dimensional plane-stress analysis. Solutions for steady-state channelling have been presented by Beuth [14]. Based on these solutions, Xia and Hutchinson developed a two-dimensional model for crack propagation in thin films [15]. Using this model we calculated the stress drop corresponding to straight cracks oriented perpendicular to the loading axis. The longitudinal stress change is

$$\Delta\sigma_x = -\sigma_0 \exp(-x/l) \tag{1}$$

where σ_0 is the initial longitudinal stress and l is a reference length which characterizes the exponential decay of the stress transverse to the crack:

$$l = \frac{\pi}{2} g(\alpha,\beta) h. \tag{2}$$

Here α, β and $g(\alpha,\beta)$ describe the elastic mismatch between the film and the substrate (see [15] for definitions) and h is the film thickness. The reference length will exceed the film thickness h when the film is stiff compared to the substrate ($\alpha \approx 1$) and will be of the order of h when the film is compliant relative to the substrate ($\alpha \approx -1$). If x is the distance normal to the crack, the average longitudinal stress σ_x in the film can be calculated:

$$\sigma_x = \sigma_0 + \frac{\int_0^{H/2} \Delta\sigma_x dx}{\int_0^{H/2} dx} = \sigma_0 \left(1 + \frac{(\exp(-H/2l) - 1)}{H/2l}\right) \tag{3}$$

where H is the mean crack distance. In Fig. 4 the normalized average stress σ_x/σ_0 is plotted against $H/2l$. In the special case of our material system, the elastic mismatch between the Ta and Cu layers and the compliant polyimide substrate is high.

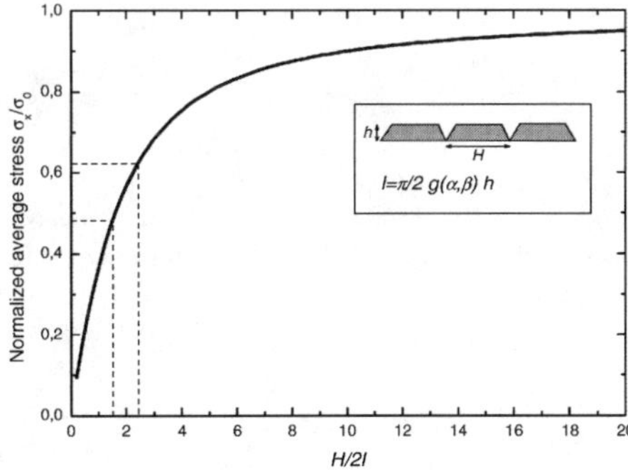

Figure 4. Calculated normalized average stress σ_x/σ_0 versus $H/2l$, following the model of Xia and Hutchinson for a semi-infinite straight crack. A stress decrease of 50% is reasonable for a film thickness of 100 nm (area bounded by the dashed lines).

If we insert the material parameters, we obtain $\alpha = 0.95$ and $g(\alpha,\beta) = 10$ (see [15] for the definition of α and $g(\alpha,\beta)$). Therefore the value for $H/2l$ lies between 1.5 and 2.5 if we consider a mean crack distance H between 6 and 10 µm (see Fig.3) and a film thickness h = 100 nm. The corresponding normalized average stress σ_x/σ_0 is approximately 0.48 to 0.62 (dashed lines in Fig. 4). This correlates quite well with the measured stress decrease of 50% for Cu/Ta and Ta/Cu/Ta samples with a Cu film thickness in that range.

In contrast to metallic films on stiff substrates, where stress relaxation due to cracking is strongly reduced, the elastic mismatch between metallic layers and compliant substrate in these samples causes stress relaxation far from the crack. On a stiff substrate, the crack density that we have observed here would only cause a slightly reduced normalized average stress σ_x/σ_0 of 0.95 ($H/2l > 20$, see Fig. 4). On the other hand, for the same stress decrease of 50 %, the crack distance would have to be between 0.5 and 1 µm.

Although the model by Xia and Hutchinson [15] describes the situation in our film systems well, there are still some open questions concerning the influence of the Cu film thickness. What defines the transition from cracking to plastic deformation, and how does the size of the plastic zone correlate with the Cu film thickness? These questions will be the subject of further investigations.

CONCLUSIONS AND OUTLOOK

We have demonstrated that layered structures of Ta and Cu thin films, which are widely used in microelectronics, cannot be directly transferred onto compliant substrates for flexible electronic circuits due to cracking of the Ta layers at relatively low total strains of 2.5%. Below 300 nm Cu film thickness the cracks propagate through the entire thin film system. The situation can be described by a two-dimensional model from Xia and Hutchinson [15]. Flexible electronic device applications will require new diffusion barriers or clever design solutions to withstand high strains. Further studies will focus on the size of the plastic zone and the thickness dependence of the crack density. By further improving the synchrotron measurements, the stress in the extremely thin and low symmetry Ta layers may also be obtained.

ACKNOWLEDGEMENTS

The authors would like to thank the beamline team of the MPI-MF-Surface-Diffraction-Beamline at ANKA (R. Weigel, N. Kaspar, A. Rühm and A. Stierle) for their excellent technical support as well as T. Wagner and F. Thiele of the MPI Thin Film Laboratory for sample preparation.

REFERENCES

1. W.D. Nix, Metall. Trans. A **20A**, 2217-45 (1989).
2. E. Arzt, Acta Mater. **46**, 5611-26 (1998).
3. R. Venkatraman and J.C. Bravman, J. Mater. Res. **7** (8), 2040-48 (1992).
4. R.P. Vinci, E.M. Zielinski and J.C. Bravman, Thin Solid Films **262**, 142-53 (1995).

5. Y.L. Shen, S. Suresh, M.Y. He, A. Bagchi, O. Kienzle, M. Rühle and A.G. Evans, J. Mater. Res. **13** (7), 1928-37 (1998).
6. T.J. Balk, G. Dehm and E. Arzt, Acta Mater. **51**, 4471-85 (2003).
7. K. Holloway and P.M. Fryer, Appl. Phys. Lett. **57** (17), 1736-38 (1990).
8. J.A. Rogers, Z. Bao, K. Baldwin, A. Dodabalapur, B. Crone, V.R. Raju, V. Kuck, H. Katz, K. Amundson, J. Ewing and P. Drzaic, Proc. Natl. Acad. Sci. USA **98** (9), 4835-40 (2001).
9. E.R. Post, M. Orth, P.R. Russo and N. Gershenfeld, IBM Syst. J. **39**, 840-60 (2000).
10. J. Böhm, P. Gruber, R. Spolenak, A. Stierle, A. Wanner and E. Arzt, Rev. Sci. Instrum. **75** (4), 1110-19 (2004).
11. J. Böhm, P. Gruber, R. Spolenak, A. Wanner and E. Arzt, these proceedings, P1.6.
12. L.G. Feinstein and R.D. Huttemann, Thin Solid Films **16**, 129-45 (1973).
13. Gruber et al., work in progress.
14. J.L. Beuth, Int. J. Solids Structures **29**, 1657-75 (1992).
15. Z.C. Xia and J.W. Hutchinson, J. Mech. Phys. Solids **48**, 1107-31 (2000).

Mat. Res. Soc. Symp. Proc. Vol. 821 © 2004 Materials Research Society P7.8

Thermomechanical properties and fatigue of nanocrystalline Ni/Cu electrodeposits

Olivier Arnould[1], Olivier Hubert and François Hild
LMT-Cachan, ENS Cachan/CNRS UMR 8535/Université Paris VI
61, avenue du Président Wilson
94235 Cachan Cedex, France

ABSTRACT

This study deals with the long-term reliability of a high precision pressure sensor using bellows mainly made of electroplated Ni. Thermomechanical properties of this deposit are obtained by several experiments and compared to theoretical models, computations and other authors' results. Bellows are expected to stay in service for many decades, thus their high cycle fatigue behavior has to be known. Stress-life fatigue curve for crack initiation and fatigue crack growth in the electroplated Ni are measured and identified using numerical computations. Results are compared with other results obtained on similar Ni electrodeposits. Normalized stress-life fatigue curve shows no specific nanosize effects.

INTRODUCTION

This papers deals with the study of aging modes of a high precision pressure sensor. It uses bellows to convert pressure variations into a displacement (figure 1a). The stiffness and airtightness of bellows require to use a Ni/Cu/Ni layered material (overall thickness ~50μm) obtained by electroforming (figure 1b). The main aging mode of this layered material [1] that affects the bellows airtightness is due to high cycle fatigue crack initiation at the surface of the nickel layers and, to a lesser extent, on its propagation through them. The fatigue behavior of the Ni deposits is linked to their microstructure and the associated (thermo)mechanical properties that depend on the electrodeposition process.

NI DEPOSIT CHARACTERIZATIONS

Microstructure analysis

Nickel is electrodeposited in a sulfamate bath ($[Ni(SO_3NH_2)_2{\cdot}4H_20]$ = 600g/l, pH = 4, θ = 50°C) with a low DC-current density ($4 \leq i \leq 10$mA/cm^2). Deposits are electroplated on a rotating aluminum mandrel first coated with a copper layer. Both are removed at the end of the electrodeposition process to obtain the bellows. Under these conditions, nickel deposits have a fine columnar structure parallel to the electroplating direction (figure 1b) with clusters (figure 2a, [2]) made of micrometer domains of low-misoriented grains (figure 2b). XRD measurements and TEM observations (figure 2c) lead to a mean grain size of about 60nm, small second order residual stresses and a very low dislocation density. Moreover, the plating conditions induce a specific inhibited growth of the nickel layers associated with hydrogen absorption [3]. This yields a (relatively low) $\langle 110 \rangle$ fiber texture along the electroplating direction (figure 1c) and a high (nano)porosity, $f > 5\%$ (density measurements), of the deposits.

[1] now at LMGC, Université de Montpellier II, cc 048, Place Eugène Bataillon, 34095 Montpellier Cedex 05, France
E-mail : arnould@lmgc.univ-montp2.fr

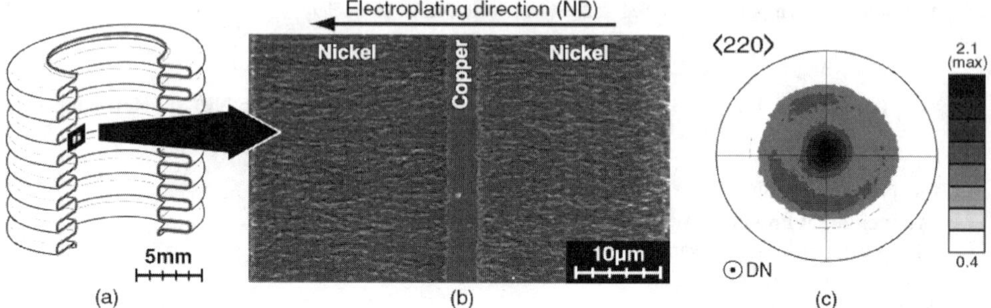

Figure 1. a) General view of electroplated bellows. b) Micrograph of a cross-section obtained in an SEM after chemical etching: fine columnar microstructure of the nickel layers. c) $\langle 220 \rangle$ pole figure obtained by DRX at the surface of the deposit.

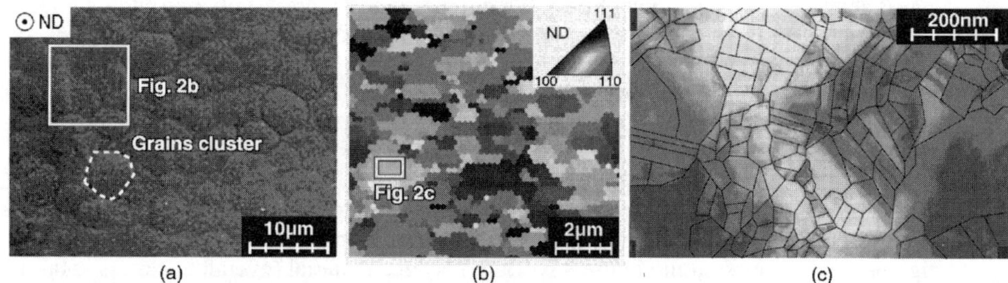

Figure 2. a) SEM micrograph of an electropolished nickel layer surface: detail of a micrometric grains cluster boundary. b) Low angle misorientation domains of (nanometer) grains obtained by EBSD mapping (measurement step: 0.2µm, grey level corresponds to a crystallographic orientation given in the inverse pole figure) at the surface of the Ni deposit. c) Bright field TEM micrograph showing the bimodal nanometer grains. (Sub)Grain boundaries are informatically drawn for a sake of clarity.

The nickel layers give the needed stiffness for the bellows but their porosity is too high to provide the required airtightness. A continuous, *i.e.,* airtight, copper layer (thickness ~3µm) is added between the nickel layers to this aim. This configuration does not modify the effective stiffness of bellows as it allows for the minimization of the *softening* effect of copper since bending is the predominant loading pattern. These microstructural characteristics induce specific thermomechanical properties that have been evaluated and compared to micromechanical models.

Thermomechanical properties

An atypical isotropic thermoelastic behavior, due to the weak fiber texture, has been shown experimentally [1, 4]. The coefficient of thermal expansion (CTE), α, is found to be greater than its usual value for nickel (*e.g.*, $\alpha(100°C) \approx 17.10^{-6}$ instead of about 14.10^{-6}). The origin of this behavior is still questionable in the present case, *i.e.*, thermoelastic homogenization shows that the porosity does not induce a change in the CTE and the (nano)size effect of the grain is still unclear [5, 6]. The grain size is "too high" to yield a significant change in the elastic properties [7]. The porosity is the main factor that yields a Young's modulus, $E = 160$GPa, less than the theoretical value of 230GPa (obtained by self-consistent elastic homogenization) for the given fiber texture [8]. Similar Young's moduli have been obtained by Buchheit *et al.* [9] for nickel deposits with a similar $\langle 110 \rangle$ fiber texture obtained with Watt's baths.

Tensile tests (figure 3a), on a specific nickel electrodeposited sheet (figure 3b), with digital image correlation measurements to evaluate the strains, are used to identify a yield strength at 0.02% plastic strain, σ_y, of the order of 600MPa, an ultimate tensile strength, σ_u, of 1.2GPa and a low plastic strain to failure of 1.6%. This is a usual behavior for this kind of material [7, 9]. Figure 3c shows the ductile fracture surface observed on a Ni sample. The measured yield stress is in good agreement with the Hall-Petch relationship [10] when considering the coefficients of the annealed nickel [11] and the mean grain size determined herein. Note that the present mean grain size is too high to lead to a reverse Hall-Petch behavior of the Ni deposits [7].

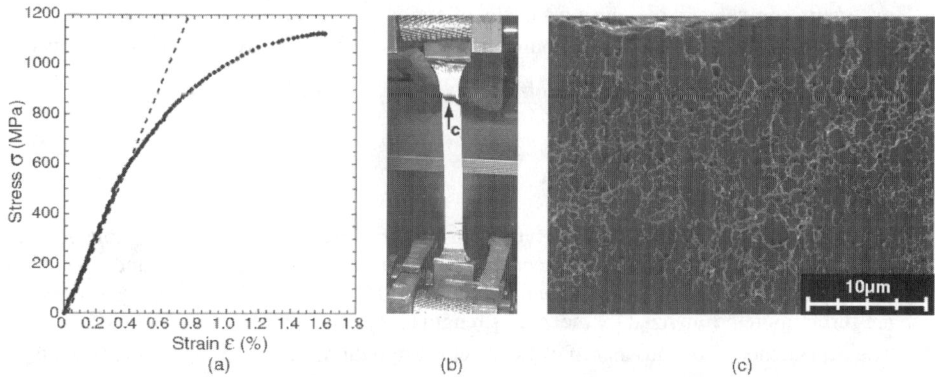

Figure 3. a) Typical stress-strain curve obtained on a Ni deposit sheet. b) View of the *post-mortem* Ni sheet sample in the tensile test device. c) SEM micrograph of the fracture surface.

FATIGUE PROPERTIES

As long-term bellows airtightness is concerned, a specific high cycle fatigue device, put in a resonant testing machine, has been designed [12]. It allows us to carry out high cycle fatigue tests of ten bellows at the same time. Tests are controlled in displacement, u, with a zero mean value, $\overline{u}$, and $\Delta u/2$ its maximum reached during a cycle (*i.e.*, half displacement range). The bellows airtightness is continuously monitored with a pressure sensor. Airtightness loss corresponds to the number of cycles for crack initiation, N_I, as the number of cycle for the crack propagation through the bellows thickness is negligible.

Crack initiation

Stress-life fatigue, *S-N*, curve for crack initiation consists in plotting an equivalent damage stress range $\Delta\sigma^*$ [12] as a function of the number of cycles for the loss in airtightness. $\Delta\sigma^*$ is computed at the most loaded point(s) of bellows by using bi-dimensional (axisymmetric) finite element (FE) analyses for a given cycle. The results are scattered. During a fatigue test, structure and material heterogeneities lead to localized microplasticity and damage phenomena in various sites whereas the whole structure remains elastic. If we assume that these sites are randomly distributed, independent and that crack initiation occurs at the weakest one, the cumulative crack initiation probability, p, within bellows Ω, can be described, at endurance, by a two-parameter Weibull model [13, 14]

$$p = 1 - \exp\left[-\frac{V_{eff}}{V_0}\left(\frac{\hat{\sigma}}{\sigma_0}\right)^m\right], \tag{1}$$

where m is the Weibull modulus, V_0 a reference volume and σ_0 the scaling stress. $\hat{\sigma}$ is the maximum equivalent stress during a cycle, of period T, over the whole structure Ω, *i.e.*, $\hat{\sigma} = \max\limits_{\Omega}(\max\limits_{t\in[0,T]} \sigma^*)$. This model accounts for the stress heterogeneity and volume effects observed during fatigue testing with V_{eff}, the effective volume, defined by

$$V_{eff} = \int_\Omega \left(\max_{t\in[0,T]} \sigma^* \big/ \hat{\sigma}\right)^m \mathrm{d}V \,. \tag{2}$$

This allows us to use other authors' fatigue results obtained with a different sample geometry and scale. Our stress ranges are corrected to have the same effective volume and stress loading ratio, *i.e.*, R_σ= -1 (using the Goodman relationship [15]), as Boyce *et al.* [16]. All stress ranges are furthermore normalized by the yield strength to take the material microstructure into account. The dependence of the number of cycle for crack initiation, N_I, to the stress range, σ_r^*, is modeled with a simple two-parameter, A and n, Stromeyer law [15]

$$N_I(p) = A \big/ \left[\Delta\sigma^* - 2\sigma_\infty(p)\right]^n, \tag{3}$$

where $\sigma_\infty(p)$ is the fatigue limit, for a given crack initiation probability, p, equal to $\hat{\sigma}$ at endurance. The parameters of the whole modeling are identified on our results and those of Boyce *et al.* [16]. The corresponding curves are reported in figure 4a where an *a priori* estimation of the mean fatigue stress threshold $\bar{\sigma}_\infty$ has been evaluated by using an infrared thermographic technique [17]. A good agreement between the different results and the identified model is shown. The mean fatigue limit, $\bar{\sigma}_\infty$, is such that $\bar{\sigma}_\infty/\sigma_y \approx 2/3$ and $\bar{\sigma}_\infty/\sigma_u \approx 1/3$ that is in agreement with results obtained with usual grain sized nickel [16]. The minimum number of cycles needed to reach the endurance domain is of the order of 10^8-10^9. That means that no specific nanosize effect seems to occur on the fatigue life of the present electroplated nickel.

Crack propagation

Tests have been carried on during some thousands of cycles, N_P, after the airtightness loss to study the crack propagation in electroplated Ni. Results are reported in figure 4b. The fatigue crack growth rate is roughly constant (about 100nm/cycle). Stress intensity factors at the crack tip for different crack lengths have been computed by a three-dimensional FE analysis [12]. A stress intensity factor proportional to the crack length, similar to the case of an orthoradially cracked hollow cylinder [18], is observed. The crack propagation rate is constant while the stress intensity factor increases for values ranging from 5 to 12 $MPa.m^{1/2}$ [12]. This behavior seems to be linked to nickel oxidation, at room temperature, at the crack tip as observed by Boyce *et al.* [16]. Constant crack growth rates for a given stress intensity factor range in the subcritical regime are, for example, observed in glassy materials in corrosive environments [19]. In this case the corrosive species diffusion to the crack tip yields a *constant* crack propagation velocity. SEM observations of the fatigue crack (figures 5a and b) show a transgranular fracture with a two-scale roughness well correlated, for the first one, with the grains cluster size and, for the second one, with the crack growth rate (close to one grain size per cycle).

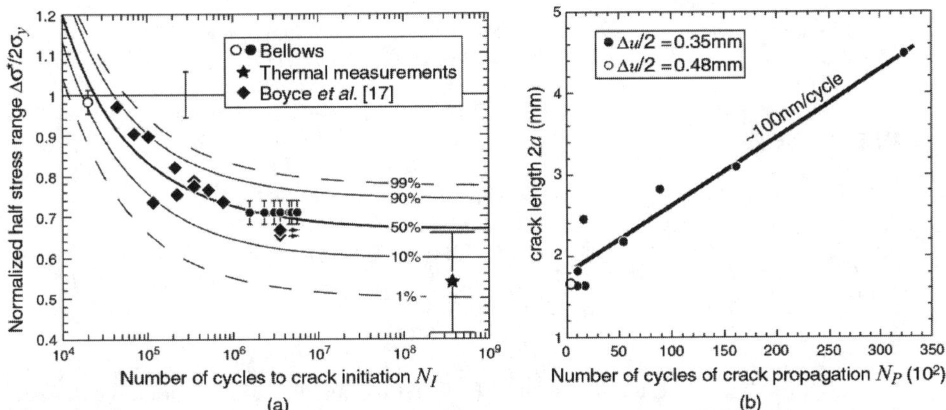

Figure 4. a) Stress-life fatigue curve for crack initiation with $R_\sigma = -1$ and $V_{eff} \approx 4.10^{-4}mm^3$. b) Fatigue crack growth for different displacement loading ranges and crack lengths.

CONCLUSIONS

Long-term airtightness of Ni/Cu/Ni electroplated bellows is studied. It is caused by fatigue crack initiation in the nickel layers. Nickel deposits are made of porous micrometer clusters of low-misoriented nanometer grains showing a weak fiber texture along the plating direction. Fatigue crack initiation curves have been measured and modeled using a probabilistic approach. The fatigue behavior of nickel electrodeposits exhibits no specific nanosize effect. Furthermore, constant fatigue crack growths have been measured for different values of the stress intensity factor. This phenomenom seems to be linked to an oxidation mechanism at the crack tip. This assumption must be confirmed by additional investigations.

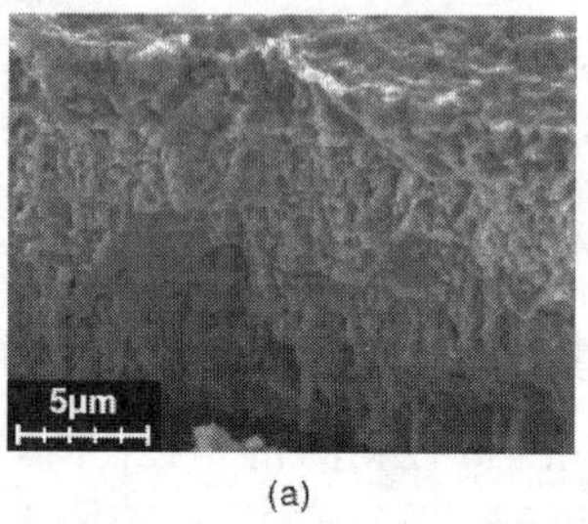

(a)

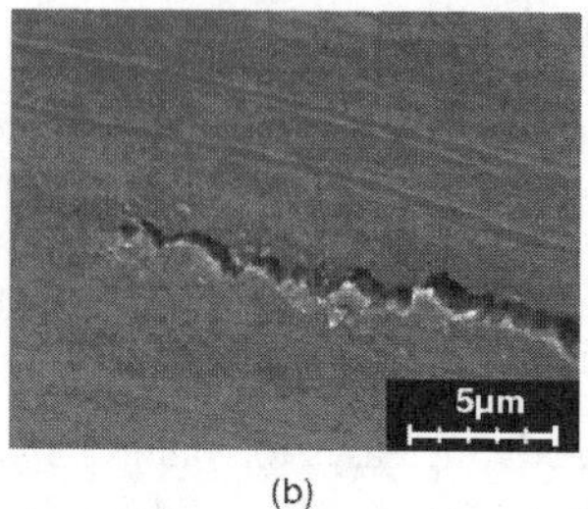

(b)

Figure 5. a) SEM micrograph of a fatigue fracture surface showing a roughness length scale close to 100nm. b) SEM micrograph of a fatigue crack tip: highlights of numerous crack deviations with a mean length scale close to the size of the low-misoriented grains domains (figure 2b).

ACKNOWLEDGEMENTS

The authors wish to thank I. Guillot (UTC Compiègne, France) for the TEM observations and V. Ji & W. Seiler (LM3 ENSAM Paris, France) for the XRD characterizations.

REFERENCES

1. O. Arnould and F. Hild, *Defect and Diffusion Forum* **203-205**, 61-80 (2002).
2. S.W. Banovic, K. Barmak and A.R. Marder, *J. Mater. Sci.* **33**, 639-645 (1998).
3. C.S. Lin, P.C. Hsu, L. Chang and C.H. Chen, *J. Appl. Electrochem.* **31**, 925-933 (2001).
4. J. Jumel, F. Taillade and F. Lepoutre, *Eur. Phys. J. – Appl. Phys.* **23**, 217-225 (2003).
5. H. Gleiter, *Prog. Mat. Sci.* **33**, 223-315 (1989).
6. T. Turi and U. Erb, *Mat. Sci. Eng.* **A204**, 34-38 (1995).
7. N. Wang, Z. Wang, K.T. Aust and U. Erb, *Mater. Sci. Eng.* **A237**, 150-158 (1997).
8. G.E. Fougere, L. Riester, M. Ferber, J.R. Weertman and R.W. Siegel, *Mater. Sci. Eng.* **A204**, 1-6 (1995).
9. T.E. Buchheit, D.A. La Van, J.R. Michael, T.R. Christenson and S.D. Leith, *Metall. Mater. Trans.* **33A**, 539-554 (2002).
10. Hall, E.O., *Proc. Phys. Soc.* **B64**, 747-753, (1951) and Petch, N.J., *J. Iron Steel Inst.* **174**, 25-28 (1953).
11. Website: neons.mems.cmu.edu/rollett/rollett.html
12. O. Arnould, PhD thesis, University of Paris VI (2003). Website: tel.ccsd.cnrs.fr
13. W. Weibull, *Roy. Swed. Inst. Eng. Res.* **151** (1939).
14. F. Hild, R. Billardon and A.S. Béranger, *Mech. Mat.* **22**, 11-21 (1996).
15. S. Suresh, *Fatigue of Materials*, Cambridge University Press (1991).
16. B.L. Boyce, J.R. Michael and P.G. Kotula, *Acta Mater.* **52(6)**, 1609-1619 (2004).
17. M.P. Luong, *Mech. Mat.* **28**, 155-163 (1998).
18. Murakami, Y., *Stress intensity factors handbook*, vol. 2, Pergamon Press, Oxford (1981).
19. A.K. Varshneya, *Fundamental of inorganic glasses*, Academic Press, Boston (1994).

Mat. Res. Soc. Symp. Proc. Vol. 821 © 2004 Materials Research Society P3.13

Debonding Under Fatigue Loading at Polymer/Inorganic Interfaces

Bree M. Sharratt[1] and Reinhold H. Dauskardt[2]
[1]Department of Aeronautics and Astronautics, Stanford University
[2]Department of Materials Science and Engineering, Stanford University
Stanford, CA 94305-2205

ABSTRACT

The mechanisms associated with cycle-by-cycle damage accumulation resulting in fatigue crack propagation between a highly constrained polymer layer and an adjacent elastic substrate are explored. Specifically, cyclic fatigue-induced crack growth between a bisphenol F model epoxy system and a passivated silicon substrate under Mode I loading is reported. Preliminary findings regarding the effects of fatigue load ratio on interfacial crack growth rates are presented. While intermediate crack growth rates were significantly accelerated under cyclic loading, the near-threshold crack growth behavior under cyclic and monotonic loading was surprisingly similar.

INTRODUCTION

The effect of fatigue loading on the subcritical interfacial crack growth between a polymer layer and an adjacent substrate is of significant interest for the integrity of a variety of components. In the case of high-density flip-chip packaging, underfill epoxies are used to support the thermo-mechanical stresses that develop in the closely-packed solder ball grid array. By filling in the standoff between the chip and the board with an epoxy, the reliability of the solder connections is improved [1,2]. However, pervasive challenges persist associated with interfacial crack growth driven by significant residual stresses from thermal expansion mismatch between the epoxy and the chip. In addition, accelerated crack growth under alternating stresses due to thermomechanical cycling may occur and is the subject of the present study.

The interface selected for study was formed between an unfilled bisphenol F epoxy bonded to a silicon substrate with a SiO_2 or Si_3N_4 passivation. Previous work examined the subcritical crack growth behavior of this interface under monotonic loading [3]. While the interface was only marginally susceptible to stress corrosion, an unexpected plateau region was observed at growth rates on the order of 1 nm/sec and was attributed to the disruption of secondary interfacial bonds by absorbed water molecules. Although Mode II loading, which may be generated under more complex thermomechanical conditions, is known to affect fatigue crack growth rates [4], the goal of the present study was to characterize the subcritical crack growth behavior resulting from Mode I cyclic fatigue.

The application of cyclic loads results in a marked acceleration in crack growth rates due to mechanically induced damage accumulation. Behavior resulting from variations in the fatigue load ratio is presented along with unique near-threshold growth rate behavior where evidence of the low growth rate plateau was again observed.

EXPERIMENTAL PROCEDURE

Specimens were comprised of a model underfill polymer, diglycidyl ether of bisphenol F epoxy cured with 2-ethyl-4-methyl-imidazole hardener, sandwiched between two SiO_2 or Si_3N_4

passivated silicon substrates in a double cantilever beam (DCB) configuration with dimensions of ~4.5 mm x 1.52 mm x 42 mm (Fig. 1).

To initiate an interfacial crack, specimens were loaded in tension at a displacement rate of 1 μm/sec until a critical load, characterized by a nonlinearity in the load-displacement curve, was reached. From a series of loadings and subsequent unloading (Fig. 2) multiple measurements for the critical strain energy release rate, G_c, and the final crack length were obtained. The critical strain energy release rate can be related to the crack length and the corresponding critical load by [5]:

$$G_c = \frac{12P_c^2a^2}{B^2h^3E'}\left(1+0.64\frac{h}{a}\right) \tag{1}$$

where P_c is the critical load, a the crack length determined from the elastic compliance of the specimen, B the thickness, h the DCB half-height, and E' the plane strain substrate modulus. Although various other relationships have been derived in this vein [6,7], this simplified expression was used in order to maintain consistency with results from previous studies [3,8]. Table 1 summarizes the critical strain energy release rates for the two interface systems.

Interface	G_c (J/m^2)	Std. Dev.
Bisphenol F/SiO_2	10.8	0.6
Bisphenol F/Si_3N_4	11.2	0.5

Table 1: Critical strain energy release rates for the bisphenol F/passivation interfaces.

Subcritical cyclic fatigue tests were performed at 20 Hz with load ratios, $R = P_{min}/P_{max}$, of 0.1 and 0.3 under a load-shedding scheme whereby the stress intensity range, ΔK, progresses according to:

$$\Delta K = \Delta K_0 e^{Q(a-a_0)} \tag{2}$$

where Q is the ΔK-gradient, set to –0.1 mm^{-1}, ΔK_0 is the initial stress intensity range, and a_0 is the initial crack length. This produced crack growth rates from 10^{-5} m/cycle to 10^{-11} m/cycle in ambient laboratory air conditions, with an average temperature and relative humidity of ~23°C and ~36%, respectively. Note that due to the highly cross-linked nature of the epoxy, we do not anticipate any effects of the loading frequency on crack growth rates. A power law relation was used to fit data in the mid-growth range:

$$\frac{da}{dN} = C(\Delta K)^m \tag{3}$$

where C is the proportionality constant and m the crack growth exponent.

Under cyclic fatigue, the effects of premature contact of the fatigue fracture surfaces on crack growth rates can be obtained by re-plotting the experimental crack growth rates as a function of an effective stress intensity range, ΔK_{eff}, defined as:

$$\Delta K_{eff} = K_{max} - K_{cl} \tag{4}$$

where K_{max} is the maximum applied stress intensity and K_{cl} is the closure stress intensity, computed from the closure load. The closure load was selected from the point of first nonlinearity of the load-displacement compliance curve.

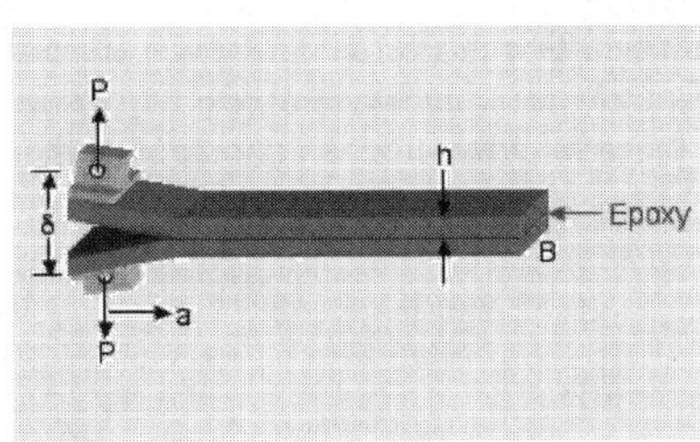

Figure 1: Schematic diagram of the double cantilever beam (DCB) specimen where the epoxy is sandwiched between two passivated silicon beams.

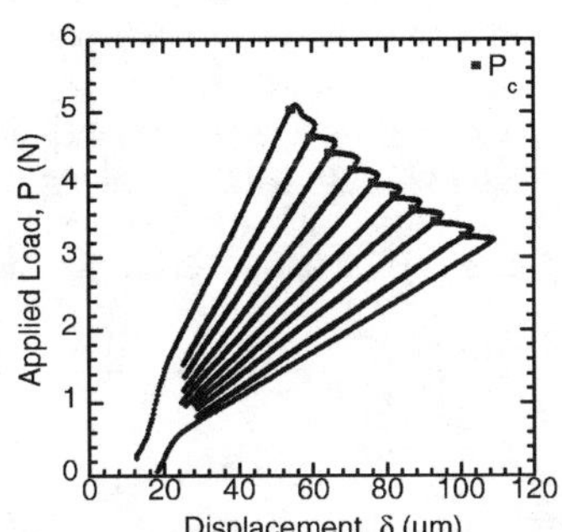

Figure 2: Load versus displacement curve resulting from critical adhesion tests of the bisphenol F/Si_3N_4 interface. The critical loads are indicated.

RESULTS

Results from cyclic fatigue tests, conducted at load ratios, *R*, of 0.1 and 0.3, are presented in Fig. 3. Characterization of the fracture surfaces indicated that debonding occurred within ~5 nm of the bisphenol F/passivation interface leaving a residual amount of epoxy on the wafer side of the fracture path.

Mid-range growth rates (10^{-7} to $2 \cdot 10^{-9}$ m/cycle) were fitted to equation (3), the results of which are summarized in Table 2. The similarity of the crack growth parameters suggests that a similar mechanism is at work, regardless of load ratio, for both interfacial systems. This was supported by atomic force microscopy (AFM) studies of the fracture surfaces (Fig. 5). In both cases, the wafer side of the fracture path was characterized by the apparent presence of residual bisphenol F epoxy asperities. Note that RMS roughness values (~1 nm) were similar for both interfacial systems and did not correlate with local debond growth rate. The epoxy residue could be the result of a craze-like damage process, the proposed existence of which is based on observations of a visible lightening of the epoxy in the vicinity of the crack front.

Load Ratio, R	Passivation	$C\left(\frac{m}{cycle}\left(MPa\sqrt{m}\right)^{-m}\right)$	m
0.1	SiO_2	$2.8 \cdot 10^{-8}$	15
0.1	Si_3N_4	$3.4 \cdot 10^{-8}$	14
0.3	SiO_2	$5.3 \cdot 10^{-7}$	13
0.3	Si_3N_4	$5.8 \cdot 10^{-8}$	16

Table 2: Constants for the power law growth rate relationship in equation (3) obtained by fitting the mid-growth range of the cyclic fatigue data curves.

The existence of a mechanically-induced fatigue effect can be verified by comparing the cyclic fatigue data with growth rates measured under monotonic loading (a load-relaxation technique where loads decrease monotonically as a result of debond extension under fixed displacement). In the case where there is no mechanically-induced fatigue process, such as in glasses [9,10], predicted growth rates for cyclic fatigue with respect to the maximum stress intensity are generally slower, as indicated by the dashed line in Fig. 4. However, experimental results for crack propagation along the bisphenol F/passivation interface reveal a dramatic acceleration of crack growth rates, verifying that a mechanically-induced damage mechanism dominates cyclic fatigue crack propagation.

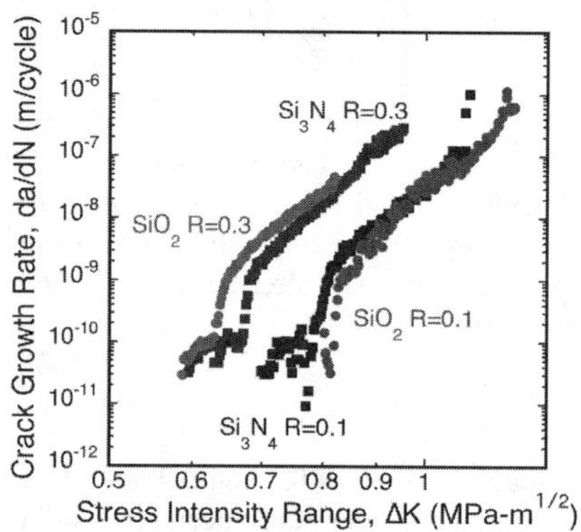

Figure 3: Crack growth rate as a function of applied stress intensity range for the bisphenol F/Si_3N_4 and bisphenol F/SiO_2 interfaces.

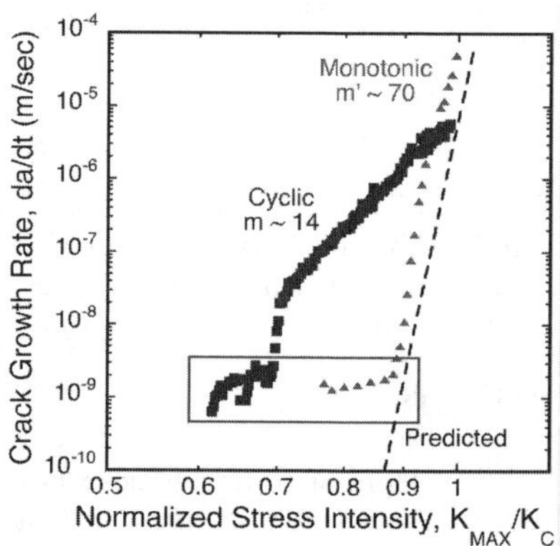

Figure 4: Comparison of crack growth rates measured under cyclic and monotonic loading. Note the similarity at low growth rates boxed in red.

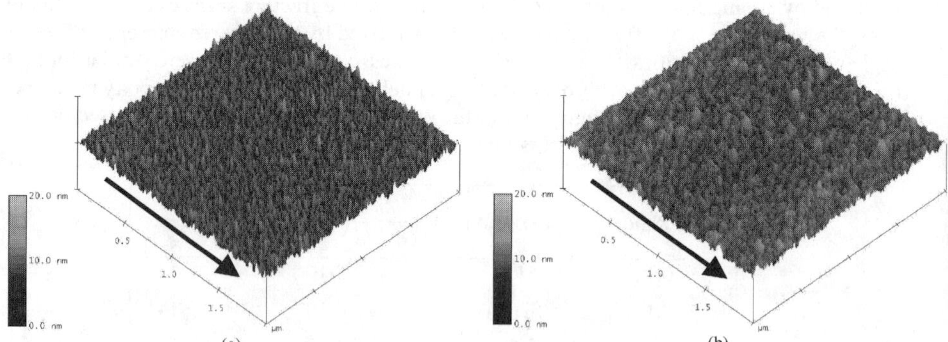

Figure 5: 2 µm x 2 µm AFM image of residual bisphenol F on the wafer side of the fracture path for (a) bisphenol F/SiO_2 interface and (b) bisphenol F/Si_3N_4 interface. Arrow indicates debond growth direction.

Fatigue crack closure

The effects of premature contact of the fracture surfaces were examined by replotting the crack growth rate as a function of the effective stress intensity, ΔK_{eff}. Crack closure due to asperity contact and epoxy layer plasticity behind the crack tip reduces the stress intensity range at the crack tip, thus lowering growth rates. Although the mechanisms associated with crack propagation appear to be similar for both passivation types, some differences were observed upon examining closure effects. While the crack growth rates obtained for the bisphenol F/Si_3N_4 interface are similar when corrected for closure (Fig. 6a), those obtained for the bisphenol F/SiO_2 interface remain accelerated with increased load ratio (Fig. 6b). This difference may be related to the differing aspect ratios of the surface asperities that exist on the wafer side of the debond path (Fig. 5).

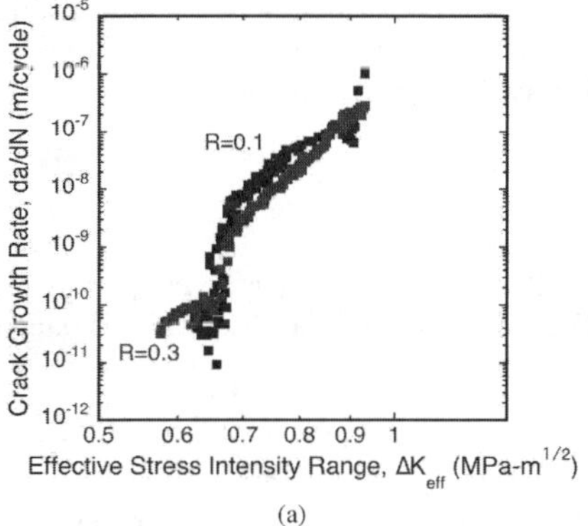

(a)

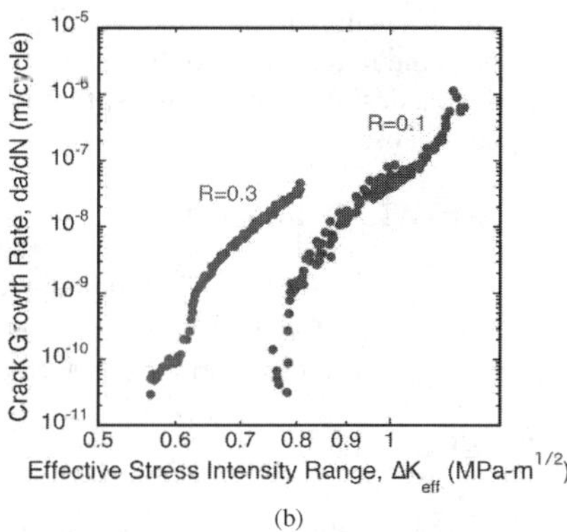

(b)

Figure 6: Crack growth rate as a function of effective stress intensity for (a) the bisphenol F/Si_3N_4 interface and (b) the bisphenol F/SiO_2 interface.

Near-threshold crack growth

The near-threshold crack growth behavior of both the bisphenol F/SiO_2 and the bisphenol F/Si_3N_4 interfaces under cyclic fatigue exhibited a remarkable similarity to that observed under monotonic loading; a surprising plateau in crack growth rates was observed at $\sim 1 \cdot 10^{-10}$ m/cycle ($2 \cdot 10^{-9}$ m/sec) (Fig. 4, boxed in red). This behavior could not be reasonably correlated to the diffusion rate of water through air or epoxy as in a transport model [8]. Rather, a reaction rate model resulted in a good fit to the experimental data under monotonic loading. Further, the resulting activation energy, low observed G_c values, and knowledge of the epoxy chemistry, indicated that the primary mechanism involved in near-threshold debond propagation was the displacement of weak secondary interfacial bonds. Cyclic fatigue observations indicate that a similar mechanism was involved in producing plateaued growth rates. However, additional factors such as residual stresses ahead of the crack tip [11], crack closure, and mechanically-induced damage accumulation may play an important role in this near-threshold region.

CONCLUSIONS

Experimental results from cyclic fatigue crack propagation between an unfilled bisphenol F epoxy and a SiO_2 or Si_3N_4 passivated silicon substrate indicate the existence of a significant mechanical fatigue effect. Surface characterization indicated that failure occurred very close to the interface, leaving residual epoxy asperities on the wafer side of the fracture path. Due to the high correlation of the crack growth rate exponents, irrespective of load ratio, the mechanism associated with crack propagation appears to be similar for both the bisphenol F/SiO_2 and bisphenol F/Si_3N_4 interfaces.

Comparison with previously reported results under monotonic loading reiterates the dominance of a mechanical fatigue effect at mid-range growth rates. However, near-threshold crack growth rates were surprisingly similar to those produced under monotonic loading. In both cases, interfacial debond growth rates plateaued at $\sim 1 \bullet 10^{-10}$ m/cycle ($2 \bullet 10^{-9}$ m/sec) and were largely insensitive to the applied stress intensity. Since low-growth rate behavior is a dominant factor in the reliability of polymer/inorganic interfaces, these results have important implications for device design as they suggest that interfacial crack growth will continue even at minimal applied loads.

ACKNOWLEDGMENTS

The authors wish to thank the Semiconductor Research Corporation for supporting this work and Zymet Inc. for providing the materials. The Perimeter Aviation Education Foundation of Winnipeg, Canada supports Bree M. Sharratt.

REFERENCES

[1] D. Suryanarayana, R. Hsiao, T.P. Gall, and J.M. McCreary, IEEE Trans. Compon. Hybrids Manuf. Technol., **14** (1), 218-223 (1991).
[2] D. Suryanarayana, T.Y. Wu, and J.A. Varloe, IEEE Trans. Compon. Hybrids Manuf. Technol., **16** (7), 858-862 (1993).
[3] L. C. Wang and R. H. Dauskardt, Mater. Res. Soc. Symp. Proc., **778**, 227-232 (2001).
[4] J. M. Snodgrass, D. Pantelidis, M. L. Jenkines, J. C. Bravman, and R. H. Dauskardt, Acta Mater., **50**, 2395-2411, (2002).
[5] M. F. Kanninen, Int. J. Fract., **9** (1), 83-92 (1978).
[6] W. B. Fichter, Int. J. Fract., **22**, 133-143, (1983).
[7] F. E. Penado, J. Comp. Mater., **27** (4), 383-407 (1993).
[8] L. C. Wang, Ph.D. Dissertation, Stanford University, 2003.
[9] A. G. Evans and E. R. Fuller, Metall. Trans., **5**, 27-33 (1974).
[10] B. Asoo, J. M. McNaney, Y. Mitamura, and R. O. Ritchie, J. Biomed. Mater. Res., **52** (3), 488-491 (2000).
[11] L. Lawson, E. Y. Chen, M. Meshii, Int. J. Fatigue, **21**, 515-534, (1999).

Mat. Res. Soc. Symp. Proc. Vol. 821 © 2004 Materials Research Society P8.18

Analysis of Residual Stress and Damage Durability with Thermal Fatigue Behavior in Thermal Barrier Coatings

Dong-Ho Park[1], Hyung-Jun Jang[1], Yeon-Gil Jung[1], Jung-Chel Chang[2], and Ungyu Paik[3]

[1]Dept. of Ceramic Science and Engineering, Changwon National University
9 Sarim-dong, Changwon, Kyungnam 641-773, Korea
[2]Power Generation Lab, Korea Electric Power Research Institute
103-16 Munji-dong, Yusung-ku, Daejon 305-380, Korea
[3]Dept. of Ceramic Engineering, Hanyang University
17 Haengdang-dong, Seongdong-gu, Seoul 133-791, Korea

ABSTRACT

Static thermal fatigue tests for thermal barrier coatings (TBCs) were conducted to observe effects of temperature and holding time on its mechanical properties, hardness and modulus, and damage durability. For which three TBCs samples with different thickness in bonding layer (0.08, 0.14, and 0.28 mm) were prepared using an air plasma spraying (APS) method. Temperature of 950 and 1100°C and holding time of 10 and 100 hr were selected for the thermal fatigue tests. The TBCs with thin bonding layer (0.08 mm) maintain sound condition for all the thermal fatigue tests, even showing an evidence of cracking at the interface between coating and bonding layers. However, the TBCs with intermediate (0.14 mm) and thick (0.28 mm) bonding layers show delamination at interface and fracture of coating layer after the thermal fatigue tests at 1100 °C for 100 hr. Thermal growth oxide (TGO) layer is created at the interface between coating and bonding layers in all the TBCs after the thermal fatigue tests, and the TGO layer thickness is mainly affected by temperature. Modulus and hardness of coating layer are increased with an increase of temperature in the thermal fatigue tests, due to the re-sintering of coating layer during the thermal fatigue tests. Effects of bonding layer thickness and thermal fatigue condition on mechanical properties, residual stresses, damage durability of the TBCs are discussed extensively.

INTRODUCTION

The advanced multi-layered ceramic coatings applied on metallic substrates are generally used as thermal barrier coatings (TBCs) in gas turbine and diesel engine components. Demands for improving performance of the turbine and the engine are met in part by increasing and decreasing the combustion and cooling temperatures, respectively [1]. However, an increase of temperature difference within the turbine and the engine leads to service limits in current TBCs and superalloys [2,3]. Therefore, estimation and understanding for properties of TBCs, such as thermal behavior, residual stresses, and damage tolerance, are essential for improving its reliability and lifetime. TBCs consisting of yttria-doped stabilized zirconia (YSZ) top coating layer and Ni-based alloy bonding layer are being widely used to enhance fuel efficiency in internal combustion engines, which are subjected to severe working conditions in elevated temperature environments.

In present work, in order to evaluate and compare mechanical properties and thermal durability of TBCs in thermal environments, static thermal fatigue tests were conducted as functions of temperature and holding time. For which three kinds of TBCs, with different thickness in bonding layer, were prepared by an air plasma spraying (APS) method [4,5]. Effects of bonding layer thickness on mechanical properties and thermal durability of the TBCs are investigated and discussed with the thermal fatigue conditions.

METERIALS AND EXPERIMENTAL PROCEDURE

Flat rectangular test specimens were machined from the shank section of a fully heat-treated GTD-111 turbine blade using an electro discharge machining process. The specimens were ground and polished to remove the recast layer. GTD-111 has a nominal composition by weight: C 0.1%, Cr 14.0%, W 3.8%, Co 9.5%, Mo 1.5%, Al 3.0%, Ti 4.9% and Ta 2.8%. For preparation of bonding and coating layers, starting materials and fabrication parameters are indicated in table I. The bonding layer thickness prepared is shown in table II.

For static thermal fatigue tests, temperature of 950 and 1100°C and holding time of 10 and 100 hr were selected. The thickness of thermally grown oxide (TGO) layer at the interface (ceramic coating/ metal alloy bonding layers) was measured using scanning electron microscopy (SEM, S2700, Hitachi, Japan), which is compared with the thermal fatigue conditions. Mechanical properties of modulus (*E*) and hardness (*H*) in the TBCs before and after the thermal fatigue tests were measured using nano-indentation (Nanoinstruments, MTS Corp, Eden Prairie), which are evaluated with bonding layer thickness [6]. The stress-strain behavior depending on bonding layer thickness before the thermal fatigue tests was determined through Hertzian-indentation, with tungsten carbide spheres of radius r = 3.18mm, using Instron testing machine (Model 1122, Instron Corp, Canton) [7,8]. The TBCs for nano-indentation and Hertzian-indentation were prepared by grinding to a 10 μm finish followed by polishing to a 1 μm surface finish, and then polished with a 10 nm SiO_2 colloidal sol. All experiments were conducted in air.

Table I. Materials and fabrication parameters for bonding and ceramic coating layers in APS

	Bond layer	Ceramic layer
Materials	AMDRY 995C (Co32Ni21Cr8Al0.5Y)	METCO 204NS-G (ZrO_2-8Y_2O_3)
Gun	Metco 9MB	Metco 9MB
Primary & secondary gas	Nitrogen & Hydrogen	Argon & Hydrogen
Powder carrier gas	Nitrogen (40±2psi)	Argon (60±2psi)
Feed rate	44±5 g/min	60±5 g/min
Gun-to-Work distance	3''±0.5	4''±0.5
Gun traverse speed	5mm/sec±1	5mm/sec±1

Table II. Bonding layer thickness prepared for static thermal fatigue tests

Test 1	Test 2	Test 3
0.05mm ~ 0.10mm	0.15mm ~ 0.20mm	0.30mm ~ 0.35mm

RESULTS AND DISCUSSION

Micrographs before the thermal fatigue tests are shown in figure 1. The bonding layer thickness indicated in figure 1 is a nominal thickness after preparation of the TBCs. Ceramic coating layer is relatively porous and the interface between coating and bonding layers is discontinuous in microstructure. However, the interface between bonding layer and substrate shows relatively continuous microstructure. There is no TGO layer at the interface between coating and bonding layers before the thermal fatigue tests. Micrographs after the thermal fatigue tests are shown in figure 2. TGO layer formed during the thermal fatigue tests can be observed. The thickness of TGO layer is increased with an increase of temperature and holding time in the thermal fatigue tests, showing main effect of temperature. The TBCs with bonding layers of 0.14 and 0.28 mm (Figs. 2(B-4) and 2(C-4)) show delamination at interface and fracture of coating layer after the thermal fatigue tests at 1100 °C for 100hr. Also, the TBCs with thin bonding layer of 0.08 mm (Fig. 2(A-4)) shows a certain evidence for delamination at interface, but there is no cracking at coating layer and fracture of coating layer. Consequently, the evolution of TGO layer in the TBCs is more dependent on thermal fatigue temperature. It can be postulated that these TGO layers will affect residual stresses and mechanical properties of the TBCs.

Indentation stress-strain behavior, $P_0(\alpha/\gamma)$, for the each bonding layer thickness before the thermal fatigue tests is compared. The change of indentation stress-strain is not observed largely with bonding layer thickness. It can be postulated that indentation stress-strain behavior is not much affected by bonding layer thickness.

The modulus before and after the thermal fatigue tests is shown in figure 3, which are represented as modulus versus distance from the interface between substrate and bonding layer. Dashed lines indicate the interface between bonding and coating layers in all the TBCs and solid lines are empirical data fits. In proportion to bonding layer thickness, the modulus before the thermal fatigue tests shows modestly higher values, but the modulus of the TBCs with thick bonding layer is increased surely. The modulus is changed sharply at the interface between bonding layer and substrate, showing relatively low values in bonding layer and continuously decrease into coating layer with increasing the distance from the interface between bonding layer and substrate. The low modulus will be helpful to absorb mechanical stresses or less loading at the outside. The modulus measured after the thermal fatigue tests at 950 °C for 100 hr and 1100 °C for 10 hr are shown in figures 3(B) and 3(C), respectively, for two bonding layers of 0.08 mm and 0.28 mm. Discontinuity of the modulus at the interface between substrate and bonding layer disappears after the thermal fatigue tests. However, there is large modulus drop at the interface between bonding and coating layers. In the TBCs after the thermal fatigue tests at higher temperature (1100 °C), the modulus are gradually changed with an increase of the distance from

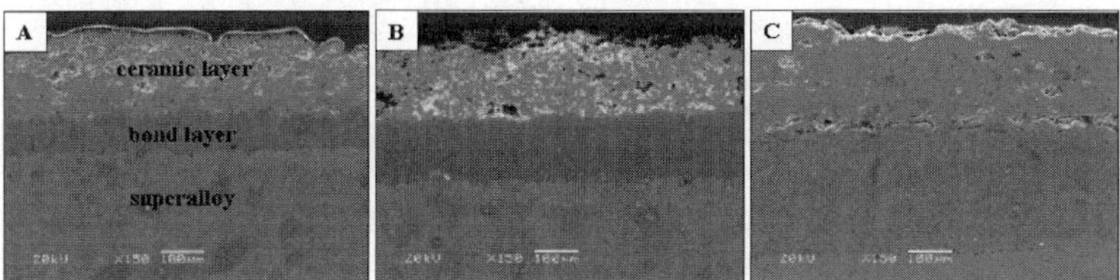

Figure 1. Micrographs of TBCs with different bonding layer thickness: (A) 0.08mm, (B) 0.14mm, and (C) 0.28mm) before thermal fatigue tests.

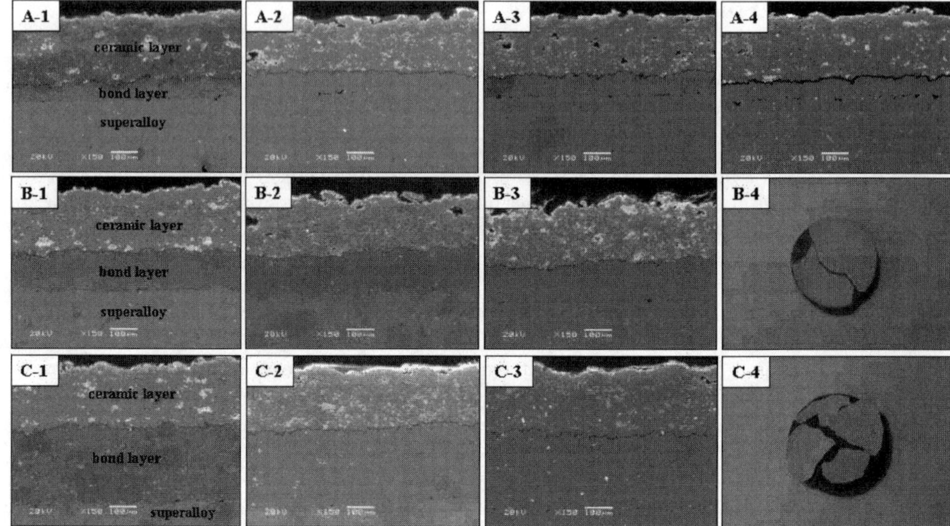

Figure 2. Micrographs of TBCs with different bonding layer thickness after thermal fatigue tests at 950 °C 10hr (A-1: 0.08 mm, B-1: 0.14 mm, and C-1: 0.28 mm), 950 °C 100hr (A-2: 0.08 mm, B-2: 0.14 mm, and C-3: 0.28 mm), 1100 °C 10hr (A-3: 0.08 mm, B-3: 0.14 mm, and C-3: 0.28 mm), and 1100 °C 100hr (A-4: 0.08 mm, B-4: 0.14 mm, and C-4: 0.28 mm).

the interface between bonding layer and substrate. The modulus is increased in bonding and coating layers after the thermal fatigue tests. The hardness before and after the thermal fatigue tests is represented in figure 4. The hardness before the thermal fatigue tests shows similar tends with the modulus of figure 3(A). After the thermal fatigue tests, a sudden change in the hardness is relieved at the interface between bonding layer and substrate. In lower temperature (950 °C), the hardness of substrate and bonding layer is not much changed. However, the hardness of coating layer is affected by the thermal fatigue tests, showing much higher hardness compared to that before the thermal fatigue tests. In higher temperature (1100 °C), the hardness of the TBCs with thin bonding layer is lower than that with thick bonding layer in all regions. Especially, the hardness in coating layer of the TBCs with thick bonding layer is much higher than those with

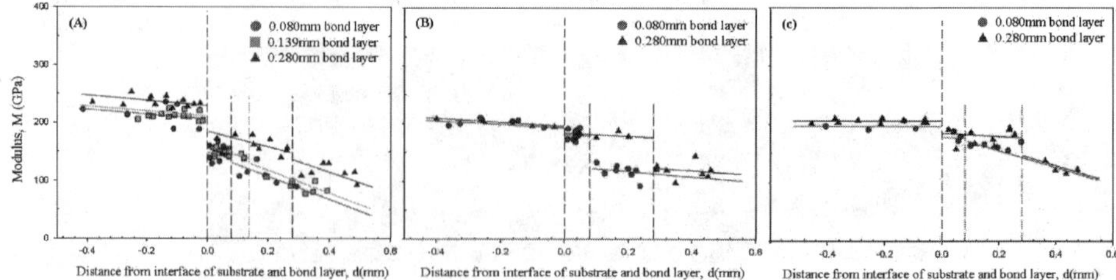

Figure 3. Modulus from nano-indentation (A: before thermal fatigue tests, B: 950 °C for100 hr, and C: 1100 °C for10 hr).

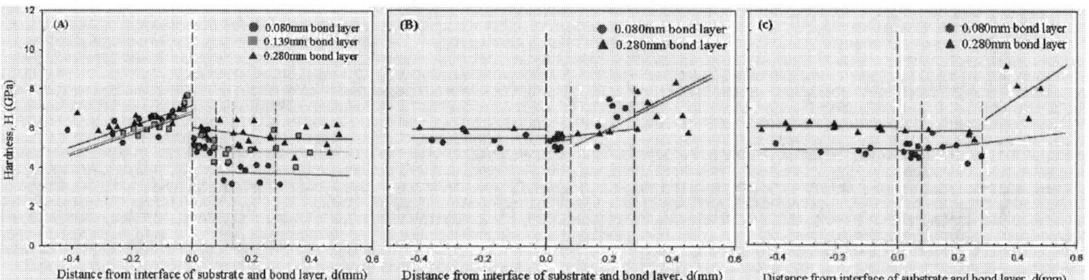

Figure 4. Hardness from nano-indentation (A: before thermal fatigue test, B: 950 °C for100 hr, and C: 1100 °C for10 hr).

thin boning layer and before the thermal fatigue tests. While, the TBCs with thin bonding layer after the thermal fatigue tests indicate similar hardness in all ranges, but higher hardness than before the thermal fatigue tests. It can be considered as one of reasons for different micrograph results of figures 2(A-4), 2(B-4), and 2(C-4). Also, this means that coating layer is re-sintered during the thermal fatigue tests. It is indicated that temperature is more effective for changing hardness than holding time.

CONCLUSIONS

The thickness of TGO layer formed during static thermal fatigue tests is dependent on temperature and holding time, showing dominant effect of temperature. The TBCs with intermediate and thick bonding layers show delamination at the interface between coating and bonding layers and fracture of coating layer after the thermal fatigue tests of 1100 °C for 100 hr, showing sound conditions below this temperature and holding time. The indentation stress-stain behavior with bonding layer thickness is not much changed. The modulus and hardness are increased through the thermal fatigue tests in both bonding and coating layers due to the re-sintering of ceramic coating layer, without changing in substrate. The discontinuity of modulus and hardness at the interfaces (substrate/bonding layer and bonding/coating layers) is decreased through the thermal fatigue tests, showing slightly lower modulus and higher hardness in coating layer than substrate and bonding layer after the thermal fatigue tests. Temperature in the thermal fatigue tests is more important parameter for mechanical properties of the TBCs and thin bonding layer is better than thick one for thermal durability.

ACKNOWLEDGMENT

This work was financially supported by Korea Electric Power Research Institute(KEPRI) and program for cultivating graduate students in regional strategic industry.

REFERENCES

1. S. Bose, and DeMasi-Marcin, *J. Therm. Spray. Technol.* **6**, 99(1997).
2. A. Bennett, *Mater. Sci. Technol.* **30**, 1(1987).
3. J.R. Brandon, and R. Taylor, *Surf. Coat. Technol.* **39**, 143(1989)
4. J.A. Haynes, M.K. Ferber, and W.D. Porter, *J. Therm. Spray. Technol.* **9**, 38(1997).
5. V. Teixeira, M. Andritschky, W. Fischer, H.P. Buchkremer, and D. Stover, *J. Mater. Proce. Technol.* **92-93**, 209(1999).
6. W.C. Oliver, G.M. Pharr, *J. Mater. Res.* **7**, 1564(1992).
7. F. Guiberteau, N.P. Padture, H. Cai and B.R. Lawn, *Philos. Mag.* **A68**, 1003(1993).
8. M.V. Swain and B.R. Lawn, *Phys. Stat. Solid.* **35**, 909(1969).

B. Modeling

Mat. Res. Soc. Symp. Proc. Vol. 821 © 2004 Materials Research Society P6.3

Hyperelastic effects in brittle materials failure

Markus J. Buehler[1], Farid F. Abraham[2] and Huajian Gao[1]
[1]Max Planck Institute for Metals Research, 70569 Stuttgart, Germany
[2]IBM Almaden Research Center, San Jose, CA 95120-6099, USA

ABSTRACT

A fact that has been neglected in most theories of brittle fracture is that the elasticity of a solid clearly depends on its state of deformation. Metals will weaken, or soften, and polymers stiffen as the strain approaches the state of materials failure. It is only for infinitesimal deformation that the elastic moduli can be considered constant and the elasticity of the solid linear. We show by large-scale atomistic simulations that hyperelasticity, the elasticity of large strains, can play a governing role in the dynamics of fracture and that linear theory is incapable of capturing all phenomena. We introduce a new concept of a characteristic length scale χ for the energy flux near the crack tip and demonstrate that the local hyperelastic wave speed governs the crack speed when the hyperelastic zone approaches this energy length scale. The new length scale χ, heretofore missing in the existing theories of dynamic fracture, helps to form a comprehensive picture of crack dynamics, explaining super-Rayleigh and supersonic fracture. We further address the stability of cracks, and show agreement of the Yoffe criterion with the dynamics of cracks in harmonic systems. We find that softening hyperelastic effects lead to a decrease in critical instability speed, and stiffening hyperelastic effects leads to an increase in critical speed. The main conclusion is that hyperelasticity plays a critical role in forming a complete picture of dynamic fracture.

INTRODUCTION

We show by large-scale atomistic simulation that hyperelasticity, the elasticity of large strains, can play a governing role in the dynamics of brittle fracture. This is in contrast to many existing theories of dynamic fracture where the linear elastic behavior of solids is assumed sufficient to predict materials failure [1]. Some experiments [2-4] as well as many computer simulations [5, 6] have shown a significantly reduced crack propagation speed in comparison with the theoretical predictions. Such discrepancies between theories, experiment and simulations can not solely be attributed to the fact that real solids have imperfections, as similar discrepancies also appear in molecular-dynamics simulations of cracks traveling in perfect atomic lattices. It was recently proposed that hyperelastic effects at the crack tip play an important role in the dynamics of fracture [3, 7]. In contrast, it is not generally accepted that hyperelasticity should play a significant role in dynamic fracture. This is because the zone of large deformation is highly confined to the crack tip, so that the region where linear elastic theory does not hold is extremely small compared to the extensions of the specimen [1].

Here we use large-scale molecular-dynamics simulations [5, 8-10] in conjunction with continuum mechanics concepts [1] to prove that hyperelasticity is crucial for understanding dynamic fracture. The first part of studies show that local hyperelasticity around the crack tip can significantly influence the limiting speed of cracks by enhancing or reducing local energy flow, even if the zone of hyperelasticity is small compared to the specimen dimensions. The hyperelastic theory drastically changes the concept of the maximum crack velocity in the

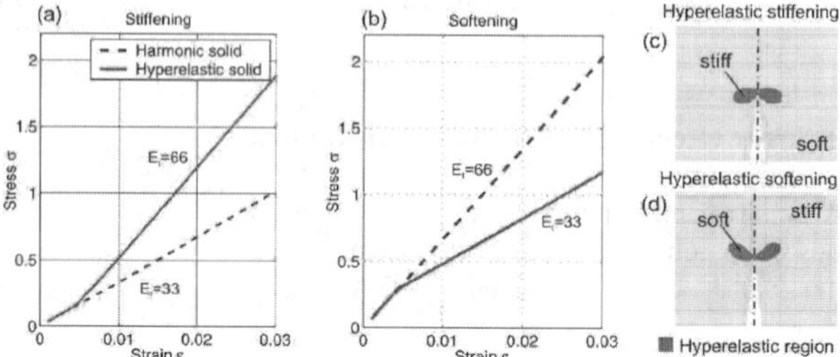

Figure 1: Elastic properties associated with the biharmonic stiffening (a) and biharmonic softening (b) potential. Plots (c) and (d) show the shape of the hyperelastic region.

classical theories. For example, the classical theories predict that mode I cracks are limited by the Rayleigh-wave speed and mode II cracks are limited by longitudinal wave speed. In contrast, both super-Rayleigh mode I and supersonic mode II cracks are allowed by hyperelasticity [8]. We find that there exists a characteristic length scale associated with energy flow near the crack tip that explains supersonic crack motion: Hyperelasticity completely dominates crack dynamics if the size of hyperelastic region approaches this characteristic length. In the second part of the paper, we report studies on the crack tip instability, and we show that hyperelasticity also plays a governing role.

MODELING

We consider a crack in a two-dimensional simulation geometry with slab width l_x propagating in a triangular hexagonal lattice (details see [11]). To avoid crack branching, a weak fracture layer is introduced so that atomic bonds across the prospective crack path snap at a critical atomic snapping distance r_{break}, while those in the rest of the slab never break. The snapping distance is used to adjust the fracture surface energy γ [11]. We adopt a biharmonic interatomic potential composed of two spring constants k_0 and $k_1 = 2k_0$ serving as a simplistic model of hyperelasticity common to a large class of real materials (see Fig. 1a and b) [11]. We consider two "model materials", one with elastic stiffening (Fig. 1a) and the other with elastic softening (Fig. 1b) behavior. The spring constant k_0 is associated with small perturbations from the equilibrium distance r_0, and the second spring constant k_1 is associated with large bond stretching ($r > r_{on}$). Harmonic systems are obtained if r_{on} is chosen to be larger than r_{break}.

CRACK SPEED AND ENERGY FLOW

We show that a localized, small hyperelastic region around the crack tip can have significant effects on the dynamics of crack propagation. In the simulations, the slab is loaded with 0.32 percent strain in mode I. The strain energy density far ahead of the crack tip is given by $S = \varepsilon_{xx}^2 l_x E/(1-\nu^2)/2$, where E is the Young's modulus at small strain. The linear elastic expression of strain energy density is valid because material far ahead of the crack is strained always below the onset threshold of the bilinear law, that is, it remains in the linear elastic regime of material response. According to the linear theory of fracture [1], the crack speed should satisfy the dynamic energy release rate equation $A(v/c_R) = 2\gamma/S$ where the function

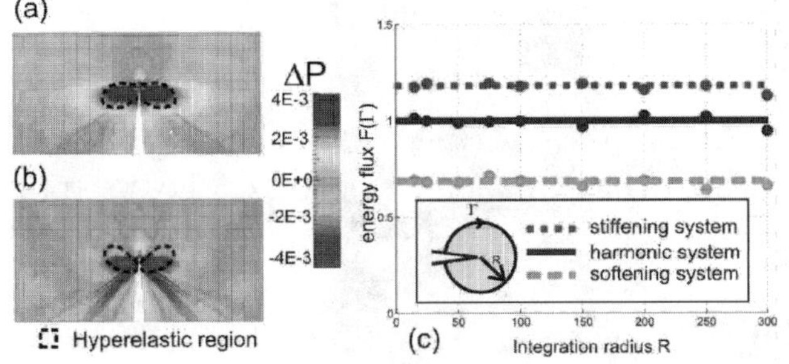

Figure 2: Subplots (a) and (b): Change of energy flow due to hyperelastic stiffening (a) and softening (b) effect. Subplot (c) shows a J-integral analysis for different choices of integration path Γ, clearly suggesting path independency and thus confirming steady-state crack motion [1].

$A(v/c_R)$ is a universal function of crack velocity. Linear theory predicts that crack velocity depends only on the ratio S/γ. Our strategy is to focus on this prediction by linear theory. In the harmonic systems, since $S \propto E$ and $\gamma \propto E$ [11], we choose the parameter r_{break} to be identical in all cases. In the biharmonic systems, we adjust the parameter r_{break}, at given values of r_{on}, E_0 and E_1, to always keep S/γ constant.

We choose $r_{break} = 1.17$ for the harmonic systems, and the crack achieves the same propagation velocity around 80 percent of the Rayleigh wave speed, consistent with linear theory. For the biharmonic systems, we choose $r_{on} = 1.1275$ and $r_{break} = 1.1558$ in the stiffening case, and $r_{break} = 1.1919$ in the softening case (always S/γ=const.). In contrast to the linear theory prediction, we find that the crack speed is about 20 % larger in the stiffening system and 30 % smaller in the softening system. These deviations can not be explained by the linear theory: The fact that we change the large-strain elasticity while keeping the small-strain elasticity constant indicates that hyperelasticity is affecting crack dynamics! Linear theory would predict the same crack limiting speed in all cases.

The region occupied by atoms having a local maximum principal strain $\varepsilon_1 > (r_{on} - r_0)/r_0$ defines the hyperelastic area A [11, 12]. Fig. 1c shows the hyperelastic area in the case of a stiffening material, and Fig. 1d shows the hyperelastic area in the case of an elastically softening material. The hyperelastic effect is highly localized to the crack tip (the pictures show a portion of the simulation slab near the crack tip).

A measure for the direction and magnitude of energy flow in the vicinity of the crack tip is the magnitude of the dynamic Poynting vector [11, 13]. A measure for the change in energy flow is obtained by subtracting the magnitude of the dynamic Poynting vector in the harmonic case from that in the biharmonic case at every point in the slab, $\Delta P = P_{biharm} - P_{harm}$. If the difference is negative, energy flow is reduced, and if the difference is positive, energy flow is enhanced. Fig. 2a and b shows the energy flow enhancement and reduction in the vicinity of the crack tip for the elastically stiffening bilinear system (a) and for the elastically softening system (b). In each plot, the local hyperelastic zone is indicated by a dotted line. The energy flow in the vicinity of the crack tip is enhanced in the stiffening case and reduced in the softening case. The change in crack velocity in the stiffening and softening system are due to enhancement or reduction of the energy flow. Steady state crack motion is confirmed by the path-independency of the dynamic J-integral (Fig. 2c), also showing change in energy flow in different cases (normalized with respect to the value of the harmonic system).

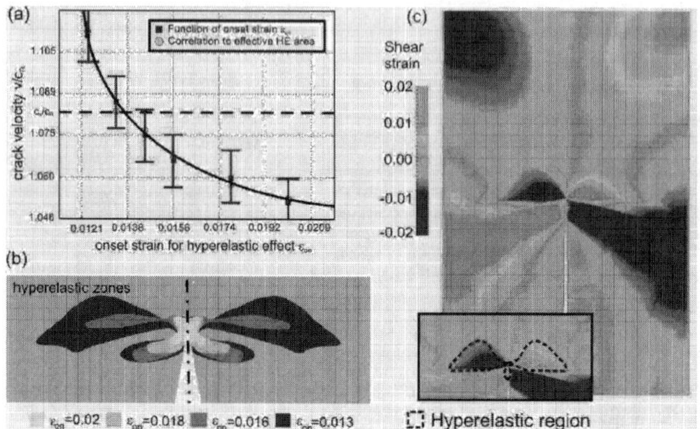

Figure 3: (a) Shape of hyperelastic region for different choices of ε_{on}, (b) crack speed as a function of ε_{on}, and (c) shear strain field during intersonic mode I crack propagation.

An important result is that when sufficient loading is applied to the system, mode I cracks in bilinear stiffening solids can reach speeds beyond the Rayleigh-wave speed. Figure 3a shows how the size of the hyperelastic region can be correlated with the crack speed, for different choices of the potential parameter ε_{on}. The larger the hyperelastic region, the higher the crack speed, and for purely harmonic systems the terminal crack velocity is the Rayleigh-wave speed. Figure 3b plots the shape of the hyperelastic region for different choices of ε_{on}. In another simulation we observe intersonic mode cracks (Fig. 3c), when the stiffening is stronger than before (here $k_1 = 4k_0$). This phenomenon has, motivated by MD simulation results [11], recently been verified in experiment [14].

Our observations can be explained by the concept of the characteristic energy length scale χ. The problem of a super-Rayleigh mode I crack in an elastically stiffening material is somewhat analogous to Broberg's [15] problem of a crack propagating in a stiff elastic strip embedded in a soft matrix (Fig. 4a). In contrast to the previous problem, the Broberg problem can be solved analytically and the characteristic energy length scale χ can be determined from numerical calculation. It was shown that the energy release rate can be expressed in the form $G = \sigma^2 h / E f(v, c_1, c_2)$, where σ is the applied stress, h is the half width of the stiff layer and f is a non-dimensional function of crack velocity and wave speeds in the strip and the surrounding matrix. The dynamic Griffith energy balance requires $G = 2\gamma$, indicating that crack propagation velocity is a function of the ratio h/χ where $\chi \propto \gamma E / \sigma^2$ is defined as a characteristic length scale for local energy flux. By dimensional analysis, the energy release rate of our hyperelastic stiffening material is expected to have similar features except that the strip width h should be replaced by a characteristic size of the hyperelastic region r_H. Therefore, we introduce the concept of a characteristic length $\chi = \beta \gamma E / \sigma^2$ for local energy flux near a crack tip. We find that the mode I crack speed reaches the local Rayleigh wave speed as soon as h/χ gets sufficiently large and verify that the scaling law holds by changing γ, E and σ independently (Fig. 4b).

In addition the studies on crack speed and energy flow, we report preliminary studies focusing on the effect of hyperelasticity on the instability dynamics of cracks. It is known that cracks in

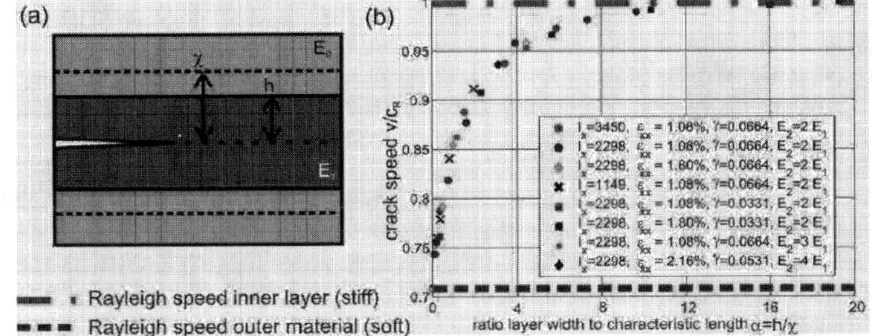

Figure 4: Numerical verification of the characteristic energy length scale in dynamic fracture (Broberg problem of crack motion in a thin stiff strip embedded in a soft matrix, $E_1 > E_0$).

softening materials become unstable at reduced speeds compared to the theoretical prediction of 73 % c_R by the Yoffe criterion [1-3, 7].

Our simulations of cracks in homogeneous materials with linear elastic properties (harmonic potential) show that the instability occurs at about 73 % c_R, in perfect agreement with the prediction by theory (Fig. 5a). The branching angle of 60 degree also matches the theoretical prediction [1], and the instability can be correlated with the development of a bimodal hoop stress as originally proposed by Yoffe (Fig. 5b) [1].

As previously, the linear systems serve as reference system when we probe crack dynamics in nonlinear materials. We use a similar biharmonic potential as in the previous sections, with the difference that it smoothens at bond breaking, leading to smoother crack acceleration (details on the potential will be published elsewhere). The results show that cracks become unstable at much lower speeds when softening materials are used (Fig. 5c), and that cracks are stable up to velocities close to the Rayleigh-wave speed in stiffening materials (Fig. 5d). When the stiffening effect is stronger ($k_1 = 4k_0$), the instability speed is even super-Rayleigh! Such observations can only be understood from the hyperelasticity point of view, as linear theory would predict the instability to always occur at 73 % c_R. Preliminary analyses suggest that the critical instability speed can be approximated by Gao's model of local limiting speed [7] in softening materials.

DISCUSSION AND CONCLUSIONS

Our studies suggest that hyperelasticity plays a dominating role in the dynamics of cracks. The effect of nonlinearities close to the crack tip is two-fold: Firstly, it has strong impact on energy flow, and secondly, it governs the instability dynamics. Figure 6 shows the classical viewpoints of length scales around a dynamic crack tip, compared to the new picture featuring the energy length scale [11]. Under a particular experimental or simulation condition, the relative importance of hyperelasticity is determined by the ratio r_H / χ. For small r_H / χ, the crack dynamics is dominated by the global linear elastic properties since much of the energy transport necessary to sustain crack motion occurs in the linear elastic region. However, when r_H / χ approaches unity, as is the case in some of our molecular dynamics simulations, the dynamics of the crack is dominated by local elastic properties because the energy transport required for crack motion occurs within the hyperelastic region. The concept of energy characteristic length χ immediately provides an explanation how the classical barrier for transport of energy over large distances can be undone by rapid transport near the tip. Our results suggest that the classical, linear elastic theories should be replaced by nonlinear theories of fracture in order to form a

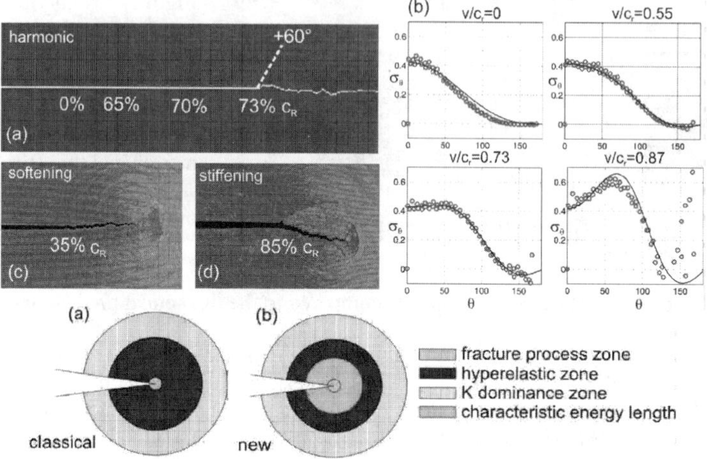

Figure 5: Dynamic crack tip instabilities (a) harmonic system, (b) softening system and (c) stiffening system. Subplot (d) shows the hoop stress σ_Θ in the harmonic systems for different crack speeds v.

Figure 6: The characteristic energy length scale in dynamic fracture, in comparison with classical length scales

comprehensive picture of crack dynamics. The viewpoint of hyperelasticity explains several experimental and computational observations, as for instance reduced critical crack speed for instability and reduced maximum crack speed [2, 3, 5].

ACKNOWLEDGMENTS

The simulations were carried out at the Max Planck Society Supercomputer Center in Munich. We gratefully acknowledge their support.

REFERENCES

1. Freund, L.B. Dynamic Fracture Mechanics (Cambridge University Press, 1990).
2. Fineberg, J. Gross, S.P., Marder, M., and Swinney, H.L. Phys. Rev. Lett. **67**, 141-144 (1991).
3. Ravi-Chandar, K. Int. J. of Fract. **90**, 83-102 (1998).
4. Cramer T., Wanner A., Gumbsch P., Phys. Rev. Lett. **85** (4): 788-791 (2000).
5. Abraham, F.F., Brodbeck, D., Rudge, W.E., Xu, X. Phys. Rev. Lett. **73**, 272-275 (1994).
6. Falk, M.L., Needleman, A. and Rice, J.R. Journal de Phys. IV France **11**, 5-43-50 (2001).
7. Gao, H. Mech. Phys. Solids **44**, 1453-1474 (1996).
8. Abraham F.F. et al. Proc. Nat. Acad. Sci. **99**, 5777-5782 (2002).
9. Abraham, F.F., Gao, H., Phys. Rev. Lett. **84**, 3113-3116 (2000).
10. Rountree, C.L. et al. Annu. Rev. Mater. Res. **32**, 377-400 (2002).
11. Buehler, M.J. Abraham, F.F., Gao, H. Nature **49**, 441-446 (2003)
12. Zimmermann, J. Continuum and atomistic modeling of dislocation nucleation at crystal surface ledges. PhD Thesis, Stanford University (1999).
13. Fratini S., Pla O., Gonzalez P., Guinea F., and Louis E. Phys. Rev. B. **66**, 104104 (2002).
14. Petersan, P.J., Deegan, R.D., Marder, M., Swinney, H.L., under submission, preprint available at: http://arxiv.org/abs/cond-mat/0311422.
15. Broberg, K.B. Int. J. Sol. Struct. **32** (6-7), 883-896 (1995).

Mat. Res. Soc. Symp. Proc. Vol. 821 © 2004 Materials Research Society P3.7

Testing Multiwall Carbon Nanotubes on Ion Erosion for Advanced Space Propulsion

Yoke Khin Yap [1,*], Jitendra Menda [1], Lakshman Kumar Vanga [1], Vijaya Kayastha [1], Jiesheng Wang [1], Lyon B. King [2,**], Svetlana Dimovski [3], Yury Gogotsi [3]

[1]Department of Physics, Michigan Technological University, Houghton, MI 49931, USA.
[2]Department of Mechanical Engineering-Engineering Mechanics, Michigan Technological University, Houghton, MI 49931, USA.
[3]Department of Materials Science and Engineering and A. J. Drexel Nanotechnology Institute, Drexel University, Philadelphia, PA 19104, USA.
[*]Email: ykyap@mtu.edu, [**]Email: lbking@mtu.edu

ABSTRACT

Are carbon nanotubes more resistant than diamonds against ion erosion? Here, we report an evaluation of multiwall carbon nanotubes (MWNTs) as the protective coating against plasma erosion in advanced space propulsion systems. We have compared polycrystalline diamond films with MWNTs, amorphous carbon (a-C) and boron nitride (BN) films. Two types of MWNTs were investigated including vertically aligned (VA) MWNTs, and those horizontally laid on the substrate surfaces. Only diamond films and VA-MWNTs survived erosion by 250 eV krypton ions of a flight-quality Hall-effect thruster. VA-MWNTs are found to bundle at their tips after ion erosion.

INTRODUCTION

The space exploration program faces enormous challenges as it seeks to achieve dramatic improvements in safety, cost, and speed of missions to the frontiers of space. Plasma propulsion systems have been recognized as far more efficient than chemical thrusters. This recognition has led to the development of highly efficient electric propulsion (EP) thrusters that are currently the only feasible technology for many deep space missions. However, these EP devices have in common electrode sputter erosion as a life-limiting process. To facilitate long thruster life, critical surfaces in EP thrusters are fabricated from sputter-resistant materials such as molybdenum (Mo). Carbon-based materials have shown nearly an order-of-magnitude improvement in sputter erosion resistance over Mo [1]. Among the tested carbon-based materials, diamond films prepared by chemical vapor deposition (CVD diamond) provide improvement by a factor of 1.5 in volumetric sputter erosion rate over others [2]. For thruster surfaces that are subject to sputter damage, yet must be electrical insulators, boron nitride ceramic has traditionally been used to increase the lifetime [3]. Recently, Meezan, et. al. found that polycrystalline diamond plates had 25% better resistance to sputtering that the traditional boron nitride ceramic [4].

On the other hand, unique mechanical properties of carbon nanotubes (CNTs) have triggered tremendous curiosities on their applications. CNTs are predicated to have extremely high Young's modulus values, similar to that of in-plane modulus of graphite (~1000GPa). This is much higher than the bulk modules of diamond (~443 GPa). Thus, it is interesting to find out the resistance of CNTs to ion erosion. In this paper, we discuss our preliminary observation on ion erosion of MWNTs by Kr ions generated by a Hall effect thruster. Similar erosion is also tested for CVD diamond films, boron nitride (BN), and amorphous-carbon (a-C) films. MWNTs are tested because they can be grown in high density with their axes either horizontally or vertically oriented to the substrate surface. The effect of tubes orientation can thus be examined.

EXPERIMENTAL DETAILS

Two types of MWNTs were used. First, MWNTs were grown vertically aligned to the substrate surface by a dual-RF-plasma enhanced CVD system [5]. Secondly, MWNTs were grown by thermal CVD so that their tube axes were aligned parallel to the substrate surface. All BN and a-C films were grown using a pulsed-laser deposition (PLD) system [6, 7]. The tested BN film was sp^2- bonded with typical IR absorption band at ~780 cm^{-1} and 1380 cm^{-1}. The a-C films were deposited at room temperature in vacuum by using a UV laser (λ=266 nm) at an intensity of ~5 GW/cm^2. Both BN and a-C films were thinner than 200 nm. The CVD diamond films are standard products of Sumitomo Electric Industries, Ltd (Japan). They are randomly oriented polycrystalline films, 800 μm thick on Si substrate. These films are mechanically polished to have a surface roughness ~ 1 nm rms.

Erosion test was conducted in a 2-meter-diameter by 4-meter-long space simulation chamber, as shown in figure 1. All the samples were placed 0.5 m downstream on the plume centerline of the exhaust beam of a 1.35-kW Hall-effect thruster. Each sample was partially masked in such a way that a portion of the material was exposed to ion erosion for 80 min, while another portion was protected. The thruster was operated on krypton propellant with a discharge voltage of 300 V producing nominally 250 Kr^+ ions at a total beam current of 1.8 A. Post-test analyses of the samples included: scanning electron microscopy (SEM) images taken at the masked and the exposed areas, profilometry across the masking boundary in an attempt to determine the eroded depth, Raman spectroscopy, Fourier transformed infrared (FTIR) spectroscopy, back-scattered electron (BSE) imaging, energy dispersive X-ray (EDX) analysis. In this paper, we focus our discussion based on the SEM images, profilometry, Raman, and FTIR spectroscopies. Our results indicate that only CVD diamond films and vertically aligned MWNTs survived the ion erosion. Furthermore, formation of vertically aligned MWNT bundles was discovered in these experiments.

Figure 1. Space simulation chamber at Michigan Tech used for the erosion test.

DISCUSSION

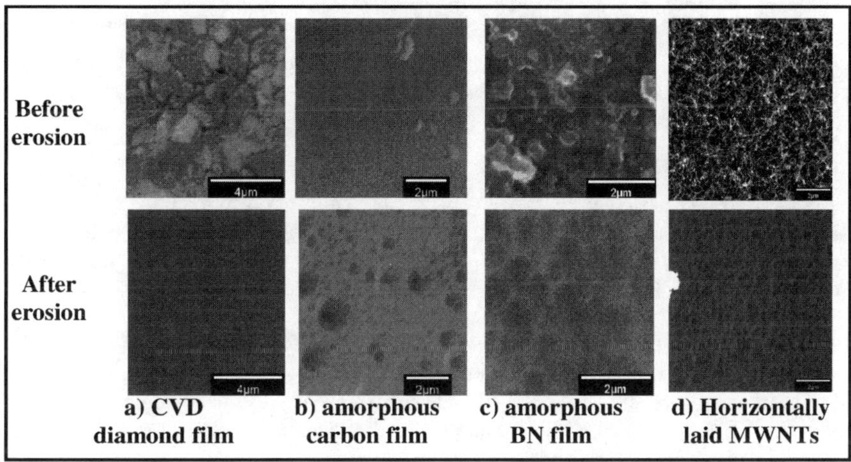

Figure 2. SEM images of CVD diamond films, amorphous carbon (a-C) films, amorphous boron nitride (BN) films, and horizontally laid multiwall carbon nanotubes (MWNTs) before and after ion erosion.

As shown by the SEM images in figure 2, all samples are eroded by the krypton ions from the Hall effect thruster. The grains of CVD diamond films are visible before erosion, but they disappear after. The surface of the diamond films could have been polished by the ions as observed in optical materials [8]. There is no significant change in Raman spectra from the masked and eroded regions. Thus, diamond film remained on the substrate after ion erosion. Profilometry analysis by a non-contact inteferometer shows that the CVD diamond film is eroded at a depth of 45 nm.

The SEM images of BN and a-C films after erosion indicate etch pits on the surface. Such morphologies do not appear by etching of amorphous materials. Thus, we conclude that this is the etched Si surface of the samples' substrate and both the BN and a-C films are totally sputtered away by the Kr ions. Complete sputtering also happened on the horizontally laid MWNTs as shown in figure 2d. As shown, no MWNTs are found on the substrate surface.

On the other hand, VA-MWNTs remained on the substrate after ion erosion. The typical image of "as-grown" VA-MWNTs is shown in figure 3a. These VA-MWNTs are about one micrometer long with an average diameter of ~70 nm, and density of ~5 x 10^9 per square centimeter. The Ni nanoparticles used for the growth of MWNTs remained on the tips of nanotubes as indicated in a brighter image contrast (figure 3a) and in backscattered electron (BSE) images (inset). Figure 3b shows both the masked and eroded regions of the VA-MWNTs sample are shown. Clearly, MWNTs survived sputtering. The boundary of the mask is seen as a bright stripe.

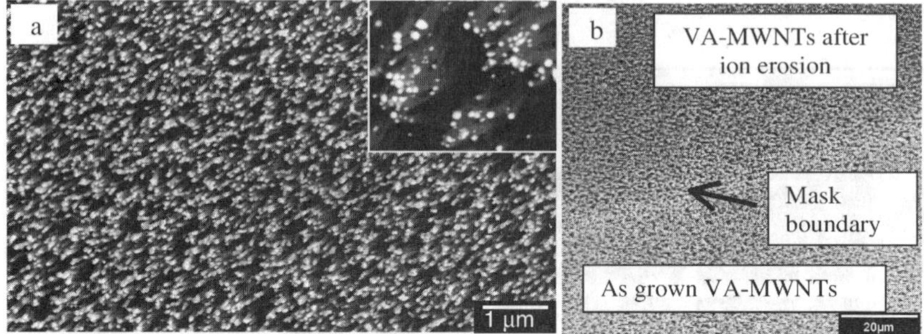

Figure 3. Image of (a) VA-MWNTs in the masked region. (b) VA-MWNTs are clearly seen in the masked and eroded regions.

High-resolution SEM images, however, suggest that VA-MWNTs were eroded by ~0.5 micron. According to Raman spectroscopy, the intensity of both the G- and D-bands at ~1580 and ~1360 cm^{-1} are reduced and broaden. These results coincide with observation by SEM that the MWNTs are shortened as shown in figure 4a. As shown, the appearance of these MWNTs has changed from a regular cylindrical tube (diameter ~70nm) to a flat-top irregular cones. Some of these cones are formed by merging and welding of several adjacent nanotubes. The change of morphology is related to the broadening of the Raman bands.

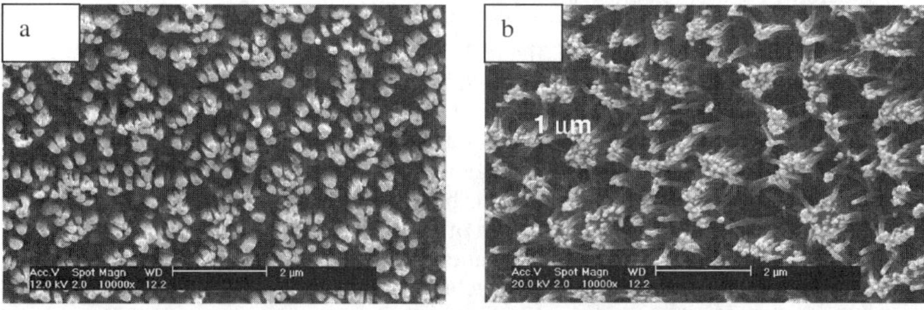

Figure 4. (a) Images of VA-MWNTs after ion erosion. The MWNTs appear as flat-top cone-like structures. (b) At the mask boundary, VA-MWNTs are found to bundle-up at their nickel tips.

In order to understand the formation of these flat-top nanostructures, we carefully mapped the transition of morphology across the mask boundary by a high-resolution field emission SEM. From this mapping, we found that individual VA-MWNTs merged with the surrounding nanotubes and formed MWNT bundles as shown in figure 4b. The only reason of bundling is the swinging of the tips of VA-MWNTs so that they can get close to each other. Under the ion sputtering, both erosion and heat are generated and caused surface melting of the tips. When these nickel tips are swinging, there are chances that the tips will stick to the adjacent tips and form the bundle. At present, the reason of swinging tips is not clear.

At the unmasked region, similar mechanism should have occurred at a faster rate. Due to a higher sputter yield at higher ionic energy and density, at one point these bundles will start to erode. We speculate that the removal of the nickel tips will avoid bundling of the VA-MWNTs and enhance the resistance of VA-MWNTs against ion erosion. This speculation will be tested by future experiments.

CONCLUSIONS

CVD diamond films have higher resistance than VA-MWNTs against ion erosion. However, VA-MWNTs are more resistant than horizontally laid MWNTs. We found that VA-MWNTs bundled at their tips as irradiated by krypton ions. The removal of nickel particles from the tips could lead to higher resistance of VA-MWNTs against ion erosion. Beside CVD diamond films, VA-MWNTs are another prospective protective coating for advanced propulsive engines.

ACKNOWLEDGEMENTS

Y.K.Y. acknowledges supports from the Michigan Tech Research Excellence Fund and the Army Research Office (W911NF-04-1-0029, through the City College of New York). S.D. and Y.G. were supported by DOE grant BSE- DE-FJ02-01ER45932.

REFERENCES

[1]. T. Haag, M. Patterson, V. Rawlin and G. Soulas, 2001 International Electric Propulsion Conference, Paper No. IEPC-01-94.

[2]. J. J. Blandino, D. G. Goodwin and C. E. Garner, Diamond & Relat. Mater. **9**, 1992 (2000).

[3]. V. Kim, V. Kozlov, A. Semenov, and I. Shkarban, 2001 International Electric Propulsion Conference, Paper No. IEPC-01-073.

[4]. N. B. Meezan, N. Gascon, and M. Capelli, 2001 International Electric Propulsion Conference, Paper No. IEPC-01-39.

[5]. T. Hirao, K. Ito, H. Furuta, Y. K. Yap, T. Ikuno, S. Honda, Y. Mori, T. Sasaki, and K. Oura, Jpn. J. Appl. Phys. **40**, L631 (2001).

[6]. Y. K. Yap, S. Kida, T. Aoyama, Y. Mori, and T. Sasaki, Appl. Phys. Lett. **73**, 915 (1998).

[7]. Y. K. Yap, M. Yoshimura, Y. Mori, and T. Sasaki, Appl. Phys. Lett. **80**, 2559 (2002).

[8]. T. Kamimura, S. Fukumoto, R. Ono, Y. K. Yap, M. Yoshimura, Y. Mori, T. Sasaki and K. Yoshida, Opt. Lett. **27**, 616 (2002).

Mat. Res. Soc. Symp. Proc. Vol. 821 © 2004 Materials Research Society P3.2

Determination of Ion Irradiation Influence on π-plasmon Properties of Carbon Nanotubes

Mariya M. Brzhezinskaya [1], Eugen M. Baitinger [1], Vladimir V. Shnitov [2] and Aleksey Smirnov [2]
[1] Physics Department of Chelyabinsk State Pedagogical University,
69 Lenin Ave., Chelyabinsk 454080, Russia
[2] Ioffe Physico-Technical Institute of the Russian Academy of Sciences,
St. Petersburg 194021, Russia

ABSTRACT

In this paper, the results of experiments on irradiation of the singlewall (SWNT) and multiwall carbon nanotube (MWNT) samples by argon ions are presented. They were obtained by reflection electron energy loss spectroscopy and Auger electron spectroscopy.

Results indicate the π-plasmon energy E_π and the full width at half maximum (FWHM) of the plasma peak were sensitive to the dose of ion irradiation. In particular, the π-plasmon energy E_π decreases and the plasma peak broadens with the increase of the dose of Ar^+ irradiation.

The π-plasmon peak broadening is associated with damage of carbon nanotubes under ion irradiation. Possible causes of the carbon nanotubes deformation and influence of deformation on π-band structure of carbon nanotubes are discussed.

INRODUCTION

The progress in synthesis of singlewall and multiwall carbon nanotubes (SWNTs and MWNTs) is now enriched with intense study of carbon nanotube (CNT) modification. The modification of structure of the carbon nanotubes using irradiation by high- and low-energy ions [1] and electrons [2] is the important part of the technology. The energy and the irradiation dose, parameters of irradiation (for instance, chemical composition of residual gases inside the chamber [3]) are the primary conditions, which in many respects determine the properties of CNTs after the modification. The study of the primary steps of modification and destruction of the CNT surface under ion flux is very important. At the primary stage, the intramolecular structure of CNTs does not change much, but CNTs begin deforming [2]. The formed metastable structures can have useful properties.

The earlier studies [4,5] indicated that energy electron loss spectroscopy (EELS) is very sensitive to the change of CNT structure, including a degree of defects and disorder of the carbon nanotubes under argon ion (Ar^+) irradiation [6,7].

EXPERIMENTAL DETAILS

The samples of SWNTs and MWNTs were used in these experiments. MWNTs were prepared by the arc-discharge evaporation [8]. All samples were produced by the closed joint-stock company «Astrin», Saint-Petersburg, Russia. MWNTs were approximately 10 nm in diameter and approximately 10 μm in length. SWNTs were prepared by the electric-arc-discharge synthesis [9,10]. The SWNTs were 1.2 - 1.6 nm in diameter and 1 - 10 μm in length.

The chemical composition of the sample surface was determined by Auger electron spectroscopy. The original samples, prepared for the experiment, contained 98% of carbon and 1 – 1.7% of oxygen. The MWNT sample contained a little nitrogen and sulfur also.

Irradiation by Ar ions (energy of 1 keV and current density of $j = 0.75$ μC/cm^2·s) and characteristic electron energy loss spectra measurement (energy of primary electrons was 1 keV, error was 0.5 eV) were carried out in ultra-high vacuum better than 10^{-9} Torr after annealing and dwell for 70 hours.

Energy loss spectroscopy of reflected electrons was used: the incident angle of the primary electron beam and the registration angle were 45°, the diameter of the beam on the sample was approximately 1 mm, the analyzer aperture was 12°. Absolute energy resolution (ΔE=Const) was 0.2 eV, when the transmission energy was 10 - 25 eV.

Each irradiation of a sample was followed by Auger spectra measurement in the mode of constant absolute energy resolution $\Delta E = 0.6$ eV in order to determine the concentration of argon, absorbed in the near surface region of the sample. The primary electron beam energy was chosen to be 2.5 keV for Auger spectra measurement. The Ar concentration was determined as [11]:

$$C_{Ar} = \frac{I_{Ar}}{S_{Ar}} \Bigg/ \left(\frac{I_C}{S_C} + \frac{I_{Ar}}{S_{Ar}} \right) \qquad (1)$$

where

S_{Ar}, S_C were coefficient of element sensitivity for argon and carbon,

I_{Ar}, I_C were amplitudes of Auger signal of argon and carbon, determined as "peak-to-peak Auger amplitudes" [11].

The samples under the test were periodically irradiated by flux of Ar^+ for 0.5 min. The total irradiation dose Q was determined as $Q = j \cdot t$, where t was the irradiation time.

RESULTS

Fig. 1 shows the dependence of the π-plasmon energy E_π on the dose of ion irradiation Q for SWNTs.

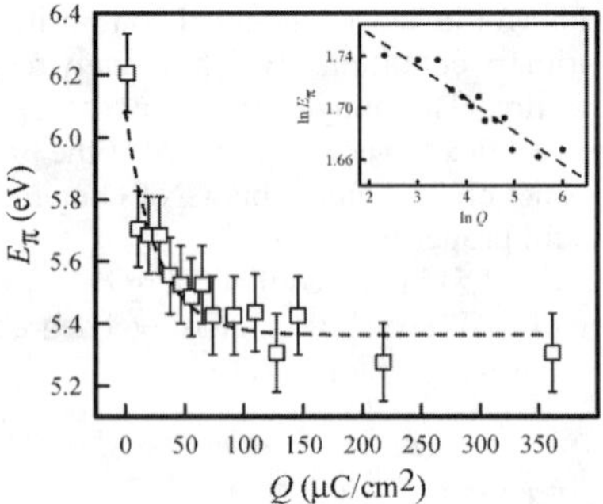

Figure 1. The π-plasmon energy dependence on the dose of ion irradiation Q for SWNTs. The inset shows the π-plasmon energy dependence on the dose of ion irradiation Q in the logarithmic scale.

The E_π value decreases significantly ($\Delta E_\pi = 0.9$ eV) with increasing dose of Ar^+ irradiation at small values of Q (< 100 μC/cm^2). The inset shows the same dependence at the double logarithmic scale. This dependence is linear.

Therefore, the determined dependence of the π-plasmon energy E_π on the dose of ion irradiation Q can be approximated by hyperbola:

$$E_\pi(Q) = Const\,/\,Q\,, \qquad (2)$$

The π-plasmon peak broadening is accompanied by its intensity decrease. The full width at half maximum (FWHM) δE_π does not change significally during the ion irradiation. It increases from 2.7 to 3.4 eV (approximately 15% only). However, this broadening reveals also nonlinear dependence on Q. The peak broadens more quickly at small values of Q.

The same experiments with MWNTs indicated that the π-plasmon energy E_π decreases from 6.4 to 5.2 eV mainly at the primary irradiation stage. The relation between E_π and Q for MWNTs can be approximated by exponential, as was shown in the previous papers (for instance, see [6]):

$$E_\pi(Q) = Const \cdot e^{-kQ}\,, \qquad (3)$$

where $k \approx 0.074$ cm^2/μC ($Q_0 = 1/k = 13.5$ μC/cm^2).

Fig. 2 and Fig. 3 show the experimental dependence of the concentration of argon, absorbed on the sample surface on the dose of ion irradiation Q (Fig. 2) and on the full width at half maximum (FWHM) δE_π (Fig. 3). Rate of absorption is high at small values of Q (< 100 μC/cm^2). However, this process slows then down and it has a tendency to saturation. The dependence of the concentration of the absorbed argon on δE_π is a linear increasing function for SWNTs (Fig. 3, line 1) and for MWNTs (Fig. 3, line 2). The slope of the line for MWNTs is 1.4 times greater that it is than for SWNTs.

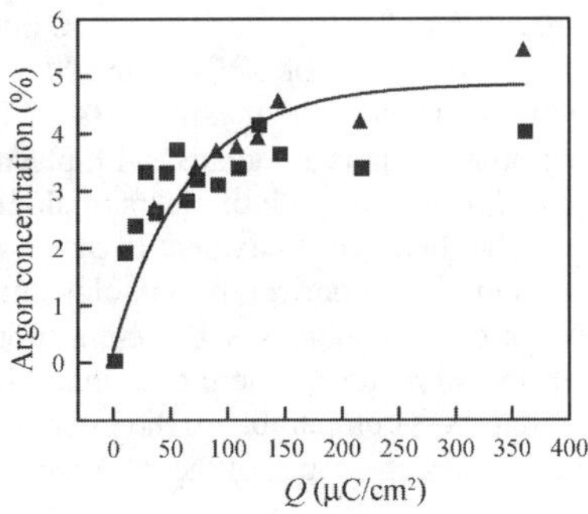

Figure 2. Dependence of Argon concentration on the surface of samples on the dose of ion irradiation Q for MWNTs (triangles) and SWNTs (squares).

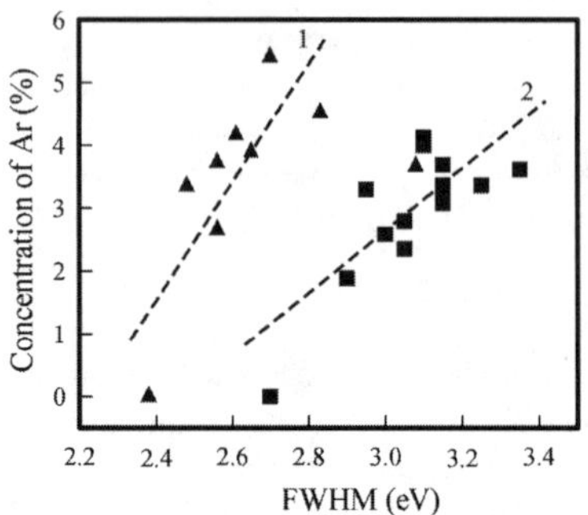

Figure 3. The relation between the full width at half maximum (FWHM) of the plasma peak δE_{π} and concentration of interstitial Argon atom for MWNTs and SWNTs. (1) – MWNTs; (2) - SWNTs.

DISCUSSION

The irradiation by moderate-energy ions primarily generates vacancy defects in the CNT walls, resulting in: a) reconstruction of hexagons to the couple of 5- and 7-atom segments [12]; b) formation of clusters of driven-out carbon atoms inside the CNTs and in the intertube space [1] and c) interstitial Ar atoms. Thus, inelastic deformation appears, causing CNT bending and change of their diameter. Perhaps, Ar ions attach to the introduced defects. In this case, the Ar concentration near surface should be equal to the concentration of defects, generated by irradiation. Distance between defects was estimated based on the value of maximum concentration of absorbed Ar (4% for SWNTs and 5.4% for MWNTs). This distance was calculated to be approximately 1.5 to 2 nm, what is about 10 interatomic distances. The experiment indicated that the modification of SWNTs and MWNTs under ion flux becomes slower when the Ar concentration reaches the maximum (see Fig. 2).

The π-plasmon peak broadening is associated with plasmon relaxation as a result of decrease of mean free time and mean free path of plasmon along CNT. Phenomenologically, the decrease of resonance plasma frequency is equivalent to plasmon relaxation, what is illustrated in Fig. 1 and Eqs. 2 and 3. Lines in Fig. 3 confirm this conclusion: When the Ar concentration, and correspondingly the number of defects, increases, the π-plasmon mean free path decreases pro rata, and the spectrum of plasmon broadens. Therefore, under the progressive amorphization, when the distance between defects is comparable to the interatomic distance, the significant broadening of the π-plasmon will be the cause of the π-plasmon disappearance.

CONCLUSIONS

Detailed analysis of the EEL spectra acquired after different doses of Ar^{+} irradiation of carbon MWNTs and SWNTs was done.

The π-plasmon energy E_π decreases significantly at the beginning of Ar^+ irradiation, when the dose Q was less than 100 μC/cm^2.

The dependence of E_π on the dose of ion irradiation Q is approximated by a hyperbolic function for SWNTs and by an exponential function for MWNTs.

The dependence π-plasmon peak broadening on the dose of ion irradiation Q is nonlinear: the π-plasmon peak broadens more quickly at small values of Q.

The dependence of δE_π on concentration of the absorbed argon is a linear function for SWNTs and for MWNTs. The slope of the line for MWNTs is 1.4 times greater that it is than for SWNTs.

The π-plasmon peak broadening is explained by plasmon relaxation as a result of decrease of mean free time and mean free path of plasmon along CNT under ion irradiation.

ACKNOWLEDGMENTS

This material is based on work supported by the Russian Ministry of Education under grant No.PD02-1.2-170.

REFERENCES

1. A.V. Krasheninnikov, K. Nordlund and J. Keinonen, Phys. Rev. **B65**, 65423 (2002).
2. B.W. Smith and D.E. Luzzi, AIP Conference Proceedings **486**, 360 (1999).
3. K. El-Hami and K. Matsushige, Intern. Journ. of Nanoscience **2**, 125 (2003).
4. M.S. Dresselhaus, G. Dresselhaus and Ph. Avouris, Carbon Nanotubes: Synthesis, Structure, Properties, and Applications, Series: Topics in Applied Physics, **80**, (2001), XV, 447 pp.
5. P.M. Ajayan, S. Iijima and T. Ichihashi, Phys.Rev.B. **47**, 6859 (1993).
6. M. Brzhezinskaya, E. Baitinger and V. Shnitov, Phisica B (2004), (in press).
7. M. Brzhezinskaya, E. Baitinger and V. Shnitov, MRS Fall Meeting Proceedings 2003 (2004), (in press).
8. S. Iijima, Nature **354**, 56 (1991).
9. E.D. Obraztsova, J.-M. Bonard, V.L. Kuznetsov, V.I. Zaikovskii, S.M. Pimenov, A.S. Pozarov, S.V. Terekhov, V.I. Konov, A.N. Obraztsov and A.P. Volkov, NanoStructured Materials **12**, 567 (1999).
10. A.S. Lobach, N.G. Spitsina, S.V. Terekhov and E.D. Obraztsova, Physics of the Solid State **44**, 475 (2002).
11. Handbook of Auger electron spectroscopy, edited by L. Davis, P.W. Palmberg, et all. Physical Electronic Division, Perkin-Elmer Corp., Second edition, (1978).
12. P.L. Langinie, J. Conard, in book: Molecular Nanostructures, edited by H. Kuzmany, J. Fink and M. Mering, World Scientific Publising Company, (1998) pp. 427-430.

AUTHOR INDEX

SUBJECT INDEX